T0189625

LONDON MATHEMATICAL SOCIETY LECTURE NOTE SERIES

Managing Editor: Professor M. Reid, Mathematics Institute,
University of Warwick, Coventry CV4 7AL, United Kingdom

The titles below are available from booksellers, or from Cambridge University Press at www.cambridge.org/mathematics

London Mathematical Society Lecture Note Series: 359

Moduli Spaces and Vector Bundles

Edited by

STEVEN B. BRADLOW

University of Illinois, Urbana-Champaign

LETICIA BRAMBILA-PAZ

Centro de Investigación en Matemáticas, Guanajuato, Mexico

OSCAR GARCÍA-PRADA

Consejo Superior de Investigaciones Científicas, Madrid

S. RAMANAN

Chennai Mathematical Institute, India

CAMBRIDGE
UNIVERSITY PRESS

Shaftesbury Road, Cambridge CB2 8EA, United Kingdom

One Liberty Plaza, 20th Floor, New York, NY 10006, USA

477 Williamstown Road, Port Melbourne, VIC 3207, Australia

314–321, 3rd Floor, Plot 3, Splendor Forum, Jasola District Centre, New Delhi – 110025, India

103 Penang Road, #05–06/07, Visioncrest Commercial, Singapore 238467

Cambridge University Press is part of Cambridge University Press & Assessment, a department of the University of Cambridge.

We share the University's mission to contribute to society through the pursuit of education, learning and research at the highest international levels of excellence.

www.cambridge.org
Information on this title: www.cambridge.org/9780521734714

First published 2009

A catalogue record for this publication is available from the British Library

Library of Congress Cataloging-in-Publication data
Moduli spaces and vector bundles / edited by Steve Bradlow . . . [et al.].
p. cm.
Includes bibliographical references.
ISBN 978-0-521-73471-4 (pbk.)
1. Vector bundles. 2. Moduli theory. I. Bradlow, Steve, 1957– II. Title.
QA612.63.M63 2009
514´.224 – dc22 2008053181

ISBN 978-0-521-73471-4 Paperback

Contents

Preface

Vector bundles and their associated moduli spaces are of fundamental importance in algebraic geometry. In recent decades the richness of the subject has been enhanced by its relationship to other areas of mathematics, including differential geometry (which has always played a fundamental role), topology (where the moduli spaces not only present interesting topological features but turn out to be useful in settling topological problems) and, perhaps most surprising of all, theoretical physics, in particular gauge theory, quantum field theory and string theory.

The concept of stability for vector bundles, and the ensuing construction of the moduli spaces which classify these bundles, goes back more than 40 years to the fundamental work of Mumford and of Narasimhan and Seshadri. Since then, there has been much progress in describing the detailed structure of these spaces using methods from pure algebraic geometry, topology, differential geometry, number theory, representation theory and theoretical physics.

Peter E. Newstead has been a leading figure in this field almost from its inception and has made many seminal contributions to our understanding of moduli spaces of stable bundles. His "Tata lecture notes" have helped an entire generation of algebraic geometers learn the foundations of moduli theory and vector bundles. He is Chair of the international research group "Vector Bundles on Algebraic Curves" (VBAC), which he initiated in 1993 and whose purpose was to encourage the development of this area of research, particularly among young mathematicians. It has been successful in bringing together researchers in this area from many countries. He has published 51 papers in this area, with several more in preparation. Most of his recent work has involved international collaboration.

His 65th birthday was thus a fitting occasion on which to survey the accomplishments leading up to the present, to showcase promising developments of the day, and to bring together the past, present and future leaders of the field. With this as motivation, the VBAC annual workshop for 2006 was organized as a "Newstead Fest" consisting of a winter college followed by a conference, and held at CIMAT in Guanajuato, Mexico in December 2006.

This volume is a tribute to Peter E. Newstead and to his contribution to algebraic geometry. It includes lecture notes (from the courses given in the college) and surveys covering both foundational material as well as the main topics of current research. It also contains research papers by participants at the conference and by colleagues of Peter. Peter Newstead's influence is prominent in all of the above. We hope that this tribute will be helpful to early-stage and experienced researchers.

Finally, the editors are delighted to add their personal tribute to Peter and to dedicate this volume to him on the occasion of his birthday.

The editors
Editor-in-Chief: L. Brambila-Paz
S. B. Bradlow
O. García-Prada
S. Ramanan

Acknowledgments

The lecture notes, surveys and research articles published here were refereed. We are indebted to the referees for their enormous cooperation and to the authors for their outstanding work. We thank the speakers and participants to "Newstead Fest", all of whom helped make it a rewarding and enjoyable event.

We thank the "Newstead Fest" local Committee, in particular Xavier Gomez-Mont (CIMAT), Jésus Muciño-Raymundo (IM-Mor-UNAM), Angela Ortega (IM-Mor-UNAM) and Claudia Reynoso (FAMAT) for their unstinting and enthusiastic support. Our thanks also to CIMAT administrative staff for their collaboration during "Newstead Fest". We also acknowledge the CUP Editorial Board for their help in publishing this volume. We take this opportunity to thank the organizations that helped make the "Newstead Fest" possible. In particular, we thank the

- Academia Mexicana de Ciencias, México, (AMC)
- Centre International de Mathématiques Pures et Appliquées, France, (CIMPA)
- Centro de Investigación en Matemáticas, México, (CIMAT)
- Clay Mathematics Institute, USA, (CMI)
- Commission on Development and Exchanges at the International Mathematical Union, Italy, (CDE-IMU)
- Consejo Nacional de Ciencia y Tecnología de Guanajuato, México, (CONCYTEG)
- Consejo Nacional de Ciencia y Tecnología, México, (CONACYT)
- Instituto de Matemáticas-Unidad Morelia, Universidad Nacional Autónoma de México, México, (IMATE-Mor-UNAM)
- International Center for Theoretical Physics, Italy, (ICTP)

- Office of External Activities at the Abdus Salam International Center for Theoretical Physics, Italy, (OEA-ICTP)
- Sociedad Matemática Mexicana, México, (SMM)

for their support.

Finally, the other three editors insist on acknowledging the extraordinary efforts of the editor-in-chief, Leticia. Without her none of these activities would have taken place, nor would they have been as memorable or enjoyable as they turned out to be.

The editors

Part I
Lecture Notes

1

Lectures on Principal Bundles

V. Balaji

Chennai Mathematical Institute
H1, SIPCOT Information Technology Park
Padur Post, Siruseri-603 103
Tamil Nadu, India.
e-mail: balaji@cmi.ac.in

To Peter Newstead on his 65th birthday

1 Introduction

The aim of these lectures (†) is to give a brief introduction to principal bundles on algebraic curves towards the construction of the moduli spaces of semistable principal bundles. The second section develops the basic machinery on principal bundles and their automorphisms. At the end of the second section, we give a proof of theorem of Grothendieck on orthogonal bundles. The third section, after developing the notions of semistability and stability, gives a modern proof of the main part of Grothendieck's theorem on classification of principal bundles on the projective line. The last section gives an outline of the construction of the moduli space of principal bundles on curves. The moduli space was constructed by A.Ramanthan in 1976. The method outlined here is from a construction in [BS]. These notes are a transcription of the lectures given in Mexico and therefore have an air of informality about them. I have consciously retained this informality despite criticism from a learned referee on the lack of rigor in some places. Indeed(‡) *"these notes are almost exactly in the form in which they were first written and distributed: as class notes, supplementing and working out my oral lectures. As such, they are far from polished and ask a lot of the reader. ...Be that as it may, my hope is that a well-intentioned reader will still be able to penetrate these notes and learn something of the subject"*.

---

† These are notes of five lectures given in Mexico in November 2006 at CIMAT, Guanajuato in the "College on Vector Bundles" which was held in honour of Peter Newstead to celebrate his 65th birthday.

‡ Lifted from D.Mumford's introduction to his "Lectures on curves on an algebraic surface"

2 Basic notions and definitions

Throughout these notes, unless otherwise stated, we have the following notations and assumptions:

(a) We work over an algebraically closed field k of characteristic zero and without loss of generality we can take k to be the field of complex numbers $\mathbb{C}$.

(b) G will stand for a reductive algebraic group often the general linear group $GL(n)$ and H a subgroup of G. Their representations are finite dimensional and rational.

(c) X is a smooth projective curve almost always in these notes.

2.1 Generalities on principal bundles

Definition 2.1. A principal G bundle $\pi : E \longrightarrow X$ with structure group G (or a G-bundle for short) is a variety E with a right G-action, the action being free, such that π is G-equivariant, X being given the trivial action. Further, the bundle π is locally trivial in the étale topology. (In other words, for every $x \in X$, there exists a neighbourhood U and an étale morphism $U' \longrightarrow U$ such that, when E is pulled back to U', it is *trivial* as G-bundle).

Remark 2.2. We remark that in our setting this definition of a principal bundle is related to the definition of a principal bundle being *locally isotrivial*, ([S]) i.e. for every $x \in X$, there exists a neighbourhood U and a finite and unramified morphism $U' \longrightarrow U$ such that, when E is pulled back to U', it is *trivial* as G-bundle. In fact, these two definitions coincide here because of the affineness of the group G. One could see this as follows: since G is affine, we have an inclusion of G in $GL(V)$ as a closed subgroup. This implies that $GL(V) \longrightarrow GL(V)/G$ is *locally isotrivial* in the sense of Serre ([S]). We now observe that any principal G–bundle can be obtained from a principal $GL(V)$–bundle by a *reduction of structure group* (see (2.5)). Further, any principal $GL(V)$–bundle is actually locally trivial in the Zariski topology (see [S]). The local isotriviality of the G–bundle now follows from that of the quotient map $GL(V) \to GL(V)/G$. In this context cf ([M]).

Remark 2.3. In general in the literature, a principal homogeneous space is defined to be locally trivial in the so-called fppf topology if the base is not smooth. Again, we remark that in our setting this is not needed even when the base is not smooth because we work with affine groups as structure groups. This follows from arguments similar to the one given above since any principal homogeneous space for the group $GL(V)$ is locally trivial in the Zariski topology.

2.1.1 Some notations and conventions

(a) By a family of H bundles on X parametrised by T we mean a principal H-bundle on $X \times T$, which we also denote by $\{E_t\}_{t \in T}$. We note that in general we may not have T to be smooth and the definition of principal bundles should be seen in the light of Remark 2.3.

(b) We recall the definitions of semisimple and reductive algebraic groups. Given a linear algebraic group G, we define the radical $R(G)$ to be $(\bigcap B)^0$ where the intersection runs over all Borel subgroups. Equivalently, $R(G)$ is the maximal, *normal*, connected, solvable subgroup of G. If $R(G) = (e)$ then G is called *semisimple*, and if $R_u(G)$, the unipotent radical of G which consists of the unipotent elements in $R(G)$, is trivial, then G is called *reductive*. (In this case $R(G)$ will be a torus). Equivalently, (by considering the derived subgroup), G is *semisimple* (resp. *reductive*) if and only if it has no connected abelian (resp. unipotent abelian), normal subgroup other than (e).

(c) Let Y be any quasi projective G-variety and let E be a G-principal bundle. For example Y could be a G-module. Then we denote by $E(Y)$ the associated bundle with fibre type Y which is the following object: $E(Y) = (E \times Y)/G$ for the twisted action of G on $E \times Y$ given by $g.(e, y) = (e.g, g^{-1}.y)$.

(d) Any G-equivariant map $\phi : F_1 \longrightarrow F_2$ will induce a morphism $E(\phi) : E(F_1) \longrightarrow E(F_2)$.

(e) A section $s : X \longrightarrow E(F)$ is given by a morphism

$$s' : E \longrightarrow F$$

such that, $s'(e.g) = g^{-1}.s'(e)$ and $s(x) = (e, s'(e))$, where $e \in E$ is such that $\pi(e) = x$, where $\pi : E \longrightarrow X$.

Definition 2.4. If $\rho : H \longrightarrow G$ is a homomorphism of groups the associated bundle $E(G)$, for the action of H on G by left multiplication through ρ, is naturally a G–bundle. We denote this G–bundle often by $\rho_*(E)$ and we say this bundle is obtained from E by extension of structure group.

Definition 2.5. A pair (E, ϕ), where E is an H–bundle and ϕ : $E(G) \longrightarrow F$ is a G–bundle isomorphism, is said to give a reduction of structure group of the bundle F to H. For convenience, we often omit ϕ and simply say E is obtained from F by reduction of structure group.

Two H–reductions of structure group (E_1, ϕ_1) and (E_2, ϕ_2) are equivalent or isomorphic if there is an H–bundle isomorphism $f : E_1 \longrightarrow E_2$ such that the following diagram commutes:

$$
\begin{array}{ccc}
E_1(G) & \xrightarrow{\ f(G)\ } & E_2(G) \\
\phi_1 \downarrow & & \downarrow \phi_2 \\
F & \xrightarrow{\quad = \quad} & F
\end{array}
\tag{1.1}
$$

Remark 2.6. A principal G-bundle E on X has an H-structure or equivalently a reduction of structure group to H if we are given a section $\sigma : X \longrightarrow E(G/H)$, where $E(G/H) \simeq E \times^G G/H$. To see this, we note the identification of the spaces $E(G/H) \simeq E/H$. Then by pulling back the principal H–bundle $E \longrightarrow E/H$ by σ, we get an H–bundle $E_H \subset E$, giving the required H-reduction. In other words, there is a natural isomorphism $E_H(G) \simeq E$. Thus, we get a correspondence between sections of $E(G/H)$ and H–reductions of E.

To see the other direction, let E_H be an H–bundle and consider the natural inclusion $E_H \hookrightarrow E_H(G) = E$ given by $z \longrightarrow (z, 1_G)$, where of course $(z, 1_G)$ is identified with (zh, h^{-1}) for $h \in H$. Going down by an action of H we get a map $X \longrightarrow E_H(G)/H = E(G/H)$. This gives the required section of $E(G/H)$.

Remark 2.7. Note that a reduction of structure group of E to $H \subset G$ can be realised by giving a G-map $s : E \longrightarrow G/H$ satisfying the property $s(e.g) = g^{-1}s(e)$. In this sense, the reduction E_H defined above can be seen to be the inverse image of the identity coset in G/H by the map s.

Remark 2.8. In the case of $G = GL(n)$, when we speak of a principal G-bundle we identify it often with the associated vector bundle by taking the associated vector bundle for the standard representation.

Remark 2.9. A $GL(n)$–bundle is completely determined by the associated vector bundle $E(V)$ (where V is the canonical n-dimensional space on which $GL(n)$ acts) as its bundle of frames. Let $\mathcal{V}$ be a vector bundle on X with fibre the vector space V. Then consider the union

$$
\bigcup_{x \in X} Isom(V, \mathcal{V}_x)
$$

where $Isom(V, V_x)$ are simply isomorphisms between the vector spaces V and V_x.

Note that there is a natural action of $GL(V)$ on the right which is easily seen to be free. This forms the total space of the principal $GL(V)$–bundle E whose associated vector bundle $E(V)$ is isomorphic to $\mathcal{V}$.

Similarly, a $PGL(n)$–bundle is equivalent to a projective bundle, i.e. an isotrivial bundle with $\mathbf{P^n}$ as fibre.

Proposition 2.10. Let E_1 and E_2 be two H–bundles. Giving an isomorphism of the H–bundles E_i is equivalent to giving a reduction of structure group of the principal $H \times H$–bundle $E_1 \times_X E_2$ to the diagonal subgroup $\Delta_H \subset H \times H$.

Proof: A reduction of structure group to the diagonal Δ gives a Δ–bundle E_Δ. Now observe that the projection maps on $H \times H$, when restricted to the diagonal, give isomorphisms of $\Delta \simeq H$. Viewing the bundle $E_\Delta \subset E_1 \times_X E_2$ as included in the fibre product, and using the two projections to E_1 and E_2, we get isomorphisms from $E_1 \simeq E_\Delta \simeq E_2$. The converse is left as an exercise.

Definition 2.11. Let P be a G–bundle. Consider the canonical adjoint action of G on itself, i.e $g \cdot g' = gg'g^{-1}$. Then we denote the associated bundle $P \times^G G$ by Ad(P).

Observe that because of the presence of an identity section, the associated fibration $Ad(P)$ is in fact a group scheme over X.

Proposition 2.12. The sections $\Gamma(X, Ad(P))$ are precisely the G–bundle automorphisms of P.

Proof: Let $\sigma : X \longrightarrow Ad(P)$ be a section. We view the section σ as remarked above as an equivariant map $\sigma : P \longrightarrow G$. Then by its definition, we have the following equivariance relation:

$$\sigma(p.g) = g^{-1} \cdot \sigma(p) = g^{-1}\sigma(p)g$$

Define the morphism:

$$f_\sigma : P \longrightarrow P$$

given by

$$f_\sigma(p) = p.\sigma(p)$$

$\forall p \in P$. The equivariance property for σ implies that $f_\sigma(p.g) = f_\sigma(p).g$ and hence its an H–morphism. Clearly it gives an automorphism of P.

For the converse, let $f : P \longrightarrow P$ be an H–bundle automorphism. Define σ as follows:

$$f(p) = p.\sigma(p)$$

Note that the H–equivariance property of f implies the equivariance property of σ which therefore defines a section of $Ad(P)$.

Remark 2.13. Let E be a reduction of structure group of the principal G–bundle P to the subgroup H. Then as we have remarked above, we can represent E as a pair (P, ϕ) where $\phi : X \longrightarrow P(G/H)$ is a section of the associated fibration. Let σ be an automorphism of the G–bundle P. Then, σ also acts as an automorphism on the associated fibration $P(G/H)$. This gives an action of the group $Aut(P)$ on the *set of all H–reductions of P.*

Two H–reductions E and F of a principal G–bundle are equivalent (i.e give isomorphic H–bundles) if and only if there exists an automorphism σ of P which takes E to F in the above sense.

To illustrate this phenomenon I give below a theorem due to Grothendieck ([G]).

Theorem 2.14. Let X be a smooth projective complex variety and let $H = O(n) \subset G = GL(n)$ be the standard inclusion. Then the canonical map induced by extension of structure groups

$$\{Isom \ classes \ of \ H-bundles\} \longrightarrow \{Isom \ classes \ of \ G-bundles\}$$

is injective. In other words, a G–bundle P has, if any, a unique reduction of structure group to H upto equivalence.

Proof. The proof of this theorem is quite beautiful and I will give it in full. Its also of importance to observe that the theorem is false for the inclusion $SO(n) \subset SL(n)$!

Let S be the space of symmetric $n \times n$–matrices which are non-singular. Then G acts on S as follows:

$$A.X := AXA^t$$

The action is known (by the Spectral theorem for non-degenerate quadratic forms) to be transitive and the isotropy subgroup at I is the standard orthogonal group $H = O(n)$. i.e $S \simeq G/H$ as a G–space. The more important fact is that there is a canonical inclusion of S in G. If $q : G \longrightarrow G/H$ is the canonical quotient map then identifying the quotient with S, the map q is given by

$$q(A) = AA^t$$

and then the restriction of the map to $S \hookrightarrow G$ is given by the map $q_S(A) = A^2$ on the space of symmetric matrices.

$$
\begin{array}{ccc}
S & \xrightarrow{\text{inclusion}} & G \\
\downarrow{\scriptstyle q_S} & & \downarrow{\scriptstyle q} \\
S & \xrightarrow{=} & G/H
\end{array}
\qquad (1.2)
$$

Let P_H be a fixed reduction of structure group of P to H and let E be any other reduction of P which is therefore given by a section $\sigma : X \longrightarrow P(G/H)$. Since we already have a reduction, we can express the new reduction σ as:

$$\sigma : X \longrightarrow P_H(G/H)$$

Consider the group scheme $Ad(P) = P(G)$. Since P has a H–reduction, we can view this group scheme as $P_H(G)$, where $H \hookrightarrow G$ acts on G by conjugation i.e $h \cdot g = h.g.h^{-1}$.

We also observe that the associated fibration $P_H(G/H)$ taken with the natural left action of H on G via its inclusion or by the conjugation action of H on G is identical. In other words, we can view the morphism associated to the canonical quotient map $G \longrightarrow G/H$ for the associated fibrations $P_H(G)$ and $P_H(G/H)$ as being induced by the conjugation action of H on G and G/H. Note that this is special to our situation since we have a H–reduction P_H to start with.

We thus get the map:

$$\phi : P_H(G) \longrightarrow P_H(G/H)$$

Observe again that the inclusion $S \hookrightarrow G$ is an H–morphism for the conjugation action of H. (Since $H = O(n)$, $A^t = A^{-1}$).

Viewing the spaces in the diagram above as a diagram of H–spaces for the conjugation action we have the following diagram of associated spaces:

$$
\begin{array}{ccc}
P_H(S) & \xrightarrow{\text{inclusion}} & P_H(G) \\
\downarrow{\scriptstyle q_S} & & \downarrow{\scriptstyle \phi} \\
P_H(S) & \xrightarrow{=} & P_H(G/H)
\end{array}
\qquad (1.3)
$$

We now note that to prove that the reduction of P given by σ is equivalent to the one given by P_H, we need to give an automorphism which takes one to the other. An automorphism is giving a section of $P(G)$ or equivalently of $P_H(G)$. Its easy to check that, giving such an automorphism is giving a section $\gamma : X \longrightarrow P_H(G)$ such that the following diagram commutes:

$$(1.4)$$

We now recall the following *interpolation* statement:

Lemma 2.15. The characteristic polynomial of a non-singular matrix A can be used to get a square root of A.

Proof. By an interpolation exercise, we can construct a polynomial $h(t)$ such that $h(t)^2 - t$ is divisible by $f(t)$, i.e $h^2(t) - t = f(t)g(t)$ for some polynomial $g(t)$. Since $f(A) = 0$ by the Cayley-Hamilton theorem, we get $h(A)^2 = A$, i.e. $h(A)$ provides a square root of A.

Now identify $P_H(G/H)$ with $P_H(S)$, which is a bundle of non-singular symmetric matrices. The section σ gives a characteristic polynomial $f(t)$ with coefficents being regular functions on X. Since X is projective these coefficients are therefore *constant*. Since these coefficients are constant, we can use the characteristic polynomial to get the *square root* of the section σ.

Take $h(t)$ as above. Then define

$$\eta(x) := h(\sigma(x))$$

This provides a section of $P_H(S)$ such that $\phi \circ \eta = \sigma$ since ϕ on S is the *squaring* operation. The composition $\eta : X \longrightarrow P_H(S) \hookrightarrow P_H(G)$ gives the required γ.

Remark 2.16. The reader should try and understand the proof in Grothendieck's paper. The main idea, as suggested by Prof Ramanan, is to compare a pair of equivalent non-degenerate quadratic forms on a vector space. When this is carried out over a family, together with choosing a square root (over X) it is essentially the classical proof. The reader can note that the proof given above works for all characteristics different from 2. The proof given here applies the definitions developed here and also naturally generalises the problem in the following sense.

In general one could work with a symmetric space G/H, where the subgroup H is given by the fixed points of an involution on G. For example, the proof generalises immediately to the case of the symplectic group. It also suggests that one could look for natural conditions for the map to be an inclusion.

Remark 2.17. It fails for $SO(n) \subset SL(n)$. In fact, it fails for $n = 2$. This can be seen as follows: $SO(2) \simeq \mathbf{G}_m$. Hence $SO(2)$ principal bundles can be identified with $\mathbf{G}_m$-bundles and hence with line bundles. Extension of structure group of a $SO(2)$–bundle to $SL(2)$ is equivalent to taking a line bundle L to $L \oplus L^*$, which is an $SL(2)$–bundle. This has clearly two inequivalent reductions L and L^*.

Remark 2.18. Olivier Serman ([O]) has shown that the map studied in the above theorem actually extends to an embedding of the moduli spaces of S–equivalence class of semistable orthogonal principal bundles into the moduli space of semistable principal $GL(n)$–bundles (see Section 4 below for the definitions).

3 Principal bundles, basic properties

Definition 3.1. A vector bundle V is said to be *semistable* (resp *stable*) if for every sub-bundle $W \subset V$,

$$\frac{deg(W)}{rk(W)} \leq \frac{deg(V)}{rk(V)}.$$

Lemma 3.2. Let V and W be semistable vector bundles on X of degree zero. Then $V \otimes W$ is semistable of degree zero.

Proof. Any semistable bundle on X of degree zero has a Jordan-Holder filtration such that its associated graded is a direct sum of stable bundles of degree zero. Note that the filtration is not unique but the associated graded is so. Hence the tensor product $V \otimes W$ gets a filtration such that its associated graded is a direct sum of tensor products of stable bundles of degree zero. We see easily that this reduces to proving the lemma when V and W are stable of degree zero. Then by the Narasimhan-Seshadri theorem, $V \otimes W$ is defined by a unitary representation of the fundamental group (namely the tensor product of the irreducible unitary representations which define V and W respectively), which implies that $V \otimes W$ is semistable (see [NS]).

Remark 3.3. Note that this theorem, as it stands, is false in positive characteristics (see [BP2]).

Remark 3.4. We remark that there is a purely algebraic proof of this theorem due to Ramanan and Ramanathan (see [RR]).

Proposition 3.5. Let E be a principal H-bundle on X with H a semisimple algebraic group. Then the following are equivalent:

(a) There exists a faithful representation $H \hookrightarrow GL(V)$ such that the induced bundle $E(V) = E \times^H V$ is semistable (of degree zero).
(b) For every representation $H \longrightarrow GL(W)$, the bundle $E(W)$ is semistable (of degree zero).

Proof. The implication (b) $\Longrightarrow$ (a) is obvious. To see the implication (a) $\Longrightarrow$ (b), we first observe that since the structure group H is semisimple, the associated vector bundle $E(V)$ is semistable of degree 0. Consider the natural tensor representation $T^{a,b}(V) = \otimes^a V \otimes \otimes^b V^*$. Then by Lemma 3.2, the bundle $E(T^{a,b}(V)) = \otimes^a E(V) \otimes \otimes^b E(V)^*$ is semistable of degree 0.

It is well-known that any H–module W is a sub-quotient of a suitable $T^{a,b}(V)$.

This can be seen as follows: Consider the action map $H \times W \longrightarrow W$. Taking the dual maps at the algebra level we have the map $W \longrightarrow k[H] \otimes W(0)$ where $W(0)$ is simply W but with the trivial H-action (namely $h.w = w \ \forall h \in H$). This follows by writing the action condition, where the action on the right is by translation on the algebra of functions. Hence, the H-module is a sub-module of $k[H]^{\dim W}$. In other words, it is enough to handle the H–module $k[H]$. For this we embed $H \hookrightarrow GL(V)$ and embed $GL(V) \hookrightarrow End(V) \times End(V^*)$ (i.e. inside matrices (A, B), such that $A.B^t = I$). Then, by dualising, we see that

$$Sym(V) \otimes Sym(V^*) \longrightarrow k[GL(V)] \longrightarrow k[H]$$

the last two maps being H–equivariant surjections. Now put W in copies of $k[H]$ and we are done. Notice that since we are in char 0 we can actually choose a splitting and embed W in a finite direct sum of tensor representations.

Thus, $E(W)$ is a sub-quotient of $E(T^{a,b}(V))$ of degree zero and this implies that $E(W)$ is also semistable of degree zero.

Definition 3.6. An H–bundle E is said to be *semistable* if it satisfies the equivalent conditions in Proposition 3.5.

3.0.2 Tannakian definitions

Definition 3.7. Another definition of a principal G-bundle is the Tannakian one: Let X be a connected smooth projective variety over **C**. Denote by $\mathrm{Vect}(X)$ the category of vector bundles over X. The category $\mathrm{Vect}(X)$ is equipped with an algebra structure defined by the tensor product operation

$$\mathrm{Vect}(X) \times \mathrm{Vect}(X) \quad \longrightarrow \quad \mathrm{Vect}(X)\,,$$

which sends any pair (V, W) to $V \otimes W$, and the direct sum operation $\oplus$, making it an additive tensor category. (See [DM, Definition 1.15]).

M.V.Nori (see [No]) gives an alternative description of principal G-bundles, which I briefly recall. For Nori however, X is allowed to be a much more general space (to be precise, a prescheme!). However we restrict ourselves to the situation where X is a smooth variety since the applications here will be in this generality.

Let $\mathrm{Rep}(G)$ denote the category of all finite dimensional complex left representations of the group G, or equivalently, left G-modules. By a G-module (or representation) we shall always mean a left G-module (or a left representation).

Given a principal G-bundle E over X and a left G-module V, the associated fiber bundle $E \times_G V$ has a natural structure of a vector bundle over X. Consider the functor

$$F(E) : \mathrm{Rep}(G) \quad \longrightarrow \quad \mathrm{Vect}(X)\,, \qquad (2.2)$$

which sends any V to the vector bundle $E \times^G V$ and sends any homomorphism between two G-modules to the naturally induced homomorphism between the two corresponding vector bundles. The functor $F(E)$ enjoys several natural abstract properties. For example, it is compatible with the algebra structures of $\mathrm{Rep}(G)$ and $\mathrm{Vect}(X)$ defined using direct sum and tensor product operations. Furthermore, $F(E)$ takes an exact sequence of G-modules to an exact sequence of vector bundles, it also takes the trivial G-module **C** to the trivial line bundle on X, and the dimension of V also coincides with the rank of the vector bundle $F(E)(V)$.

Nori proves that the collection of principal G-bundles over X is in bijective correspondence with the collection of functors from

$$F : \mathrm{Rep}(G) \longrightarrow \mathrm{Vect}(X)$$

satisfying the following properties: *strict, exact, faithful, tensor functor such that F_x (defined by $F_x(V) = F(V)_x$) is a fibre functor on the category* Rep(G).

(a) Strict: a morphism of vector bundles is said to be strict if the cokernel is also locally free. Let $u : V \longrightarrow W$ be a G-module map. Then we need the induced morphism $F(u) : F(V) \longrightarrow F(W)$ to be strict. In particular, $ker F(u)$ and $Im F(u)$ are also locally free.

(b) Exact: $ker F(u) = F(ker(u))$, $coker F(u) = F(coker(u))$.

(c) Faithful: $F(Hom(V, W)) \hookrightarrow Hom(F(V), F(W))$.

(d) Tensor functor: $F(V \otimes W) = V(V) \otimes F(W)$ and $F(trivial) = \mathcal{O}_X$.

(e) The functor F_x is a fibre functor, which says that using the pair (Rep(G), F_x), by the easy half of Grothendieck-Tannaka theorem, one recovers the group G from the Tannakian category (Rep(G), F_x) (see [DM, Page 130]).

Given F, there is a E unique upto a unique isomorphism such that $F \simeq F(E)$ (i.e. naturally equivalent).

Remark 3.8. More generally, this definition allows us to talk about torsors in any Tannaka category. For example, we could work with the category of semiharmonic Higgs bundles on a smooth projective variety, or the category of parabolic bundles with quasi-parabolic structure prescribed on a fixed divisor with simple normal crossings and the weights satisfying some natural conditions. Then a principal Higgs bundle or principal parabolic bundle could be defined as a functor satisfying Nori's axioms with values in the Higgs category or parabolic category (cf.[Si]). This method can be used to construct moduli spaces of these objects.

Remark 3.9. The Narasimhan-Seshadri theorem illustrates in a natural manner the philosophy underlying Grothendieck's Tannakian formalism wherein "fiber functors" play the role of points. Let V be a stable bundle of degree 0; it is known that it is a flat bundle associated to the representation ρ. Let $\mathcal{C}(V)$ be the full sub-category consisting of subquotients of tensor powers of V and V^*. Then, this gives an example of a Tannaka category. The Grothendieck-Tannaka theorem realises this category as a category of representations of a pro-algebraic group scheme $M(V)$. The image of $M(V)$, i.e $Im(M(V)) \in GL(V_x)$ is the monodromy group of the representation ρ i.e the Zariski closure of $\rho(\pi(X, x))$ in $GL(n)$ (See [BK] for generalizations to higher dimensional varieties).

3.0.3 Semistability and stability of principal bundles

Definition 3.10. Fix a Borel subgroup B and a maximal torus $T \subset B$. Let P be a parabolic subgroup of the reductive group G. A character $\chi : P \longrightarrow \mathbf{C}^*$ is said to be *dominant* if it is trivial on the connected component of the centre of G and such that with respect to positive roots $\alpha \in X^*(T)$, we have $(\alpha, \chi|_T) \geq 0$. This notion does not depend on the choice of B. Here $(\ ,\ )$ denotes the natural W–invariant pairing on $X^*(T) \otimes \mathbf{Q}$.

Remark 3.11. Let χ be a character of P. Then, we consider the P–bundle $G \longrightarrow G/P$. The character χ then defines a line bundle L_χ on G/P. The property of dominance of χ is equivalent to the ampleness of the line bundle L_χ^*.

Definition 3.12. (A. Ramanathan)([R1], [R2]) E is *semistable* if $\forall$ parabolic subgroup P of H, $\forall$ reduction $\sigma_P : X \longrightarrow E(H/P)$ and $\forall$ dominant character χ of P, the bundle $\sigma_P^*(L_\chi)$ has degree ≤ 0.

Theorem 3.13. (A.Ramanathan) Let H be semisimple. Then E is semistable in the above sense if and only if it is so in the sense of Definition 3.6.

Remark 3.14. The proof of this theorem is quite non-trivial. I will simply refer the reader to the paper ([RR]) which in fact gives an algebraic proof of this result. We note that the assumption of semisimplicity of H can be replaced by reductivity of H in Ramanathan's definition. This extension to reductive H will be needed below when we handle Levi subgroups of parabolic subgroups.

Remark 3.15. On comparing the definition of semistability and stability of vector bundles (Definition 3.1) with Ramanathan's definition given above, it seems as if one is only interested in imposing Ramanathan's condition for maximal parabolic subgroups. This is indeed correct. It is not too hard to see that by expressing the dominant character of a given parabolic in terms of the characters of the maximal parabolic subgroups containing it, one can always reduce to the case of maximal parabolic subgroups. Once we reduce to checking the conditions for a maximal parabolic P, since $Pic(G/P) = \mathbf{Z}$ it follows that there is only one inequality to be checked as in the case of vector bundles (where again the data coming from all characters of the parabolic subgroups do not figure in the inequalities).

Remark 3.16. For the most part we will be working with curves of $g \geq 2$. If $g = 0$ we have the Grothendieck-Harder theorem (G. Harder proved it over fields of positive characteristic) which says that all G–bundles (G, connected and reductive) have reduction of structure group to the maximal torus upto action by the Weyl group.

3.0.4 The Harder-Narasimhan filtration

Let V be a vector bundle on X. Then recall that there is a unique filtration of V

$$0 = V_0 \subset V_1 \subset V_2 \subset ... \subset V_{l-1} \subset V_l = V$$

by subbundles and such that the filtration is characterised by the two conditions, namely, that all the quotients V_i/V_{i-1} are *semistable* for $i \in [1, l]$ and further, the slopes $deg(V_i/V_{i-1})/rank(V_i/V_{i-1})$ are strictly decreasing as i increases. This filtration was introduced by Harder and Narasimhan (see [HN]) and is called the HN-filtration of V. This allows us to reduce, in some sense, all questions on vector bundles to semistable vector bundles. In the principal bundle setting we have the following:

Definition 3.17. (Harder-Narasimhan Reduction) Let E be a principal G bundle on X and (P, σ_P) be a reduction of structure group of E to a parabolic subgroup P of G, then this reduction is called H-N reduction if the following two conditions hold:

(a) If L is the Levi quotient of P then the principal L–bundle $E_P(L) = E_P \times^P L$ over X is a semistable L–bundle.

(b) For any dominant character χ of P with respect to some Borel subgroup $B \subset P$ of G, the associated line bundle L_χ over X has degree > 0.

Remark 3.18. For $G = GL(n, k)$ a reduction E_P gives a filtration of the rank n vector bundle associated to the standard representation. With a bit of work one can check that E_P is an H-N reduction in the above sense if and only if the corresponding filtration of the associated vector bundle filtration coincides with its Harder Narasimhan filtration (see [AB, Chapter 10]).

Theorem 3.19. (Grothendieck) (See [G]) Let E be a principal G–bundle on $\mathbf{P}^1$, where G is connected reductive. Then E has a reduction of structure group to the maximal torus of G unique upto an action of the Weyl group.

Proof: The proof proceeds as follows. Case 1: E is a semistable G–bundle.

Now we reduce further to the case when G is semisimple, in which case Grothendieck's theorem would read:

If E is a semistable bundle with semisimple structure group, then E is trivial.

For this, consider the quotient group $\overline{G}$ associated to G by going modulo the identity component of the center. Then, since E is semistable, it follows that $E(\overline{G})$ is also semistable. This needs a proof but is not difficult. (see [R2])

Now since $\overline{G}$ is semisimple, we get, by the *case for semisimple groups* of Grothendieck's theorem, that $E(\overline{G})$ is *trivial.* This implies that the structure group of E reduces to the identity component of the center of G proving the theorem.

To complete the proof we need to prove the result if the group G is *semisimple* then the only semistable bundle will be the trivial one. Let $G \hookrightarrow SL(V)$ be a faithful representation. Then consider $E(V)$ the extension of structure group to $SL(V)$. Since $E(V)$ is semistable with trivial determinant, by Riemann-Roch, $E(V)$ has a section, which will be nowhere vanishing (because of the semistability and degree 0 property of $E(V)$). Since we are on $\mathbf{P}^1$, we get a splitting of $E(V)$ with a trivial factor. Now an induction on rank of $E(V)$ proves that $E(V)$ is *trivial.* Since $E(V)$ comes with a reduction of structure group to G, we get a section of $E(SL(V)) \times^{SL(V)} SL(V)/G$. Since $E(SL(V))$ is trivial, this section gives a map from $\mathbf{P}^1 \longrightarrow SL(V)/G$ which will be constant, since $SL(V)/G$ is an affine variety. This implies that E is trivial, completing the proof.

Case 2: E is not semistable. Then we reduce to the semistable case as follows:

Suppose that E is unstable. Then by the existence of HN reduction, we have a parabolic subgroup $P \subset G$ and a reduction of structure group E_P to P such that the associated Levi bundle $E_P(L)$ (via the quotient $P \to P/U \simeq L$) is *semistable.* It follows by the earlier case that $E_P(L)$ has a reduction of structure group to the maximal torus.

Let E' be the P–bundle obtained from $E_P(L)$ by extension of structure group $L \hookrightarrow P$ (choose "a" Levi subgroup of P). Clearly, E' has a reduction of structure group to the maximal torus.

We will therefore be done if we show that $E' \simeq E_P$ as P–bundles.

We observe that the parabolic subgroup is a semi-direct product of L and the unipotent radical U. The point to note is that U has a filtration by the additive group $\mathbf{G_a}$ and hence on $\mathbf{P^1}$, one can prove without much difficulty that $H^1(\mathbf{P^1}, U) = 0$. Observe that we have an exact sequence of cohomology sets:

$$H^1(\mathbf{P^1}, U) \longrightarrow H^1(\mathbf{P^1}, P) \longrightarrow H^1(\mathbf{P^1}, L)$$

and the vanishing of $H^1(\mathbf{P^1}, U)$ and the exactness of the sequence should be interpreted as saying that two P–bundles are isomorphic if they are so as L–bundles, when the structure group is extended via $P \to P/U \simeq L$. This implies that $E' \simeq E_P$ as P–bundles.

As regards the orbit under the Weyl group action we note that $W \simeq N(T)/T$ and $N(T)$ acts as inner automorphisms in G and T is stable under the operations. This gives an action of $N(T)$ on $H^1(\mathbf{P^1}, T)$ compatible with the action on $H^1(\mathbf{P^1}, G)$. The uniqueness of the reduction upto the action of W is the content of [G, Section 2] and we refer the reader to the paper for the complete proof.

Remark 3.20. In the case of elliptic curves the same reduction statement is true for semistable bundles with vanishing Chern classes (true when G is semisimple!) and the moduli is simply the X^k/W. In the case of vector bundles it is the symmetric product of the curve (Atiyah). For the more general principal bundle case see [La] or [FMW].

Definition 3.21. A reduction of structure group of E to a parabolic subgroup P is called *admissible* if for any character χ on P which is trivial on the center of H, the line bundle associated to the reduced P-bundle E_P has degree zero.

Definition 3.22. An H-bundle E is said to be *polystable* if it has a reduction of structure group to a Levi R of a parabolic P such that the reduced R-bundle E_R is stable and the extended P bundle $E_R(P)$ is an admissible reduction of structure group for E.

Remark 3.23. One can define the associated graded $gr(E)$ of a semistable E as follows: firstly one shows that there exists an admissible reduction of structure group to P with the added property that $E_P(R)$ is stable as an R-bundle. Note that in general R is only a reductive group. We define $gr(E) := E_P(R)(G)$. This will be the unique closed orbit (in the GIT construction) in the orbit closure of E.

In the case $H = GL(n)$, if E is the associated vector bundle, an example of an admissible reduction is writing E as an extension, namely:

$$0 \longrightarrow E_1 \longrightarrow E \longrightarrow E_2 \longrightarrow 0$$

where $deg(E_i) = 0$. One knows by the theory of semistable bundles that in general E has a filtration $E_k \subset ... \subset E_1 \subset E$ of bundles such that the successive quotients E_i/E_{i+1} are stable of degree 0 and then we define $gr(E) \simeq \oplus E_i/E_{i+1}$.

See Lemma 4.9 for the corresponding notion for principal bundles and also see [R1] for the precise definition of the associated graded of a semistable principal bundle.

4 Moduli spaces of principal bundles

The moduli space of principal G-bundles (for a reductive algebraic group G) on a smooth projective curve X was constructed by A. Ramanathan over fields of characteristic zero (cf. [R1] [R2]). His method using Geometric Invariant Theory followed the basic lines of the construction for the case of vector bundles (cf. [Ses]). The properness (and hence the projectivity) of the moduli space is an end product of this method of construction. One knows that this property (for the case of vector bundles) could be proved a priori, before constructing the moduli spaces and is referred to as the *semistable reduction theorem* (cf Langton [L]). In this section I outline a construction from a paper of mine with Seshadri (see [BS]). This method has certain advantages in that it gives strategies for generalisation to curves over fields of positive characteristic (see [BP2]) as well as to some moduli constructions over more general smooth projective varieties (see [B]). Due to contingencies of space and also since these notes essentially presented a single perspective of the subject I have not devoted any part of the notes to many other papers in the field where the study of principal bundles on curves have been carried out. Among many papers, I will draw the attention of the reader to the following: The papers of Kumar-Narasimhan-Ramanathan ([KNR]) and the papers of Beauville-Laszlo ([BL]) study the moduli stack of principal bundles over curves and compute the Picard groups of these stacks. This approach follows the classic paper of A.Weil ([We]) which initiated the adelic method of construction of these stacks. The reader should refer to the notes by C.Sorger ([So]) for a comprehensive survey of these approaches. The moduli spaces of principal bundles have been generalized along with the notion which generalizes the Gieseker-Maruyama stability. There has been much work on this in recent years. See for example the papers of A.Schmitt ([Sch0]), Gomez-Sols ([GS]); the very recent preprint by Gomez-Langer-Schmitt-Sols ([GLSS]) carries out the construction over fields of positive characteristics. See also the survey by Alexander Schmitt appearing in this volume.

4.1 Construction of the moduli space

For the present purpose, we take $G = SL(n, \mathbb{C})$ and $H \subset G$ a semisimple subgroup. We will be interested in constructing a moduli space of H-bundles. Roughly, we wish to give the set M_H of isomorphism classes of H-bundles the structure of an algebraic scheme in a *natural way*. Natural here could mean representing (in a certain *coarse* sense) the functor F_H or equivalently, given a family of H-bundles $F \longrightarrow X \times T$, parametrised by an algebraic scheme T, the natural set map

$$t \mapsto (F_t)$$

from

$$T \longrightarrow M_H$$

is a **morphism**, where by F_t we mean $F|_{X \times t}$ and by (F_t) we mean the isomorphism class of the bundle F_t. This is almost okay, except that we would need to take the normalisation of the space M_H to get the moduli space. We note that there are *bad* bundles forcing us to refine our equivalence from isomorphism classes to S-equivalence. Indeed one can have a family of H-bundles $\{E_t\}_{t \in \mathbf{A}^1}$, on $\mathbf{A}^1 \times X$, parametrised by the affine line, such that $\forall\, t \neq 0 \in \mathbf{A}^1$ $E_t \simeq E$ and $E_0 \simeq gr(E)$. See the last part of construction below (Proposition 4.8) where this phenomenon arises while defining the points of the moduli space. The S-equivalence is in some sense identifying the general points on this family and the points in the limit so as to make the resultant moduli a scheme.

We recall very briefly the Grothendieck Quot scheme used in the construction of the moduli space of vector bundles (cf.[N]).

Let $\mathcal{F}$ be a coherent sheaf on X and let $\mathcal{F}(m)$ be $\mathcal{F} \otimes \mathcal{O}_X(m)$ (where following the standard notation $\mathcal{O}_X(1)$ is a fixed very ample line bundle on X and $\mathcal{O}_X(m)$ its m-fold tensor power).

Choose an integer $m_0 = m_0(n, d)$ ($n = $ rk, $d = $ deg) such that for any $m \geq m_0$ and any semistable bundle V of rank n and deg d on X and we have $h^1(V(m)) = 0$ and $V(m)$ is generated by its global sections.

Let $\chi = h^0(V(m))$ and consider the Quot scheme Q consisting of coherent sheaves $\mathcal{F}$ on X which are quotients of $\mathbb{C}^\chi \otimes_{\mathbb{C}} \mathcal{O}_X$ with a fixed Hilbert polynomial P. The group $\mathcal{G} = GL(\chi, \mathbb{C})$ canonically acts on Q

and hence on $X \times Q$ (trivial action on X) and lifts to an action on the universal sheaf $\mathcal{E}$ on $X \times Q$.

Let R denote the $\mathcal{G}$-invariant locally closed subset of Q defined by

$$R = \left\{ q \in Q \mid \begin{array}{l} \mathcal{E}_q = \mathcal{E} \mid_{X \times q} \text{ is locally free s.t. the canonical map} \\ \mathbb{C}^\chi \longrightarrow H^0(\mathcal{E}_q) \text{ is an } \textit{isomorphism, } \det \mathcal{E}_q \simeq \mathcal{O}_X \end{array} \right\}.$$

We denote by Q^{ss} the $\mathcal{G}$-invariant open subset of R consisting of semistable bundles and let $\mathcal{E}$ continue to denote the restriction of $\mathcal{E}$ to $X \times Q^{ss}$.

Henceforth, 'by abuse of notation', we shall write Q for Q^{ss}.

Definition 4.1. Recall the definition of a *good quotient*. Let $H \times T \longrightarrow T$ be an action of H on T. $p : T \longrightarrow Y = T//H$ is called a good quotient if:

- (a) p is onto, affine and H-invariant.
- (b) $p_*(\mathcal{O}_T^H) = \mathcal{O}_Y$ where $\mathcal{O}_T^H$ is the sheaf of H-invariant sections.
- (c) Closed H–invariant subsets of T map to closed subsets of Y and disjoint closed H–invariant subsets go to disjoint closed sets by p.

If all the orbits are closed then p is called a geometric quotient. If further the action map $(H \times T) \longrightarrow (T \times T)$ is a closed embedding, in which case we say the action is *free* (in the strong sense!), then, p is an H-torsor or principal bundle. There can be situations where the action is set theoretically free and a *geometric quotient* exists but with p not being a principal bundle (see [GIT, Example 0.4, Page 11])). However, in char 0 (as in our case), by Luna's theorem a geometric quotient by a *free* action is always a principal bundle (see Newstead [N]).

4.2 The construction of the moduli space for principal bundles

Fix a base point $x \in X$ (cf. Remark 2.3). Let $q'' : (\text{Sch}) \longrightarrow (\text{Sets})$ be the following functor:

$$q''(T) = \left\{ (V_t, s_t) \mid \begin{array}{l} \{V_t\} \text{ is a family of semistable} \\ \text{principal } G\text{-bundles} \\ \text{parametrised by } T \text{ and} \\ s_t \in \Gamma(X, V(G/H)_t) \ \forall \, t \in T \end{array} \right\}.$$

i.e. $q''(T)$ consists pairs of rank n vector bundles (or equivalently principal G-bundles) together with a reduction of structure group to H.

By appealing to the general theory of Hilbert schemes, one can show without much difficulty (cf. [R1, Lemma 3.8.1]) that q'' is representable by a Q-scheme, which we denote by Q''.

The universal sheaf $\mathcal{E}$ on $X \times Q$ is in fact a vector bundle. Denoting by the same $\mathcal{E}$ the associated principal G-bundle, set $Q' = (\mathcal{E}/H)_x$. Then in our notation $Q' = \mathcal{E}(G/H)_x$ i.e. we take the bundle over $X \times Q$ associated to $\mathcal{E}$ with fibre G/H and take its restriction to $x \times Q \approx Q$. Let $f : Q' \longrightarrow Q$ be the natural map. Then, since H is reductive, f is an *affine morphism*.

Observe that Q' parametrises semistable vector bundles together with *initial values at x of possible reductions to H*.

Define the "evaluation map" of Q-schemes as follows:

$$\phi_x : Q'' \longrightarrow Q'$$

$$(V, s) \longmapsto (V, s(x)).$$

We then have the following two technical lemmas. The reader may refer to [BS, Lemma 8.1] for a proof of the first one.

Lemma 4.2. The evaluation map $\phi_x : Q'' \longrightarrow Q'$ is *proper*.

Lemma 4.3. The evaluation map ϕ_x is *injective*.

Proof. Since H is semisimple, it follows that G/H is affine and by a classical result of Chevalley, we have a G–embedding $G/H \hookrightarrow W$ in a finite dimensional G–module W. Let (E, s) and $(E', s') \in Q''$ such that $\phi_x(E, s) = \phi_x(E', s')$ in Q'. i.e. $(E, s(x)) = (E', s'(x))$. So we may assume that $E \simeq E'$ and that s and s' are two different sections of $E(G/H)$ with $s(x) = s'(x)$.

Using $G/H \hookrightarrow W$, we may consider s and s' as sections in

$$\Gamma(X, E(W)).$$

Observe that since E is semistable of degree 0, so is the associated bundle $E(W)$.

Recall the following fact:

If E and F are semistable vector bundles with $\mu(E) = \mu(F)$, then the evaluation map

$$\phi_x : \mathrm{Hom}(E, F) \longrightarrow \mathrm{Hom}(E_x, F_x) \qquad (*)$$

is *injective*.

In our situation, since $s, s' \in H^0(\mathcal{O}_X, E(W))$, by $(*)$ since $\phi_x(s) = \phi_x(s')$, we get $s = s'$, proving *injectivity*.

Remark 4.4. It is immediate that the $\mathcal{G}$-action on Q lifts to an action on Q''.

Recall the commutative diagram

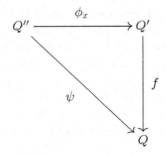

By Lemma 4.2 and Lemma 4.3, ϕ_x is a proper injection and hence affine. One knows that f is affine (with fibres G/H). Hence ψ is a $\mathcal{G}$–*equivariant affine morphism.*

Remark 4.5. Let (E, s) and (E', s') be in the same $\mathcal{G}$-orbit of Q''. Then we have $E \simeq E'$. Identifying E' with E, we see that s and s' lie in the same orbit of $\mathrm{Aut}_{G}E$ on $\Gamma(X, E(G/H))$. Then using Remark 2.13 we see that the reductions s and s' give isomorphic H-bundles.

Conversely, if (E, s) and (E', s') are such that $E \simeq E'$ and the reductions s, s' give isomorphic H-bundles, then using once again Remark 2.13 we see that (E, s) and (E', s') lie in the same $\mathcal{G}$-orbit.

Consider the $\mathcal{G}$-action on Q'' with the linearisation induced by the *affine $\mathcal{G}$-morphism* $Q'' \longrightarrow Q$. It is seen without much difficulty that, since a good quotient of Q by $\mathcal{G}$ exists and since $Q'' \longrightarrow Q$ is an affine $\mathcal{G}$–equivariant map, a good quotient $Q''/\mathcal{G}$ exists (cf. [N] Lemma 4.1).

Moreover by the universal property of categorical quotients, the canonical morphism

$$\overline{\psi} : Q''//\mathcal{G} \longrightarrow Q//\mathcal{G}$$

is also *affine.*

Theorem 4.6. Let $M_X(H)$ denote the scheme $Q''//\mathcal{G}$. Then this scheme is the coarse moduli scheme of semistable H-bundles. Further $M_X(H)$ is projective and moreover if $H \hookrightarrow GL(V)$ is a faithful representation, then the canonical morphism $\overline{\psi} : M_X(H) \longrightarrow M_X(GL(V))$ is *finite.*

Proof. We need only check the last statement. Theorem 4.10, which is termed the *semistable reduction theorem* then implies that the moduli space $M_X(H)$ is projective, and therefore $\overline{\psi}$ is *proper*. By the remarks made above $\overline{\psi}$ is also *affine*, therefore it follows that $\overline{\psi}$ is *finite*.

Remark 4.7. We have supposed that H is semisimple; however, it is not difficult to treat the more general case when H is *reductive*. Let H be then reductive and $\overline{H} = H$ mod centre, its adjoint group. Let P be a principal H-bundle and $\overline{P}$ the $\overline{H}$-bundle, obtained by extension of structure groups. *We define P to be semistable if $\overline{P}$ is semistable.*

Recall that over a curve the topological type of the bundle is completely determined by $\pi_1(G)$. For example, consider the exact sequence (assume the G is semisimple)

$$1 \longrightarrow \pi_1(G) \longrightarrow G^{sc} \longrightarrow G \longrightarrow 1$$

Then one observes that for the simply connected group, principal bundles are of one topological type (namely trivial!). Then follow the cohomology map

$$H^1(X, \mathbf{G}) \longrightarrow H^2(X, \pi_1(G)) \simeq \pi_1(G)$$

where, $\mathbf{G}$ is the sheaf of continuous functions with values in G. If we fix a topological isomorphism class c for principal H-bundles, this fixes a topological isomorphism $\overline{c}$ for principal $\overline{H}$-bundles. Then the moduli space $M_X(H)_c$ is "essentially" $M_X(\overline{H})_{\overline{c}} \times$ (product of Jacobians). This can be made rigorous and leads to the construction of $M_X(H)_c$.

4.3 Points of the moduli

In this subsection we will briefly describe the k-valued points of the moduli space $M_X(H)$. The general functorial description of $M_X(H)$ as a coarse moduli scheme follows by the usual process.

Proposition 4.8. The "points" of $M_X(H)$ are given by isomorphism classes of polystable principal H-bundles.

We firstly remark that since the quotient $q : Q'' \longrightarrow M_X(H)$ obtained above is a good quotient, it follows that each fibre $q^{-1}(E)$ for $E \in M_X(H)$ has a unique closed $\mathcal{G}$-orbit. Let us denote an orbit $\mathcal{G} \cdot E$ by $O(E)$. The proposition is a consequence of the following:

Lemma 4.9. If $O(E)$ is closed then E is polystable.

Proof. (Idea of proof) Recall the definition of a polystable bundle Def 3.22 and the definition of *admissible reductions* Def 3.21. If E has no admissible reduction of structure group to a parabolic subgroup then it is polystable and there is nothing to prove.

Suppose then that E has an admissible reduction E_P, to $P \subset H$. Recall by the general theory of parabolic subgroups that there exists a 1-PS $\xi : \mathbf{G}_m \longrightarrow H$ such that $P = P(\xi)$. Let $L(\xi)$ and $U(\xi)$ be its canonical Levi subgroup and unipotent subgroup respectively. The Levi subgroup will be the centraliser of this 1-PS ξ and one knows $P(\xi) = L(\xi) \cdot U(\xi) = U(\xi) \cdot L(\xi)$. In particular, if $h \in P$ then $lim_{t \to 0} \xi(t) \cdot h \cdot \xi(t)^{-1}$ exists. From these considerations one can show that there is a morphism

$$f : P(\xi) \times \mathbf{A}^1 \longrightarrow P(\xi)$$

such that $f(h, 0) = m \cdot u$, where $h \in P$ and $h = m \cdot u$, $m \in L$ and $u \in U$. (see [R1, Lemma 3.5.12])

Consider the P-bundle E_P. Then, using the natural projection $P \longrightarrow L$ where $L = L(\xi)$, we get an L-bundle $E_P(L)$. Again, using the inclusion $L \hookrightarrow P \hookrightarrow H$, we get a new H-bundle $E_P(L)(H)$. Let us denote this H-bundle by $E_P(L, H)$. It follows from the definition of admissible reductions and polystability that $E_P(L, H)$ is *polystable*.

Further, from the family of maps f defined above, and composing with the inclusion $P(\xi) \hookrightarrow H$ we obtain a family of H-bundles $E_P(f_t)$ for $t \neq 0$ and all these bundle are isomorphic to the given bundle E. Following ([R1, Prop.3.5 pp 313]), one can prove that the bundle $E_P(L, H)$ is the limit of $E_P(f_t)$. It follows that $E_P(L, H)$ is in the $\mathcal{G}$-orbit $O(E)$ because $O(E)$ is closed. This implies that $E \simeq E_P(L, H)$, implying that E is polystable. Q.E.D.

To complete the projectivity of the moduli space, we need the following result. (cf [BS] and [F] for proofs, as well as the recent paper [H]).

Theorem 4.10. Let P_K be a family of semistable principal H-bundles on $X \times \operatorname{Spec} K$, or equivalently, if H_K denotes the group scheme $H \times \operatorname{Spec} K$, a semi-stable H_K-bundle P_K on X_K. Then there exists a finite extension L/K, with the integral closure B of A in L, such that, P_K, after base change to $\operatorname{Spec} B$, extends to a semistable H_B-bundle P_B on X_B.

Remark 4.11. We outline in brief the classical Langton for vector bundles (cf [L]): We are given V_K a semistable family on X_K. Firstly, there exists an extension of V_K to a V_A which we may choose as reflexive since we are on a family of curves. (the choice is quite non-canonical as can be seen from Proposition 6 of Langton's paper). If the special fibre V_k is not semistable then we let $F_i(V_k)$ be the first non-trivial term in the HN-filtration. We may assume that the extension V_A has been chosen that $\mu(F_i)$ is minimal and also among these with the least $rk(F_i)$. We get a new model $V_A' = ker(V_A \longrightarrow V_k/F_i(V_k))$ and an extension

$$0 \longrightarrow V_k/F_i(V_k) \longrightarrow V_k' \longrightarrow F_i(V_k) \longrightarrow 0$$

From this, after some work, one shows that $F_i(V_k')$ injects into $F_i(V_k)$. Therefore by minimality, $F_i(V_k') = F_i(V_k)$. Continuing, one gets a decreasing family of models for V_K. Taking their intersection (making sense of it!) one gets a sub $W_A \subset V_A$ such that $W_k \simeq F_i(V_k)$. Thus, $\mu(W_K) = \mu(F_i)$, which is greater than $\mu(V_K)$, since V_k is not semistable. This implies that V_K itself is not semistable which gives a contradiction!

Remark 4.12. The first crucial difficulty in proving our theorem is even at the very first step. Namely, there is no choice of extension of the family E_K as an H_A family or even as a torsor with a modified group scheme. If one chooses a representation then the associated vector bundle family via Langton's theorem has no canonical extension (be it the limit or the family!). Indeed, we cannot simply use Langton's theorem as we require the limiting bundle to be polystable, which can be ensured only by the GIT construction of the moduli space of vector bundles. We refer the reader to some recent work by Heinloth ([H]), who gives a Langton–type proof of the semistable reduction theorem which handles even positive characteristics. Heinloth, using a theorem of Drinfeld and Simpson ([DS]) as well as the ind-projectivity of the affine grassmannian gets around our initial difficulty of extending the family of principal bundles across the puncture without changing the structure group. The problem then is exactly as in Langton's proof, namely to modify the possibly unstable extension using principal bundle analogues of "Hecke" correspondences to get a semistable extension. We note that Heinloth's arguments work only over curves.

Remark 4.13. Let $H \subset G$, where G is a linear group. In the notation of §2 let F_H and F_G stand for the functors associated to families of semistable bundles of degree zero. The inclusion of H in G induces a morphism of functors $F_H \longrightarrow F_G$. We remark that the semistable reduction theorem for principal H-bundles **need not** imply that the induced morphism $F_H \longrightarrow F_G$ is a proper morphism of functors. Indeed, this does not seem to be the case. However, it does imply that the associated morphism at the level of moduli spaces is indeed proper (see Theorem 4.6).

Remark 4.14. We need to go to a ramified cover to extend our family unlike the vector bundle case. This can be seen in the following example: Let (V_A, q_A) be a family of trivial vector bundles equipped with a non-degenerate quadratic form thus making it a family of orthogonal bundles. Consider the family $(V_A \oplus V_A, q_A \oplus t.q_A)$ where t is the uniformising parameter. This is generically a family of orthogonal bundles the limit being the trivial bundle with degenerate quadratic form. The vector bundle family has trivially extended but the quadratic form fails to extend as a non-degenerate form. By going to the quadratic extension (totally ramified) $K(\sqrt{t})$ and letting B be the integral closure of A in this extension, we see that by modifying the quadratic form in its equivalence, namely $Q.X.Q^t$ using the matrix $Q = (q_{ij})$ where

$$Q = \begin{pmatrix} 1 & 0 \\ 0 & \sqrt{(t)}^{-1} \end{pmatrix}$$

we can extend the family to the closed fibre.

Acknowledgments I am grateful to Alexander Schmitt for his numerous suggestions and corrections of an earlier manuscript of these notes. I thank both the referees, one for the criticisms and the other for the creative suggestions. Both have helped make the notes more polished than before.

Bibliography

[AB] M.F. Atiyah and R. Bott: *The Yang-Mills equations over Riemann Surfaces*, Phil. Trans. Roy. Soc. Lond. A **308** (1983), 523–615.

[BS] V.Balaji and C.S.Seshadri: *Semistable Principal Bundles-I (characteristic zero)* Journal of Algebra, **258** (2002), pp 321–347.

[BK] V.Balaji and J.Kollar: *Holonomy groups of stable vector bundles*, (to appear in RIMS Journal, Kyoto) (2008) (archiv:math.AG/0601120)

[BP2] V.Balaji and A.J.Parameswaran: *Semistable principal bundles-II (in positive characteristics)*, Transformation Groups, **18**, No 1, (2003), pp 3–36.

[B] V.Balaji: *Principal bundles on projective varieties and the Donaldson-Uhlenbeck compactification*, Journal Differential Geometry **76**, (2007), (arXiv:math.AG/0505106).

[BL] A.Beauville and Y.Laszlo: *Conformal blocks and generalized theta functions*, Journal Communications in Mathematical Physics **164**, Number 2, (1994) pp 385–419.

[DM] P.Deligne and D.Mumford: *The irreducibility of the space of curves of a given genus*, Publ I.H.E.S. **36** (1969) pp 75–109.

[DS] V.Drinfeld and C.Simpson: *B-Structures on G−bundles and local triviality*, Math. Res. Lett. **2**, (1995) 823–829.

[F] G.Faltings: *Stable G-bundles and projective connections*, J. Alg. Geom., **2** (1993), 507–568.

[FMW] R. Friedman, J.W. Morgan, E. Witten: *Principal G−bundles over elliptic curves*, Math. Res. Lett, **5** (1998), 97–118.

[GIT] D. Mumford, J. Fogarty and F. Kirwan: *Geometric invariant theory*, Ergebnisse der Mathematik und ihrer Grenzgebiete (2), Springer-Verlag, Berlin, 1994.

[GS] T.Gómez and I.Sols: *Moduli spaces of principal sheaves over projective varieties*, Annals of Math. **161** (2005) no 2, 1037–1092.

[GLSS] T.L. G'omez, A. Langer, A.H.W. Schmitt, I. Sols: *Moduli Spaces for Principal Bundles in Arbitrary Characteristic*, (math.AG/0506511)

[G] A.Grothendieck: *Sur la classification des fibrés holomorphes sur la sphère de Riemann*, Amer.J.Math. **79** (1956), 121–138.

[HN] G. Harder and M. S. Narasimhan: *On the cohomology groups of moduli spaces of vector bundles on curves*, Math. Ann., 212 (1975), 215–248.

[H] J. Heinloth : *Semistable reduction for G−bundles on curves*, J. Algebraic Geom. (to appear),http : //www.ams.org/distribution/jag/

[KNR] S.Kumar, M.S.Narasimhan and A.Ramanathan: *Infinite Grassmanians and moduli spaces of G−bundles*, Math Annalen, **300**(1994), 41–75.

[L] S.Langton: *Valuative criterion for families of vector bundles on algebraic varieties*, Annals of Mathematics (2) **101** (1975) pp 88–110.

[La] Y. Laszlo :*About G−bundles over elliptic curves*, Annales de l'institut Fourier, 48 no. 2 (1998), p. 413–424

[M] J.Milne : *Étale Cohomology*, Princeton University Press,

[NS] M.S. Narasimhan and C.S.Seshadri: *Stable and unitary vector bundles on a compact Riemann surface*, Annals of Mathematics (2)**82** (1965) pp 540–567.

[N] P.Newstead: *Introduction to moduli problems an orbit spaces*, Tata Inst. of Fundamental Research, Vol 51, Springer Verlag, 1978.

[No] M.V. Nori: *The fundamental group scheme*, Proc. Ind. Acad. Sci. Math. Sci. **91** (1982), 73–122.

[R1] A. Ramanathan: *Stable principal bundles on a compact Riemann surface - Construction of moduli space* (Thesis), Bombay University 1976, Proc.Ind.Acad.Sci., **106**, (1996), 301–328 and 421–449.

[R2] A. Ramanathan: *Stable principal bundles on a compact Riemann surface*, Math. Ann. **213** (1975), 129–152.

[RR] S.Ramanan and A.Ramanathan: *Some remarks on the instability flag*, Tohoku Math. Journal **36** (1984), 269–291.

[Sch0] A.Schmitt: *Singular Principal Bundles over higher dimensional manifolds and their moduli spaces*, IMRN, **23** (2002), 1183–1209.

[O] Olivier Serman: *Moduli spaces of orthogonal and symplectic bundles over an algebraic curve* (preprint).

[S] J.P.Serre: *Espaces fibrés algébriques*, Anneaux de Chow et applications, Séminaire Chevalley, (1958), also Documents Mathématiques **1** (2001), 107–140.

[Ses] C.S. Seshadri: *Fibrés vectoriels sur les courbes algébriques*, Asterisque., **96** (1982).

[Si] C. Simpson: *Higgs bundles and Local systems*, Pub. I.H.E.S. **75** (1992), 5–95.

[So] C.Sorger: *Lectures on moduli of principal G−bundles over algebraic curves*, ICTP Lecture Notes (1995).

[We] A.Weil: *Généralisation des fonctiones abeliennes*, J.Math. pures. appl. **17**, (1938), 47–87.

2

Brill-Noether Theory for stable
vector bundles

Ivona Grzegorczyk

Department of Mathematics,
California State University Channel Islands,
Camarillo, CA 91320
e-mail ivona.grze@csuci.edu;

Montserrat Teixidor i Bigas

Mathematics Department,
Tufts University,
Medford MA 02155
e-mail: montserrat.teixidoribigas@tufts.edu

Abstract

This paper gives an overview of the main results of Brill-Noether Theory for vector bundles on algebraic curves.

1 Introduction

Let C be a projective, algebraic curve of genus g which will be non-singular most of the time. Let $U(n, d)$ be the moduli space of stable bundles on C of rank n and degree d. A Brill-Noether subvariety $B_{n,d}^k$ of $U(n, d)$ is a subset of $U(n, d)$ whose points correspond to bundles having at least k independent sections (often denoted by $W_{n,d}^{k-1}$). Brill-Noether theory describes the geometry of $B_{n,d}^k$ and in this paper we present the foundations of this theory and an overview of current results. Even the basic questions such as when $B_{n,d}^k$ is non-empty, what is its dimension and whether it is irreducible are subtle and of great interest. While the topology of $U(n, d)$ depends only on g, the answers to the above questions turn out to depend on the algebraic structure of C. In various cases we have definite answers which are valid for a *general* curve, but fail for certain *special* curves. This allows to define subvarieties of the moduli space of curves and contribute to the study of its geometry. For instance, the first counterexample to the Harris-Morrison slope conjecture can be defined in such a way ([FP]).

Classical Brill-Noether Theory deals with line bundles ($n = 1$). The natural generalization of most of the results known in this case turn out to be false for some $n \geq 2$ and particular values of d, k: the Brill-Noether locus can be empty when the expected dimension is positive and non-empty when it is negative even on the generic curve, it can be reducible on the generic curve with components of different dimensions and the singular locus may be larger than expected.

Moreover, there are very few values of d, n, k, $n \geq 2$ for which there is a complete picture of the situation. Nevertheless, for a large set of values of n, d, k, there are partial (positive) results.

We will start with a scheme theoretic description of $B_{n,d}^k$ and the study of its tangent space. This justifies its expected dimension and singular locus. We will describe the complete solution for small slopes ($\mu \leq 2$) (see [BGN], [Me1]). We will present the results for higher slope from [T1] and [BMNO] and include some stronger results for special curves. In the last section, we concentrate on the case of rank two and canonical determinant.

2 Scheme structure on $B_{n,d}^k$

We start by giving a scheme structure to $B_{n,d}^k$. Assume first for simplicity that the greatest common divisor of n and d is one. There exists then a Poincaré bundle $\mathcal{E}$ on $C \times U(n,d)$. Denote by $p_1 : C \times U(n,d) \rightarrow C$, $p_2 : C \times U(n,d) \rightarrow U(n,d)$ the two projections. Let D be a divisor of large degree on C, $deg(D) \geq 2g - 1 - \frac{d}{n}$. Consider the exact sequence on C

$$0 \rightarrow \mathcal{O}_C \rightarrow \mathcal{O}(D) \rightarrow \mathcal{O}_D(D) \rightarrow 0.$$

Taking the pull back of this sequence to the product $C \times U(n,d)$, tensoring with $\mathcal{E}$ and pushing down to $U(n,d)$, one obtains

$$0 \rightarrow p_{2*}\mathcal{E} \rightarrow p_{2*}\mathcal{E} \otimes p_1^*(\mathcal{O}(D)) \rightarrow p_{2*}(\mathcal{E} \otimes p_1^*(\mathcal{O}_D(D))) \rightarrow \dots$$

Note now that $p_{2*}(\mathcal{E} \otimes p_1^*(\mathcal{O}_D(D)))$ is a vector bundle of rank $n \deg D$ whose fiber over the point $E \in U(n,d)$ is $H^0(C, E \otimes \mathcal{O}_D(D))$. As E is stable and $K \otimes E^*(-D)$ has negative degree, $h^0(K \otimes E^*(-D)) = 0$. Then, $h^0(C, E(D)) = d + n \deg D + n(1 - g) = \alpha$ is independent of the element $E \in U(n,d)$. Hence $p_{2*}(\mathcal{E} \otimes p_1^*(\mathcal{O}_D(D)))$ is a vector bundle of rank α. Define $B_{n,d}^k$ as the locus where the map

$$\varphi : p_{2*}\mathcal{E} \otimes p_1^*(\mathcal{O}(D)) \rightarrow p_{2*}(\mathcal{E} \otimes p_1^*(\mathcal{O}_D(D)))$$

has rank at most $\alpha - k$.

From the theory of determinantal varieties, the dimension of $B^k_{n,d}$ at any point is at least $\dim U(n,d) - k(n\deg D - (\alpha - k)) = \rho$ with expected equality. Moreover, provided $B^{k+1}_{n,d} \neq U(n,d)$, the locus where the rank of the map of vector bundles goes down by one more (namely $B^{k+1}_{n,d}$) is contained in the singular locus, again with expected equality.

Definition 2.1. *The expected dimension of $B^k_{n,d}$ is called the Brill-Noether number and will be denoted by*

$$\rho^k_{n,d} = n^2(g-1) + 1 - k(k - d + n(g-1)).$$

When the rank and the degree are not relatively prime, a Poincaré bundle on $C \times U(n,d)$ does not exist. One can however take a suitable cover on which the bundle exists and make the construction there.

We want to study now the tangent space to $B^k_{n,d}$ at a point E (see [W], [BR]). We first show how the tangent space to $U(n,d)$ at E is identified to $H^1(C, E^* \otimes E)$. Consider a tangent vector to $U(n,d)$ at E, namely a map from $\mathrm{Spec}(\mathbf{k}[\epsilon]/\epsilon^2)$ to $U(n,d)$ with the closed point going to E. This is equivalent to giving a vector bundle $\mathcal{E}_\epsilon$ on $C_\epsilon = C \times \mathrm{Spec}(\mathbf{k}[\epsilon]/\epsilon^2)$ extending E. Then $\mathcal{E}_\epsilon$ fits in an exact sequence

$$0 \to E \to \mathcal{E}_\epsilon \to E \to 0.$$

The class of this extension gives the corresponding element in the cohomology $H^1(C, E^* \otimes E)$.

The bundle $\mathcal{E}_\epsilon$ can be described explicitly: Let $\varphi \in H^1(C, E^* \otimes E)$. Choose a suitable cover U_i of C, $U_{ij} = U_i \cap U_j$. Represent φ by a coboundary (φ_{ij}) with $\varphi_{ij} \in H^0(U_{ij}, Hom(E, E))$. Consider the trivial extension of E to $U_i \times Spec\mathbf{k}(\epsilon)/\epsilon^2$, namely $E_{U_i} \oplus \epsilon E_{U_i}$. Take gluings on U_{ij} given by

$$\begin{pmatrix} Id & 0 \\ \varphi_{ij} & Id \end{pmatrix}.$$

This gives the vector bundle $\mathcal{E}_\epsilon$.

Assume now that a section s of E can be extended to a section s_ϵ of the deformation. Hence there exist local sections $s'_i \in H^0(U_i, E_{|U_i})$ such that $(s_{|U_i}, s'_i)$ define a section of $\mathcal{E}_\epsilon$.

By construction of $\mathcal{E}_\epsilon$ this means that

$$\begin{pmatrix} Id & 0 \\ \varphi_{ij} & Id \end{pmatrix} \begin{pmatrix} s_{|U_i} \\ s'_i \end{pmatrix} = \begin{pmatrix} s_{|U_j} \\ s'_j \end{pmatrix}.$$

The first condition $(s_{|U_i})_{|U_{ij}} = (s_{|U_j})_{|U_{ij}}$ is automatically satisfied as s is a global section. The second condition can be written as $\varphi_{ij}(s) = s'_j - s'_i$. Equivalently, $\varphi_{ij}(s)$ is a cocycle, namely

$$\varphi_{ij} \in Ker \quad (H^1(C, E^* \otimes E) \quad \to \quad H^1(C, E))$$
$$\nu_{ij} \quad \to \quad \nu_{ij}(s) \quad .$$

This result can be reformulated as follows: the set of infinitessimal deformations of the vector bundle E that have sections deforming a certain subspace $V \subset H^0(C, E)$ consists of the orthogonal to the image of the following map

Definition 2.2. *The natural cup-product map of sections*

$$P_V : V \otimes H^0(C, K \otimes E^*) \to H^0(C, K \otimes E \otimes E^*).$$

is called the Petri map. When $V = H^0(C, E)$, we write P for $P_{H^0(C,E)}$.

Note now that if E is stable, $h^0(C, E^* \otimes E) = 1$ as the only automorphisms of E are multiples of the identity. Hence,

$$h^1(C, E^* \otimes E) = h^0(K \otimes E \otimes E^*) = n^2(g-1) + 1 = \dim U(n, d).$$

Take $V = H^0(C, E)$ above and assume that it has dimension k. By Riemann-Roch Theorem, $h^0(C, K \otimes E^*) = k - d + n(g-1)$. Hence,

$$\rho^k_{n,d} = h^1(C, E^* \otimes E) - h^0(C, E)h^1(C, E)$$

Equivalently, $B^k_{n,d}$ is non-singular of dimension $\rho^k_{n,d}$ at E if and only if the Petri map above is injective. This result allows us to present the first example of points in the singular locus of $B^k_{n,d}$ that are not in $B^{k+1}_{n,d}$ (see [T1]).

Example 2.3. Let C be a generic curve of genus at least six. Let L be a line bundle of degree $g - 2$ with at least two independent sections and E a generic extension

$$0 \to L \to E \to K \otimes L^{-1} \to 0.$$

Then, E has two sections and the Petri map is not injective.

Note first that from the condition on the genus and Brill-Noether Theory for rank one, such an L with two sections exists.

It is easy to check with a moduli count that for a generic extension the vector bundle is stable (see [RT]) and has precisely two sections. Dualizing the sequence above and tensoring with K, one obtains an analogous sequence

$$0 \to L \to K \otimes E^* \to K \otimes L^{-1} \to 0.$$

Hence $H^0(C, L)$ is contained both in $H^0(C, E)$ and in $H^0(C, K \otimes E^*)$. Let now s, t be two independent sections of L. Then, $s \otimes t - t \otimes s$ is in the kernel of the Petri map.

For this particular case of rank two and determinant canonical, one can check that these are essentially the only cases in which the Petri map is not injective (see [T3]) but not much is known in general.

3 Classical Results and Brill-Noether geography

We start by looking at low genus and rank.

$g = 0$. Grothedieck proved that all bundles on a projective line are decomposable into direct sums of line bundles (see [Gr]), i.e. $E = L_1 \oplus L_2 \oplus ... \oplus L_n$ and there are no stable vector bundles for $n > 1$. If $d = nd_1 + d_2$, the least unstable vector bundle of rank n and degree d is

$$\mathcal{O}(d_1 + 1)^{d_2} \oplus \mathcal{O}(d_1)^{n-d_2}.$$

This vector bundle has an n^2-dimensional space of endomorphisms.

The space of pairs (E, V) where V is a subspace of dimension k of $H^0(C, E)$ is parameterized by the grassmannian of subspaces of dimension k of $H^0(C, E)$ which has dimension $\rho + n^2 - 1$.

$g = 1$. Atiyah (see [A]) classified the bundles over elliptic curves: if $(n, d) = 1$, the moduli space of stable vector bundles is isomorphic to the curve itself. If $(n, d) = h > 1$, there are no stable vector bundles of this rank and degree and a generic semistable bundle is a direct sum of h stable vector bundles all of the same rank and degree. From Riemann-Roch, if $d > 0$, $h^0(C, E) = d$. Then, the dimension of the set of pairs (E, V) with V a k-dimensional space of sections of $H^0(C, E)$ is the expected dimension $\rho_{n,d}^k$ of the Brill-Noether locus.

$n = 1$. Line bundles of degree d are classified by the Jacobian variety $J^d(C) = Pic^d(C)$, $\dim J^d(C) = g$. All line bundles are stable and $B_{1,d}^k$ are subvarieties of $J^d(C)$ and their expected dimension equals $\rho_{1,d}^k = \rho = g - k(k - d + g - 1)$. To sketch the region of interest for Brill-Noether theory in the (d, k) plane we consider the bounds given by the following classical results ([ACGH]):

Riemann-Roch theorem for line bundles. *Let L be a line bundle on C. Then $h^0(C, L) - h^1(C, L) = d - (g - 1)$.*

Clifford theorem. *Let D be an special effective divisor of degree d and $L = \mathcal{O}(D)$, $d \leq 2g$. Let $h^0(C, \mathcal{O}(D)) = k$. Then*

$$k \leq \frac{1}{2} \deg D + 1 = \frac{1}{2} \deg L + 1.$$

Note, that when $\rho = 0$ we can solve for d

$$d = (k-1)(1 + \frac{g}{k})$$

i.e. $\rho = 0$ is a hyperbola in the d, k-plane passing through $(0,1)$ (corresponding to $\mathcal{O}_C$) and through $(2g-2, g)$ (corresponding to K_C). Points below $\rho = 0$ in the region between the Clifford and the Riemann-Roch lines correspond to $B_{1,d}^k$ of expected positive dimensions. Serre duality for line bundles on a curve C given by $h^0(C, \mathcal{O}(D)) - h^0(C, \mathcal{O}(K-D)) = d + g + 1$ makes two subregions dual to each other, see Picture 1a.

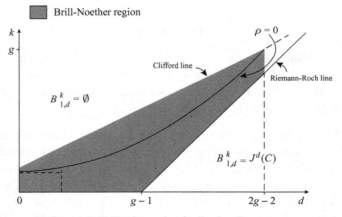

Picture 1a. Brill-Noether region for line bundles.

The following results give answers to the basic questions of Brill-Noether theory in this case (see [ACGH1]).

Connectedness Theorem. *Let C be a smooth curve of genus g. Let $d \geq 1$, $k \geq 0$. Assume $\rho = g - k(g - d + k - 1) \geq 1$. Then $B_{1,d}^k$ is connected on any curve and irreducible on the generic curve.*

Dimension Theorem. *Let C be a general curve of genus g. Let $d \geq 1$, $k \geq 0$. Assume $\rho = g - k(g - d + k - 1) \geq 0$. Then $B_{1,d}^k$ is non-empty reduced and of pure dimension ρ. If $\rho < 0$, $B_{1,d}^k$ is empty.*

Smoothness Theorem. *Let C be a general curve of genus g. Let $d \geq 1$, $k \geq 0$. Assume $\rho = g - k(g - d + k - 1) \geq 0$. Then $B^k_{1,d}$ is non-singular outside of $B^{k+1}_{1,d}$.*

In summary, for rank one the answers to the basic questions are affirmative: if $\rho \geq 0$, the Brill-Noether locus is non-empty on any curve and it is connected as soon as $\rho > 0$. It may be reducible, but each component has dimension at least ρ.

Moreover, for the generic curve, the dimension of the Brill-Noether locus is exactly ρ (with the locus being empty for $\rho < 0$), it is irreducible and of dim ρ and Sing $B^k_{1,d} = B^{k+1}_{1,d}$.

4 Problem in higher ranks (and genus)

We want to study the non-emptiness of $B^k_{n,d}$. We will make use of two parameters; the slope $\mu = \frac{d}{n}$ and a sort of "dimensional slope" $\lambda = \frac{k}{n}$.

To define the area of interest for non-emptiness of $B^k_{n,d}$ in the (μ, λ)-plane is shown

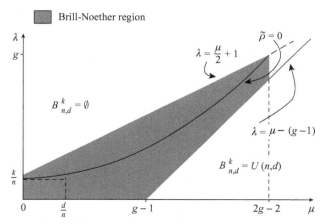

Picture 1b. Brill-Noether region for $n > 1$.

This picture makes use of the Riemann-Roch and Generalized Clifford Theorems.

Riemann-Roch Theorem *The dimension of the space of sections of a vector bundle of rank n and degree d is given by $h^0(E) = d + n(1 - g) + h^1(E)$.*

The Riemann-Roch line in the (k, d)-plane $k + n(g - 1) = d$ determines a half plane $k + n(g - 1) \leq d$ where all the possible values of k, d necessarily lie. Dividing by n, we obtain $\lambda + (g - 1) \leq \mu$.

Clifford's Theorem (see [BGN]) *If E is special semi-stable* $2k - 2n \leq$
d.

Dividing by n, this gives $2\lambda - 2 \leq \mu$.

Using the expected dimension equation $\rho_{n,d}^k = n^2(g-1) + 1 - k(k-d+n(g-1))$ we define a new curve $\widetilde{\rho}_{n,d}^k := \frac{\rho-1}{n^2} = 0$. This easily translates to μ and λ parametization as

$$\widetilde{\rho}_{n,d}^k = g - 1 - \lambda(\lambda - \mu + g - 1).$$

The Brill-Noether region for rank at least two is bounded by $\widetilde{\rho}_{n,d}^k \geq 0$ and the Riemann Roch and Clifford lines. Serre duality describes a "symmetry" line between the subregions on the left and right, see Picture 1b.

For $\mu(E) = \frac{d}{n} \leq 1$ the non-emptiness problem of Brill-Noether loci was solved completely by Brambila-Paz, Grzegorczyk, Newstead (see Picture 2 and [BGN1]) and their results and methods were extended to $\mu(E) = \frac{d}{n} \leq 2$ by Mercat (see [M1], [M3]).

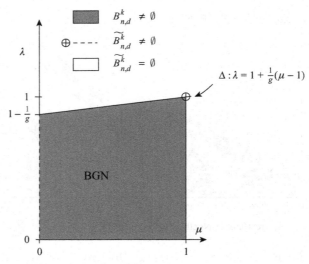

Picture 2. Bundles of slope ≤ 1

The main results for $\mu < 2$, C non-singular are stated below.

Theorem 4.1. *Assume* $\mu < 2$. *Then,* $B_{n,d}^k$ *is non-empty if and only if* $d > 0$, $n \leq d + (n-k)g$ *and* $(n,d,k) \neq (n,n,n)$. *If* $B_{n,d}^k$ *is not empty, then it is irreducible, of dimension* $\rho_{n,d}^k$ *and Sing* $B_{n,d}^k = B_{n,d}^{k+1}$.

Theorem 4.2. *Denote by $\widetilde{B}^k_{n,d}$ the subscheme of the moduli space of S-equivalence classes of semistable bundles whose points correspond to bundles having at least k independent sections. Then, $\widetilde{B}^k_{n,d}$ is non-empty if and only if either $d = 0$ and $k \leq n$ or $d > 0$ and $n \leq d + (n - k)g$. If $\widetilde{B}^k_{n,d}$ is not empty, then it is irreducible.*

Note that the equation $n = d + (n-k)g$ corresponds to $1 = \mu + (1-\lambda)g$ in (μ, λ) coordinates, i.e. $\lambda = 1 + \frac{1}{g}(\mu - 1)$. This is a line which we will denote by Δ. The point $(1, 1)$ is on the line and the Δ line is tangent to $\widetilde{\rho} = 0$ at the point $(1,1)$, see Picture 3 (note that E_K is a vector bundle dual to the kernel of the valuation morphism $H^0(K) \otimes \mathcal{O}_C \to K$).

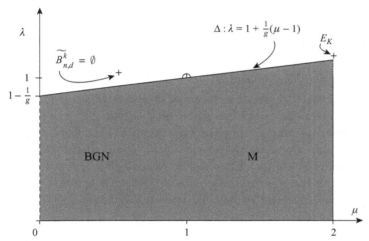

Picture 3. Vector bundles of slope ≤ 2

The main point of the proof in [BGN] is to show that all vector bundles $E \in B^k_{n,d}$, $\mu(E) \leq 1$, can be presented as extensions

$$0 \to O^k \to E \to F \to 0.$$

Such an extension corresponds to an element

$$e = (e_1, ..., e_k) \in H^1(F^*)^{\oplus k}.$$

The condition $n \leq d + (n - k)g$ is equivalent to $k \leq h^1(F^*)$. If this condition fails, the elements $e_1, ..., e_k$ are linearly dependent in $H^1(F^*)$

and the extension partially split. Therefore, the corresponding E cannot be stable.

For $1 < \mu < 2$ consider the vector bundle defined as the kernel of the evaluation map of sections of the canonical map

$$(*) \quad 0 \to M_K \to H^0(K) \otimes \mathcal{O}_C \to K \to 0,$$

where K is a canonical divisor generated by its global sections. As $E \otimes K$ is a stable vector bundle of slope larger than $2(g-1)$, $H^1(C, E \otimes K) = 0$. As $\mu(E) < \mu(M_K^*) = 2$, there are no non-trivial maps from M_K^* to E, hence $h^0(C, M_K^* \otimes E) = 0$. Then, using the long exact sequence in cohomology of $(*)$ tensored with E, one can see that the inequality $n \leq d + (n-k)g$ still holds in this case.

These results were extended to $\mu = 2$ as follows:

Theorem 4.3. *If C is nonhyperelliptic then $B_{n,d}^k$ is non-empty if and only if $n \leq d + (n-k)g$ or $E \simeq M_K^*$.*

If C is hyperelliptic, then $B_{n,d}^k$ is non-empty if and only if $k \leq n$ or $E \simeq g_k^1$. There is at least one component of the right dimension equal to $\rho_{n,d}^k$.

Remark 4.4.   (a) The material in this section provides many examples with $\rho_{n,d}^k > 0$ and $B_{n,d}^k = \phi$ against the natural expectations.

(b) Note that $\widetilde{B}_{n,0}^k$ is not a closure of $B_{n,0}^k = \phi$, i.e semistable bundles with sections exist when stable ones do not.

(c) Recent results on coherent systems show that when $1 \leq \mu \leq 2$ there are no other components of $B_{n,d}^k$, (see [BGMMN]).

(d) For $\mu(E) \leq 2$, the Δ line that is tangent to $\widetilde{\rho}_{n,d}^k = 0$ at the point $(1,1)$ is the nonemptiness boundary (even though it is completely in the Brill-Noether region). Does this indicate existence of similar boundaries for other μ's?

5 Non-emptiness of Brill-Noether loci for large number of sections

In this section, we explain the main known results about existence of vector bundles with sections mostly when $k > n$, as in the case $k \leq n$ better results can be obtained with the methods of the previous section.

Theorem 5.1. *Let C be a generic curve of genus $g \geq 2$. Let d, n, k be positive integers with $k > n$. Write*

$$d = nd_1 + d_2, \quad k = nk_1 + k_2, \quad d_2 < n, k_2 < n.$$

Then, $B_{n,d}^k$ is non-empty and has one component of the expected dimension ρ if one of the conditions below is satisfied:

$$g - (k_1 + 1)(g - d_1 + k_1 - 1) \geq 1, \ 0 \neq d_2 \geq k_2$$
$$g - k_1(g - d_1 + k_1 - 1) > 1, \ d_2 = k_2 = 0$$
$$g - (k_1 + 1)(g - d_1 + k_1) \geq 1, \ d_2 < k_2.$$

See Picture 5 for graphic presentation of this result in the Brill-Noether region.

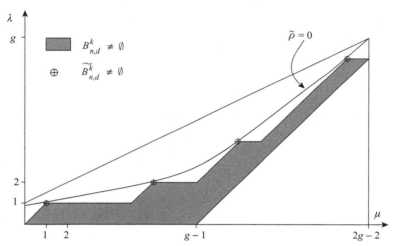

Picture 5. Teixidor's parallelograms.

We sketch here two methods of proof. The first one is based on degeneration techniques. Assume given a family of curves

$$\mathcal{C} \to T$$

and construct the Brill-Noether locus for the family in a way similar to what was done in section 2 for a single curve

$$\mathcal{B}_{n,d}^k = \{(t, E) | t \in T, \ E \in B_{n,d}^k(\mathcal{C}_t)\}.$$

The dimension of this scheme at any point is at least $\dim T + \rho_{n,d}^k$. Assume that we can find a point (t_0, E) such that $\dim(B_{n,d}^k(\mathcal{C}_{t_0}))_E = \rho_{n,d}^k$. Then $\dim \mathcal{B}_{n,d}^k \leq \rho_{n,d}^k + \dim T$ and therefore we have equality.

The dimension of the generic fiber of the projection $\mathcal{B}_{n,d}^k \to T$ is at most the dimension of the fiber over t_0, namely $\rho_{n,d}^k$. But it cannot be any smaller, as the fiber over a point $t \in T$ is $B_{n,d}^k(\mathcal{C}_t)$ which has dimension at least $\rho_{n,d}^k$. Hence there is equality which is the result we are looking for.

The main difficulty of course is finding a particular curve and vector bundle for which the result works. The curves that seem to work best are obtained by taking g elliptic curves $C_1, ..., C_g$ each glued to the next by a chain of rational components. There are three technical problems that need to be solved with this approach:

1) The limit of a vector bundle on a non-singular curve is a torsion-free, not necessarily locally-free sheaf on a nodal curve.

2) The definition of stability cannot be applied directly to the reducible situation to produce moduli spaces of sheaves on reducible curves.

3) It is not clear what one should mean by the limit of the k-dimensional space of sections.

The first problem can be solved by blowing up some of the nodes and replacing the curve by a curve with some more components (see [G], [EH]).

The second problem was solved in two different ways by Gieseker and Seshadri (see [G], [S]). Gieseker considered the concept of Hilbert stability: a vector bundle and a space of sections that generate it give rise to a map from the curve to the Grassmannian. He then took the Hilbert scheme of non-singular curves in the Grassmannian and considered its closure.

Seshadri on the other hand, generalized the concept of stability to torsion-free sheaves on reducible curves: given a curve C with components C_i, take a positive weight a_i for each component with $\sum a_i = 1$. Define the a_i-slope of a torsion-free sheaf E as

$$\frac{\chi(E)}{\sum a_i rank(E_{|C_i})}.$$

As usual, a sheaf is stable if its slope is larger than the slope of any of its subbundles. Seshadri then constructed a moduli space for a_i-stable sheaves.

The third problem is probably the trickiest and requires a generalisation of the techniques of Eisenbud and Harris for limit linear series ([EH]) to higher rank (see [T1]).

Once these three obstacles have been overcome, one can use the knowledge of vector bundles on elliptic curves to obtain the results. The main advantage of this method is that it can still be used for numerical conditions different from those in 5.1 to improve the results in particular situations (see next section).

A second approach was taken by Mercat ([Me2]). He proved the following

Lemma 5.2. . *Let $L_1, ...L_n$ be line bundles of the same degree d' not isomorphic to each other and F either a vector bundle of slope higher than d' or a torsion sheaf. Then, there exists a stable E that can be written as an extension*

$$0 \to L_1 \oplus ... \oplus L_n \to E \to F \to 0.$$

One can then deduce 5.1 by taking line bundles L_i with lots of sections. For example, if $d_2 < k_2$, take $L_i \in B_{1,d_1}^{k_1+1}$ (which are known to exist by classical Brill-Noether Theory) and F a torsion sheaf of degree d_2.

This method does not give the fact that one can get a component of dimension ρ. In fact the set of E as described above when the L_i have plenty of sections form a proper subset of any component of $B_{n,d}^k$. On the other hand, using this method, for certain choices of d, n, k one can obtain an E with k sections for which $\rho_{n,d}^k < 0$. Namely, one obtains **non-empty** Brill-Noether loci with **negative** Brill-Noether number for every curve of a given genus.

6 Further results

The results for small slopes can be used to obtain non-emptiness for larger slopes (and small number of sections). Results can be summarized in the following picture:

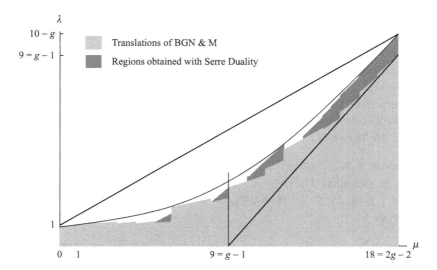

Picture 6. Non-existence results for $g = 10$.

The "translation method" based on twisting and dualizing vector bundles with known properties used in [BMNO] leads to extension of the previous results for a general curve C to give the following:

Theorem 6.1. *([BMNO]) $B_{n,d}^{ks}$ is not empty if the following conditions are satisfied $d = nd' + d''$ with $0 < d'' < 2n$, $1 < s < g$, $d' \geq \frac{(s-1)(s+g)}{s}$, $n \leq d'' + (n-k)g$, $(d'', k) \neq (n,n)$.*

The above gives polygonal regions below the $\tilde{\rho}_{n,d}^k = 0$ curve that sometimes go beyond Teixidor parallelograms (see Picture 6). Note though that these are only valid for small values of k ($k \leq \frac{d''}{g} + \frac{n(g-1)}{g} < 2n$).

More can be said of course when we restrict the values of the parameters. For example, in rank two the best result so far seems to be the following:

Theorem 6.2. *(see [T5]) If $k \geq 2$ and $k - d + 2(g-1) \geq 2$, then, $B_{2,d}^k$ is non-empty and has one component of the right dimension on the generic curve for $\rho \geq 1$ and odd d or $\rho \geq 5$ and d even.*

If $g \geq 2$ and $3 \leq d \leq 2g-1$ then $B_{2,d}^2 \neq \phi$, and the locus is irreducible, reduced of dimension $\rho = 2d - 3$. Further, if $d > g+1$, then the singular locus of $B_{2,d}^2$ is the union of $B_{2,d}^3$ and a set F with $\dim F = 2g - 6$.

If $g \geq 3$ and $0 < d < 2g$ then $B_{2,d}^3 \neq \phi$ if and only if $\rho_{2,d}^3 \geq 0$. Moreover if $\rho > 0$, $B_{2,d}^3$ is irreducible, reduced of dimension ρ.

It is well known that special curves do not behave as generic curves from the point of view of classical Brill-Noether Theory. The same is of course true for higher rank. Here is a sampling of results:

Results on hyperelliptic curves (see Picture 7a below for a summary and [BGMNO] for details) give examples of non-empty loci for values of λ, μ very close to the Clifford line.

Bielliptic curves also behave differently from generic curves and known results are presented on Picture 7b below and stated in the following

Theorem 6.3. *(Ballico) Let C be a bielliptic curve (i.e. there exists a two to one map $f : C \to C'$ where C' is an elliptic curve). Then if $h^0(E) \leq \frac{d}{2}$ ($\lambda \leq \frac{\mu}{2}$) then $\tilde{B}_{n,d}^k \neq \phi$. If $h^0(E) < \frac{d}{2}$ ($\lambda < \frac{\mu}{2}$) then $B_{n,d}^k \neq \phi$.*

Denote by $B_{n,L}^k$ the locus of stable vector bundles of rank n and determinant L with at least k sections. One has the following result (see [V]):

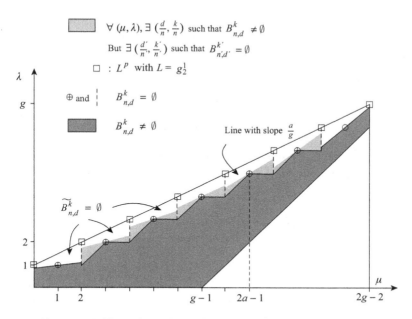

Picture 7a. Brill-Noether region for hyperelliptic curves.

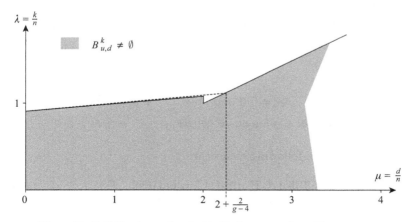

Picture 7b. Brill-Noether region for bi-elliptic curves of $g = 12$.

Theorem 6.4. *Let C be a curve lying on a $K3$ surface. Then $B_{2,K}^k \neq \phi$ if $k \leq \left[\frac{g}{2}\right] + 2$.*

The $B_{2,K}^{\left[\frac{g}{2}\right]+2}$ locus is known to be empty on the generic curve. Hence curves lying on $K3$ surfaces are exceptional from the point of view of

Brill-Noether Theory of rank two and determinant canonical. This is particularly relevant as generic curves on $K3$'s are well-behaved from the point of view of classical Brill-Noether Theory. Therefore, higher rank Brill-Noether provides a new tool to study the geometry of $\mathcal{M}_g$ (see also some very interesting related results in [AN]).

7 Generalized Clifford bounds

Clifford's Theorem for line bundles was generalized to semistable vector bundles by Xiao (see 4). The first extension of this bound was given by Re in the following :

Theorem 7.1. *(see [R]) Let C be a nonhyperelliptic curve, E a stable vector bundle on C such that $1 \le \mu \le 2g - 1$ then $h^0(E) \le \frac{d+n}{2}$, (or $\lambda \le \frac{\mu+1}{2}$).*

This inequality moved the Clifford bound lower in the (μ, λ) plane towards the $\tilde{\rho}$ curve, giving more non-existence of stable vector bundles results on nonhyperelliptic curves.

To extend this idea one makes the following definition:

Definition of Clifford index. The Clifford index, denoted by γ_C, is the maximal integer such that $h^0(C, L) \le \frac{\deg L - \gamma_C}{2} + 1$ for line bundles L such that $h^0(L) \ge 2$ and $h^1(L) \ge 2$.

Remark 7.2.   (a) $\gamma_C = 0$ if and only if C is hyperelliptic.

(b) $\gamma_C = \lceil \frac{g-1}{2} \rceil$ for C of genus g generic.

Mercat made the following conjecture

Generalized Clifford Bound Conjecture. *For $g \ge 4$ and E semistable, $h^0(E) \le \frac{d - \gamma_C n}{2} + n$ for $n\gamma_C + 2 \le \mu(E) \le 2g - 4 - n\gamma_C$.*

At this time he can only prove the above result for $\gamma_C \le 2$ and gives the following upper bound for non-existence (see also Picture 8 for graphic interpretation and [M4] for details).

Theorem 7.3. *If $2 \le \mu(E) \le 2 + \frac{2}{g-4}$, then $h^0(E) \le n + \frac{1}{g-2}(d - n)$. If $2 + \frac{2}{g-4} \le \mu(E) \le 2g - 4 - \frac{2}{g-4}$, then $h^0(C, E) \le \frac{d}{2}$.*

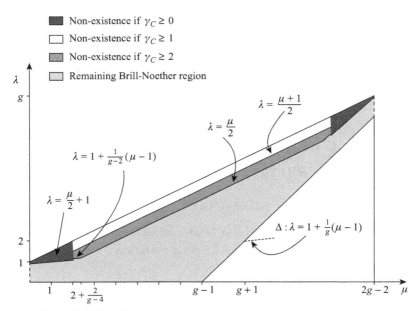

Picture 8. Clifford bounds.

As the Clifford index of the generic curve of genus 10 is 4, one would expect from 7.2 for vector bundles of rank 2 $h^0(E) \leq \frac{d-8}{2} + 2$ for $6 \leq \mu(E) \leq 12$. For example for $d = 12$, the point corresponding to $B^4_{2,12}$ represents a "corner point" for the Clifford bound inequalities. One has in fact (see [GMN]):

Theorem 7.4. . *Let C be a general curve of genus 10. Then $B^4_{2,d} \neq \phi$ if and only if $d \geq 13$. However, there exists a special curve for which $B^4_{2,12} \neq \phi$*

This result shows again, that even the basic properties of $B^k_{n,d}$ may depend on the curve structure, (see [GMN]). For further results on special curves see [BO] in this volume.

8 The case of rank two and canonical determinant

Assume now that E is a vector bundle of rank two and canonical determinant. In this section, we shall assume that the characteristic of the ground field is different from two. We want to consider the set

$$B_{2,K}^k = \{E \in U(2,K) | h^0(C,E) \ge k\}.$$

Proposition 8.1. *(see [BF], [M1]) The set $B_{2,K}^k$ can be given a natural scheme structure. Its expected dimension is*

$$\rho_{2,K}^k = \dim\ U(2,K) - \binom{k+1}{2}.$$

The tangent space to $B_{2,K}^k$ at a point E is naturally identified to the orthogonal to the image of the symmetric Petri map

$$S^2(H^0(C,E)) \to H^0(C,S^2(E)).$$

Proof The proof uses the following fact:

Lemma 8.2. *(see [H, Mu]). Let α be an antisymmetric non-degenerate form on a vector bundle $\mathcal{F}$ of rank $2n$ on a variety X. Let A and A' be isotropic subbundles of rank n. Then,*

$$\{x \in X | \dim(A_x \cap A'_x) \ge k,\ A_x \cap A'_x \equiv k\ (mod\ 2)\}$$

has codimension at most $\binom{k+1}{2}$ in X.

The construction that follows is a direct generalization of the one in [H]. As in section 2, let us assume for simplicity that there exists a Poincaré bundle $\mathcal{E}$ on $C \times U(2,K)$. Denote by $p_1 : C \times U(2,K) \to C$, $p_2 : C \times U(2,K) \to U(2,K)$ the two projections. Let D be a divisor of large degree on C. Define

$$\mathcal{F} = p_{2*}((\mathcal{E} \otimes p_1^*\mathcal{O}_C(D))/(\mathcal{E} \otimes p_1^*\mathcal{O}_C(-D))).$$

This is a vector bundle on $U(2,K)$ of rank 4deg$D$. Define $N = 2$deg$D$.

Define an antisymmetric form α in $\mathcal{F}$ by $\alpha(s_i, t_j) = \sum_{P \in D} Res(s_i \wedge t_j)$ where $s_i \wedge t_j$ is taken as a section of $K(D)$ and Res denotes the residue.

Note now that

$$A = p_{2*}(\mathcal{E}/(\mathcal{E} \otimes p_1^*\mathcal{O}_C(-D)))$$

is a subbundle of rank N that is isotropic (as we are considering holomorphic sections only, the residues are zero). Also

$$A' = p_{2*}(\mathcal{E} \otimes p_1^*\mathcal{O}_C(D))$$

is an isotropic subbundle: by the residue Theorem, the sum of the residues at all the poles is zero. The intersection of the fibers of these two subbundles at a point E is $H^0(C, E)$. Then the locus

$$B_{2,K}^k = \{x \in U(2, K) \mid \dim(A_x \cap A_x') \geq k\}.$$

Therefore, by 8.2, the expected dimension of $B_{2,K}^k$ is as claimed.

Let us now turn to the computation of the tangent space. We first need to identify the tangent space to $U(2, K) \subset U(2, 2g-2)$ as a subset of $H^1(C, E^* \otimes E)$. Denote by $Trl(E)$ the sheaf of traceless endomorphisms of E. We have a natural decomposition of $E^* \otimes E$ into direct sum

$$
\begin{array}{ccc}
E^* \otimes E & \to & \mathcal{O}_C \oplus Trl(E) \\
\varphi & \to & (\frac{Tr(\varphi)}{2} Id, \varphi - \frac{Tr(\varphi)}{2} Id)
\end{array}
$$

the inverse map being given by the two natural inclusions. We can identify the tangent space to $U(2, K)$ with $H^1(C, Trl(E))$. This can be seen with an infinitesimal computation as in section 2. With the notations in section 2, let φ_{ij} be a representative of an element in $H^1(C, E^* \otimes E)$. Take a local basis e_1, e_2 for the fiber of E on the open set U_i. Then, $e_1 \wedge e_2$ is a basis for $\wedge^2 E$ in U_i. We want to see which deformations $\mathcal{E}_\epsilon$ of E preserve the determinant. Note that $e_1 \wedge e_2$ glues in U_j with

$$(e_1 + \epsilon\varphi_{ij}(e_1)) \wedge (e_2 + \epsilon\varphi_{ij}(e_2)) = e_1 \wedge e_2 + \epsilon(\varphi_{ij}(e_1) \wedge e_2 + e_1 \wedge \varphi_{ij}(e_2)).$$

Writing $\varphi_{i,j}$ as a square matrix in terms of the basis e_1, e_2,

$$\varphi_{i,j} = \begin{pmatrix} a_{11} & a_{12} \\ a_{21} & a_{22} \end{pmatrix}.$$

One can check that

$$\varphi_{ij}(e_1) \wedge e_2 + e_1 \wedge \varphi_{ij}(e_2) = (a_{11}e_1 + a_{12}e_2) \wedge e_2 + e_1 \wedge (a_{21}e_1 + a_{22}e_2) =$$
$$(a_{11} + a_{22})(e_1 \wedge e_2) = Tr(\varphi_{i,j})(e_1 \wedge e_2)$$

Therefore, the deformation preserves the determinant if and only if the trace is zero. Hence, the tangent space to $U(2, K)$ at E is $H^1(C, Trl(E))$.

From the computations in section two, the deformations that preserve a certain subspace of sections correspond to the orthogonal to the image of the Petri map. We need to compute the intersection of this orthogonal with $H^1(C, Trl(E))$.

From $\wedge^2(E) \cong K$, we obtain the identification $E \cong E^* \otimes K \cong Hom(E, K)$ given by

$$
\begin{array}{ccc}
E & \to & Hom(E, K) \\
e & \to & \psi_e : E \to K \\
& & v \to e \wedge v
\end{array}
$$

Similarly, one has the identification

$$E \otimes E \cong E^* \otimes K \otimes E \cong Hom(E, K \otimes E).$$

Consider the composition of this identification with the direct sum decomposition of $E^* \otimes E$ tensored with K

$$E \otimes E \cong E^* \otimes K \otimes E \cong K \oplus (K \otimes Trl(E)).$$

The projection on the first summand is (up to a factor of two)

$$e \otimes v \to Tr(\psi_e \otimes v).$$

Computing locally, one checks that this is the determinant

$$e \otimes v \to e \wedge v.$$

Therefore, $H^0(C, K \otimes Trl(E))$ can be identified to the kernel of this map, namely $H^0(C, S^2(E))$.

With these identifications, the composition of the Petri map with the projection onto $H^0(C, S^2(E))$ becomes

$$H^0(C, E) \otimes H^0(C, E) \to H^0(C, E \otimes E) \to H^0(C, S^2(E)).$$

This map vanishes on $\wedge^2 H^0(C, E)$ and therefore gives rise to the symmetric Petri map

$$S^2(H^0(C, E)) \to H^0(C, S^2(E)).$$

Then, the orthogonal to the image of this map is identified to the tangent space at E to $B_{2,K}^k$. $\qquad \square$

As in the case of non-fixed determinant, the map being injective is equivalent to E being a non-singular point on a component of the expected dimension of $B_{2,K}^k$. This is proved in [T4]. Therefore we have the following

Theorem 8.3. *For a generic curve of genus g, any component of $B_{2,K}^k$ has the expected dimension $\rho_{2,K}^k$ (and the locus is empty when $\rho_{2,K}^k < 0$).*

In terms of existence, one has the following result

Theorem 8.4. *(see [T2]) Let C be a generic curve of genus at least two. If $k = 2k_1$, then $B_{2,K}^k$ is non-empty and has a component of the expected dimension if $g \geq k_1^2$, if $k_1 > 2$; $g \geq 5$, if $k_1 = 2$; $g \geq 3$, if $k_1 = 1$. If $k = 2k_1 + 1$, $B_{2,K}^k$ is non-empty and has a component of the expected dimension if*

$$g \geq k_1^2 + k_1 + 1.$$

This result was proved using degeneration techniques. Very similar results were obtained in [P] computing the cohomology class of the locus inside $U(2, K)$ and showing that it is non-zero.

Note that for small values of k, the expected dimension of $B_{2,K}^k$ is larger than the expected dimension of $B_{2,2g-2}^k$. This allows us to produce some more counterexamples to the expected results. Here we state only the case of an odd number of sections:

Example 8.5. Let C be a generic curve of genus $g \geq 2$. If $k_1^2 + k_1 + 1 \leq g < 2k_1^2 + k_1$, then $B_{2,2g-2}^{2k_1+1}$ is reducible and has one component of dimension larger than expected.

Bibliography

[A] M.F. Atiyah, *Vector bundles over an elliptic curve.* Proc. London Math. Soc. **7** (1957), (3), 414–452.

[ACGH] E. Arbarello, M. Cornalba, P.A. Griffiths, J. Harris *Geometry of algebraic curves.* Vol **1** Springer-Verlag, New York, 1985.

[AN] M.Aprodu, J.Nagel, *Non-vanishing for Koszul cohomology of curves.* Comment. Math. Helv. **82 n3** (2007), 617–628.

[BF] A.Bertram, B.Feinberg, *On stable rank two bundles with canonical determinant and many sections.* in "Algebraic Geometry", ed P.Newstead, Marcel Dekker 1998, 259–269.

[BR] I.Biswas, S.Ramanan, *An infinitesimal study of the moduli space of Hitchin pairs.* J.London Math.Soc. **(2)49** (1994), 219–231.

[BGMMN] S. Bradlow, O. Garcia-Prada, V.Mercat, Munoz, P.Newstead, *Moduli spaces of coherent systems of small slope on algebraic curves.* Comm. Alg. (is correct.)

[BGN] L. Brambila-Paz, I. Grzegorczyk, P.Newstead, *Geography of Brill-Noether loci for small slopes.* J. Algebraic Geom. **6 n4** (1997), 645–669.

[BMNO] L.Brambila-Paz, V.Mercat, P.Newstead, F.Ongay, *Nonemptiness of Brill-Noether loci.* Internat. J. Math. **11** (2000), no. 6, 737–760.

[BO] L. Brambila-Paz and A. Ortega, *Brill-Noether bundles and coherent systems on special curves* Moduli spaces and vector bundles, ed. L. Brambila-Paz, S.Bradlow, O. Garcia-Prada, and S. Ramanan, Cambridge University Press. London Math. Soc. Lecture Note Series 359.

[DN] J.M.Drezet, M.S. Narasimhan, *Groupe de Picard des varietes de modules de fibres semi-stables sur les courbes algebriques.* Invent. Math. **97 no. 1** (1989), 53–94.

[EH] D.Eisenbud, J.Harris, *Limit linear series: basic theory.* Invent. Math. **85, n2** (1986), 337–371.

[FP] G.Farkas, M. Popa, *Effective divisors on $\overline{\mathcal{M}}_g$, curves on K3 surfaces, and the slope conjecture.* J. Algebraic Geom. **14 n2** (2005), 241–267.

[H] J.Harris, *Theta-characteristics on algebraic curves.* Trans. Amer. Math. Soc. **271 n2** (1982), 611–638.

[G] D.Gieseker, *A degeneration of the moduli space of stable bundles.* J. Differential Geom. **19 V no. 1** (1984), 173–206.

[Gr] A.Grothendieck, *Sur la classification des fibres holomorphes sur la sphere de Riemann.* Amer. J. Math. **79 no. 1** (1957), 121–138.

[GMN] I. Grzegorczyk, V.Mercat, P.Newstead, *Stable bundles of rank 2 and degree 12 with 4 sections on a curve of genus 10.* Preprint.

[Me1] V.Mercat, *Le probleme de Brill-Noether pour des fibres stables de petite pente.* J. Reine Angew. Math. **506** (1999), 1–41.

[Me2] V.Mercat, *Le probleme de Brill-Noether et le theoreme de Teixidor.* Manuscripta Math. **98 n1** (1999), 75–85.

[Me3] V.Mercat, *Fibres stable de pente* 2. Bull. London Math. Soc. **33** (2001), 535–542.

[Me4] V.Mercat, *Clifford's Theorem on higher rank vector bundles.* Internat. J. Math (2002), 785–796.

[M1] S.Mukai, *Curves and Brill-Noether Theory.* MSRI publications **28**, Cambridge University Press 1995, 145–158.

[M2] S.Mukai, *Non-abelian Brill-Noether Theory and Fano threefolds.* Sugaku Expositions **14** (2001), no. 2, 125–153.

[Mu] D.Mumford, *Theta characteristics of an algebraic curve.* Ann. Sci. cole Norm. Sup. (4) 4 (1971), 181–192.

[P]   S.Park, *Non-emptiness of Brill-Noether loci in* $M(2, K)$. Preprint.

[R]   R.Re, *Multiplication of sections and Clifford bounds for stable vector bundles on curves.* Comm. Algebra **26 n6** (1998), 1931–1944.

[RT] B.Russo, M.Teixidor i Bigas, *On a conjecture of Lange.* J. Algebraic Geom. **8 n3** (1999), 483–496.

[S]   C.S.Seshadri, *Fibres vectoriels sur les courbes algebriques.* Asterisque 96, Societe Matematique de France 1982.

[T1] M.Teixidor i Bigas, *Brill-Noether Theory for stable vector bundles.* Duke Math **J62 N2**(1991), 385–400.

[T2] M.Teixidor i Bigas, *Rank two vector bundles with canonical determinant.* Math. Nachr. **265** (2004), 100–106.

[T3] M.Teixidor i Bigas, *On the Gieseker-Petri map for rank* 2 *vector bundles.* Manuscripta Math. **75 n4** (1992), 375–382.

[T4] M.Teixidor i Bigas, *Petri map for rank* 2 *bundles with canonical determinant.* Compositio Mathematica, in press.

[T5] M.Teixidor i Bigas, *Existence of vector bundles of rank two with sections.* AdvGeom. **5** (2005), 37–47.

[V]   C. Voisin, *Sur l'application de Wahl des coubres satisfaisant le condition de Brill-Noether-Petri.* Acta.Math. **168** (192), 249–272.

[W]   G.Welters, *Polarised abelian varieties and the heat equations.* Comp. Math. **49** (1983), 173–194.

3

Introduction to Fourier-Mukai and Nahm transforms with an application to coherent systems on elliptic curves

U. Bruzzo

Scuola Internazionale Superiore di Studi Avanzati, Trieste, Italy, and Istituto Nazionale di Fisica Nucleare, Sezione di Trieste
e-mail: bruzzo@sissa.it

D. Hernández Ruipérez

Departamento de Matemáticas and
Instituto Universitario de Física Fundamental y Matemáticas (IUFFYM)
Universidad de Salamanca, Spain
e-mail: ruiperez@gugu.usal.es

C. Tejero Prieto

Departamento de Matemáticas and
Instituto Universitario de Física Fundamental y Matemáticas (IUFFYM)
Universidad de Salamanca, Spain
e-mail: carlost@usal.es

These notes record, in a slightly expanded way, the lectures given by the first two authors at the College on Moduli Spaces of Vector Bundles that took place at CIMAT in Guanajuato, Mexico, from November 27th to December 8th, 2006. The college, together with the ensuing conference on the same topic, was held in occasion of Peter Newstead's 65th anniversary. It has been a great pleasure and a privilege to contribute to celebrate Peter's outstanding achievements in algebraic geometry and his lifelong dedication to the progress of mathematical knowledge. We warmly thank the organizers of the college and conference for inviting us, thus allowing us to participate in Peter's celebration.

The main emphasis in these notes is on the Fourier-Mukai transforms as equivalences of derived categories of coherent sheaves on algebraic varieties. For this reason, the first Section is devoted to a basic (but we hope, understandable) introduction to derived categories. In the second Section we develop the basic theory of Fourier-Mukai transforms.

Another aim of our lectures was to outline the relations between Fourier-Mukai and Nahm transforms. This is the topic of Section 3. Finally, Section 4 is devoted to the application of the theory of Fourier-Mukai transforms to the study of coherent systems.

This is a review paper. Most of the material is taken from [BBH08] and [HT08], although the presentation is different in some places. We refer the reader to those works for further details and for a systematic treatment.

1 Derived categories

Introduction

We start with an introduction to derived categories, especially in connection with Fourier-Mukai transforms. A more comprehensive treatment may be found in [BBH08]. As a witness to the relevance of derived categories in this theory, one may mention the title of the paper where Mukai introduced the transform now known as Fourier-Mukai's: "Duality between $D(X)$ and $D(\hat{X})$ with its application to Picard sheaves" [Muk81].

Let us describe the original Mukai transform from a naive point of view. Let X be a complex abelian variety and $\mathcal{E}$ a vector bundle on X. We consider here only algebraic (or holomorphic) vector bundles, so we can also think of $\mathcal{E}$ as a smooth hermitian bundle E endowed with a Hermitian connection ∇ which is compatible with the complex structure. We fix an index i and look for the various cohomology spaces $H^i(X, \mathcal{E} \otimes \mathcal{P}_\xi)$ where $\mathcal{P}_\xi$ varies in the space $\hat{X}$ of all flat line bundles on X (the dual abelian variety of X). A natural question is whether the collection of vector spaces $H^i(X, \mathcal{E} \otimes \mathcal{P}_\xi)$ define a vector bundle on $\hat{X}$. In some cases this happens, for instance if one has $H^j(X, \mathcal{E} \otimes \mathcal{P}_\xi) = 0$ for any $j \neq i$ and $\xi \in \hat{X}$.

In general, one cannot expect to be so lucky, and such a vector bundle (or more generally sheaf) may not exist. What one can do is to mimic the construction of the cohomology groups to get objects that play a similar role. On the product $X \times \hat{X}$ there is a *universal* line bundle $\mathcal{P}$, called the Poincaré bundle, whose restriction to the fibre $\hat{\pi}^{-1}(\xi)$ over ξ of the projection $\hat{\pi} \colon X \times \hat{X} \to \hat{X}$ is the line bundle $\mathcal{P}_\xi$; we normalise $\mathcal{P}$ so that it restricts to the trivial line bundle on the fibre of the origin x^0 of X for the other projection $\pi \colon X \times \hat{X} \to X$. In analogy with the construction of the cohomology groups of a sheaf, we take the sheaf $\mathcal{F} = \pi^* \mathcal{E} \otimes \mathcal{P}$ (whose restriction to $\hat{\pi}^{-1}(\xi)$ is precisely $\mathcal{E} \otimes \mathcal{P}_\xi$), a resolution

$$0 \to \mathcal{F} \to \mathcal{R}^0 \to \mathcal{R}^1 \to \cdots \to \mathcal{R}^n \to \ldots$$

by injective sheaves, and define the *higher direct images* of $\mathcal{F}$ under $\hat{\pi}$ as the cohomology sheaves $R^i\hat{\pi}_*\mathcal{F} = \mathcal{H}^i(\hat{\pi}_*\mathcal{R}^\bullet)$ of the complex

$$0 \to \hat{\pi}_*\mathcal{F} \to \hat{\pi}_*\mathcal{R}^0 \to \hat{\pi}_*\mathcal{R}^1 \to \cdots \to \hat{\pi}_*\mathcal{R}^n \to \dots .$$

The relationship between the sheaves $R^i\pi_*(\pi^*\mathcal{E}\otimes\mathcal{P})$ and the cohomology groups $H^i(X, \mathcal{E}\otimes\mathcal{P}_\xi)$ is given by some "cohomology base change" theorems [Hart77, III.12]. This shows that the sheaves $R^i\hat{\pi}_*(\pi^*\mathcal{E}\otimes\mathcal{P})$ encode more information than the cohomology groups of the fibres. Another classical fact is that the higher direct images are independent of the resolution $\mathcal{R}^\bullet$ of $\mathcal{F}$, that is, if $0 \to \mathcal{F} \to \tilde{\mathcal{R}}^\bullet$ is another acyclic resolution of $\mathcal{F}$ (meaning that the higher direct images $R^i\hat{\pi}^i\tilde{\mathcal{R}}^j$ of the sheaves $\tilde{\mathcal{R}}^j$ are zero for every $i > 0$, $j \geq 0$), then the complexes of sheaves $\hat{\pi}_*\mathcal{R}^\bullet$ and $\hat{\pi}_*\tilde{\mathcal{R}}^\bullet$ have the same cohomology sheaves. If we identify two complexes of sheaves when they have the same cohomology sheaves (we say that they are *quasi-isomorphic*), and write $\mathbf{R}\hat{\pi}_*\mathcal{F}$ for the "class" of any of the complexes $\hat{\pi}_*\mathcal{R}^\bullet$, the information about the cohomology groups $H^i(X, \mathcal{E}\otimes\mathcal{P}_\xi)$ is encoded in the single object $\Phi(\mathcal{E}) = \mathbf{R}\hat{\pi}_*\mathcal{F} = \mathbf{R}\hat{\pi}_*(\pi^*\mathcal{E}\otimes\mathcal{P})$.

To make good sense of all this, we need to construct, out of any abelian category, another category, which is called the *derived category*, where quasi-isomorphic complexes become isomorphic and we can define "derived functors" (such as $\mathbf{R}\hat{\pi}_*$) and also some derived versions of the pullback functor π^* and of the tensor product.

1.1 Categories of complexes

A *complex* $(\mathcal{K}^\bullet, d_{\mathcal{K}^\bullet})$ in an abelian category $\mathfrak{A}$ is a sequence

$$\cdots \to \mathcal{K}^{n-1} \xrightarrow{d^{n-1}} \mathcal{K}^n \xrightarrow{d^n} \mathcal{K}^{n+1} \to \cdots$$

of morphisms in $\mathfrak{A}$ such that $d^{n+1} \circ d^n = 0$ for all $n \in \mathbb{Z}$. The family of morphisms $d_{\mathcal{K}^\bullet}$ is called the *differential* of the complex $\mathcal{K}^\bullet$.

The *category of complexes* $\mathbf{C}(\mathfrak{A})$ is the category whose objects are complexes $(\mathcal{K}^\bullet, d_{\mathcal{K}^\bullet})$ in $\mathfrak{A}$ and whose morphisms $f\colon (\mathcal{K}^\bullet, d_{\mathcal{K}^\bullet}) \to (\mathcal{L}^\bullet, d_{\mathcal{L}^\bullet})$ are collections of morphisms $f^n : \mathcal{K}^n \to \mathcal{L}^n$, $n \in \mathbb{Z}$, in $\mathfrak{A}$ such that the diagrams

$$
\begin{array}{ccccccccc}
\cdots & \longrightarrow & \mathcal{K}^{n-1} & \xrightarrow{d^{n-1}} & \mathcal{K}^n & \xrightarrow{d^n} & \mathcal{K}^{n+1} & \xrightarrow{d^{n+1}} & \cdots \\
& & \downarrow{f^{n-1}} & & \downarrow{f^n} & & \downarrow{f^{n+1}} & & \\
\cdots & \longrightarrow & \mathcal{L}^{n-1} & \xrightarrow{d^{n-1}} & \mathcal{L}^n & \xrightarrow{d^n} & \mathcal{L}^{n+1} & \xrightarrow{d^{n+1}} & \cdots
\end{array}
$$

commute.

The direct sum $\mathcal{K}^\bullet \oplus \mathcal{L}^\bullet$ of two complexes $\mathcal{K}^\bullet$ and $\mathcal{L}^\bullet$ is defined in the obvious way. One can also describe in a natural way the kernel and the cokernel of a morphism of complexes, and readily check that the category $\mathbf{C}(\mathfrak{A})$ of complexes of an abelian category is abelian as well.

We define the complex of homomorphisms $\mathrm{Hom}^\bullet(\mathcal{K}^\bullet, \mathcal{L}^\bullet)$ by setting

$$\mathrm{Hom}(\mathcal{K}^\bullet, \mathcal{L}^\bullet)^n = \prod_i \mathrm{Hom}_{\mathfrak{A}}(\mathcal{K}^i, \mathcal{L}^{i+n})$$

for each $n \in \mathbb{Z}$, together with a differential defined by

$$
\begin{aligned}
d^n \colon \mathrm{Hom}(\mathcal{K}^\bullet, \mathcal{L}^\bullet)^n &\to \mathrm{Hom}(\mathcal{K}^\bullet, \mathcal{L}^\bullet)^{n+1} \\
f^i &\mapsto d_{\mathcal{L}^\bullet}^{i+n} \circ f^i + (-1)^{n+1} f^{i+1} \circ d_{\mathcal{K}^\bullet}^i
\end{aligned}
\tag{1.1}
$$

When $\mathfrak{A}$ has tensor products and arbitrary direct sums, we can define the tensor product of complexes by letting $(\mathcal{K}^\bullet \otimes \mathcal{L}^\bullet)^n = \bigoplus_{p+q=n} (\mathcal{K}^p \otimes \mathcal{L}^q)$ with the differential $d_{\mathcal{K}^\bullet}^p \otimes \mathrm{Id} + (-1)^p \mathrm{Id} \otimes d_{\mathcal{L}^\bullet}^q$ over $\mathcal{K}^p \otimes \mathcal{L}^q$. If $\mathfrak{A}$ has tensor products but not arbitrary direct sums, $\mathcal{K}^\bullet \otimes \mathcal{L}^\bullet$ is defined whenever for every n there are only a finite number of summands in $\bigoplus_{p+q=n} (\mathcal{K}^p \otimes \mathcal{L}^q)$.

The shift $\mathcal{K}^\bullet[n]$ of a complex $\mathcal{K}^\bullet$ by an integer number n, is defined by setting $\mathcal{K}[n]^p = \mathcal{K}^{p+n}$ with the differential $d_{\mathcal{K}^\bullet[n]} = (-1)^n d_{\mathcal{K}^\bullet}$. A morphism of complexes $f \colon \mathcal{K}^\bullet \to \mathcal{L}^\bullet$ induces another morphism of complexes $f[n] \colon \mathcal{K}^\bullet[n] \to \mathcal{L}^\bullet[n]$ given by $f[n]^p = f^{p+n}$. In this way, $\mathcal{K}^\bullet \mapsto \mathcal{K}^\bullet[n]$ is an additive functor. Sometimes we shall denote $\tau(\mathcal{K}^\bullet) = \mathcal{K}^\bullet[1]$, so that $\tau^n(\mathcal{K}^\bullet) = \mathcal{K}^\bullet[n]$ for any integer n. The n-th cohomology of a complex $\mathcal{K}^\bullet$ is the object

$$\mathcal{H}^n(\mathcal{K}^\bullet) = \ker d^n / \mathrm{Im} d^{n-1}.$$

We say that $\mathcal{Z}^n(\mathcal{K}^\bullet) = \ker d^n$ are the n-cycles of $\mathcal{K}^\bullet$ and $\mathcal{B}^n(\mathcal{K}^\bullet) = \mathrm{Im} d^{n-1}$ are the n-boundaries. A morphism of complexes $f \colon \mathcal{K}^\bullet \to \mathcal{L}^\bullet$ maps cycles to cycles and boundaries to boundaries, so that it yields for every n a morphism

$$\mathcal{H}^n(f) \colon \mathcal{H}^n(\mathcal{K}^\bullet) \to \mathcal{H}^n(\mathcal{L}^\bullet).$$

One has $\mathcal{H}^n(\mathcal{K}^\bullet[m]) \simeq \mathcal{H}^{n+m}(\mathcal{K}^\bullet)$ and $\mathcal{H}^n(f[m]) \simeq \mathcal{H}^{n+m}(f)$.

A complex $\mathcal{K}^\bullet$ is said to be *acyclic* or *exact* if $\mathcal{H}(\mathcal{K}^\bullet) = 0$; a morphism of complexes $f \colon \mathcal{K}^\bullet \to \mathcal{L}^\bullet$ is a *quasi-isomorphism* if $\mathcal{H}(f) \colon \mathcal{H}(\mathcal{K}^\bullet) \to \mathcal{H}(\mathcal{L}^\bullet)$ is an isomorphism. The composition of two quasi-isomorphisms is a quasi-isomorphism.

A morphism of complexes $f \colon \mathcal{K}^\bullet \to \mathcal{L}^\bullet$ is *homotopic to zero* if there is a collection of morphisms $h^n \colon \mathcal{K}^n \to \mathcal{L}^{n-1}$ (a homotopy) such that $f^n = h^{n+1} \circ d_{\mathcal{K}^\bullet}^n + d_{\mathcal{L}^\bullet}^{n-1} \circ h^r$ for every n. A complex $\mathcal{K}^\bullet$ is *homotopic to zero* if its identity morphism is homotopic to zero. Two morphisms $f, g \colon \mathcal{K}^\bullet \to \mathcal{L}^\bullet$ are *homotopic* if $f - g$ is homotopic to zero.

The sum of two morphisms homotopic to zero is homotopic to zero as well. Furthermore, $f \circ g$ is homotopic to zero if either f or g is homotopic to zero.

The *homotopy category* $K(\mathfrak{A})$ is the category whose objects are the objects of $\mathbf{C}(\mathfrak{A})$ and whose morphisms are

$$\mathrm{Hom}_{K(\mathfrak{A})}(\mathcal{K}^\bullet, \mathcal{L}^\bullet) = \mathrm{Hom}_{\mathbf{C}(\mathfrak{A})}(\mathcal{K}^\bullet, \mathcal{L}^\bullet) / \mathrm{Ht}(\mathcal{K}^\bullet, \mathcal{L}^\bullet) \,,$$

where $\mathrm{Ht}(\mathcal{K}^\bullet, \mathcal{L}^\bullet)$ is the set of morphisms which are homotopic to zero.

One can see from Equation (1.1) that the n-cycles of the complex of homomorphisms $\mathrm{Hom}^\bullet(\mathcal{K}^\bullet, \mathcal{L}^\bullet)$ are the morphisms of complexes $\mathcal{K}^\bullet \to \mathcal{L}^\bullet[n]$, while the n-boundaries are the morphisms homotopic to zero. Thus,

$$\mathcal{H}^n\left(\mathrm{Hom}^\bullet(\mathcal{K}^\bullet, \mathcal{L}^\bullet)\right) = \mathrm{Hom}_{K(\mathfrak{A})}(\mathcal{K}^\bullet, \mathcal{L}^\bullet[n]) \,.$$

If $f \colon \mathcal{K}^\bullet \to \mathcal{L}^\bullet$ is homotopic to zero, then $\mathcal{H}(f) = 0$; hence, *two homotopic morphisms induce the same morphism in cohomology* and a *complex $\mathcal{K}^\bullet$ which is homotopic to zero is acyclic.*

The homotopy category $K(\mathfrak{A})$ does not have neither kernels nor cokernels. This can be overcome by introducing the notion of cone of a morphism.

Definition 1.1. The cone of a morphism of complexes $f \colon \mathcal{K}^\bullet \to \mathcal{L}^\bullet$ is the complex $\mathrm{Cone}(f)$ such that $\mathrm{Cone}(f)^n = \mathcal{K}^{n+1} \oplus \mathcal{L}^n$, equipped with differential

$$d^n_{\mathrm{Cone}(f)} = \begin{pmatrix} -d^{n+1}_{\mathcal{K}^\bullet} & 0 \\ f^{n+1} & d^n_{\mathcal{L}^\bullet} \end{pmatrix}$$

$\triangle$

$\mathrm{Cone}(f)$ is not isomorphic to the direct sum $\mathcal{K}^\bullet[1] \oplus \mathcal{L}^\bullet$ because it has another differential. There are functorial morphisms $\beta \colon \mathrm{Cone}(f) \to \mathcal{K}^\bullet[1]$, $(k, l) \mapsto k$, and $\alpha \colon \mathcal{L}^\bullet \to \mathrm{Cone}(f)$, $l \mapsto (0, l)$.

The sequence

$$\mathcal{K}^\bullet \xrightarrow{f} \mathcal{L}^\bullet \xrightarrow{\alpha} \mathrm{Cone}\, f \xrightarrow{\beta} \mathcal{K}^\bullet[1] \,,$$

in $K(\mathfrak{A})$ is called *a distinguished (or exact) triangle* in $K(\mathfrak{A})$ and is also written in the form

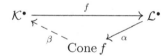

where the dashed arrow stands for a morphism $\mathrm{Cone}\, f \to \mathcal{K}^\bullet[1]$. Notice that $\alpha \circ f = 0$ and $\beta \circ \alpha = 0$.

Proposition 1.1. *Given an exact triangle* $\mathcal{K}^\bullet \xrightarrow{f} \mathcal{L}^\bullet \xrightarrow{\alpha} \mathrm{Cone}\, f \xrightarrow{\beta} \mathcal{K}^\bullet[1]$ *in* $K(\mathfrak{A})$, *for every integer n there is an exact sequence of cohomology groups*

$$\mathcal{H}^n(\mathcal{K}^\bullet) \xrightarrow{\mathcal{H}^n(f)} \mathcal{H}^n(\mathcal{L}^\bullet) \xrightarrow{\mathcal{H}^n(\alpha)} \mathcal{H}^n(\mathrm{Cone}\, f)$$

$$\xrightarrow{\mathcal{H}^n(\beta)} \mathcal{H}^n(\mathcal{K}^\bullet[1]) \simeq \mathcal{H}^{n+1}(\mathcal{K}^\bullet).$$

$\square$

One also obtains the so-called *cohomology long exact sequence*:

$$\cdots \xrightarrow{\mathcal{H}^{n-1}(\beta)} \mathcal{H}^n(\mathcal{K}^\bullet) \xrightarrow{\mathcal{H}^n(f)} \mathcal{H}^n(\mathcal{L}^\bullet)$$

$$\xrightarrow{\mathcal{H}^n(\alpha)} \mathcal{H}^n(\mathrm{Cone}\, f) \xrightarrow{\mathcal{H}^n(\beta)} \mathcal{H}^{n+1}(\mathcal{K}^\bullet) \cdots$$

Proposition 1.1 tells us that the functors $\mathcal{H}^n \colon K(\mathfrak{A}) \to \mathfrak{A}$ are cohomological. This means the following: if $\mathfrak{A}$, $\mathfrak{B}$ are abelian categories, an additive functor $F \colon K(\mathfrak{A}) \to \mathfrak{B}$ is *cohomological* if for every exact triangle $\mathcal{K}^\bullet \xrightarrow{f} \mathcal{L}^\bullet \xrightarrow{\alpha} \mathrm{Cone}\, f \xrightarrow{\beta} \mathcal{K}^\bullet[1]$ the sequence $F(\mathcal{K}^\bullet) \xrightarrow{F(f)} F(\mathcal{L}^\bullet) \xrightarrow{\alpha} F(\mathrm{Cone}\, f) \xrightarrow{F(\beta)} F(\mathcal{K}^\bullet)[1]$ is exact.

Corollary 1.2. *A morphism of complexes* $f \colon \mathcal{K}^\bullet \to \mathcal{L}^\bullet$ *is a quasi-isomorphism if and only if* $\mathrm{Cone}(f)$ *is acyclic.* $\square$

If $0 \to \mathcal{K}^\bullet \xrightarrow{f} \mathcal{L}^\bullet \xrightarrow{g} \mathcal{N}^\bullet \to 0$ is an exact sequence of complexes (in $\mathbf{C}(\mathfrak{A})$), then there is a morphism of complexes $\mathrm{Cone}(f) \to \mathcal{N}^\bullet$ defined in degree n by $(a_{n+1}, b_n) \in \mathcal{K}^{n+1} \oplus \mathcal{L}^n \mapsto g(b_n) \in \mathcal{N}^n$. One easily checks that it is a quasi-isomorphism. Combining this with the cohomology long exact sequence we obtain the more customary form of the latter, i.e., there exist functorial morphisms $\delta^n \colon H^n(\mathcal{N}^\bullet) \to H^{n+1}(\mathcal{L}^\bullet)$ such that one has an exact sequence

$$\cdots \xrightarrow{\delta^{n-1}} H^n(\mathcal{L}^\bullet) \to H^n(\mathcal{M}^\bullet) \to H^n(\mathcal{N}^\bullet) \xrightarrow{\delta^n} H^{n+1}(\mathcal{L}^\bullet) \to$$

$$\to H^{n+1}(\mathcal{M}^\bullet) \to H^{n+1}(\mathcal{N}^\bullet) \xrightarrow{\delta^{n+1}} \cdots$$

1.2 Derived Category

In our route toward the definition of a category where quasi-isomorphic complexes are actually isomorphic, we have first identified homotopic morphisms, and then we have moved from the category of complexes $\mathbf{C}(\mathfrak{A})$ to the homotopy category $K(\mathfrak{A})$. A second step is to "localise" by quasi-isomorphims. This localisation is a fraction calculus for categories. Recall that given a ring A (e.g., the integer numbers) and $S \subset A$ which is a multiplicative system (that is, it contains the unity and is closed under products), then one can define the localised ring $S^{-1}A$; elements in $S^{-1}A$ are equivalence classes a/s of pairs $(a, s) \in A \times S$ where $(a, s) \sim (a', s')$ (or $a/s = a'/s'$) if there exists $t \in S$ such that $t(as' - a's) = 0$. The elements $s \in S$ become invertible in $S^{-1}A$ because $s/1 \cdot 1/s = 1$. One can proceed in a similar way with morphisms of complexes, since quasi-isomorphisms verify the conditions for being a multiplicative system: the identity is a quasi-isomorphism and the composition of two quasi-isomorphisms is a quasi-isomorphism. We can then define a "fraction" f/ϕ as a diagram of (homotopy classes of) morphisms of complexes

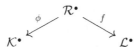

where ϕ is a quasi-isomorphism. Two diagrams f/ϕ and g/ψ of the same type are said to be equivalent if there are quasi-isomorphisms $\mathcal{R}^\bullet \leftarrow \mathcal{T}^\bullet \to \mathcal{S}^\bullet$ such that the diagram

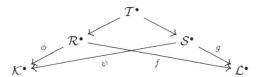

commutes in $K(\mathfrak{A})$. Equivalence of fractions is actually an equivalence relation; this follows from the next Proposition, whose proof is based on the properties of the cone of a morphism.

Proposition 1.3. *Given morphisms of complexes* $\mathcal{M}^\bullet \xrightarrow{f} \mathcal{N}^\bullet \xleftarrow{g} \mathcal{R}^\bullet$ *in* $K(\mathfrak{A})$, *there are morphisms of complexes* $\mathcal{M}^\bullet \xleftarrow{g'} \mathcal{Z}^\bullet \xrightarrow{f'} \mathcal{R}^\bullet$ *such that the diagram*

$$\begin{array}{ccc} \mathcal{Z}^\bullet & \xrightarrow{f'} & \mathcal{R}^\bullet \\ g' \downarrow & & \downarrow g \\ \mathcal{M}^\bullet & \xrightarrow{f} & \mathcal{N}^\bullet \end{array}$$

is commutative in $K(\mathfrak{A})$. *Moreover,* f' *(respectively,* g'*) is a quasi-isomorphism if and only if* f *(respectively,* g*) is so.* $\qquad \square$

Definition 1.2. The derived category $D(\mathfrak{A})$ of $\mathfrak{A}$ is the category with the same objects as $K(\mathfrak{A})$ (i.e., complexes of objects of $\mathfrak{A}$), and whose morphisms are the equivalence classes $[f/\phi]$ of diagrams as above. $\triangle$

In order for this definition to make sense we need to say how to compose morphisms. Given two morphisms $[f/\phi]$ and $[g/\psi]$ in $D(\mathfrak{A})$, their composition is defined by the diagram

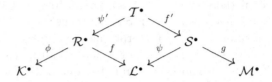

which exists by Proposition 1.3. Hence, we set $[g/\psi] \circ [f/\phi] = [(g \circ f')/(\phi \circ \psi')]$; one can readily see that this definition makes sense and that $D(\mathfrak{A})$ is an additive category.

A morphism $f \colon \mathcal{K}^\bullet \to \mathcal{L}^\bullet$ in $K(\mathfrak{A})$ defines the morphism $f/\mathrm{Id}_{\mathcal{K}^\bullet} \colon \mathcal{K}^\bullet \to \mathcal{L}^\bullet$ in the derived category, which we denote simply by f. This defines a functor $K(\mathfrak{A}) \to D(\mathfrak{A})$ which is additive.

A morphism $f/\phi \colon \mathcal{K}^\bullet \to \mathcal{L}^\bullet$ in the derived category induces a morphism in cohomology $\mathcal{H}(f/\phi) \colon \mathcal{H}(\mathcal{K}^\bullet) \xrightarrow{\mathcal{H}(\phi)^{-1}} \mathcal{H}(\mathcal{R}^\bullet) \xrightarrow{\mathcal{H}(f)} \mathcal{H}(\mathcal{L}^\bullet)$, which is independent of the representative f/ϕ of the class and is compatible with compositions.

Definition 1.3. Two complexes $\mathcal{K}^\bullet$ and $\mathcal{L}^\bullet$ are quasi-isomorphic if there is a complex $\mathcal{Z}^\bullet$ with quasi-isomorphisms $\mathcal{K}^\bullet \leftarrow \mathcal{Z}^\bullet \to \mathcal{L}^\bullet$. $\triangle$

Lemma 1.3 implies that the notion of quasi-isomorphism induces an equivalence relation between complexes. Eventually, we have the result we were looking for:

Proposition 1.4. *A morphism of complexes $f \colon \mathcal{K}^\bullet \to \mathcal{L}^\bullet$ is a quasi-isomorphism if and only if the induced morphism in the derived category is an isomorphism. Moreover, two complexes are quasi-isomorphic if and only if they are isomorphic in $D(\mathfrak{A})$.* $\square$

Proposition 1.5. *Let $\mathfrak{C}$ be an additive category. An additive functor $F \colon K(\mathfrak{A}) \to \mathfrak{C}$ factors through an additive functor $D(\mathfrak{A}) \to \mathfrak{C}$ if and only if it maps quasi-isomorphisms to isomorphisms. If $\mathfrak{B}$ is an abelian category, an additive functor $G \colon K(\mathfrak{A}) \to K(\mathfrak{B})$ mapping quasi-isomorphisms into quasi-isomorphisms induces an additive functor $G \colon$*

$D(\mathfrak{A}) \to D(\mathfrak{B})$ *such that the diagram*

$$
\begin{array}{ccc}
\mathbf{C}(\mathfrak{A}) & \xrightarrow{\ G\ } & \mathbf{C}(\mathfrak{B}) \\
\downarrow & & \downarrow \\
D(\mathfrak{A}) & \xrightarrow{\ G\ } & D(\mathfrak{B})
\end{array}
$$

is commutative. $\qquad\qquad\square$

We can also define derived categories out of some subcategories of $\mathbf{C}(\mathfrak{A})$; the only condition we need is that all the operations we have done can be performed in the new situation. More precisely, we should be able to construct the corresponding homotopy category and to localise by quasi-isomorphims; to this end one needs to define the cone of a morphism inside the new category. Some examples are the following:

Example 1.4. A complex $\mathcal{K}^\bullet$ is *bounded below* (resp. *bounded above*) if there is an integer n_0 such that $\mathcal{K}^{\bullet n} = 0$ for all $n \leq n_0$ (resp. $n \geq n_0$). A complex is *bounded* if it is bounded on both sides.

Bounded below complexes form a category $\mathbf{C}^+(\mathfrak{A})$. We can define its homotopy category $K^+(\mathfrak{A})$ and a "derived" category $D^+(\mathfrak{A})$ as we did before. By Proposition 1.5, the natural functor $K^+(\mathfrak{A}) \to D(\mathfrak{A})$ induces a functor $\gamma\colon D^+(\mathfrak{A}) \to D(\mathfrak{A})$. The latter is fully faithful, that is,

$$
\mathrm{Hom}_{D^+(\mathfrak{A})}(\mathcal{K}^\bullet, \mathcal{M}^\bullet) \simeq \mathrm{Hom}_{D(\mathfrak{A})}(\gamma(\mathcal{K}^\bullet), \gamma(\mathcal{M}^\bullet))
$$

for any pair of objects $\mathcal{K}^\bullet, \mathcal{M}^\bullet$ in $D^+(\mathfrak{A})$, and its essential image is the faithful subcategory of $D(\mathfrak{A})$ consisting of complexes in $\mathfrak{A}$ with bounded below cohomology. (The essential image of the functor γ is the subcategory of objects which are isomorphic to objects of the form $\gamma(\mathcal{K}^\bullet)$ for some $\mathcal{K}^\bullet$ in $D^+(\mathfrak{A})$). One can also define the categories $\mathbf{C}^-(\mathfrak{A})$ of *bounded above* complexes and $\mathbf{C}^b(\mathfrak{A})$ of complexes *bounded* on both sides, giving rise to "derived" categories $D^-(\mathfrak{A})$ and $D^b(\mathfrak{A})$. These are characterised as faithful subcategories of $D(\mathfrak{A})$ as above. $\qquad\triangle$

Example 1.5. An abelian subcategory $\mathfrak{A}'$ of $\mathfrak{A}$ is *thick* if any extension in $\mathfrak{A}$ of two objects of $\mathfrak{A}'$ is also in $\mathfrak{A}'$. If $\mathfrak{A}'$ is a thick abelian subcategory of $\mathfrak{A}$, we denote by $\mathbf{C}_{\mathfrak{A}'}(\mathfrak{A})$ the category of complexes whose cohomology objects are in $\mathfrak{A}'$. We can construct its homotopy category $K_{\mathfrak{A}'}(\mathfrak{A})$ and its derived category $D_{\mathfrak{A}'}(\mathfrak{A})$. The functor $K_{\mathfrak{A}'}(\mathfrak{A}) \to D(\mathfrak{A})$ induces a fully faithful functor $D_{\mathfrak{A}'}(\mathfrak{A}) \to D(\mathfrak{A})$ (cf. Proposition 1.5), whose essential image is the subcategory of $D(\mathfrak{A})$ whose objects are the complexes with cohomology objects in $\mathfrak{A}'$. $\qquad\triangle$

Example 1.6. We can also introduce the homotopy categories $K^+_{\mathfrak{A}'}(\mathfrak{A})$, $K^-_{\mathfrak{A}'}(\mathfrak{A})$ and $K^b_{\mathfrak{A}'}(\mathfrak{A})$ of complexes bounded below, above and on both sides, respectively, whose cohomology objects are in the subcategory $\mathfrak{A}'$ of $\mathfrak{A}$. The corresponding derived categories $D^+_{\mathfrak{A}'}(\mathfrak{A})$, $D^-_{\mathfrak{A}'}(\mathfrak{A})$ and $D^b_{\mathfrak{A}'}(\mathfrak{A})$ can be defined as well.   △

Let us write $\star$ for any of the symbols $+$, $-$, b, or for no symbol at all. The natural functor $K^\star(\mathfrak{A}') \to D(\mathfrak{A})$ maps quasi-isomorphisms to isomorphisms, so that induces a functor $D^\star(\mathfrak{A}') \to D^\star_{\mathfrak{A}'}(\mathfrak{A})$. In general, it may fail to be an equivalence of categories. There are special notations for the derived categories we are most interested in:

- If $\mathfrak{A}$ is the category of modules over a commutative ring A, we simply write $D(A)$, $D^+(A)$, $D^-(A)$, and $D^b(A)$.

- If $\mathfrak{A} = \mathfrak{Mod}(X)$ is the category of sheaves of $\mathcal{O}_X$-modules on an algebraic variety X, we write $D(X)$, $D^+(X)$, $D^-(X)$, and $D^b(X)$.

- If $\mathfrak{A} = \mathfrak{Mod}(X)$ and $\mathfrak{A}' = \mathfrak{Qco}(X)$ is the category of quasi-coherent sheaves of $\mathcal{O}_X$-modules, the derived category $D_{\mathfrak{A}'}(\mathfrak{A})$ of complexes of $\mathcal{O}_X$-modules with quasi-coherent cohomology sheaves is denoted $D_{qc}(X)$. In a similar way, we have the categories $D^+_{qc}(X)$, $D^-_{qc}(X)$ and $D^b_{qc}(X)$.

- If $\mathfrak{A} = \mathfrak{Mod}(X)$ and $\mathfrak{A}' = \mathfrak{Coh}(X)$ is the category of coherent sheaves of $\mathcal{O}_X$-modules, the derived category $D_{\mathfrak{A}'}(\mathfrak{A})$ of complexes of $\mathcal{O}_X$-modules with coherent cohomology sheaves is denoted $D_c(X)$. One also has the derived categories $D^+_c(X)$, $D^-_c(X)$ and $D^b_c(X)$.

- If $\mathfrak{A} = \mathfrak{Qco}(X)$ and $\mathfrak{A}' = \mathfrak{Coh}(X)$, we have the derived categories $D_c(\mathfrak{Qco}(X))$ $D^+_c(\mathfrak{Qco}(X))$, $D^-_c(\mathfrak{Qco}(X))$ and $D^b_c(\mathfrak{Qco}(X))$.

One has equivalences of categories $D^+_{qc}(X) \simeq D^+(\mathfrak{Qco}(X))$ and $D^b_{qc}(X) \simeq D^b(\mathfrak{Qco}(X))$. The first equivalence is a consequence of the fact that every quasi-coherent sheaf on an algebraic variety can be embedded as a subsheaf of an *injective* quasi-coherent sheaf. One also has $D^+_c(X) \simeq D^+_c(\mathfrak{Qco}(X))$ and $D^b_c(X) \simeq D^b_c(\mathfrak{Qco}(X))$. When X is smooth, the same is true for unbounded complexes as well, so that $D^\star_{qc}(X) \simeq D^\star(\mathfrak{Qco}(X))$ and $D^\star(\mathfrak{Coh}(X)) \simeq D^\star_c(\mathfrak{Qco}(X)) \simeq D^\star_c(X)$ for any value of $\star$.

1.2.1 *The derived category as a triangulated category*

Derived categories are examples of triangulated categories. We shall not give here a formal definition but shall just point out some of the features of the derived category that make it into a triangulated category.

The first is the existence of a shift functor $\tau \colon D(\mathfrak{A}) \to D(\mathfrak{A})$, $\tau(\mathcal{K}^\bullet) = \mathcal{K}^\bullet[1]$, which is an equivalence of categories. The second is the existence of "triangles", and among them a class of "distinguished triangles" satisfying some properties we do not describe here. A *triangle* in $D(\mathfrak{A})$ is a sequence of morphisms

$$\mathcal{K}^\bullet \xrightarrow{u} \mathcal{L}^\bullet \xrightarrow{v} \mathcal{M}^\bullet \xrightarrow{w} \mathcal{K}^\bullet[1]$$

which we also write in the form

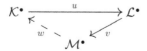

where the dashed arrow stands for the morphism $\mathcal{M}^\bullet \xrightarrow{w} \mathcal{K}^\bullet[1]$. A morphism of triangles is defined in the obvious way, and we say that a triangle is *distinguished* or *exact* if it is isomorphic to the triangle defined by the cone of a morphism $f \colon \mathcal{K}^\bullet \to \mathcal{L}^\bullet$, which is the triangle $\mathcal{K}^\bullet \xrightarrow{f} \mathcal{L}^\bullet \xrightarrow{\alpha} \mathrm{Cone}(f) \xrightarrow{\beta} \mathcal{K}^\bullet[1]$. From Proposition 1.1, an exact triangle in $D(\mathfrak{A})$ induces a long exact sequence in cohomology

$$\cdots \to \mathcal{H}^i(\mathcal{A}^\bullet) \xrightarrow{\mathcal{H}^i(u)} \mathcal{H}^i(\mathcal{B}^\bullet) \xrightarrow{\mathcal{H}^i(v)} \mathcal{H}^i(\mathcal{C}^\bullet) \xrightarrow{\mathcal{H}^i(w)}$$

$$\mathcal{H}^{i+1}(\mathcal{A}^\bullet) \xrightarrow{\mathcal{H}^{i+1}(u)} \mathcal{H}^{i+1}(\mathcal{B}^\bullet) \xrightarrow{\mathcal{H}^{i+1}(v)} \mathcal{H}^{i+1}(\mathcal{C}^\bullet) \xrightarrow{\mathcal{H}^{i+1}(w)} \cdots$$

Definition 1.7. If $\mathfrak{B}$ is another abelian category, an additive functor $F \colon D^\star_{\mathfrak{A}'}(\mathfrak{A}) \to D(\mathfrak{B})$ is said to be *exact* if it commutes with the shift functor, $F(\mathcal{K}^\bullet[1]) \simeq F(\mathcal{K}^\bullet)[1]$, and maps exact triangles to exact triangles. △

Then, for any exact triangle $\mathcal{K}^\bullet \xrightarrow{u} \mathcal{L}^\bullet \xrightarrow{v} \mathcal{M}^\bullet \xrightarrow{w} \mathcal{K}^\bullet[1]$ we have a long exact sequence

$$\cdots \to \mathcal{H}^i(F(\mathcal{K}^\bullet)) \to \mathcal{H}^i(F(\mathcal{L}^\bullet)) \to \mathcal{H}^i(F(\mathcal{M}^\bullet)) \to$$

$$\mathcal{H}^{i+1}(F(\mathcal{K}^\bullet)) \to \mathcal{H}^{i+1}(F(\mathcal{L}^\bullet)) \to \mathcal{H}^{i+1}(F(\mathcal{M}^\bullet)) \to \cdots$$

1.3 Derived Functors

The cohomology groups of a sheaf $\mathcal{F}$ on an algebraic variety X are the cohomology objects of the complex of global sections $\Gamma(X, \mathcal{I}^\bullet)$ of a resolution $\mathcal{I}^\bullet$ of $\mathcal{F}$ by injective sheaves; the resulting groups do not depend on the injective resolution, due to a result known as abstract de Rham theorem. In order to define derived functors on the derived category we shall mimic this construction. Let $\mathfrak{A}$ be an abelian category with enough injectives. Thus, any object $\mathcal{M}$ in $\mathfrak{A}$ has an injective resolution $\mathcal{M} \to I^0(\mathcal{M}) \to I^1(\mathcal{M}) \to \ldots$ which can be chosen to be functorial in $\mathcal{M}$. One can prove (by using bicomplexes, a notion we have not introduced in these notes) that for any complex $\mathcal{M}^\bullet$ there is a complex of injective objects $I(\mathcal{M}^\bullet)$ and a quasi-isomorphism

$$\mathcal{M}^\bullet \to I(\mathcal{M}^\bullet),$$

which defines a functor $I\colon K(\mathfrak{A}) \to K(\mathfrak{A})$. Let $\mathfrak{B}$ be another abelian category and $F\colon \mathfrak{A} \to \mathfrak{B}$ a left-exact functor. Then F induces a functor $\mathbf{R}F\colon K^+(\mathfrak{A}) \to D^+(\mathfrak{B})$ defined by $\mathbf{R}F(\mathcal{M}^\bullet) = F(I(\mathcal{M}^\bullet))$. Moreover, if $\mathcal{J}^\bullet$ is an acyclic complex of injective objects then $F(\mathcal{J}^\bullet)$ is acyclic, because $\mathcal{J}^\bullet$ splits. This implies that $\mathbf{R}F$ maps quasi-isomorphisms to isomorphisms and thus (cf. Proposition 1.5) yields a functor

$$\mathbf{R}F\colon D^+(\mathfrak{A}) \to D^+(\mathfrak{B}),$$

which is the *right derived functor* of F.

We can also derive on the right functors from $K(\mathfrak{A})$ to $K(\mathfrak{B})$ that are not induced by a left-exact functor. We shall give some examples in Subsection 1.3.3.

As it is customary for the "classical" right i-th derived functor of F, we use the notation $\mathbf{R}^i F(\mathcal{M}^\bullet) = \mathcal{H}^i(\mathbf{R}F(\mathcal{M}^\bullet))$. The right derived functor $\mathbf{R}F$ is exact. In particular, an exact triangle in $K(\mathfrak{A})$, $\mathcal{M}'^\bullet \to \mathcal{M}^\bullet \to \mathcal{M}''^\bullet \to \mathcal{M}'^\bullet[1]$ induces a long exact sequence

$$\cdots \to \mathbf{R}^i F(\mathcal{M}'^\bullet) \to \mathbf{R}^i F(\mathcal{M}^\bullet) \to \mathbf{R}^i F(\mathcal{M}''^\bullet) \to$$
$$\mathbf{R}^{i+1} F(\mathcal{M}'^\bullet) \to \mathbf{R}^{i+1} F(\mathcal{M}^\bullet) \to \mathbf{R}^{i+1} F(\mathcal{M}''^\bullet) \to \cdots$$

For any bounded below complex $\mathcal{M}^\bullet$ there is a natural morphism $F(\mathcal{M}^\bullet) \to \mathbf{R}F(\mathcal{M}^\bullet)$ in the derived category. $\mathcal{M}^\bullet$ is said to be F-*acyclic* if this morphism is an isomorphism in $D^+(\mathfrak{B})$.

The right derived functor $\mathbf{R}F$ satisfies a version of the de Rham theorem, namely, if a complex $\mathcal{M}^\bullet$ is isomorphic in the derived category $D^+(\mathfrak{A})$ to an F-acyclic complex $\mathcal{J}^\bullet$, then $\mathbf{R}F(\mathcal{M}^\bullet) \simeq F(\mathcal{J}^\bullet)$ in $D^+(\mathfrak{B})$.

Let $\mathfrak{C}$ be a third abelian category and $G\colon \mathfrak{B} \to \mathfrak{C}$ another left-exact functor.

Proposition 1.6 (Grothendieck's composite functor theorem). *If F transforms complexes of injective objects into G-acyclic complexes, one has a natural isomorphism $\mathbf{R}(G \circ F) \overset{\sim}{\to} \mathbf{R}G \circ \mathbf{R}F$.* $\qquad\square$

The theory of right derived functors can be applied when $\mathfrak{A}$ is one of the categories $\mathfrak{Mod}(X)$ or $\mathfrak{Qco}(X)$ because both have enough injectives.

One can develop a parallel theory of derived left exact functors if one assumes that $\mathfrak{A}$ has enough projectives, so that any object $\mathcal{M}$ has a functorial projective resolution $\cdots \to P^1(\mathcal{M}) \to P^0(\mathcal{M}) \to \mathcal{M} \to 0$. Then for every bounded above complex $\mathcal{M}^\bullet$ there exists a bounded above complex $P(\mathcal{M}^\bullet)$ of projective objects which defines a functor $P \colon K^-(\mathfrak{A}) \to K^-(\mathfrak{A})$. The functor $\mathbf{L}F \colon K^-(\mathfrak{A}) \to K^-(\mathfrak{B})$ given by $\mathbf{L}F(\mathcal{M}^\bullet) = F(P(\mathcal{M}^\bullet))$ defines as above a *left derived functor*

$$\mathbf{L}F \colon D^-(\mathfrak{A}) \to D^-(\mathfrak{B}).$$

Analogous properties to those stated for right derived functors hold for left derived functors.

One should note that the categories $\mathfrak{Mod}(X)$, $\mathfrak{Qco}(X)$ and $\mathfrak{Coh}(X)$ do not have enough projectives. However if X is a (quasi-)projective, any quasi-coherent sheaf has a (possibly infinite) resolution by locally free sheaves which may have infinite rank, and the problem is circumvented by considering complexes $\mathcal{P}^\bullet$ of locally free sheaves. We shall come again to this point in Subsection 1.3.2.

1.3.1 Derived Direct Image

Let $f \colon X \to Y$ be a morphism of algebraic varieties. Since the direct image functor $f_* \colon \mathfrak{Mod}(X) \to \mathfrak{Mod}(Y)$ is left-exact, it induces a right derived functor

$$\mathbf{R}f_* \colon D^+(X) \to D^+(Y)$$

described as $\mathbf{R}f_*\mathcal{M}^\bullet \simeq f_*(\mathcal{I}^\bullet)$ where $\mathcal{I}^\bullet$ is a complex of injective $\mathcal{O}_X$-modules quasi-isomorphic to $\mathcal{M}^\bullet$. When f is quasi-compact and locally of finite type (as it often happens), the direct image of a quasi-coherent sheaf is also quasi-coherent; in this case, $\mathbf{R}f_*$ defines a functor $\mathbf{R}f_* \colon D^+_{qc}(X) \to D^+_{qc}(Y)$. When f is proper, so that the higher direct images of a coherent sheaf are coherent as well (cf. [EGAIII-1, Thm.3.2.1] or [Hart77, Thm. 5.2] in the projective case), we also have a functor $\mathbf{R}f_* \colon D^+_c(X) \to D^+_c(Y)$. Moreover, the dimension of X bounds the number of higher direct images of a sheaf of $\mathcal{O}_X$-modules, so that $\mathbf{R}f_*$ induces also a functor $\mathbf{R}f_* \colon D^b_{qc}(X) \to D^b_{qc}(Y)$. In this case $\mathbf{R}f_*$ extends to a functor $\mathbf{R}f_* \colon D_{qc}(X) \to D_{qc}(Y)$ which maps $D^b_c(X)$ to $D^b_c(Y)$.

If Y is a point, $\mathfrak{A}_Y$ is the category of abelian groups and f_* is the functor of global sections $\Gamma(X,\ )$. In this case, $\mathbf{R}f_*\mathcal{M}^\bullet = \mathbf{R}\Gamma(X, \mathcal{M}^\bullet)$

and $\mathbf{R}^i f_* \mathcal{M}^\bullet$ is called the i-th hypercohomology group $\mathbb{H}^i(X, \mathcal{M}^\bullet)$ of the complex $\mathcal{M}^\bullet$. It coincides with the cohomology group $H^i(X, \mathcal{M})$ when the complex reduces to a single sheaf.

1.3.2 The derived inverse image

Let $f: X \to Y$ be a morphism of algebraic varieties. One can prove that any sheaf of $\mathcal{O}_X$-modules $\mathcal{M}$ is a quotient of a *flat* sheaf of $\mathcal{O}_X$-modules $P(\mathcal{M})$ and that one can choose $P(\mathcal{M})$ depending functorially on $\mathcal{M}$. One then shows that for any bounded above complex $\mathcal{M}^\bullet$, there is a complex $P(\mathcal{M}^\bullet)$ of flat sheaves, and a quasi-isomorphism $P(\mathcal{M}^\bullet) \to \mathcal{M}^\bullet$ which defines a functor $K^-(\mathfrak{Mod}(Y)) \to K^-(\mathfrak{Mod}(Y))$. Moreover $\mathbf{L}f^*(\mathcal{M}^\bullet) = f^*(P(\mathcal{M}^\bullet))$ gives a left derived functor

$$\mathbf{L}f^* : D^-(Y) \to D^-(X) ,$$

which induces functors $\mathbf{L}f^* : D_{qc}^-(X) \to D_{qc}^-(Y)$ and $\mathbf{L}f^* : D_c^-(X) \to D_c^-(Y)$. In some cases it also induces a functor $\mathbf{L}f^* : D_c(X) \to D_c(Y)$ which maps $D_c^b(X)$ to $D_c^b(Y)$; this happens for instance when Y is smooth. Another case is when f is *of finite homological dimension*, that is, when for every coherent sheaf $\mathcal{G}$ on Y there are only a finite number of nonzero derived inverse images $\mathbf{L}_j f^*(\mathcal{G}) = \mathcal{H}^{-j}(\mathbf{L}f^*(\mathcal{G}))$; in particular, flat morphisms are of finite homological dimension.

1.3.3 Derived homomorphism functor and derived tensor product

We wish to construct a "derived functor" of the functor of global homomorphisms for complexes. Although this functor is not induced by a left-exact functor between the original abelian categories, we still can derive the complex of homomorphisms by mimicking the procedure used so far. We are not detailing here the entire process (see [BBH08, Appendix A]); let us just mention that one eventually obtains a bifunctor

$$\mathbf{R}\mathrm{Hom}_X^\bullet : D(X)^0 \times D^+(X) \to D(k)$$

described as $\mathbf{R}\mathrm{Hom}_X^\bullet(\mathcal{K}^\bullet, \mathcal{L}^\bullet) \simeq \mathrm{Hom}^\bullet(\mathcal{K}^\bullet, \mathcal{I}^\bullet)$, where $\mathcal{I}^\bullet$ is a complex of injective sheaves quasi-isomorphic to $\mathcal{L}^\bullet$. We use the notation

$$\mathrm{Ext}_X^i(\mathcal{K}^\bullet, \mathcal{L}^\bullet) = \mathbf{R}^i\mathrm{Hom}_X^\bullet(\mathcal{K}^\bullet, \mathcal{L}^\bullet) = \mathcal{H}^i(\mathbf{R}\mathrm{Hom}^\bullet(\mathcal{K}^\bullet, \mathcal{L}^\bullet)) .$$

Proposition 1.7. (Yoneda's formula) *If* $\mathcal{M}^\bullet \in D(X)$ *and* $\mathcal{N}^\bullet \in D^+(X)$, *one has*

$$\mathrm{Ext}_X^i(\mathcal{M}^\bullet, \mathcal{N}^\bullet) \simeq \mathrm{Hom}_{D(X)}^i(\mathcal{M}^\bullet, \mathcal{N}^\bullet) := \mathrm{Hom}_{D(X)}(\mathcal{M}^\bullet, \mathcal{N}^\bullet[i]) .$$

$\square$

Let Y be another algebraic variety and $\Psi \colon D(X) \to D(Y)$ an exact functor which maps bounded complexes to bounded complexes.

Corollary 1.8. (Parseval's formula) *If Ψ is fully faithful, there are isomorphisms*

$$
\begin{array}{ccc}
\mathrm{Ext}^i_X(\mathcal{M}^\bullet, \mathcal{N}^\bullet) & \xrightarrow{\;\sim\;} & \mathrm{Ext}^i_Y(\Psi(\mathcal{M}^\bullet), \Psi(\mathcal{N}^\bullet)) \\
\Big\downarrow{\scriptstyle\simeq} & & \Big\downarrow{\scriptstyle\simeq} \\
\mathrm{Hom}^i_{D(X)}(\mathcal{M}^\bullet, \mathcal{N}^\bullet) & \xrightarrow{\;\sim\;} & \mathrm{Hom}^i_{D(Y)}(\Psi(\mathcal{M}^\bullet), \Psi(\mathcal{N}^\bullet))
\end{array}
$$

$\square$

One can define the complex $\mathcal{H}om^\bullet_{\mathcal{O}_X}(\mathcal{M}^\bullet, \mathcal{N}^\bullet)$ of sheaves of homomorphisms; this is given by $\mathcal{H}om^n(\mathcal{M}^\bullet, \mathcal{N}^\bullet) = \prod_i \mathcal{H}om_{\mathcal{O}_X}(\mathcal{M}^i, \mathcal{N}^{i+n})$ with the differential $df = f \circ d_{\mathcal{M}^\bullet} + (-1)^{n+1} d_{\mathcal{N}^\bullet} \circ f$. There is also a derived sheaf homomorphism

$$
\mathbf{R}\mathcal{H}om^\bullet_{\mathcal{O}_X} \colon D(X)^0 \times D^+(X) \to D(X).
$$

By applying Grothendieck's composite functor theorem, we obtain for every open $U \subseteq X$ an isomorphism in the derived category $D(U)$:

$$
\mathbf{R}\Gamma(U, \mathbf{R}\mathcal{H}om^\bullet_{\mathcal{O}_X}(\mathcal{K}^\bullet, \mathcal{L}^\bullet)) \simeq \mathbf{R}\mathrm{Hom}^\bullet_{\mathcal{O}_U}(\mathcal{K}^\bullet{}_{|U}, \mathcal{L}^\bullet{}_{|U}).
$$

By following a similar procedure, one can derive the functor "tensor product of complexes": there exists a bifunctor, called derived tensor product,

$$
\overset{\mathbf{L}}{\otimes} \colon D(X) \times D(X) \to D(X),
$$

whose description is $\mathcal{M}^\bullet \overset{\mathbf{L}}{\otimes} \mathcal{N}^\bullet = \mathcal{M}^\bullet \otimes P(\mathcal{N}^\bullet)$, where $P(\mathcal{N}^\bullet) \to \mathcal{N}^\bullet$ is a quasi-isomophism and $P(\mathcal{N}^\bullet)$ is a complex of flat sheaves.

1.3.4 Base change in the derived category

There are some remarkable formulas in the derived category that relate the various derived functors (cf. [Hart77]). Here we describe only the following strengthened version of the *derived category base change formula*.

Proposition 1.9. [BBH08, Prop.A.74] *Let us consider a cartesian diagram of morphisms of algebraic varieties*

$$
\begin{array}{ccc}
X \times_Y \widetilde{Y} & \xrightarrow{\;\tilde{g}\;} & X \\
{\scriptstyle\tilde{f}}\Big\downarrow & & \Big\downarrow{\scriptstyle f} \\
\widetilde{Y} & \xrightarrow{\;g\;} & Y
\end{array}
$$

For any complex $\mathcal{M}^\bullet$ of $\mathcal{O}_X$-modules there is a natural morphism

$$\mathbf{L}g^*\mathbf{R}f_*\mathcal{M}^\bullet \to \mathbf{R}\tilde{f}_*\mathbf{L}\tilde{g}^*\mathcal{M}^\bullet.$$

Moreover, if $\mathcal{M}^\bullet$ has quasi-coherent cohomology and either f or g is flat, the above morphism is an isomorphism.   □

2 Integral functors and Fourier-Mukai transforms

2.1 Definitions

We start this section with a general definition of integral functor; Fourier-Mukai transforms will provide interesting examples. The varieties involved can be quite general algebraic varieties, though in the applications they will be mainly smooth and projective. Most of the results described here are valid over an arbitrary algebraically closed field $\Bbbk$, possibly requiring that $\mathrm{ch}\,\Bbbk = 0$. However, in order to simplify our treatment, we shall assume that $\Bbbk$ is the field $\mathbb{C}$ of complex numbers.

Let us consider the diagram

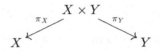

involving the projections of the cartesian product $X \times Y$ onto its factors. Any object $\mathcal{K}^\bullet$ in the derived category $D_c^-(X \times Y)$ is the *kernel* of an *integral functor*:

$$\Phi_{X \to Y}^{\mathcal{K}^\bullet} : D_c^-(X) \to D_c^-(Y)$$

$$\mathcal{E}^\bullet \mapsto \Phi_{X \to Y}^{\mathcal{K}^\bullet}(\mathcal{E}^\bullet) = \mathbf{R}\pi_{Y*}(\pi_X^*\mathcal{E}^\bullet \overset{\mathbf{L}}{\otimes} \mathcal{K}^\bullet).$$

The integral functor $\Phi_{X \to Y}^{\mathcal{K}^\bullet}$ is said to be a *Fourier-Mukai functor* if it is an equivalence of categories, and a *Fourier-Mukai transform* is a Fourier-Mukai functor whose kernel $\mathcal{K}^\bullet$ reduces to a single sheaf $\mathcal{K}$.

If $\mathcal{K}^\bullet$ is of *finite Tor-dimension over X*, that is, if it is isomorphic in the derived category to a bounded complex of sheaves that are flat over X, then the integral functor $\Phi_{X \to Y}^{\mathcal{K}^\bullet}$ maps $D_c^b(X)$ to $D_c^b(Y)$ and can be extended to a functor $\Phi_{X \to Y}^{\mathcal{K}^\bullet} : D(X) \to D(Y)$ between the entire derived categories. Note that the condition of finite Tor-dimensionality is always fulfilled if X and Y are *smooth* and $\mathcal{K}^\bullet$ is an object of $D_c^b(X \times Y)$, because in this case $\mathcal{K}^\bullet$ is isomorphic in the derived category to a bounded complex of locally free sheaves.

Since an integral functor is the composition of the functors $\mathbf{L}\pi_X^*$ (which is isomorphic to π_X^* because π_X is a flat morphism), $(\ )\overset{\mathbf{L}}{\otimes}\mathcal{K}^\bullet$, and $\mathbf{R}\pi_{Y*}$, and these functors are exact, *any integral functor is exact as well*. In particular, for any exact sequence $0 \to \mathcal{F} \to \mathcal{E} \to \mathcal{G} \to 0$ of sheaves in X we obtain an exact sequence

$$\cdots \to \Phi^{i-1}(\mathcal{G}) \to \Phi^i(\mathcal{F}) \to \Phi^i(\mathcal{E}) \to \Phi^i(\mathcal{G}) \to \Phi^{i+1}(\mathcal{F}) \to \cdots$$

where we have written $\Phi^i(\ ) = \mathcal{H}^i(\Phi_{X\to Y}^{\mathcal{K}^\bullet}(\ ))$.

Definition 2.1. Let $\Phi_{X\to Y}^{\mathcal{K}^\bullet}$ be an integral functor. A complex $\mathcal{F}^\bullet$ in $D_c^-(X)$ is WIT$_i$ if $\Phi_{X\to Y}^{\mathcal{K}^\bullet}(\mathcal{F}^\bullet) \simeq \mathcal{G}[-i]$ in $D(Y)$, for a coherent sheaf $\mathcal{G}$ on Y. If in addition $\mathcal{G}$ is locally free, we say that $\mathcal{F}^\bullet$ is IT$_i$. $\triangle$

In Section 3 we shall see a connection between integral functors and index theory. This will make it clear that the "IT" condition is related with the "Atiyah-Singer index theorem", and that the "W" of "WIT" stands for "weak".

Using the cohomology base change theorem [Hart77, III.12.11] one proves the following criterion for the WIT condition to hold.

Proposition 2.1. *Assume that the kernel $\mathcal{Q}$ is a locally free sheaf on the product $X \times Y$. A coherent sheaf $\mathcal{F}$ on X is IT$_i$ if and only if $H^j(X, \mathcal{F} \otimes \mathcal{Q}_y) = 0$ for all $y \in Y$ and for all $j \neq i$, where $\mathcal{Q}_y$ denotes the restriction of $\mathcal{Q}$ to $X \times \{y\}$. Furthermore, $\mathcal{F}$ is WIT$_0$ if and only if it is IT$_0$.* $\square$

Let us list the simplest examples of integral functors.

Example 2.2. Let $\delta\colon X \hookrightarrow X \times X$ denote the diagonal immersion, and write Δ for its image. The structure sheaf $\mathcal{O}_\Delta = \delta_* \mathcal{O}_X$ of Δ defines an integral functor $\Phi_{X\to X}^{\mathcal{O}_\Delta}\colon D_c^-(X) \to D_c^-(X)$, which is isomorphic to the identity functor $\Phi_{X\to X}^{\mathcal{O}_\Delta} \simeq \mathrm{Id}$. Thus, the identity functor is a Fourier-Mukai transform. $\triangle$

Example 2.3. If $\mathcal{L}$ is a line bundle on X, the functor $\Phi_{X\to X}^{\delta_*\mathcal{L}}$ consists of the twist by $\mathcal{L}$. It is also a Fourier-Mukai transform, whose quasi-inverse is the twist by $\mathcal{L}^{-1}$, that is, the Fourier-Mukai transform with kernel $\delta_*\mathcal{L}^{-1}$. $\triangle$

Example 2.4. If $f\colon X \to Y$ is a proper morphism and $\mathcal{K}^\bullet$ is the structure sheaf of the graph $\Gamma_f \subset X\times Y$, one has isomorphisms of functors $\Phi_{X\to Y}^{\mathcal{K}^\bullet} \simeq \mathbf{R}f_*\colon D_c^-(X) \to D_c^-(Y)$ and $\Phi_{Y\to X}^{\mathcal{K}^\bullet} \simeq \mathbf{L}f^*\colon D_c^-(Y) \to D_c^-(X)$. $\triangle$

2.1.1 The abelian Fourier-Mukai transform

Let X be an abelian variety; we can think of X as a torus. Let $\hat{X}$ be its dual abelian variety; a closed point $\xi \in \hat{X}$ corresponds to a zero degree line bundle $\mathcal{P}_\xi$ over X (or a flat line bundle). $\hat{X}$ is a "fine moduli space", in the sense that there exists a line bundle $\mathcal{P}$ over $X \times \hat{X}$, whose restriction to the fibre $X \simeq \pi_{\hat{X}}^{-1}\xi$ of the projection $\pi_{\hat{X}}$ is precisely the flat line bundle $\mathcal{P}_\xi$ corresponding to ξ. This *universal Poincaré line bundle* is uniquely characterised by this property up to twisting by pull-backs $\pi_{\hat{X}}^* \mathcal{N}$ of line bundles $\mathcal{N}$ on $\hat{X}$. To avoid any ambiguity, $\mathcal{P}$ is normalised so that its restriction to the fibre $\hat{X} \simeq \pi_X^{-1}(0)$ of the origin $0 \in X$, is trivial. This fixes $\mathcal{P}$ uniquely. An explicit description of $\mathcal{P}$ in differential-geometric terms will be given in Section 3.

The first example of a Fourier-Mukai transform was introduced by Mukai in this setting [Muk81]. Mukai's seminal idea was to use the normalised Poincaré bundle $\mathcal{P}$ to define an integral functor

$$\mathbf{S} = \Phi_{X \to \hat{X}}^{\mathcal{P}} : D_c^b(X) \to D_c^b(\hat{X}),$$

which turns out to be an equivalence of triangulated categories, that is, a *Fourier-Mukai transform*. We shall give a short proof of this fact in Theorem 2.8. Moreover, the functor $\Phi_{\hat{X} \to X}^{\mathcal{P}^*[g]}$, where $g = \dim X$, is quasi-inverse to $\Phi_{X \to \hat{X}}^{\mathcal{P}}$. We shall call $\mathbf{S} = \Phi_{X \to \hat{X}}^{\mathcal{P}}$ the *abelian Fourier-Mukai transform* and $\widehat{\mathbf{S}} = \Phi_{\hat{X} \to X}^{\mathcal{P}^*}$ the *dual abelian Fourier-Mukai transform*. Actually, instead of $\widehat{\mathbf{S}}$, Mukai considers the functor $\widetilde{\mathbf{S}} = \Phi_{\hat{X} \to X}^{\mathcal{P}} : D_c^b(\hat{X}) \to D_c^b(X)$. One has $\widetilde{\mathbf{S}} \circ \mathbf{S} \simeq \iota_{\hat{X}}^* \circ [-g]$, where $\iota_{\hat{X}} : \hat{X} \to \hat{X}$ is the involution which maps a line bundle to its dual.

2.1.2 Orlov's representation theorem

We have seen a few examples of integral functors. A natural problem is the characterisation of the exact functors $D_c^b(X) \to D_c^b(Y)$ that are integral. The most important result in this direction is due to Orlov [Or97]:

Theorem 2.2. *Let X and Y be smooth projective varieties. Any fully faithful exact functor $\Psi \colon D_c^b(X) \to D_c^b(Y)$ is an integral functor.*   □

Orlov's original statement assumed that the exact functor had a right adjoint; however, Bondal and Van den Bergh proved that any exact functor $D_c^b(X) \to D_c^b(Y)$ satisfies this property [BvdB03].

2.2 General properties of integral functors

The first property of integral functors we would like to describe is that the composition of two of them is again an integral functor.

If Z is another proper variety, we consider the diagram

$$\begin{array}{ccc} & X \times Y \times Z & \\ {\scriptstyle \pi_{XY}} \swarrow & \downarrow{\scriptstyle \pi_{Y,Z}} & \searrow {\scriptstyle \pi_{XZ}} \\ X \times Y & Y \times Z & X \times Z \end{array}$$

Given kernels $\mathcal{K}^\bullet$ in $D_c^-(X \times Y)$ and $\mathcal{L}^\bullet$ in $D_c^-(Y \times Z)$, the composition of the integral functors defined by them is given by the following Proposition:

Proposition 2.3. *There is a natural isomorphism of functors* $\Phi_{Y \to Z}^{\mathcal{L}^\bullet} \circ \Phi_{X \to Y}^{\mathcal{K}^\bullet} \simeq \Phi_{X \to Z}^{\mathcal{L}^\bullet * \mathcal{K}^\bullet}$, *where* $\mathcal{L}^\bullet * \mathcal{K}^\bullet = \mathbf{R}\pi_{XZ*}(\pi_{XY}^* \mathcal{K}^\bullet \overset{\mathbf{L}}{\otimes} \pi_{YZ}^* \mathcal{L}^\bullet)$ *in* $D_c^-(X \times Z)$. $\qquad\square$

2.2.1 Action of integral functors on cohomology

Integral functors act on cohomology, and the study of this action allows one to determine the topological invariants of the transform of a complex in terms of the topological invariants of the complex. This is very useful in studying the effect of integral functors on moduli spaces of sheaves.

We need to recall the notion of Mukai vector and pairing. The Mukai vector of a complex $\mathcal{E}^\bullet$ in $D_c^b(X)$ for a smooth projective variety X is defined as

$$v(\mathcal{E}^\bullet) = \mathrm{ch}(\mathcal{E}^\bullet) \cdot \sqrt{\mathrm{td}(X)} \,,$$

where $\mathrm{td}(X) \in A^\bullet(X) \otimes \mathbb{Q}$ is the Todd class of X. We can define a symmetric bilinear form $\langle \cdot, \cdot \rangle$ on the rational Chow group $A^\bullet(X) \otimes \mathbb{Q}$ by setting

$$\langle v, w \rangle = - \int_X v^* \cdot w \cdot \exp(\tfrac{1}{2} c_1(X)) \,.$$

The Mukai pairing naturally extends to the even rational cohomology $\bigoplus_j H^{2j}(X, \mathbb{Q})$.

If X and Y are smooth proper varieties, and $\mathcal{K}^\bullet$ is a kernel in $D_c^b(X \times Y)$, by the Grothendieck-Riemann-Roch theorem the integral functor $\Phi_{X \to Y}^{\mathcal{K}^\bullet} : D_c^b(X) \to D_c^b(Y)$ gives rise to the commutative diagram

$$\begin{array}{ccccc} D_c^b(X) & \overset{v}{\longrightarrow} & A^\bullet(X) \otimes \mathbb{Q} & \longrightarrow & H^\bullet(X, \mathbb{Q}) \\ \downarrow{\scriptstyle \Phi_{X \to Y}^{\mathcal{K}^\bullet}} & & \downarrow{\scriptstyle f^{\mathcal{K}^\bullet}} & & \downarrow{\scriptstyle f^{\mathcal{K}^\bullet}} \\ D_c^b(Y) & \overset{v}{\longrightarrow} & A^\bullet(Y) \otimes \mathbb{Q} & \longrightarrow & H^\bullet(Y, \mathbb{Q}) \end{array}$$

where both $f^{\mathcal{K}^\bullet}$ are the $\mathbb{Q}$-vector space homomorphisms defined by

$$f^{\mathcal{K}^\bullet}(\alpha) = \pi_{Y*}(\pi_X^* \alpha \cdot v(\mathcal{K}^\bullet)).$$

The map $f^{\mathcal{K}^\bullet}$ sends $H^{2\bullet}(X, \mathbb{Q})$ to $H^{2\bullet}(Y, \mathbb{Q})$ and depends functorially on the kernel, i.e., $f^{\mathcal{L}^\bullet * \mathcal{K}^\bullet} = f^{\mathcal{L}^\bullet} \circ f^{\mathcal{K}^\bullet}$. This implies the following result.

Corollary 2.4. *Let X and Y be smooth proper varieties and $\Phi_{X \to Y}^{\mathcal{K}^\bullet}$: $D_c^b(X) \to D_c^b(Y)$ a Fourier-Mukai functor. Then the morphisms $f^{\mathcal{K}^\bullet}$: $A^\bullet(X) \otimes \mathbb{Q} \to A^\bullet(Y) \otimes \mathbb{Q}$ and $f^{\mathcal{K}^\bullet} : H^\bullet(X, \mathbb{Q}) \to H^\bullet(Y, \mathbb{Q})$ are isomorphisms. Moreover, the latter induces an isomorphism of vector spaces between the even cohomology rings.* $\square$

2.2.2 Fully faithful integral functors and Fourier-Mukai functors

In this Subsection all varieties are smooth and projective unless otherwise stated. The first step in characterising the kernels $\mathcal{K}^\bullet$ in $D_c^b(X \times Y)$ that give rise to equivalences is to determine the kernels for which $\Phi_{X \to Y}^{\mathcal{K}^\bullet} : D_c^b(X) \to D_c^b(Y)$ is fully faithful. The idea is to study the effect of a fully faithful integral functor $\Phi_{X \to Y}^{\mathcal{K}^\bullet}$ on the skyscraper sheaves $\mathcal{O}_x$ of the points. Due to the Parseval formula (Corollary 1.8), for any pair of points x_1, x_2 of X one has

$$\operatorname{Hom}_{D(X)}^i(\mathcal{O}_{x_1}, \mathcal{O}_{x_2}) \simeq$$
$$\operatorname{Hom}_{D(Y)}^i(\Phi_{X \to Y}^{\mathcal{K}^\bullet}(\mathcal{O}_{x_1}), \Phi_{X \to Y}^{\mathcal{K}^\bullet}(\mathcal{O}_{x_2})) \simeq \operatorname{Hom}_{D(Y)}^i(\mathbf{L}j_{x_1}^* \mathcal{K}^\bullet, \mathbf{L}j_{x_2}^* \mathcal{K}^\bullet).$$
$$(2.1)$$

It follows that if $\Phi_{X \to Y}^{\mathcal{K}^\bullet}$ is fully faithful, the kernel $\mathcal{K}^\bullet$ fulfills the following properties:

(a) $\operatorname{Hom}_{D(Y)}^i(\mathbf{L}j_{x_1}^* \mathcal{K}^\bullet, \mathbf{L}j_{x_2}^* \mathcal{K}^\bullet) = 0$ unless $x_1 = x_2$ and $0 \leq i \leq \dim X$;

(b) $\operatorname{Hom}_{D(Y)}^0(\mathbf{L}j_x^* \mathcal{K}^\bullet, \mathbf{L}j_x^* \mathcal{K}^\bullet) = \mathbb{C}$.

Kernels $\mathcal{K}^\bullet$ satisfying these properties were called *strongly simple over* X by Maciocia [Mac96], but this notion had already been used implicitly in [BO95]. The following crucial result was originally proved by Bondal and Orlov [BO95].

Theorem 2.5. *Let X and Y be smooth projective varieties, and $\mathcal{K}^\bullet$ a kernel in $D^b(X \times Y)$. The functor $\Phi_{X \to Y}^{\mathcal{K}^\bullet}$ is fully faithful if and only if $\mathcal{K}^\bullet$ is strongly simple over X.* $\square$

Building on Bondal and Orlov's work, Bridgeland and Maciocia [Bri99, Mac96] determined the kernels that give rise to equivalences of categories. They called *special* an object $\mathcal{F}^\bullet$ in $D_c^b(X)$ such that $\mathcal{F}^\bullet \otimes \omega_X \simeq \mathcal{F}^\bullet$ in $D_c^b(X)$, where ω_X is the canonical line bundle.

Proposition 2.6. *Let X and Y be smooth projective varieties of the same dimension n, and let $\mathcal{K}^\bullet$ be a kernel in $D_c^b(X \times Y)$. Then $\Phi_{X \to Y}^{\mathcal{K}^\bullet}$ is a Fourier-Mukai functor if and only if $\mathcal{K}^\bullet$ is strongly simple over X and $\mathbf{L}j_x^* \mathcal{K}^\bullet$ is special for all $x \in X$.* □

When the canonical bundles of X and Y are trivial this takes a simpler form.

Proposition 2.7. *Let X and Y be smooth projective varieties of the same dimension with trivial canonical bundles and $\mathcal{K}^\bullet$ an object in $D_c^b(X \times Y)$ strongly simple over X. Then $\Phi_{X \to Y}^{\mathcal{K}^\bullet}$ is a Fourier-Mukai functor and $\Phi_{Y \to X}^{\mathcal{K}^{\bullet \vee}[n]}$ is a quasi-inverse to $\Phi_{X \to Y}^{\mathcal{K}^\bullet}$.* □

The characterisation of the kernels that induce fully faithful integral functors or equivalences has been generalised to the case of singular Cohen-Macaulay varieties in [HLS05, HLS08].

2.2.3 The abelian Fourier-Mukai transform revisited

Here we apply the characterisation of Fourier-Mukai functors to give a simple proof of the fact that the "abelian Fourier-Mukai transform" **S** and the "dual abelian Fourier-Mukai transform" $\widehat{\mathbf{S}}$ are truly Fourier-Mukai transforms, that is, they are equivalences of categories.

Let X be an abelian variety of dimension g and $\hat{X}$ its dual abelian variety. If $\mathcal{P}$ is the Poincaré line bundle on $X \times \hat{X}$, the restriction $\mathbf{L}j_\xi^* \mathcal{P}$ is the line bundle $\mathcal{P}_\xi$ on X corresponding to the point $\xi \in \hat{X}$. Since $\mathrm{Hom}_{D(X)}^i(\mathcal{P}_{\xi_1}, \mathcal{P}_{\xi_2})) \simeq H^i(X, \mathcal{P}_{\xi_1}^* \otimes \mathcal{P}_{\xi_2})$ for any pair of points ξ_1 and ξ_2 of $\hat{X}$, one has:

(a) $\mathrm{Hom}_{D(X)}^i(\mathcal{P}_{\xi_1}, \mathcal{P}_{\xi_2})) = 0$ unless $\xi_1 = \xi_2$ and $0 \leq i \leq g$;

(b) $\mathrm{Hom}_{D(X)}^0(\mathcal{P}_{\xi_1}, \mathcal{P}_{\xi_1}) = \mathbb{C}$ for any $\xi \in \hat{X}$.

In other words, $\mathcal{P}$ is strongly simple over $\hat{X}$. By Proposition 2.7, it is strongly simple over X as well. So we have:

Theorem 2.8. *The functors* $\mathbf{S} = \Phi_{X \to \hat{X}}^{\mathcal{P}} : D_c^b(X) \to D_c^b(\hat{X})$ *and* $\widehat{\mathbf{S}} = \Phi_{\hat{X} \to X}^{\mathcal{P}^*} : D_c^b(\hat{X}) \to D_c^b(X)$ *are equivalences of categories.* □

2.2.4 Fourier-Mukai functors on K3 and abelian surfaces

Let Y be a smooth projective surface and X a fine moduli space of *special* stable sheaves on Y with fixed Mukai vector v. Let $\mathcal{Q}$ be a universal sheaf on $X \times Y$ for the corresponding moduli problem, so that $\mathcal{Q}$ is flat over X and $\mathcal{Q}_x$ is a stable special sheaf on Y with Mukai vector v. Given closed points x and z in X, one has $\chi(\mathcal{Q}_x, \mathcal{Q}_z) = -\langle v, v \rangle = -v^2$.

Proposition 2.9. [BrM01] *Assume that X is a projective surface. Then X is smooth if and only if $v^2 = 0$. Moreover, in this case $\Phi^{\mathcal{Q}}_{X \to Y} : D^b_c(X) \to D^b_c(Y)$ is a Fourier-Mukai functor.* $\square$

With the help of Proposition 2.9 one can construct Fourier-Mukai transforms for K3 surfaces. Let Y be a K3 surface with a polarisation H and v a Mukai vector with $v^2 = 0$. Assume that $v = (r, c, s)$ is primitive (i.e., not divisible by an integer) and that the greatest common divisor of the numbers $(r, c \cdot H, s)$ is 1. If the moduli space $X = M_v(Y)$ of sheaves on Y with Mukai vector v that are (Gieseker) stable with respect to H is nonempty, it is a projective variety of dimension $v^2 + 2 = 2$ (cf. [Mar77, Mar78]). Moreover, there exists a universal sheaf $\mathcal{Q}$ on $X \times Y$. Proposition 2.9 implies that X is smooth and that $\Phi^{\mathcal{Q}}_{X \to Y} : D^b_c(X) \to D^b_c(Y)$ is a Fourier-Mukai transform. We shall also see (cf. Proposition 2.22) that X is again a K3 surface.

Example 2.5. A first instance of this situation, which has also been the first example of a nontrivial Fourier-Mukai transform on K3 surfaces, was given in [BBH97a], were a class of K3 surfaces called *strongly reflexive* was introduced. A K3 surface Y is strongly reflexive if it carries a polarization H and a divisor ℓ such that $H^2 = 2$, $H \cdot \ell = 0$, $\ell^2 = -12$, and X has no nodal curves of degree 1 or 2. Strongly reflexive K3 surfaces do exist. There is indeed a nonempty coarse moduli space of strongly reflexive K3 surfaces, which is an irreducible quasi-projective scheme of dimension 18 [BBH97b]. On a strongly reflexive K3 surface Y one may take the Mukai vector $v = (2, \ell, -3)$, which fulfils all the above requirements. One proves that $M_v(Y) \neq \emptyset$ [BBH97a], and then $\Phi^{\mathcal{Q}}_{X \to Y} : D^b_c(X) \to D^b_c(Y)$ is a Fourier-Mukai transform. This may be used to construct further examples [BM96]. $\triangle$

Example 2.6. Later Mukai provided another example [Muk99]. He considered a K3 surface Y such that there exist coprime positive integers r, s and a polarization H in Y with $H^2 = 2rs$. He proved that $X = M_v(Y)$ is nonempty (and fine) and that the universal family induces a Fourier-Mukai transform $D^b_c(X) \simeq D^b_c(Y)$. $\triangle$

2.2.5 Relative integral functors and base change

In this Subsection we generalise the notion of integral functor to the relative setting, namely, we shall deal with morphisms (or families) instead of single varieties.

We consider two (proper) morphisms of algebraic schemes $p\colon X \to B$ and $q\colon Y \to B$ and denote by $\tilde\pi_X$, $\tilde\pi_Y$ the projections of the fibre product $X \times_B Y$ onto its factors. If we set $\rho = p \circ \tilde\pi_X = q \circ \tilde\pi_Y$, we have a cartesian diagram

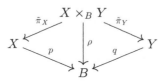

An object $\mathcal{K}^\bullet$ in the derived category $D_c^-(X \times_B Y)$ (a "relative kernel"), induces a relative integral functor $\Phi\colon D^-(X) \to D^-(Y)$ which is defined as

$$\Phi(\mathcal{E}^\bullet) = \mathbf{R}\tilde\pi_{Y*}(\mathbf{L}\tilde\pi_X^* \mathcal{E}^\bullet \overset{\mathbf{L}}{\otimes} \mathcal{K}^\bullet)\,.$$

This functor is an (ordinary) integral functor whose kernel in the derived category $D_c^-(X \times Y)$ is $j_*\mathcal{K}^\bullet$, where $j\colon X \times_B Y \hookrightarrow X \times Y$ is the natural closed immersion. Thus the results about integral functors can be applied to relative integral functors as well. In the remainder of this Subection we assume that $\mathcal{K}^\bullet$ is of finite Tor-dimension over X; then, Φ can be extended to a functor $\Phi\colon D(X) \to D(Y)$ which maps $D_c^b(X)$ to $D_c^b(Y)$.

One of the most interesting features of relative integral functors is their compatibility with base changes. Let $f\colon S \to B$ be a morphism. For any morphism $g\colon Z \to B$ we denote by $g_S\colon Z_S = Z \times_B S \to S$ and $f_Z\colon Z_S \to Z$ the induced morphisms. We consider the kernel $\mathcal{K}_S^\bullet = \mathbf{L}f_{X \times_B Y}^* \mathcal{K}^\bullet$ and the induced relative integral functor

$$\Phi_S\colon D_c^-(X_S) \to D_c^-(Y_S)\,; \qquad \Phi_S(\mathcal{E}^\bullet) = \mathbf{R}\tilde\pi_{Y_S*}(\mathbf{L}\tilde\pi_{X_S}^* \mathcal{E}^\bullet \overset{\mathbf{L}}{\otimes} \mathcal{K}_S^\bullet)\,.$$

Since $\mathcal{K}_S^\bullet$ is of finite Tor-dimension over S, Φ_S maps $D_c^b(X_S)$ to $D_c^b(Y_S)$. Base change compatibility is expressed by the following result, whose proof uses base change in the derived category (cf. Proposition 1.9).

Proposition 2.10. *Assume either that $f\colon S \to B$ or $p\colon X \to B$ is flat. Then for every object $\mathcal{E}^\bullet$ in $D^b(X)$ there is a functorial isomorphism* $\mathbf{L}f_Y^* \Phi(\mathcal{E}^\bullet) \simeq \Phi_S(\mathbf{L}f_X^* \mathcal{E}^\bullet)$ *in the derived category of Y_S.* $\qquad\square$

If the morphism $p\colon X \to B$ is flat, there is no need to assume that the base change morphism is flat, a fact which is often neglected. In this case, by denoting by j_t the immersions of both fibres $X_t = p^{-1}(t) \hookrightarrow X$ and $Y_t = q^{-1}(t) \hookrightarrow$ over a closed point $t \in B$, one has $\mathbf{L}j_t^* \Phi(\mathcal{E}^\bullet) \simeq \Phi_t(\mathbf{L}j_t^* \mathcal{E}^\bullet)$. One has the following result.

Corollary 2.11. *Assume that $p\colon X \to B$ is flat, and let $\mathcal{E}^\bullet$ be an object in $D^b(X)$. Then the derived restriction $\mathbf{L}j_t^* \mathcal{E}^\bullet$ to the fibre X_t is WIT$_i$ for every t if and only if $\mathcal{E}^\bullet$ is WIT$_i$ and $\Phi^i(\mathcal{E}^\bullet)$ is flat over B.* $\qquad\square$

The condition of being WIT$_i$ is open on the base, as the following Proposition asserts.

Proposition 2.12. *Let $p\colon X \to B$ be a flat morphism and $\mathcal{E}$ be a sheaf on X flat over B. The set U of points in B such that the restriction $\mathcal{E}_t$ of $\mathcal{E}$ to the fibre X_t is WIT$_i$ is a nonempty open subscheme of B.* $\qquad\square$

2.2.6 Fourier-Mukai functors between moduli spaces

Integral transforms define in many cases algebraic morphisms between moduli spaces. Assume that X and Y are smooth projective varieties and let $\Phi\colon D^b(X) \to D^b(Y)$ be an integral functor. We consider the functor $\mathbf{M}_{X,P}$ which associates to any variety T the set of all coherent sheaves $\mathcal{E}$ on $T \times X$, flat over T, and whose restrictions $\mathcal{E}_t = j_t^* \mathcal{E}$ to the fibres $X_t \simeq X$ of $\pi_T\colon T \times X \to T$ have Hilbert polynomial P. Let $\mathbf{M}_X$ be a subfunctor of $\mathbf{M}_{X,P}$ which parametrises WIT$_i$ sheaves for a certain index i. By Corollary 2.11, if $\mathcal{E}$ is in $\mathbf{M}_X(T)$ the sheaf $\widehat{\mathcal{E}} = \Phi_T^i(\mathcal{E})$ is flat over T, so that for a fixed i the fibres $(\widehat{\mathcal{E}})_t \simeq \widehat{\mathcal{E}}_t$ have the same Hilbert polynomial $\hat{P}$. Thus the transforms $\widehat{\mathcal{E}}$ are in $\mathbf{M}_{Y,\hat{P}}(T)$. Proposition 2.10 implies now that Φ yields a morphism of functors $\Phi_\mathbf{M}\colon \mathbf{M}_X \to \mathbf{M}_{Y,\hat{P}}$.

Proposition 2.13. *If $\mathbf{M}_X$ has a coarse moduli scheme M_X and Φ is a Fourier-Mukai functor, then $\mathbf{M}_Y = \Phi(\mathbf{M}_X)$ is coarsely representable by a moduli scheme M_Y, and Φ induces an isomorphism of schemes $M_X \simeq M_Y$. Moreover M_X is a fine moduli scheme if and only if M_Y is a fine moduli scheme.* $\qquad\square$

When $\mathbf{M}_X$ is the moduli functor of all the skyscraper sheaves $\mathcal{O}_x$, one has:

Corollary 2.14. *If Φ is a Fourier-Mukai functor, then X is a fine moduli space for the moduli functor of the sheaves $\Phi^i(\mathcal{O}_x)$ over Y. Moreover these sheaves are simple.* $\qquad\square$

2.3 Fourier-Mukai partners

The fact that two smooth algebraic varieties have equivalent derived categories entails strong constraints on their geometry. We already know that this fact is equivalent to the existence of a Fourier-Mukai functor between their derived categories (cf. Theorem 2.2). In this Section we describe some important results in this direction. All the varieties are projective.

Definition 2.7. Two varieties X and Y are Fourier-Mukai partners if there is an exact equivalence of triangulated categories

$$F \colon D^b_c(X) \xrightarrow{\sim} D^b_c(Y).$$

$\triangle$

Lemma 2.8. *Let X be a smooth variety. Every Fourier-Mukai partner of X is smooth.* $\square$

Theorem 2.15. *Let X, Y be smooth Fourier-Mukai partners, so that there is a Fourier-Mukai functor $\Phi^{\mathcal{K}^\bullet}_{X \to Y} \colon D^b(X) \to D^b(Y)$.*

(a) *X and Y have the same dimension.*

(b) *There is an isomorphism $H^0(X, \omega^i_X) \simeq H^0(Y, \omega^i_Y)$ for every integer i, so that X and Y have the same Kodaira dimension.*

(c) *ω_X and ω_Y have the same order, that is, ω^k_X is trivial if and only if ω^k_Y is trivial. Thus, ω_X is trivial if and only if ω_Y is trivial and in this case, the functor $\Phi^{\mathcal{K}^{\bullet\vee}[n]}_{Y \to X}$ is a quasi-inverse to $\Phi^{\mathcal{K}^\bullet}_{X \to Y}$. Moreover, $\omega^r_X \simeq \mathcal{O}_X$ and $\omega^r_Y \simeq \mathcal{O}_Y$ with $r = \mathrm{rk}(\mathcal{K}^\bullet)$.*

$\square$

Theorem 2.15 implies that if the kernel $\mathcal{K}^\bullet$ has positive rank a certain power ω^r_X of the canonical bundle of X is trivial, with $r \neq 0$. If X is a curve, it has to be elliptic (and then $\omega_X \simeq \mathcal{O}_X$); if X is a surface, it is either abelian, K3, Enriques or bielliptic (corresponding to the cases cases $\omega_X \simeq \mathcal{O}_X$, $\omega^2_X \simeq \mathcal{O}_X$ and $\omega^{12}_X \simeq \mathcal{O}_X$, cf. [Hart77, Thm. 6.3]). In dimension 3 the most important example is provided by Calabi-Yau varieties (for which, by definition, $\omega_X \simeq \mathcal{O}_X$).

The following result will be useful later on.

Proposition 2.16. *Let X, Y be proper smooth algebraic varieties of dimension n and $\Phi \colon D^b_c(X) \xrightarrow{\sim} D^b_c(Y)$ a Fourier-Mukai functor. For* *every (closed) point* $x \in X$ *the inequality*

$$\sum_i \dim \mathrm{Hom}^1_{D(Y)}(\Phi^i(\mathcal{O}_x), \Phi^i(\mathcal{O}_x)) \leq n$$

holds true.

Proof There is a spectral sequence

$$E_2^{p,q} = \bigoplus_i \mathrm{Hom}^p_{D(Y)}(\Phi^i(\mathcal{O}_x), \Phi^{i+q}(\mathcal{O}_x))$$

converging to $E_\infty^{p+q} = \mathrm{Hom}^{p+q}_{D(Y)}(\Phi(\mathcal{O}_x), \Phi(\mathcal{O}_x))$. The exact sequence of lower terms of the spectral sequence gives $0 \to E_2^{1,0} \to E_\infty^1$. By the Parseval formula (Corollary 1.8), one has $\mathrm{Hom}^1_{D(Y)}(\Phi(\mathcal{O}_x), \Phi(\mathcal{O}_x)) \simeq \mathrm{Hom}^1_{D(X)}(\mathcal{O}_x, \mathcal{O}_x) \simeq \mathbb{C}^n$. □

2.3.1 D-equivalence implies K-equivalence

Two smooth algebraic varieties X and Y are *K-equivalent* if there are a normal variety $\widetilde{Z}$ and birational morphisms $\tilde{p}_X : \widetilde{Z} \to X$, $\tilde{p}_Y : \widetilde{Z} \to Y$ such that $\tilde{p}_X^* K_X$ and $\tilde{p}_Y^* K_Y$ are $\mathbb{Q}$-linearly equivalent.

The next result is due to Kawamata.

Theorem 2.17. ("*D-equivalence implies K-equivalence*")
[Kaw02a, Kaw02] *Let X, Y be smooth Fourier-Mukai partners.*

(a) *The line bundle ω_X (resp. ω_X^*) is nef if and only if ω_Y (resp. ω_Y^*) is nef.*

(b) *If the Kodaira dimension $\kappa(X)$ is equal to $\dim X$ (or if $\kappa(X, \omega_X^*) = \dim X$), then X and Y are K-equivalent.*

□

A consequence of this result is Bondal and Orlov's "reconstruction theorem" [BO01].

Theorem 2.18. *Let X, Y be smooth Fourier-Mukai partners. If either ω_X or ω_Y is ample or anti-ample, there is an isomorphism $X \simeq Y$.* □

Very few results are available for Fourier-Mukai partners of singular varieties. It is known that any Fourier-Mukai partner of a Cohen-Macaulay (resp. Gorenstein) variety is Cohen-Macaulay (resp. Gorenstein) as well [HLS08].

2.3.2 Fourier-Mukai partners of curves

Here we prove that the only Fourier-Mukai partner of a smooth projective curve is the curve itself.

Theorem 2.19. *A smooth curve X has no Fourier-Mukai partners but itself.*

Proof Let Y be a Fourier-Mukai partner of X of genus g. By Theorem 2.18, one has $X \simeq Y$ if $g > 1$ or $g = 0$ because in these cases the canonical line bundle is ample or anti-ample, respectively. Assume then that X is elliptic and take a Fourier-Mukai functor $\Phi \colon D^b_c(Y) \to D^b_c(X)$. One has $\sum_i \dim \mathrm{Hom}^1_{D(X)}(\Phi^i(\mathcal{O}_y), \Phi^i(\mathcal{O}_y)) \leq 1$ for any point $y \in Y$ (cf. Proposition 2.16), and then there is a unique value of i for which $\Phi^i(\mathcal{O}_y) \neq 0$. By Proposition 2.12, i is actually independent of y, and then Y is a fine moduli space of simple sheaves over X by Corollary 2.14. If the sheaves $\Phi^i(\mathcal{O}_y)$ are torsion-free, they are stable by Corollary 4.6 and thus $Y \simeq X$ by Corollary 4.7. If they have torsion, they are skyscraper sheaves of length 1, so that $Y \simeq X$. $\qquad\square$

2.3.3 Fourier-Mukai partners of surfaces

The main result about Fourier-Mukai partners of algebraic surfaces is the following Theorem.

Theorem 2.20. *A smooth surface has a finite number of Fourier-Mukai partners.* $\qquad\square$

This was proved by Bridgeland and Maciocia [BrM01] for minimal surfaces. Kawamata [Kaw02a] completed the result by including the surfaces with (-1)-curves. His proof is actually simpler and more direct, and exploits the geometric properties of the support of the kernel of the corresponding Fourier-Mukai functor.

The case of minimal surfaces is treated with a case-by-case approach, essentially based on the Enriques-Kodaira classification, in view of the fact that Fourier-Mukai partners have the same Kodaira dimension (Theorem 2.15).

Proposition 2.21. *Two smooth surfaces X and Y that are Fourier-Mukai partners have the same Picard number, the same Betti numbers, and therefore the same topological Euler characteristic.*

The study of the Fourier-Mukai partners of K3 and abelian surfaces is particularly interesting.

Proposition 2.22. *Let X be a K3 (resp. an abelian) surface and Y a Fourier-Mukai partner of X. Then Y is a K3 (resp. an abelian) surface as well.*

Proof Since ω_X is trivial, ω_Y is trivial as well by Theorem 2.15. Then Y is either K3 or abelian. Moreover $H^\bullet(X, \mathbb{Q}) \simeq H^\bullet(Y, \mathbb{Q})$ by Corollary 2.4, then if X is K3, Y is also K3 and if X is abelian, Y is abelian as well. □

By a result of Orlov, the Fourier-Mukai partners of a K3 or an abelian surface are completely characterised in terms of isometries of the transcendental lattice $\mathbf{T}(X)$. This is defined as the orthogonal complement to $Pic(X)$ in $H^2(X, \mathbb{Z})$.

Proposition 2.23. [Or97, Thm. 3.3] *Let X, Y be two K3 or abelian surfaces. X and Y are Fourier-Mukai partners if and only if the lattices $\mathbf{T}(X)$ and $\mathbf{T}(Y)$ are Hodge isometric.* □

2.3.4 Fourier-Mukai partners for threefolds

Minimal models of threefolds are in general not characterised; at present, one is only able to prove in some special cases that birational threefolds have equivalent derived categories. Two important results in this direction are the following, both due to Bridgeland.

Theorem 2.24. [Bri02, Thm. 1.1] *Let X be a (complex) threefold with terminal singularities and $f_1 : Y_1 \to X$, $f_2 : Y_2 \to X$ crepant resolutions of singularities. Then there is an equivalence of triangulated categories $D_c^b(Y_1) \simeq D_c^b(Y_2)$.* □

Since any birational map between smooth Calabi-Yau threefolds is crepant, we deduce the following result.

Theorem 2.25. [Bri02] *Let X and Y be two birational smooth Calabi-Yau threefolds. Then X and Y are Fourier-Mukai partners.* □

The proof of Theorem 2.24 relies on the fact that any crepant birational map between threefolds with only terminal singularities can be decomposed into a sequence of particularly simple birational transformations, called *flops*. This reduces the proof to the case of flops. There are two different proofs, one due to Bridgeland [Bri02], who explicitly constructs the flop using a moduli spaces of point perverse sheaves, and another due to Van den Bergh [VdB04] who uses noncommutative techniques. The description of all the Fourier-Mukai partners of a Calabi-Yau threefold is still unknown. Cӑldӑraru [Ca1] has found some explicit models of Fourier-Mukai partners for three-dimensional Calabi-Yau threefolds.

3 The Nahm transform

This construction was introduced by Nahm in 1983 [N]. Starting from an instanton on a 4-dimensional flat torus, it yields an instanton on the dual torus. Later this was formalised by Schenk [Sc] and Braam and van Baal [BvB]. According to their picture, the Nahm transform is an index-theoretic construction, where, given a vector bundle E on a flat torus X equipped with an anti-self-dual connection ∇, the dual torus $\hat{X}$ is regarded as the parameter space of a family of Dirac operators twisted by ∇. The index of this family yields, under suitable conditions, an instanton $\hat{\nabla}$ on $\hat{X}$. A survey of some properties of the standard version of the Nahm transform was given by M. Jardim in [J5].

The connection between the Nahm and the Fourier-Mukai transform was seemingly first realised by Braam-van Baal and Schenk, and a first formalization was given by Donaldson and Kronheimer [DKr]. The link between the two constructions is a relation between index bundles and higher direct images, in accordance with Illusie's definition of the "analytical index" of a relative elliptic complex [Ill71].

Mainly following [BBH94], we shall describe here the relation between the Fourier-Mukai and Nahm transforms by considering the second as a particular case of a more general class of transforms, that we call *Kähler Nahm transforms*. We shall also introduce a special case of such transforms when the manifolds involved have a hyperkähler structure, considering a generalization of the notion of instanton (the *quaternionic instantons*) and proving that the "hyperkähler Fourier-Mukai transform" preserves the quaternionic instanton condition. This will be mainly taken from [BBH2]. However, the whole theory recalled in this section is described in more detail in [BBH08].

3.1 Line bundles on complex tori

We provide here a few basic facts about the description of line bundles on complex tori, which will be useful to introduce the Nahm transform. If V is a g-dimensional complex vector space, and Ξ a nondegenerate lattice in it, the quotient $T = V/\Xi$ comes with a natural structure of g-dimensional complex manifold, and is said to be a *complex torus* of dimension g. Any generator of the lattice Ξ corresponds to a loop in T, thus one has natural identifications $\Xi \simeq \pi_1(T) \simeq H_1(T, \mathbb{Z})$, and as a consequence, also $H^k(T, \mathbb{Z}) \simeq \Lambda^k \Xi^*$. Moreover, the space $\mathcal{H}(T)$ of hermitian forms $H: V \times V \to \mathbb{C}$ that satisfy the condition $\mathrm{Im}(H(\Xi, \Xi)) \subset \mathbb{Z}$ may be identified with the Néron-Severi group $\mathrm{NS}(T)$.

Definition 3.1. A *semicharacter* associated with an element $H \in \mathcal{H}(T)$ is a map $\chi: \Xi \to U(1)$ such that $\chi(\lambda + \mu) = \chi(\lambda)\,\chi(\mu)\,e^{iH(\lambda,\mu)}$. An element $H \in \mathcal{H}(T)$ and an associated semicharacter χ define an *automorphy factor* $a: V \times \Xi \to U(1)$ by $a(v, \lambda) = \chi(\lambda)\,e^{\pi H(v,\lambda) + \frac{\pi}{2} H(\lambda,\lambda)}$.

Proposition 3.1. [Birlan99] *The holomorphic functions s on V that satisfy the condition $s(v + \lambda) = a(v, \lambda)\,s(v)$ for all $v \in V$ and $\lambda \in \Xi$, where a is an automorphy factor associated with an element $H \in \mathcal{H}(T)$, are in a one-to-one correspondence with sections of a line bundle L on T such that $c_1(L) = H$.*

One may define the dual torus T^* as follows. Let Ω be the conjugate dual space of V, and let $\Xi^* = \{\ell \in \Omega \,|\, \ell(\Xi) \subset \mathbb{Z}\}$ be the lattice dual to Ξ. If we set $T^* = \Omega/\Xi^*$, by the natural isomorphism $T^* \simeq \mathrm{Hom}_{\mathbb{Z}}(\Xi, U(1))$ and the exact sequence

$$0 \to T^* \to Pic(T) \to \mathrm{NS}(T) \to 0,$$

we see that T^* parametrises flat $U(1)$ bundles on T.

Proposition 3.1 may be used to construct the Poincaré bundle $\mathcal{P}$ on $T \times T^*$. Let $H \in \mathcal{H}(T \times T^*)$ be given by

$$H(v, w, \alpha, \beta) = \overline{\beta(v)} + \alpha(w) \tag{3.1}$$

where $v, w \in V$, $\alpha, \beta \in \Omega$, and consider the associated semicharacter

$$\chi(\lambda, \mu) = e^{i\pi\,\mu(\lambda)}. \tag{3.2}$$

The Poincaré bundle is the line bundle $\mathcal{P}$ given by the hermitian form (3.1) and the semicharacter (3.2). This Poincaré bundle is automatically normalised as in Section 2. Moreover, it comes with a natural hermitian metric, which is expressed on the functions on the universal covering of $T \times T^*$ corresponding (via the automorphy condition) to sections of $\mathcal{P}$ in terms of the standard hermitian metric on $\mathbb{C}^n$. For any element $\xi \in T^*$ we shall denote $\mathcal{P}_\xi = \mathcal{P}_{|T \times \{\xi\}}$ the line bundle on T parametrised by ξ.

3.2 Nahm transform

Let us briefly recall Nahm's transform in its original version. Let T be a flat Riemannian 4-torus, equipped with a compatible complex structure, T^* its dual torus. As we have seen, the Poincaré bundle on $T \times T^*$ comes with a natural hermitian metric. Let $\nabla_{\mathcal{P}}$ be the corresponding Chern connection (the unique connection on $\mathcal{P}$ compatible both with the hermitian metric and the complex structure of $\mathcal{P}$). Furthermore, let E be an hermitian vector bundle on T whose Chern connection ∇ is anti-self-dual (ASD), i.e., its curvature F_∇ satisfies the ASD condition $F_\nabla = - * F_\nabla$, where $*$ is Hodge duality on forms on T. A survey of the Nahm transform and their most recent generalisations can be founded in [J5].

If $\xi \in T^*$, one has a coupled connection ∇_ξ in $E \otimes \mathcal{P}_\xi$, and correspondingly, a family of *twisted Dirac operators*

$$D_\xi : \Gamma(E \otimes \mathcal{P}_\xi \otimes S_+) \to \Gamma(E \otimes \mathcal{P}_\xi \otimes S_-) \tag{3.3}$$

where $S_\pm$ is the spinor bundle of positive/negative helicity on T. If the pair (E, ∇) satisfies an irreducibility condition (it is *without flat factors,* i.e., there is no ∇-compatible splitting $E = E' \oplus L$, where L is a flat line bundle), then for every $\xi \in T^*$ we have ker $D_\xi = 0$, and then by Atiyah-Singer's index theory (minus) the index of the family of Dirac operators is a vector bundle $\hat{E}$ on T^*, whose fibre at $\xi \in T^*$ is the vector space $coker D_\xi$.

The bundle $\hat{E}$ may be equipped with a metric and a compatible connection. Indeed, completing the spaces of sections appearing in Eq. (3.3) in the natural L^2 norms, we have for every $\xi \in T^*$ an exact sequence

$$0 \to \hat{E}_\xi \to L^2(E \otimes \mathcal{P}_\xi \otimes S_-) \xrightarrow{D_\xi^*} L^2(E \otimes \mathcal{P}_\xi \otimes S_+) \to 0.$$

By restricting the scalar product in the space in the middle in this sequence one defines an hermitian metric in the bundle $\hat{E}$. Moreover, if we regard the spaces $L^2(E \otimes \mathcal{P}_\xi \otimes S_\pm)$ as the fibres of trivial ∞-dimensional bundles on T^*, this exact sequence allows one to define a connection on $\hat{E}$ by a projection formula: one takes a section of $\hat{E}$, regards it as section of the bundle in the middle, takes the covariant derivative with respect to the trivial connection, and then projects back to $\hat{E}$ using the scalar product. One shows that the resulting connection $\hat{\nabla}$ is compatible with the metric, and is ASD.

The pair $(\hat{E}, \hat{\nabla})$ is the *Nahm transform* of (E, ∇). The Atiyah-Singer theorem for families allows one to compute the topological invariants of

$\hat{E}$, getting

$$(\hat{r}, c_1(\hat{E}), \mathrm{ch}_2(\hat{E})) = -(\mathrm{ch}_2(E), c_1(E), r),$$

where to compare the first Chern classes we use the natural identification of the groups $H^2(T, \mathbb{Z})$ and $H^2(\hat{T}, \mathbb{Z})$.

3.3 Fourier-Mukai vs. Nahm

In order to compare the Fourier-Mukai transform with Nahm's construction, it is convenient to recast the latter into a more general form. To this end, we consider a flat proper submersive holomorphic morphism of complex manifolds $f \colon Z \to Y$. We call $\mathcal{O}_{Z/Y} = f^{-1}\mathcal{C}_Y^\infty \otimes_{f^{-1}\mathcal{O}_Y} \mathcal{O}_Z$ the sheaf of "relatively holomorphic functions" (locally, if x and y are holomorphic coordinates on the fibres of f and on Y respectively, the sections of $\mathcal{O}_{Z/Y}$ are functions of the variables $x, y, \bar{y}$).

If F is a holomorphic vector bundle on Z (whose sheaf of sections we shall denote by $\mathcal{F}$), then $\mathcal{F}^r = \mathcal{F} \otimes_{\mathcal{O}_Z} \mathcal{O}_{Z/Y}$ has a relative holomorphic structure i.e., its sheaf of sections has a structure of $\mathcal{O}_{Z/Y}$-module. The relative Dolbeault complex provides a (fine) resolution of this $\mathcal{O}_{Z/Y}$-module:

$$0 \to \mathcal{F}^r \to \mathcal{F} \otimes_{\mathcal{O}_Z} \mathcal{C}_Z^\infty \xrightarrow{\bar{\partial}_{Z/Y}} \mathcal{F} \otimes_{\mathcal{O}_Z} \Omega_{Z/Y}^{0,1} \to \cdots$$

Moreover, one has $R^i f_* \mathcal{F}^r \simeq R^i f_* \mathcal{F} \otimes_{\mathcal{O}_Y} \mathcal{C}_Y^\infty$, i.e., the higher direct images of $\mathcal{F}^r$ come with a natural holomorphic structure.

Definition 3.2. The bundle F satisfies the even (odd) IT condition if $R^i f_* \mathcal{F}^r = 0$ for i odd (even), and the non-vanishing higher direct images are locally free.

Now, let us assume that the sheaf of relative differentials $\Omega_{Z/Y}^1$ has a *relative Kähler structure*, i.e., it is equipped with a Hermitian metric such that the corresponding 2-form is closed under the relative exterior differential. We define relative spinor bundles $\Sigma_\pm = \bigoplus_{k \text{ even, odd}} \bigwedge^k \Omega_{Z/Y}^{0,1}$, and a relative Dirac operator $D = \bar{\partial}_{Z/Y} + \bar{\partial}_{Z/Y}^* \colon f_*(\mathcal{F} \otimes_{\mathcal{O}_Z} \Sigma_+) \to f_*(\mathcal{F} \otimes_{\mathcal{O}_Z} \Sigma_-)$. By restricting to the fibres of f, this provides a family of Dirac operators parametrised by Y.

Theorem 3.2. *Assume that the relative canonical bundle of $f \colon Z \to Y$ is trivial. If F satisfies (e.g.) the even IT condition, then $\ker D = 0$, and*

$$-\operatorname{ind}(D) = coker(D) \simeq \bigoplus_{k \text{ odd}} R^k f_* \mathcal{F}^r.$$

If F has a Hermitian metric, let ∇ be the corresponding Chern connection. By generalizing the constructions we have seen in the case of the original Nahm transform, one can use these data to induce on $coker(D)$ a Hermitian metric and a connection $\hat{\nabla}$, which turns out to be a Chern connection, and may be regarded as a "direct image" of ∇.

Now we apply these constructions to the case when Z is a product $X \times Y$ to define a generalised Nahm transform. We denote by π_X, π_Y the projections of $X \times Y$ onto its factors. Let X be a compact Kähler manifold with trivial canonical bundle, Y a Kähler manifold, E a Hermitian holomorphic vector bundle on X, and finally, let Q be a Hermitian holomorphic vector bundle on $X \times Y$. Applying the previous construction to the (Hermitian holomorphic) bundle $\pi_X^* E \otimes Q$ on $X \times Y$ we construct a Hermitian bundle $\hat{E}$ on Y. The pair $(\hat{E}, \hat{\nabla})$, where $\hat{\nabla}$ is the Chern connection of the Hermitian holomorphic bundle $\hat{E}$, is the *generalised Nahm transform* of (E, ∇).

The formalism we have so far developed provides a direct proof of the following compatibility condition between the Fourier-Mukai and generalised Nahm transforms.

Theorem 3.3. *Assume that $F = \pi_X^* E \otimes Q$ satisfies an IT condition (say the even one). Then the sheaf of holomorphic sections of $\hat{E} = coker(D)$ is isomorphic to the Fourier-Mukai transform $\Phi_{X \to Y}^{\mathcal{Q}}(\mathcal{E})$ of the sheaf $\mathcal{E}$ of holomorphic sections of E.*

Here $\mathcal{Q}$ is the sheaf of holomorphic sections of the bundle Q.

3.4 Hyperkähler Fourier-Mukai transform

In some situations the Fourier-Mukai transform preserves the stability of the sheaves it acts on. We have seen that the original Nahm transform maps instantons to instantons. These two results are actually related by the so-called Hitchin-Kobayashi correspondence, according to which on a compact Kähler manifold, a holomorphic vector bundle is polystable (i.e., it is a direct sum of stable sheaves having the same slope) if and only if it carries an hermitian metric which satisfies a certain differential condition (it is an Hermitian-Yang-Mills metric). In the case of complex dimension 2, and for bundles of zero degree, the Hermitian-Yang-Mills condition is equivalent to saying that the Chern connection is ASD, and this establishes the link between the two "preservation" results. One might wonder if our generalised Nahm transform may provide further examples of preservation of some instanton-like condition. One such instance is provided by hyperkähler geometry.

Let X be a hyperkähler manifold, and let I_k be 3 basic complex structures. The automorphism $\Sigma_k(I_k \otimes I_k)$ acting on $\Lambda^2 T^* X$ has two eigenspaces, with eigenvalues 3 and -1 respectively: $\Lambda^2 T^* X = \mathfrak{e}_1 \oplus \mathfrak{e}_2$. One has $(\mathfrak{e}_1)_x = \bigcap_{u \in Z_x} (\Lambda^2 T^* X)_u^{1,1}$, where $Z \to X$ is the twistor space of X.

Definition 3.3. A connection ∇ on a complex vector bundle E on X is a *quaternionic instanton* if its curvature F_∇ takes values in $\mathfrak{e}_1$.

Let us say that a complex vector bundle E is *hyperstable* if it is stable with respect to any Kähler structure in the hyperkähler family of X.

Theorem 3.4. *Let X and Y be hyperkähler manifolds, (E, ∇) a quaternionic instanton on X, and $(Q, \tilde{\nabla})$ a quaternionic instanton on $X \times Y$. The Nahm transform of (E, ∇) with kernel $(Q, \tilde{\nabla})$ is a quaternionic instanton on Y.*

This may be regarded as a "stability preservation" result in view of the following extension of the Hitchin-Kobayashi correspondence, which is not difficult to prove.

Theorem 3.5. *Let E be a vector bundle on a hyperkähler manifold X which has zero degree with respect to all Kähler structures in the hyperkähler family of X. Then E is hyperstable if and only if it admits an hermitian metric such that the corresponding Chern connection is an irreducible quaternionic instanton.*

The proof of Theorem 3.4 exploits a generalization of the classical Atiyah-Ward correspondence. The latter states that there is a one-to-one correspondence between instantons on a compact, connected, orientable ASD Riemannian 4-manifold X, and holomorphic vector bundles on the twistor space Z of X that are holomorphically trivial along the fibres of the projection $Z \to X$ (as stated, this correspondence holds true for instantons whose structure group is the general linear group. Unitary instantons on X correspond to bundles on Z that carry an additional structure, called a *real form*).

The generalization of this correspondence to quaternionic instantons on higher-dimensional hyperkähler manifolds reads as follows.

Theorem 3.6. *There is a one-to-one correspondence between the following objects:*

(a) *gauge equivalence classes of (Hermitian) quaternionic instantons on a hyperkähler manifold X;*

(b) *isomorphism classes of holomorphic vector bundles on the twistor space Z of X, holomorphically trivial along the fibres of Z (carrying a positive real form).*

The proof of Theorem 3.4 is based on the natural isomorphism $Z_{X \times Y} \simeq Z_X \times_{\mathbb{P}^1} Z_Y$ and on the following commutative diagram

$$
\begin{array}{ccccc}
Z_X & \xleftarrow{\ t_1\ } & Z_{X \times Y} & \xrightarrow{\ t_2\ } & Z_Y \\
\ \downarrow{\scriptstyle p_1} & & \ \downarrow{\scriptstyle q} & & \ \downarrow{\scriptstyle p_2} \\
X & \xleftarrow{\ \pi_1\ } & X \times Y & \xrightarrow{\ \pi_2\ } & Y
\end{array}
\ .
$$

All data in the spaces in the bottom row are lifted to the first row by using the generalised Atiyah-Ward correspondence. Then one performs a (relative) Fourier-Mukai transform between Z_X and Z_Y, and descends from Z_Y to Y using the generalised Atiyah-Ward correspondence again, after several consistency checks.

4 Moduli spaces of sheaves and coherent systems on elliptic curves

Since its very first appearance the Fourier-Mukai transform has been an important tool in the study of moduli spaces of sheaves. A key feature is that, under suitable hypotheses, the Fourier-Mukai transform preserves the stability (or semistability) of sheaves and thus produces isomorphisms between different moduli spaces. Among such applications, one can list the original contributions by Mukai [Muk87a, Muk87b] and the study of moduli spaces of stable sheaves on abelian or K3 surfaces [Mac96b, FL91, BM96, BBH98a].

Since a complete account of all the applications of the Fourier-Mukai transform exceeds the scope of these notes (see [BBH08] for a comprehensive treatment), we devote this Section to describing two particularly interesting examples. The first, which can be nowadays considered as classical, is the moduli spaces of sheaves on elliptic curves. The second is the theory of coherent systems on an elliptic curve.

In the first case, the Fourier-Mukai transform provides new and easier proofs of Atiyah's classical theorems. This approach was introduced in [Bri98t, Pol03, HePl05] but our treatment, taken directly from [BBH08, Chap. 3], is somehow different.

The application of the Fourier-Mukai transform to the study of coherent systems on elliptic curves follows recent work by two of the authors [HT08]. Coherent systems on algebraic curves are "decorated objects" and their definition, notion of stability and moduli spaces have been introduced and studied by Le Potier [LeP93], King and Newstead [KN95], García-Prada, Bradlow, Muñoz and Newstead [BG02, BGMN03]. The specific case of elliptic curves has been studied by Lange and Newstead [LN05].

4.1 Application of Fourier-Mukai transforms to the moduli spaces of sheaves on elliptic curves

Let X be an elliptic curve, i.e., a smooth curve of genus 1 with a fixed point x^0. Then X can be regarded as an abelian variety of dimension one such that x^0 is the identity of the group law in X. The morphism $X \to \hat{X}$ mapping x to the line bundle $\mathcal{O}_X(x - x^0)$ is an isomorphism, so that we can identify X with its dual variety $\hat{X}$. Using this identification we can write the Poincaré bundle described in Section 2 in the form

$$\mathcal{P} \simeq \mathcal{O}_{X \times X}(\Delta_\iota) \otimes \pi_1^* \mathcal{O}_X(-x^0) \otimes \pi_2^* \mathcal{O}_X(-x^0), \qquad (4.1)$$

where Δ_ι is the graph of the isomorphism $\iota \colon X \to X$ defined as $\iota(x) = -x$. Both the abelian Fourier-Mukai transform $\mathbf{S}$ and the dual abelian Fourier-Mukai transform $\widehat{\mathbf{S}}$ are autoequivalences of $D_c^b(X)$.

Given an object $\mathcal{E}^\bullet$ of $D_c^b(X)$, we can write its Chern character as $\mathrm{ch}(\mathcal{E}^\bullet) = (n, d)$, where $n = \mathrm{ch}_0(\mathcal{E}^\bullet)$ is its rank and $d = \mathrm{ch}_1(\mathcal{E}^\bullet)$ its degree, thought as an integer number. Equation (4.1) and the Grothendieck-Riemann-Roch theorem give the following result.

Proposition 4.1. If $\mathrm{ch}(\mathcal{E}^\bullet) = (n, d)$, then $\mathrm{ch}(\mathbf{S}(\mathcal{E}^\bullet)) = (d, -n)$. $\square$

When $\mathcal{E}^\bullet$ is WIT_i, one has $\mathbf{S}(\mathcal{E}) = \widehat{\mathcal{E}}[-i]$ so that

$$\mathrm{ch}(\mathbf{S}(\mathcal{E}^\bullet)) = (-1)^i \, \mathrm{ch}(\widehat{\mathcal{E}}).$$

4.1.1 (Semi)stable sheaves on an elliptic curve

μ-semistable sheaves on an *elliptic curve* X are characterized by the following result [Pol03, Lemma 14.5].

Proposition 4.2. *Any indecomposable torsion-free sheaf on X is semi-stable.*

Proof If $\mathcal{E}$ is a torsion-free sheaf on X and $0 \subset \mathcal{E}_1 \subset \cdots \subset \mathcal{E}_n = \mathcal{E}$ is its Harder-Narasimhan filtration, the quotient sheaves $\mathcal{G}_i = \mathcal{E}_i / \mathcal{E}_{i-1}$

are μ-semistable with $\mu(\mathcal{G}_i) > \mu(\mathcal{G}_{i+1})$. Thus $\mathrm{Hom}_X(\mathcal{G}_i, \mathcal{G}_{i+1}) = 0$ so that $\mathrm{Ext}^1_X(\mathcal{G}_{i+1}, \mathcal{G}_i) = 0$ by Serre duality. This implies that the Harder-Narasimhan filtration splits, then if $\mathcal{E}$ is indecomposable it is also semistable. □

Corollary 4.3. *Let $\mathcal{E}$ be a semistable sheaf of rank n and degree d on X.*

(a) *If $d < 0$, then $\mathcal{E}$ is IT_1 for both $\mathbf{S}$ and $\widehat{\mathbf{S}}$, and both transforms are semistable.*

(b) *If $d > 0$, then $\mathcal{E}$ is IT_0 for both $\mathbf{S}$ and $\widehat{\mathbf{S}}$, and both transforms are semistable.*

(c) *If $d \neq 0$, then $\mathcal{E}$ is locally free.*

(d) *If $d = 0$ and $\mathcal{E}$ is stable, then $\mathcal{E}$ is a line bundle. Thus, any semistable sheaf of degree 0 is WIT_1 and the unique transform $\widehat{\mathcal{E}}$ is a skyscraper sheaf. Moreover a torsion-free sheaf of degree 0 is semistable if and only if it is S-equivalent to a direct sum of degree 0 line bundles: $\mathcal{E} \sim \oplus_i \mathcal{L}_i^{\oplus n_i}$, with $\sum_i n_i = n$.*

Proof 1. One has $H^0(X, \mathcal{E} \otimes \mathcal{P}_\xi) \simeq \mathrm{Hom}_X(\mathcal{P}_\xi^*, \mathcal{E}) = 0$ for every point $\xi \in \hat{X}$ since $\mathcal{E}$ is semistable and $d < 0$. Then $\mathcal{E}$ is IT_1 by Proposition 2.1. Assume that $\mathcal{E}$ is semistable. To prove that $\widehat{\mathcal{E}}$ is semistable we can assume that $\mathcal{E}$ is indecomposable; then $\widehat{\mathcal{E}}$ is indecomposable as well, and then it is semistable by Proposition 4.2. A similar argument proves the semistability of $\widehat{\mathbf{S}}(\mathcal{E})$.

2. One has $H^1(X, \mathcal{E} \otimes \mathcal{P}_\xi)^* \simeq \mathrm{Hom}_X(\mathcal{E} \otimes \mathcal{P}_\xi, \mathcal{O}_X) \simeq \mathrm{Hom}_X(\mathcal{E}, \mathcal{P}_\xi^*)$ by Serre duality. Since $\mathcal{E}$ is semistable of positive degree, the second group is zero and then $\mathcal{E}$ is IT_0. Proceeding as in the first part one proves the semistability of $\widehat{\mathcal{E}}$. The proof for $\widehat{\mathbf{S}}$ is analogous.

3. If $d \neq 0$, $\mathcal{E}$ is either IT_0 or IT_1 with respect to $\mathbf{S}$, according to whether $d > 0$ or $d < 0$, due to part 1 or 2. Moreover, $\widehat{\mathcal{E}}$ is semistable of nonzero degree, so that it is IT_1 or IT_0 with respect to $\widehat{\mathbf{S}}$. It follows that $\mathcal{E} \simeq \widehat{\mathbf{S}}^{1-i}(\mathbf{S}^i(\mathcal{E}))$ (with $i = 0$ or 1) is locally free.

4. If $\mathcal{E}$ is stable of degree 0, then $H^0(X, \mathcal{E} \otimes \mathcal{P}_\xi) \simeq \mathrm{Hom}_X(\mathcal{P}_\xi^*, \mathcal{E}) = 0$ unless $\mathcal{E} \simeq \mathcal{P}_\xi^*$. Thus if $\mathcal{E}$ is not a line bundle, it is IT_1; by Proposition 4.1 $\widehat{\mathcal{E}}$ is locally free of rank 0; thus $\widehat{\mathcal{E}} = 0$ so that $\mathcal{E} = 0$ by the invertibility of $\mathbf{S}$. For the second part, assume that $\mathcal{E}$ is semistable of degree 0; then it has a Jordan-Holder filtration $0 = \mathcal{E}_0 \subset \mathcal{E}_1 \subset \cdots \subset \mathcal{E}_n = \mathcal{E}$ whose quotients $\mathcal{G}_i = \mathcal{E}_i / \mathcal{E}_{i-1}$ are stable of degree 0, that is, they are line bundles of degree 0. Thus, $\mathcal{G}_i \simeq \mathcal{P}_{\xi_i}$ for a point $\xi_i \in \hat{X}$. Since the

sheaves $\mathcal{P}_{\xi_i}$ are WIT$_1$ and $\widehat{\mathcal{P}}_{\xi_i} \simeq \mathcal{O}_{\iota(\xi_i)}$, we see that $\mathcal{E}$ is WIT$_1$ and $\widehat{\mathcal{E}}$ is a skyscraper sheaf. Analogous arguments prove that $\mathcal{E}$ is WIT$_1$ with respect to $\widehat{\mathbf{S}}$ and that $\widehat{\mathbf{S}}^1(\mathcal{E})$ is a skyscraper sheaf.   □

4.1.2 Geometry of the moduli spaces of stable sheaves on elliptic curves

Let us consider the full subcategory $\mathfrak{Coh}^{ss}_{n,d}(X)$ of the category $\mathfrak{Coh}(X)$ of coherent sheaves on X whose objects are semistable sheaves of rank n and degree d. We also consider the category $\mathfrak{Skn}_n(X)$ of skyscraper sheaves of length n on X. Corollary 4.3 implies the following result.

Proposition 4.4. *The abelian Fourier-Mukai transform induces equivalences of categories* $\mathfrak{Coh}^{ss}_{n,d}(X) \simeq \mathfrak{Coh}^{ss}_{d,-n}(X)$ *if $d > 0$, and* $\mathfrak{Coh}^{ss}_{n,0}(X) \simeq \mathfrak{Skn}_n(X)$.   □

The Fourier-Mukai transform $\Psi = \Phi^{\delta * \mathcal{L}}_{X \to X} : D^b(X) \xrightarrow{\sim} D^b(X)$, which is nothing but the twist by $\mathcal{L} = \mathcal{O}_X(x_0)$, also induces an equivalence $\mathfrak{Coh}^{ss}_{n,d}(X) \simeq \mathfrak{Coh}^{ss}_{n,d+n}(X)$. By composing the Fourier-Mukai transforms $\mathbf{S}$ and Ψ in an appropriate way, and using Euclid's algorithm we have:

Proposition 4.5. *For every pair (n,d) of integers $(n > 0)$, there is a Fourier-Mukai functor* $\tilde{\Phi} \colon D^b(X) \xrightarrow{\sim} D^b(X)$ *which induces an equivalence of categories*

$$\mathfrak{Coh}^{ss}_{n,d}(X) \simeq \mathfrak{Coh}^{ss}_{\bar{n},0}(X) \overset{\mathbf{S}}{\simeq} \mathfrak{Skn}_{\bar{n}}(X),$$

where $\bar{n} = \gcd(n,d)$.   □

Corollary 4.6. *Let $\mathcal{E}$ be a torsion-free sheaf of rank n and degree d on X. Then $\mathcal{E}$ is stable if and only if it is simple, and if and only if it is semistable and $\gcd(n,d) = 1$. Thus, the integral functors of Proposition 4.5 map stable sheaves to stable sheaves.*   □

The structure of the coarse moduli space $\mathcal{M}^{ss}(n,d)$ of semistable sheaves of rank n and degree d on X can be obtained in a similar way. Let Sym$^n X$ be the n-th symmetric product.

Corollary 4.7. *For every pair (n,d) of integers $(n > 0)$, there is a Fourier-Mukai functor which induces an isomorphism of moduli spaces*

$$\mathcal{M}^{ss}(n,d) \simeq \mathcal{M}^{ss}(\bar{n},0) \overset{\mathbf{S}}{\simeq} \text{Sym}^{\bar{n}} X,$$

where $\bar{n} = \gcd(n,d)$. Then, if Y is a nonempty moduli space of stable torsion-free sheaves on X, there is an isomorphism $Y \simeq X$.

Proof Propositions 4.5 and 2.13 imply the first part, because $\mathrm{Sym}^{\bar{n}} X$ is a coarse moduli space for the moduli functor of skyscraper sheaves of length $\bar{n}$ on X. The last part follows now from Corollary 4.6. $\square$

4.1.3 Autoequivalences of the derived category of an elliptic curve

Let X be an elliptic curve. Since $H^{even}(X, \mathbb{Z}) \simeq \mathbb{Z} \oplus \mathbb{Z}$, if

$$\Phi = \Phi_{X \to X}^{\mathcal{K}^\bullet} : D_c^b(X) \to D_c^b(X)$$

is an integral functor $(\mathcal{K}^\bullet \in D_c^b(X \times X))$, Equation (2.2.1) yields a commutative diagram

$$
\begin{array}{ccc}
D_c^b(X) & \xrightarrow{\Phi} & D_c^b(X) \\
v \downarrow & & \downarrow v \\
\mathbb{Z} \oplus \mathbb{Z} & \xrightarrow{\Phi_*} & \mathbb{Z} \oplus \mathbb{Z}
\end{array}
$$

where $\Phi_* = f^{\mathcal{K}^\bullet}$. When Φ is a Fourier-Mukai functor, Φ_* is a matrix in $SL(2, \mathbb{Z})$. For instance, if $\mathcal{P}$ is the Poincaré line bundle on $X \times X$ and $\mathbf{S} = \Phi_{X \to X}^{\mathcal{P}}$ is the abelian Fourier-Mukai transform, one has $\mathbf{S}_* = \begin{pmatrix} 0 & 1 \\ -1 & 0 \end{pmatrix}$ (cf. Proposition 4.1).

Due to Orlov's representation Theorem 2.2, we have a representation in $SL(2, \mathbb{Z})$ of the group $\mathrm{Aut}\,(D_c^b(X))$ of derived auto-equivalences of the derived category of X. The study of such representation is due to Bridgeland [Bri98], who proved the following result.

Proposition 4.8. *Given a matrix* $A = \begin{pmatrix} \alpha & \beta \\ \gamma & \delta \end{pmatrix} \in SL(2, \mathbb{Z})$ *such that* $\beta > 0$, *there exist vector bundles on* $X \times X$ *that are strongly simple over both factors, and which restrict to give bundles of Chern character* (β, α) *on the first factor and* (β, δ) *on the second. For any such bundle* $\mathcal{Q}(A)$, *the associated integral functor* $\Phi^{\mathcal{Q}(A)}$ *is a Fourier-Mukai transform, and moreover* $\Phi_*^{\mathcal{Q}(A)} = A$. $\square$

Thus, Proposition 4.8 essentially describes all the Fourier-Mukai functors on an elliptic curve. A more precise result was proved by Hille and van den Bergh [HiVdB].

Theorem 4.9. *Let X be an elliptic curve. There is an exact sequence of groups*

$$0 \to 2\mathbb{Z} \times \mathrm{Aut}\,(X) \ltimes Pic^0(X) \to \mathrm{Aut}\,(D_c^b(X)) \xrightarrow{ch} SL(2, \mathbb{Z}) \to 0\,,$$

where $n \in \mathbb{Z}$ acts as the shift functor $[n]$, and the transform corresponding to $(\varphi, \mathcal{L}) \in \mathrm{Aut}\,(X) \ltimes Pic^0(X)$ sends $\mathcal{E}^\bullet$ to $\varphi_*(\mathcal{L} \otimes \mathcal{E}^\bullet)$. $\square$

Here we have set $\text{ch}(\Phi) = \Phi_*$. Given $A = \left(\begin{smallmatrix} \alpha & \beta \\ \gamma & \delta \end{smallmatrix}\right) \in SL(2,\mathbb{Z})$ with $\beta > 0$, by proceeding as in the proof of Corollary 4.3 one can determine the behaviour of the transforms $\Phi^{\mathcal{Q}(A)}$ [HT08].

Proposition 4.10. *Let $\mathcal{E}$ be a semistable (resp. stable) vector bundle of Chern character $\text{ch}(\mathcal{E}) = (r,d)$.*

(a) *If $\alpha r + \beta d > 0$ then $\mathcal{E}$ is IT_0 with respect to $\Phi^{\mathcal{Q}(A)}$ and the unique transform $\widehat{\mathcal{E}}$ is also semistable (resp. stable).*

(b) *If $\alpha r + \beta d = 0$ then $\mathcal{E}$ is WIT_1 with respect to $\Phi^{\mathcal{Q}(A)}$ and the unique transform $\widehat{\mathcal{E}}$ is a torsion sheaf.*

(c) *If $\alpha r + \beta d < 0$ then $\mathcal{E}$ is IT_1 with respect to $\Phi^{\mathcal{Q}(A)}$ and the unique transform $\widehat{\mathcal{E}}$ is also semistable (resp. stable).*

$\square$

If we write $\Psi^{\mathcal{Q}(A)} = \Phi^{Q(A)*}$, the functor $\Psi^{\mathcal{Q}(A)}[1]$ is a quasi-inverse of $\Phi^{\mathcal{Q}(A)}$.

We are now interested in the Fourier-Mukai functors Φ such that $\Phi(\mathcal{O}_X) \simeq \mathcal{O}_X[i]$ for some integer i; that is, $\mathcal{O}_X$ is WIT_i with respect to Φ and $\Phi^i(\mathcal{O}_X) = \mathcal{O}_X$. These Fourier-Mukai functors will be relevant in the study of coherent systems on an elliptic curve, as we will see in Section 4.2. Using Proposition 4.10 and the similar statement for Ψ one proves:

Proposition 4.11. *Let a be a positive integer. There exists a Fourier-Mukai transform $\Phi_a \colon D^b_c(X) \to D^b_c(X)$, unique up to composition with an automorphism of X, such that $(\Phi_a)_* = \left(\begin{smallmatrix} 1 & a \\ 0 & 1 \end{smallmatrix}\right)$ and $\Phi_a(\mathcal{O}_X) \simeq \mathcal{O}_X$ (that is, $\mathcal{O}_X$ is IT_0 and $\Phi^0_a(\mathcal{O}_X) = \mathcal{O}_X$).* $\square$

4.2 Coherent systems

A coherent system of type (r,d,k) on a smooth projective curve X is defined as a pair $(\mathcal{E},V)$ consisting of a vector bundle $\mathcal{E}$ (a locally free sheaf) of rank r and degree d over X and a vector subspace $V \subset H^0(X,\mathcal{E})$ of dimension k. A morphism $f \colon (\mathcal{E}',V') \to (\mathcal{E},V)$ of coherent systems is a homomorphism of vector bundles $f \colon \mathcal{E}' \to \mathcal{E}$ such that $f(V') \subset V$. If $\mathcal{E}'$ is a subbundle of $\mathcal{E}$ then we say that $(\mathcal{E}',V')$ is a *coherent subsystem* of $(\mathcal{E},V)$. Coherent systems on X form an additive category $\mathfrak{S}(X)$ (see [LeP93] §4.1).

As for many other decorated objects, the notion of stability (and semistability) for coherent systems depends on the choice of a real parameter. For any real number α, the α-slope of a coherent system $(\mathcal{E}, V)$ of type (r, d, k) is defined by

$$\mu_\alpha(\mathcal{E}, V) = \frac{d}{r} + \alpha\frac{k}{r}.$$

A coherent system $(\mathcal{E}, V)$ is called α-*stable* (α-*semistable*) if

$$\mu_\alpha(\mathcal{E}', V') < \mu_\alpha(\mathcal{E}, V) \quad (\mu_\alpha(\mathcal{E}', V') \leq \mu_\alpha(\mathcal{E}, V))$$

for every proper coherent subsystem $(\mathcal{E}', V')$ of $(\mathcal{E}, V)$.

A coherent system $(\mathcal{E}, V)$ gives rise to an evaluation map $V \otimes \mathcal{O}_X \to \mathcal{E}$. This enables us to consider coherent systems as objects of the abelian category $\mathfrak{C}(X)$ whose objects are arbitrary sheaf maps $\varphi \colon V \otimes \mathcal{O}_X \to \mathcal{E}$, where V is a finite dimensional vector space and $\mathcal{E}$ is any coherent sheaf (cf. [KN95]). A morphism from $\varphi_1 \colon V_1 \otimes \mathcal{O}_X \to \mathcal{E}_1$ to $\varphi_2 \colon V_2 \otimes \mathcal{O}_X \to \mathcal{E}_2$ in $\mathfrak{C}(X)$ is defined by a linear map $f \colon V_1 \to V_2$ and a sheaf morphism $g \colon \mathcal{E}_1 \to \mathcal{E}_2$ such that the obvious diagram commutes. With this definition, the category $\mathfrak{S}(X)$ of coherent systems is a full subcategory of $\mathfrak{C}(X)$.

One can easily see that the condition for an object $\varphi \colon V \otimes \mathcal{O}_X \to \mathcal{E}$ of $\mathfrak{C}(X)$ to represent a coherent system is that $\mathcal{E}$ is a vector bundle and the induced map $H^0(\varphi) \colon V \to H^0(X, \mathcal{E})$ is injective. The latter condition is equivalent to $H^0(X, \ker \varphi) = 0$.

One can extend to $\mathfrak{C}(X)$ the notion of α-(semi)stability of coherent systems, and this extension does not introduce new semistable objects (cf. [KN95]). Moreover, the full subcategory $\mathfrak{S}_{\alpha,\mu}(X)$ of $\mathfrak{C}(X)$ consisting of α-semistable coherent systems with fixed α-slope μ is a Noetherian and Artinian abelian category whose simple objects are precisely the α-stable coherent systems [KN95, RV94].

4.3 Moduli spaces of coherent systems

There exists a (coarse) moduli space for the α-stable coherent systems of type (r, d, k) on X. It is a quasiprojective variety which we denote by $G(\alpha; r, d, k)$. The reader is referred to [BGMN03, Section 2.1] and [B08] for further information about $G(\alpha; r, d, k)$.

Here, we only recall that α-stable coherent systems exist only for $\alpha > 0$ if $k \geq 1$. The range of the parameter α is divided into open intervals determined by a finite number of critical values

$$0 = \alpha_0 < \alpha_1 < \cdots < \alpha_L.$$

The moduli spaces for all values of α in the interval (α_i, α_{i+1}) are isomorphic; if $k \geq r$ this is also true for the interval (α_L, ∞) (see [BGMN03, Propositions 4.2 & 4.6]).

In this Section we describe some well known facts about the moduli spaces of coherent systems for the two limit cases of small and large values of α. We will always assume $d \neq 0$ and $k > 0$.

4.3.1 Small values of the parameter

We denote by $G_0(r, d, k)$ the moduli space of α-stable coherent systems of type (r, d, k) with $0 < \alpha < \alpha_1$, where α_1 is the first critical value.

Proposition 4.12. [RV94] *A coherent system $(\mathcal{E}, V)$ of type (r, d, k) is α-stable, with $0 < \alpha < \alpha_1$, if and only if $\mathcal{E}$ is semistable and $k'/r' < k/r$, for all coherent subsystems $(\mathcal{E}', V')$ of type (r', d', k') with $0 \neq \mathcal{E}' \neq \mathcal{E}$ and $\mu(\mathcal{E}') = \mu(\mathcal{E})$.*   □

4.3.2 Large values of the parameter

Let us denote by $G_L(r, d, k)$ the moduli space of α-stable coherent systems of type (r, d, k) with $\alpha_L < \alpha < \frac{d}{r-k}$ (we are assuming that $0 < k < r$). $G_L(r, d, k)$ has been described by Bradlow and García-Prada [BG02] (see also [BGMN03]) in terms of the Brambilla-Grzegorczyk-Newstead extensions, BGN extensions for short [BGN97].

BGN extensions of type (r, d, k) are defined as extensions of vector bundles

$$0 \to \mathcal{O}_X^{\oplus k} \to \mathcal{E} \to \mathcal{F} \to 0,$$

where $\mathcal{E}$ has rank $r > k$ and degree $d > 0$, which satisfy the following conditions:

(a) $H^0(X, \mathcal{F}^*) = 0$
(b) If $(e_1, \dots, e_k) \in \mathrm{Ext}_X^1(\mathcal{F}, \mathcal{O}_X^{\oplus k}) \simeq H^1(X, \mathcal{F}^*)^{\oplus k}$ denotes the class of the extension, then $e_1, \dots, e_k$ are linearly independent as vectors in $H^1(X, \mathcal{F}^*)$.

Coherent systems can be regarded as BGN-extensions due to the following result.

Proposition 4.13. [BG02, Proposition 4.1] *Let $(\mathcal{E}, V)$ be an α-semistable coherent system of type (r, d, k) with $\alpha_L < \alpha < \frac{d}{r-k}$. The evaluation map of $(\mathcal{E}, V)$ defines a BGN extension $0 \to \mathcal{O}_X^{\oplus k} \to \mathcal{E} \to \mathcal{F} \to 0$, with $\mathcal{F}$ semistable. Moreover, any BGN extension $0 \to \mathcal{O}_X^{\oplus k} \to \mathcal{E} \to \mathcal{F} \to 0$*

where the quotient $\mathcal{F}$ is stable gives rise to an α-stable coherent system, with $\alpha_L < \alpha < \frac{d}{r-k}$. □

A complete characterization of the BGN-extensions which give rise to α-stable coherent systems has been given in [HT08].

Proposition 4.14. *A BGN extension of type (r, d, k), $0 \to \mathcal{O}_X^{\oplus k} \to \mathcal{E} \to \mathcal{F} \to 0$ defines an α-stable coherent system, with $\alpha_L < \alpha < \frac{d}{r-k}$, if and only if $\mathcal{F}$ is semistable and one has $k'/r' > k/r$ for all subextensions $0 \to \mathcal{O}_X^{\oplus k'} \to \mathcal{E}' \to \mathcal{F}' \to 0$ of type (r', d', k') with $\mu(\mathcal{F}') = \mu(\mathcal{F})$.* □

We now denote by $\mathcal{BGN}(r, d, k)$, $\mathcal{BGN}^s(r, d, k)$ the families of BGN extension classes of type (r, d, k) on X in which the quotient is semistable or stable, respectively. Due to Proposition 4.13 we have inclusions $\mathcal{BGN}^s(r, d, k) \hookrightarrow G_L(r, d, k) \hookrightarrow \mathcal{BGN}(r, d, k)$.

Proposition 4.15. [LN05, Proposition 3.2 and Lemma 4.1] *Let $(\mathcal{E}, V)$ be an α-stable coherent system of type (r, d, k) on an elliptic curve X with $d \neq 0$, $k > 0$. Then every indecomposable direct summand of $\mathcal{E}$ has positive degree.* □

4.4 Fourier-Mukai transforms of coherent systems on elliptic curves

In this Subsection X is an elliptic curve. We also assume that $d \neq 0$, $k > 0$.

Recall (Proposition 4.11) that for every integer number $a > 0$ there is a Fourier-Mukai transform Φ_a such that $\mathcal{O}_X$ is IT_0, $\Phi_a^0(\mathcal{O}_X) = \mathcal{O}_X$ and $(\Phi_a)_* = \left(\begin{smallmatrix} 1 & a \\ 0 & 1 \end{smallmatrix}\right)$. If $\Psi_a[1]: D_c^b(X) \to D_c^b(X)$ is the quasi-inverse of Φ_a, the Fourier-Mukai transforms Φ_a, Ψ_a define functors $\Phi_a^0: \mathfrak{C}(X) \to \mathfrak{C}(X)$, $\Psi_a^1: \mathfrak{C}(X) \to \mathfrak{C}(X)$ that send the object $\varphi: V \otimes \mathcal{O}_X \to \mathcal{E}$ to $\Phi_a^0(\varphi): V \otimes \mathcal{O}_X \to \Phi_a^0(\mathcal{E})$ and $\Psi_a^1(\varphi): V \otimes \mathcal{O}_X \to \Psi_a^1(\mathcal{E})$, respectively.

Proposition 4.16. *Let $\varphi: V \otimes \mathcal{O}_X \to \mathcal{E}$ be a coherent system.*

(a) *The Φ_a^0-transform $\Phi_a^0(\varphi): V \otimes \mathcal{O}_X \to \Phi_a^0(\mathcal{E})$ is a coherent system.*
(b) *If $\mathcal{E}$ is IT_1 with respect to Ψ_a, $\Psi_a^1(\varphi): V \otimes \mathcal{O}_X \to \Psi_a^1(\mathcal{E})$ is a coherent system.*

□

Therefore, the functor $\Phi_a^0: \mathfrak{C}(X) \to \mathfrak{C}(X)$ preserves the subcategory $\mathfrak{S}(X)$ of coherent systems and induces a functor $\Phi_a^0: \mathfrak{S}(X) \to \mathfrak{S}(X)$.

4.4.1 Preservation of stability. Small α

In this subsection α-stability refers to a positive α which smaller than the first critical value.

As a consequence of Propositions 4.12 and 4.15, if $\varphi\colon V \otimes \mathcal{O}_X \to \mathcal{E}$ is a coherent system in the moduli space $G_0(r, d, k)$, then $\mathcal{E}$ is semistable of positive degree. Hence, $r + ad > 0$ and Proposition 4.10 implies that $\mathcal{E}$ is IT_0 with respect to Φ_a and that $\Phi_a^0(E)$ is semistable with Chern character $\mathrm{ch}(\Phi_a^0(E)) = (r + ad, d)$. This is not enough to prove that the transformed coherent system $\Phi_a^0(\varphi)\colon V \otimes \mathcal{O}_X \to \Phi_a^0(\mathcal{E})$ is stable, but one can prove that the remaining conditions required by Proposition 4.12 are also fulfilled. One then gets the following result [HT08].

Theorem 4.17. *The Fourier-Mukai transform Φ_a induces an isomorphism of moduli spaces*

$$\Phi_a^0 : G_0(r, d, k) \overset{\sim}{\to} G_0(r + ad, d, k),$$

whose inverse is induced by Ψ_a. Therefore, the isomorphism type of $G_0(r, d, k)$ depends only on the class $[r] \in \mathbb{Z}/d\,\mathbb{Z}$. $\square$

4.4.2 Preservation of stability. Large α

In this subsection we suppose that $0 < k < r$. Under this assumption, Lange and Newstead proved in [LN05, Theorem 5.2] that for an elliptic curve the moduli space $G(\alpha; r, d, k)$ is non empty if and only if $0 < \alpha < \frac{d}{r-k}$ and either $k < d$ or $k = d$ and $\gcd(r, d) = 1$. Moreover, in this case the largest critical value α_L verifies $\alpha_L < \frac{d}{r-k}$.

Recall that for large α the moduli spaces $G(\alpha; r, d, k)$ are described in terms of BGN-extensions. If $0 \to \mathcal{O}_X^{\oplus k} \to \mathcal{E} \to \mathcal{F} \to 0$ is a BGN-extension, the quotient $\mathcal{F}$ is IT_0 with respect to Φ_a and $\Phi_a^0(F)$ is semistable by Proposition 4.10. Since $\mathcal{O}_X$ is also IT_0, it follows that we have an exact sequence $0 \to \mathcal{O}_X^{\oplus k} \to \Phi_a^0(\mathcal{E}) \to \Phi_a^0(\mathcal{F}) \to 0$. Using again Proposition 4.10, one can prove that this exact sequence is actually a BGN-extension and that it is stable if the original BGN-extension is so. Then one has:

Theorem 4.18. ([HT08, Thm. 4.12]) *The Fourier-Mukai transform Φ_a induces an isomorphism $\Phi_a^0\colon \mathcal{BGN}(r, d, k) \overset{\sim}{\to} \mathcal{BGN}(r + ad, d, k)$ by sending a BGN extension $0 \to \mathcal{O}_X^{\oplus k} \to \mathcal{E} \to \mathcal{F} \to 0$ to $0 \to \mathcal{O}_X^{\oplus k} \to \Phi_a^0(\mathcal{E}) \to \Phi_a^0(\mathcal{F}) \to 0$. This restricts to an isomorphism $\Phi_a^0\colon \mathcal{BGN}^s(r, d, k) \overset{\sim}{\to} \mathcal{BGN}^s(r + ad, d, k)$. The inverse isomorphism is induced by Ψ_a.* $\square$

Theorem 4.19. [HT08] *The Fourier-Mukai transform Φ_a induces an isomorphism*

$$\Phi_a^0 \colon G_L(r,d,k) \overset{\sim}{\to} G_L(r+ad,d,k)\,.$$

Therefore, the isomorphism type of $G_L(r,d,k)$ depends only on the class $[r] \in \mathbb{Z}/d\,\mathbb{Z}$.

Proof We only sketch an idea of the proof. Given a coherent system in $G_L(r,d,k)$, we know by Proposition 4.13 that it defines an extension $0 \to \mathcal{O}_X^{\oplus k} \to \mathcal{E} \to \mathcal{F} \to 0$ which belongs to $\mathcal{BGN}(r,d,k)$. By Proposition 4.18 the transformed extension $0 \to \mathcal{O}_X^{\oplus k} \to \Phi_a^0(\mathcal{E}) \to \Phi_a^0(\mathcal{F}) \to 0$ belongs to $\mathcal{BGN}(r+ad,d,k)$. The importance of Proposition 4.14 becomes apparent here, because it tells us what conditions a BGN-extension has to fulfill in order to define an α-stable coherent system. The proof then consists in checking that the transformed extension fulfills those additional conditions. This is a rather technical issue and we shall omit it here. $\qquad\square$

One may draw a diagram which summarizes all this information.

$$
\begin{array}{ccccc}
\mathcal{BGN}^s(r,d,k) & \lhook\joinrel\longrightarrow & G_L(r,d,k) & \lhook\joinrel\longrightarrow & \mathcal{BGN}(r,d,k) \\
\Big\downarrow{\scriptstyle\Phi_a^0} & & \Big\downarrow{\scriptstyle\Phi_a^0} & & \Big\downarrow{\scriptstyle\Phi_a^0} \\
\mathcal{BGN}^s(r+ad,d,k) & \lhook\joinrel\longrightarrow & G_L(r+ad,d,k) & \lhook\joinrel\longrightarrow & \mathcal{BGN}(r+ad,d,k)
\end{array}
$$

4.4.3 Birational type of the moduli spaces $G(\alpha;r,d,k)$

Let $0 = \alpha_0 < \alpha_1 < \cdots < \alpha_L$ be the critical values for coherent systems of type (r,d,k), so that the moduli spaces $G(\alpha;r,d,k)$ for any two values of $\alpha \in (\alpha_i, \alpha_{i+1})$ coincide. Then, there is only a finite number of different moduli spaces. Moreover, one has:

Theorem 4.20. [LN05, Theorem 4.4] *The birational type of $G(\alpha;r,d,k)$ is independent of $\alpha \in (\alpha_0, \alpha_L)$.* $\qquad\square$

It follows that we can determine the birational type of any of the finitely many different moduli spaces simply by computing one of them. We then choose $G_0(r,d,k)$, which has been studied in full generality.

Theorem 4.21. ([HT08, Thm. 5.2]) *Let a be a positive integer. The birational types of $G(\alpha;r,d,k)$ and $G(\alpha;r+ad,d,k)$ are the same. Therefore, the birational type of $G(\alpha;r,d,k)$ depends only on the class $[r] \in \mathbb{Z}/d\,\mathbb{Z}$.*

Proof The birational types of $G(\alpha; r, d, k)$ and $G(\alpha; r + ad, d, k)$ are the same as those of $G_0(r, d, k)$ and $G_0(r + ad, d, k)$, respectively, by Theorem 4.20. Moreover, the Fourier-Mukai transform Φ_a induces an isomorphism $\Phi_a^0 : G_0(r, d, k) \to G_0(r + ad, d, k)$ by Theorem 4.17 and we finish by Theorem 4.20. $\qquad\square$

Bibliography

[BBH08] C. BARTOCCI, U. BRUZZO, AND D. HERNÁNDEZ RUIPÉREZ, *Fourier-Mukai and Nahm transforms in geometry and mathematical physics*. To appear in Progress in Mathematics, Birkhaüser, 2009.

[BBH94] ——, *Fourier-Mukai transform and index theory*, Manuscripta Math., 85 (1994), pp. 141–163.

[BBH97a] ——, *A Fourier-Mukai transform for stable bundles on K3 surfaces*, J. Reine Angew. Math., 486 (1997), pp. 1–16.

[BBH97b] ——, *Moduli of reflexive K3 surfaces*, in Complex analysis and geometry (Trento, 1995), Pitman Res. Notes Math. Ser., vol. 366, Longman, Harlow, 1997, pp. 60–68.

[BBH98a] ——, *Existence of μ-stable vector bundles on K3 surfaces and the Fourier-Mukai transform*, in Algebraic geometry (Catania, 1993/Barcelona, 1994), Lecture Notes in Pure and Appl. Math., vol. 200, Dekker, New York, 1998, pp. 245–257.

[BBH2] ——, *A hyper-Kähler Fourier transform*, Differential Geom. Appl., 8 (1998), pp. 239–249.

[Birlan99] C. BIRKENHAKE AND H. LANGE, *Complex tori*, vol. 177 of Progress in Mathematics, Birkhäuser Boston Inc., Boston, MA, 1999.

[BvdB03] A. BONDAL AND M. VAN DEN BERGH, *Generators and representability of functors in commutative and noncommutative geometry*, Mosc. Math. J., 3 (2003), pp. 1–36, 258.

[BO95] A. I. BONDAL AND D. O. ORLOV, *Semi orthogonal decomposition for algebraic varieties*. MPIM Preprint 95/15 (1995), `math.AG/9506012`.

[BO01] ——, *Reconstruction of a variety from the derived category and groups of autoequivalences*, Compositio Math., 125 (2001), pp. 327–344.

[BvB] P. J. BRAAM AND P. VAN BAAL, *Nahm's transformation for instantons*, Comm. Math. Phys., 122 (1989), pp. 267–280.

[BG02] S. B. BRADLOW AND O. GARCÍA-PRADA, *An application of coherent systems to a Brill-Noether problem*, J. Reine Angew. Math., 551 (2002), pp. 123–143.

[BGMN03] S. B. BRADLOW, O. GARCÍA-PRADA, V. MUÑOZ, AND P. E. NEWSTEAD, *Coherent systems and Brill-Noether theory*, Internat. J. Math., 14 (2003), pp. 683–733.

[B08] L. BRAMBILA-PAZ, *Non-emptiness of moduli spaces of coherent systems*. Internat. J. Math. 19, n.7 (2008), pp. 777–799.

[BGN97] L. BRAMBILA-PAZ, I. GRZEGORCZYK, AND P. E. NEWSTEAD, *Geography of Brill-Noether loci for small slopes*, J. Algebraic Geom., 6 (1997), pp. 645–669.

[Bri98] T. BRIDGELAND, *Fourier-Mukai transforms for elliptic surfaces*, J. Reine Angew. Math., 498 (1998), pp. 115–133.

[Bri98t] ——, *Fourier-Mukai Transforms for Surfaces and Moduli Spaces of Stable Sheaves*, PhD thesis, University of Edinburgh, 1998.

[Bri99] ——, *Equivalences of triangulated categories and Fourier-Mukai transforms*, Bull. London Math. Soc., 31 (1999), pp. 25–34.

[Bri02] ——, *Flops and derived categories*, Invent. Math., 147 (2002), pp. 613–632.

[BrM01] T. BRIDGELAND AND A. MACIOCIA, *Complex surfaces with equivalent derived categories*, Math. Z., 236 (2001), pp. 677–697.

[BM96] U. BRUZZO AND A. MACIOCIA, *Hilbert schemes of points on some K3 surfaces and Gieseker stable bundles*, Math. Proc. Cambridge Philos. Soc., 120 (1996), pp. 255–261.

[Ca1] A. CĂLDĂRARU, *Fiberwise stable bundles on elliptic threefolds with relative Picard number one*, C. R. Math. Acad. Sci. Paris, 334 (2002), pp. 469–472.

[DKr] S. K. DONALDSON AND P. B. KRONHEIMER, *The geometry of four-manifolds*, The Clarendon Press – Oxford University Press, Oxford, 1990.

[FL91] R. FAHLAOUI AND Y. LASZLO, *Transformée de Fourier et stabilité sur les surfaces abéliennes*, Compositio Math., 79 (1991), pp. 271–278.

[EGAIII-1] A. GROTHENDIECK, *Éléments de géométrie algébrique. III. Étude cohomologique des faisceaux cohérents. I*, Inst. Hautes Études Sci. Publ. Math., (1961), p. 167.

[Hart77] R. HARTSHORNE, *Algebraic geometry*, Graduate Texts in Mathematics, vol. 52, Springer-Verlag, New York, 1977.

[HePl05] G. HEIN AND D. PLOOG, *Fourier-Mukai transforms and stable bundles on elliptic curves*, Beiträge Algebra Geom., 46 (2005), pp. 423–434.

[HLS05] D. HERNÁNDEZ RUIPÉREZ, A. C. LÓPEZ MARTÍN, AND F. SANCHO DE SALAS, *Fourier-Mukai transforms for Gorenstein schemes*, Adv. in Maths., 211 (2007), pp. 594–620.

[HLS08] D. HERNÁNDEZ RUIPÉREZ, A. C. LÓPEZ MARTÍN, AND F. SANCHO DE SALAS, *Relative integral functors for singular fibrations and singular partners*, J. Eur. Math. Soc. (JEMS), (2009). To appear. Also `arXiv:math/0610319`.

[HT08] D. HERNÁNDEZ RUIPÉREZ AND C. TEJERO PRIETO, *Fourier-Mukai transforms for coherent systems on elliptic curves*, J. London Math. Soc., (2007). doi: 10.1112/jlms/jdm089. Also `arXiv:math/0603249`.

[HiVdB] L. HILLE AND M. D. VAN DEN BERGH, *Fourier-Mukai transforms*, in Handbook on tilting theory, vol. 332 of London Math. Soc. Lecture Note Series, Cambridge Univ. Press, 2007.

[Ill71] L. ILLUSIE, *Définition de l'indice analytique d'un complexe elliptique relatif, Exposé II, Appendix II*, in Théorie des intersections et théorème de Riemann-Roch (SGA6), Lecture Notes in Math., vol. 225, Springer Verlag, Berlin, 1971, pp. 199–221.

[J5] M. JARDIM, *A survey on Nahm transform*, J. Geom. Phys., 52 (2004), pp. 313–327.

[Kaw02a] Y. KAWAMATA, *Francia's flip and derived categories*, in Algebraic geometry, de Gruyter, Berlin, 2002, pp. 197–215.

[Kaw02] ——, *Equivalences of derived categories of sheaves on smooth stacks*, Amer. J. Math., 126 (2004), pp. 1057–1083.

[KN95] A. D. KING AND P. E. NEWSTEAD, *Moduli of Brill-Noether pairs on algebraic curves*, Internat. J. Math., 6 (1995), pp. 733–748.

[LN05] H. LANGE AND P. E. NEWSTEAD, *Coherent systems on elliptic curves*, Internat. J. Math., 16 (2005), pp. 787–805.

[LeP93] J. LE POTIER, *Systèmes cohérents et structures de niveau*, Astérisque, (1993), p. 143.

[Mac96] A. MACIOCIA, *Generalized Fourier-Mukai transforms*, J. Reine Angew. Math., 480 (1996), pp. 197–211.

[Mac96b] ——, *Gieseker stability and the Fourier-Mukai transform for abelian surfaces*, Quart. J. Math. Oxford Ser. (2), 47 (1996), pp. 87–100.

[Mar77] M. MARUYAMA, *Moduli of stable sheaves. I*, J. Math. Kyoto Univ., 17 (1977), pp. 91–126.

[Mar78] ——, *Moduli of stable sheaves. II*, J. Math. Kyoto Univ., 18 (1978), pp. 557–614.

[Muk81] S. MUKAI, *Duality between $D(X)$ and $D(\hat{X})$ with its application to Picard sheaves*, Nagoya Math. J., 81 (1981), pp. 153–175.

[Muk87b] ——, *Fourier functor and its application to the moduli of bundles on an abelian variety*, in Algebraic geometry, Sendai, 1985, Adv. Stud. Pure Math., vol. 10, North-Holland, Amsterdam, 1987, pp. 515–550.

[Muk87a] ——, *On the moduli space of bundles on K3 surfaces. I*, in Vector bundles on algebraic varieties (Bombay, 1984), Tata Inst. Fund. Res. Stud. Math., vol. 11, Tata Inst. Fund. Res., Bombay, 1987, pp. 341–413.

[Muk99] ——, *Duality of polarized K3 surfaces*, in New trends in algebraic geometry (Warwick, 1996), London Math. Soc. Lecture Note Ser., vol. 264, Cambridge Univ. Press, Cambridge, 1999, pp. 311–326.

[N] W. NAHM, *Self-dual monopoles and calorons*, in Group theoretical methods in physics (Trieste, 1983), Lecture Notes in Phys., vol. 201, Springer-Verlag, Berlin, 1984, pp. 189–200.

[Or97] D. O. ORLOV, *Equivalences of derived categories and K3 surfaces*, J. Math. Sci. (New York), 84 (1997), pp. 1361–1381. Algebraic geometry, 7.

[Pol03] A. POLISHCHUK, *Abelian varieties, theta functions and the Fourier transform*, Cambridge Tracts in Mathematics, vol. 153, Cambridge University Press, Cambridge, 2003.

[RV94] N. RAGHAVENDRA AND P. A. VISHWANATH, *Moduli of pairs and generalized theta divisors*, Tohoku Math. J. (2), 46 (1994), pp. 321–340.

[Sc] H. SCHENK, *On a generalised Fourier transform of instantons over flat tori*, Comm. Math. Phys., 116 (1988), pp. 177–183.

[VdB04] M. VAN DEN BERGH, *Three-dimensional flops and noncommutative rings*, Duke Math. J., 122 (2004), pp. 423–455.

4

Geometric Invariant Theory

P. E. Newstead

Department of Mathematical Sciences
University of Liverpool
Peach Street Liverpool L69 7ZL UK
e-mail: newstead@liverpool.ac.uk

These notes are based on lectures given at the CIMAT College on Vector Bundles and describe a method of constructing quotients in algebraic geometry. Geometric Invariant Theory (GIT) is due originally to Mumford [GIT], but some of the ideas go back to 19th century invariant theory, especially the work of Hilbert in the 1890s. The lecture notes are essentially unchanged from those given out when the lectures were given and are intended to be reasonably self-contained, although some proofs are omitted; further details, if needed, can be found in [Tata]. I have added an appendix describing the application of GIT to moduli problems and particularly to the construction of moduli spaces for vector bundles on algebraic curves; this material is adapted from my "Polish" notes [Pol].

The natural context for the construction is that of schemes, but for simplicity (except to a limited extent in the appendix) we work always with varieties defined over an algebraically closed field k; this field can have any characteristic. All topological terms refer to the Zariski topology.

The references are divided into two sections. The first lists works directly connected with the material described in the lectures and served as the reference list for those lectures. The second lists some other key works; it is not by any means a comprehensive list.

For the construction of quotients, we make the standard assumption that the group G is reductive. One should, however, remark that there are many interesting problems for which G is not reductive and it is possible to use variants of GIT for such groups. For recent work in this direction, see [DK]; a survey by Frances Kirwan is included in this volume.

For another introduction to GIT for reductive groups over $\mathbf{C}$, see [Sch]. This includes some more advanced concepts and describes recent

progress in the construction of moduli spaces of vector bundles and principal bundles with added structure.

1 Lecture 1 – Quotients

1.1 *Types of quotient*

Let G be an algebraic group acting on a variety X (through a morphism
$G \times X \to X : (g, x) \mapsto g.x)$. Our object is to construct a good **quotient**
for this action; what could this mean?

(i) **Orbit space.** It may happen that the set of orbits X/G has a
   natural structure of variety. In fact this rarely happens, although
   there is always a dense open G-invariant subset U of X for which
   U/G exists [Ros] (a fact which is not as well known as it might be).
   However we are looking for something more global. Examples
   1.1–1.3 below illustrate some of the problems with the orbit space
   approach.

(ii) **Categorical quotient.** We say that $\phi : X \to Y$ (or just Y) is a
   categorical quotient of X by G if every morphism $X \to Z$ which
   is constant on orbits factors uniquely through ϕ. This is uniquely
   determined if it exists and has good functorial properties, but
   not necessarily good geometric ones (see the examples below).
   The ultimate in this approach is the use of **stacks**; these allow
   quotients to exist by definition, but geometrical properties may
   not be obvious.

Example 1.1. k^* acts on k^2 by $t.(x, y) = (tx, t^{-1}y)$. The orbits are

- for each $\alpha \in k^*$, the conic $\{(x, y) : xy = \alpha\}$
- the punctured x-axis $\{(x, 0) : x \in k^*\}$
- the punctured y-axis $\{(0, y) : y \in k^*\}$
- the origin $\{(0, 0)\}$.

In order to get a separated quotient, one has to combine the three last
orbits listed and indeed one then gets a categorical quotient isomorphic
to k, the quotient morphism being given by $(x, y) \mapsto xy$. One cannot
obtain a separated orbit space (that is, as a variety), even if one deletes
the orbit $\{(0, 0)\}$, which has lower dimension than the others, since both
punctured axes are limits of the other orbits as $\alpha \to 0$.

Example 1.2. k^* acts on k^n $(n \geq 2)$ by $t.(x_1, \ldots, x_n) = (tx_1, \ldots, tx_n)$.
The origin lies in the closure of every orbit, so any morphism which is

constant on orbits is constant. Thus there is no orbit space but there is a categorical quotient consisting of a single point.

Example 1.3. k^* acts on $k^n \setminus \{(0, \dots, 0)\}$ by the same formula. This time the projective space $\mathbf{P}^{n-1}$ is a categorical quotient and also an orbit space.

Our aim is to construct categorical quotients with good geometrical properties. Our method of doing this is Geometric Invariant Theory (GIT). There are other methods using stacks or algebraic spaces or by direct construction (Example 1.3 above, for instance) etc. Here we shall concentrate on GIT, which has proved extremely useful and, when k is the complex numbers, has important and surprising connections with symplectic geometry. Proofs are not included; we will comment on these in Lecture 3.

We shall always write $G.x$ for the orbit $\{y \in X : y = g.x \text{ for some } g \in G\}$ of x.

1.2 Rings of invariants

For any variety X, let $A(X)$ denote the algebra of morphisms $X \to k$. The action of G on X determines an action of G on $A(X)$. We write $A(X)^G$ for the algebra of elements of $A(X)$ which are G-invariant. If $X \to Z$ is constant on orbits, the natural homomorphism $A(Z) \to A(X)$ has image contained in $A(X)^G$. In particular, if Y is a categorical quotient, the homomorphism $A(Y) \to A(X)^G$ is an isomorphism.

Example 1.4. In Examples 1.1, 1.2, 1.3, we have $G = k^*$ and $A(X)^G = k[\alpha], k, k$ respectively. In Examples 1.1, 1.2, this gives good geometric answers; in Example 1.3, it does not. In fact, in Example 1.3, we have a very well behaved quotient $Y = \mathbf{P}^{n-1}$ and $A(Y)$ is isomorphic to $A(X)^G = k$, but this in no way represents the geometry of the quotient.

Example 1.5. $G = \mathrm{GL}(n)$ acts on $X = \mathrm{M}(n)$ ($n \times n$ matrices) by conjugation. It is not hard to see that $A(X)^G$ is the polynomial algebra on the coefficients of the characteristic polynomial. These coefficients determine a morphism $\mathrm{M}(n) \to k^n$, which is a categorical quotient, but not an orbit space since all matrices with the same eigenvalues map to the same point of k, but not all such matrices are in the same orbit (cf. Jordan normal forms).

1.3 Affine Quotients

Now suppose that X is an affine variety and that we look for a quotient which is also an affine variety (this rules out Example 1.3, but includes Examples 1.1, 1.2 and 1.5). In this case $A(X)$ is just the affine coordinate ring of X, and, by Hilbert's Nullstellensatz, we can recover X from $A(X)$. Moreover, a k-algebra has the form $A(Z)$ for some affine variety Z if and only if it is finitely generated and has no nilpotent elements. The second property is necessarily true of $A(X)^G$, so one question to be answered is

Question. Is $A(X)^G$ finitely generated?

This is a version of Hilbert's 14th problem and the answer in general is NO [N1, N2]. However the answer is always YES if G is **reductive**. The meaning of reductive will be discussed in Lecture 3; for the time being, it is enough to remark that many important examples, including GL(p), SL(p) and PGL(p), are reductive. It turns out that this is sufficient for us to construct well behaved quotients in the affine case.

Theorem 1.1. *Let G be a reductive group acting on an affine variety X. Then there exists an affine variety Y and a morphism $\phi : X \to Y$ such that*

  (i) *ϕ is constant on orbits;*
 (ii) *ϕ is surjective;*
(iii) *if U is an open subset of Y, the induced homomorphism*

$$\phi^* : A(U) \to A(\phi^{-1}(U))^G$$

    is an isomorphism;
 (iv) *if W is a closed G-invariant subset of X, then $\phi(W)$ is closed in Y;*
  (v) *if W_1, W_2 are closed G-invariant subsets of X and $W_1 \cap W_2 = \emptyset$, then $\phi(W_1) \cap \phi(W_2) = \emptyset$.*

Corollary 1.2. *For any open subset U of Y, U is a categorical quotient of $\phi^{-1}(U)$ by G.*

Corollary 1.3. *If $\phi(x_1) = \phi(x_2)$, then $\overline{G.x_1} \cap \overline{G.x_2} \neq \emptyset$.*

Corollary 1.4. *If G acts on $\phi^{-1}(U)$ with closed orbits, then $U = \phi^{-1}(U)/G$ (i.e. U is an orbit space).*

Note that Corollary 1.3 states that ϕ separates orbits to the greatest extent possible, consistent with Y being a variety. This is perhaps more significant than the simple fact that $A(X)^G$ is finitely generated.

1.4 Projective quotients

The case in which we are most interested is when X is **projective**; this is because we would like our quotients to be **complete** (**compact**) or at least to have natural compactifications. To apply the previous section, we need to consider **affine open subsets** of X which are invariant under G. We suppose $X \subset \mathbf{P}^n$ and that G acts **linearly**, that is through a homomorphism $G \to \mathrm{GL}(n+1)$. Then G acts homogeneously on the polynomial ring $k[X_0, \ldots, X_n]$. If f is any non-constant G-invariant homogeneous polynomial for this action, then $X_f = \{x \in X : f(x) \neq 0\}$ is a G-invariant affine open subset of X.

Definition. A point $x \in X$ is

- *semistable* for the action of G if there exists f such that $x \in X_f$;
- *stable* for the action of G if x has finite stabiliser (or equivalently $\dim G.x = \dim G$) and there exists f as above such that G acts on X_f with closed orbits. (In [GIT], this is called "properly stable".)

Lemma 1.5. *The subsets X^{ss}, X^s of X of semistable (resp. stable) points of X are G-invariant open subsets of X.*

Exercise. Prove Lemma 1.5.

Definition. $\phi : X \to Y$ is

- a *good quotient* of X by G if ϕ is an affine morphism (that is, the inverse image of every affine open set in Y is affine) and (i)–(v) of Theorem 1.1 hold;
- a *geometric quotient* if it is a good quotient and also an orbit space.

It is conventional to write $Y = X /\!/ G$ when Y is a good quotient and $Y = X/G$ when Y is a geometric quotient. The concept of "good quotient" given here is due to Seshadri; the idea originates in [GIT], but the definition given there is not identical with ours. Our "geometric quotient" differs from that in [GIT] by the insistence that ϕ be an affine morphism. This is a reasonable requirement since G itself is affine.

The properties of the corollaries to Theorem 1.1 carry over to good quotients.

Proposition 1.6. *Let $\phi : X \to Y$ be a good quotient of X by G. Then*

(i) *Y is a categorical quotient of X by G;*

(ii) *$\phi(x_1) = \phi(x_2) \Leftrightarrow \overline{G.x_1} \cap \overline{G.x_2} \neq \emptyset$;*

(iii) *if G acts on X with closed orbits, then $Y = X/G$ is a geometric quotient of X by G.*

We come now to the main theorem of this lecture.

Theorem 1.7. *Let G be a reductive group acting linearly on a projective variety X. Then*

(i) *there exists a good quotient $\phi : X^{ss} \to Y$ and $Y = X^{ss}//G$ is projective;*

(ii) *there exists an open subset Y^s of Y such that $\phi^{-1}(Y^s) = X^s$ and $Y^s = X^s/G$ is a geometric quotient of X^s;*

(iii) *for $x_1, x_2 \in X^{ss}$, $\phi(x_1) = \phi(x_2) \Leftrightarrow \overline{G.x_1} \cap \overline{G.x_2} \cap X^{ss} \neq \emptyset$;*

(iv) *for $x \in X^{ss}$, x is stable if and only if x has finite stabiliser and $G.x$ is closed in X^{ss}.*

Remark 1.1. In some ways the most significant point of this theorem is that Y is **projective**. This means both that it is complete and that it can be embedded in projective space. (In fact (see 3.4), $Y = \operatorname{Proj} A(\hat{X})^G$, where $\hat{X}$ is the affine variety in k^{n+1} lying over X.)

Remark 1.2. For G not necessarily reductive, Rosenlicht's result [Ros] gives a dense open subset U of X such that U/G exists with all the properties of a geometric quotient except that the morphism ϕ need not be affine.

2 Lecture 2 – The Hilbert-Mumford criterion

It is in general a very difficult problem to calculate a ring of invariants. However, if we are simply interested in calculating basic properties of a projective quotient, it is often sufficient to compute the stable and semistable points for the action. In this lecture we will describe a usable criterion for doing this, due originally to Hilbert for semistability for $G = \operatorname{SL}(n)$ and extended by Mumford to stability and arbitrary reductive G. We begin with a definition of stability not involving projective space. We suppose throughout that G is a reductive group.

2.1 Affine definition of stability

Proposition 2.1. *Let X be a projective variety in $\mathbf{P}^n$ and G a reductive group acting linearly on X. Let $x \in X$ and $\hat{x} \in k^{n+1}$ be a point lying over x. Then*

(i) *x is semistable if and only if $0 \notin \overline{G.\hat{x}}$;*

(ii) *x is stable if and only if the stabiliser of $\hat{x}$ is finite and $G.\hat{x}$ is closed in k^{n+1}.*

Proof (i) Suppose first that x is semistable, and let f be a G-invariant homogeneous polynomial of degree ≥ 1 such that $f(x) \neq 0$. We have clearly $f(\hat{x}) \neq 0$, so $f(y)$ is equal to a *non-zero* constant for $y \in G.\hat{x}$. Hence $0 \notin \overline{G.\hat{x}}$.

Conversely, if $0 \notin \overline{G.\hat{x}}$, it is a simple consequence of geometric reductivity (see Lecture 3 and [Tata, Lemma 3.3]) that there exists a G-invariant polynomial f such that

$$f(0) = 0, \quad f\left(\overline{G.\hat{x}}\right) = 1.$$

But then f has constant term 0; so some homogeneous part of f must be non-zero at $\hat{x}$. Hence x is semistable.

(ii) Suppose that x is stable. Then x has finite stabiliser and there exists a G-invariant homogeneous polynomial f of degree ≥ 1 such that $f(x) \neq 0$ and $G.x$ is closed in X_f. It is obvious that $\hat{x}$ has finite stabiliser. Now put $\alpha = f(\hat{x}) \neq 0$ and let

$$Z_\alpha = \left\{ y \in k^{n+1} : f(y) = \alpha \right\}.$$

Clearly Z_α is closed in k^{n+1} and it is easy to see that the morphism $Z_\alpha \to \mathbf{P}^n_f$ is surjective and finite. Moreover the inverse image of $G.x$ is closed in Z_α and is the union of a finite number of orbits of G acting on Z_α. Since all these orbits have the same dimension, each of them, in particular $G.\hat{x}$, is closed in Z_α and therefore in k^{n+1}.

Conversely, suppose that $\hat{x}$ has finite stabiliser and $G.\hat{x}$ is closed in k^{n+1}. By (i), x is semistable, so there exists a G-invariant homogeneous polynomial f of degree ≥ 1 such that $f(x) \neq 0$. As above $Z_\alpha \to \mathbf{P}^n_f$ is surjective and finite. This morphism maps $G.\hat{x}$ surjectively to $G.x$; it follows that x has finite stabiliser and $G.x$ is closed in $\mathbf{P}^n_f$. This holds for all f such that $f(x) \neq 0$, so $G.x$ is closed in X^{ss}. It follows from Theorem 1.7(iv) that x is stable. $\qquad\square$

2.2 Actions of k^*

Suppose first that $G = k^*$. We identify k^* with a subset of $\mathbf{P}^1$ in the obvious way. In fact, for any $\alpha \in k$, we identify α with $(1 : \alpha) \in \mathbf{P}^1$ and write ∞ for the extra point $(0 : 1)$. Note that any morphism $k^* \to \mathbf{P}^n$ extends to a morphism $\mathbf{P}^1 \to \mathbf{P}^n$, but the corresponding statement is not true for a morphism $k^* \to k^{n+1}$. In fact, if ϕ is such a morphism and is non-constant,

(a) any point of $\overline{\phi(k^*)} \setminus \phi(k^*)$ must be equal to one of $\lim_{t \to 0} \phi(t)$ and $\lim_{t \to \infty} \phi(t)$;

(b) $\phi(k^*)$ is closed in k^{n+1} if and only if neither limit exists.

Now suppose $G = k^*$ acts linearly on a projective variety X in $\mathbf{P}^n$. The induced linear action on k^{n+1} can be diagonalised. In other words there exists a basis $e_0, \ldots, e_n$ of k^{n+1} such that

$$t.e_i = t^{r_i} e_i$$

for some integers r_i. Now choose $\hat{x} \in k^{n+1}$ lying over x and write $\hat{x} = \sum \hat{x}_i e_i$, so that

$$t.\hat{x} = \sum t^{r_i} \hat{x}_i e_i,$$

and define

$$\mu(x) = \max\{-r_i : x_i \neq 0\}.$$

Thus $\mu(x)$ is the unique integer μ such that $\lim_{t \to 0} t^{\mu}(t.\hat{x})$ exists and $\neq 0$. (It is easy to see that $\mu(x)$ is independent of the choice of $\hat{x}$ and of the basis $e_0, \ldots, e_n$.) Note that

$$\mu(x) > 0 \Leftrightarrow \lim_{t \to 0} t.\hat{x} \text{ does not exist,}$$

$$\mu(x) = 0 \Leftrightarrow \lim_{t \to 0} t.\hat{x} \text{ exists and } \neq 0.$$

Similarly we define

$$\mu^-(x) = \max\{r_i : x_i \neq 0\}.$$

Then

$$\mu^-(x) > 0 \Leftrightarrow \lim_{t \to \infty} t.\hat{x} \text{ does not exist,}$$

$$\mu^-(x) = 0 \Leftrightarrow \lim_{t \to \infty} t.\hat{x} \text{ exists and } \neq 0.$$

The following proposition now follows immediately from Proposition 2.1, (a) and (b).

Proposition 2.2. *Let* $G = k^*$ *act linearly on a projective variety* X *in* $\mathbf{P}^n$. *Then, for any* $x \in X$,

(i) x *is semistable if and only if* $\mu(x) \geq 0$ *and* $\mu^-(x) \geq 0$;

(ii) x *is stable if and only if* $\mu(x) > 0$ *and* $\mu^-(x) > 0$.

2.3 The criterion

We now consider an arbitrary linear action of a reductive group G on a projective variety X.

Definition A 1-parameter subgroup (1-PS) of G is a non-trivial homomorphism of algebraic groups $\lambda : k^* \to G$.

It is a standard fact that the image of any homomorphism of algebraic groups is closed. It now follows easily from Proposition 2.1 that a stable (semistable) point of X for the action of G is necessarily stable (semistable) for the action of k^* induced by any 1-PS.

For any 1-PS λ, we write $\mu(x, \lambda)$ for the value of $\mu(x)$ for the action of k^* on X induced by λ. (It is not necessary to define $\mu^-(x, \lambda)$ because this is just $\mu(x, \lambda^{-1})$ where λ^{-1} is the 1-PS defined by $\lambda^{-1}(t) = \lambda(t^{-1})$.) It now follows from Proposition 2.2 that

- x semistable $\Rightarrow \mu(x, \lambda) \geq 0$ for every 1-PS λ of G;
- x stable $\Rightarrow \mu(x, \lambda) > 0$ for every 1-PS λ of G.

We can now state the Hilbert-Mumford Criterion.

Theorem 2.3. *Let* G *be a reductive group acting linearly on a projective variety* X *in* $\mathbf{P}^n$. *Then the two* $\Rightarrow$ *above can be replaced by* $\Leftrightarrow$.

In order to prove this theorem, one has to prove the following assertions.

- x non-stable $\Rightarrow \exists \lambda$ such that $\mu(x, \lambda) \leq 0$,
- x non-semistable $\Rightarrow \exists \lambda$ such that $\mu(x, \lambda) < 0$.

In view of Proposition 2.2, this is equivalent to showing that, if x is non-stable (non-semistable), then there is a 1-PS λ such that x is still non-stable (non-semistable) for the induced action of k^*. These statements are at least plausible if G has a plentiful supply of 1-PS; this is the significance of the hypothesis that G is reductive.

A proof for the semistable case when $G = \mathrm{SL}(p)$ and $k = \mathbf{C}$ was given by Hilbert using convergent power series. Mumford and Seshadri obtained a proof valid for all k and all reductive G using formal power series and a theorem of Iwahori. A proof for all k but restricted to the case $G = \mathrm{SL}(p)$ is given in [Tata] and also in [Muk].

2.4 An observation

The following remark can simplify computations.

Remark 2.1. It follows at once from the definition that

$$\mu(g.x, \lambda) = \mu(x, g^{-1}\lambda g)$$

for any $g \in G$. This allows us to replace λ by a convenient conjugate in making calculations. In particular, for $G = \mathrm{SL}(p)$, it is well known that every 1-PS is conjugate to a diagonal matrix of the form

$$\lambda(t) = \begin{pmatrix} t^{r_0} & 0 & \cdots & 0 \\ 0 & t^{r_1} & \cdots & 0 \\ \vdots & \vdots & \ddots & \vdots \\ 0 & \cdots & \cdots & t^{r_{p-1}} \end{pmatrix}, \tag{2.1}$$

with

$$\sum r_i = 0, \quad r_0 \geq r_1 \geq \cdots \geq r_{p-1}, \quad \text{not all } r_i = 0.$$

(This follows almost at once from the fact (already used) that any linear action of k^* is diagonalisable.) We deduce

Proposition 2.4. *Let* $\mathrm{SL}(p)$ *act linearly on a projective variety* X. *A point* x *of* X *is stable (semistable) for this action if and only if* $\mu(g.x, \lambda) > 0$ (≥ 0) *for every* $g \in \mathrm{SL}(p)$ *and every 1-PS* λ *of* $\mathrm{SL}(p)$ *of the form (2.1).*

3 Lecture 3 – Further details

3.1 Reductivity and finite generation

The key to the proofs of the results in Lecture 1 is the concept of a *reductive* group.

Note first that any linear algebraic group G possesses a unique maximal connected normal solvable subgroup, called its *radical*.

Definition. Let G be a linear algebraic group. Then G is

- *reductive* if its radical is isomorphic to a direct product of copies of k^*;

- *linearly reductive* if, for every linear action of G on k^n, and every invariant point v of k^n, $v \neq 0$, there exists a G-invariant homogeneous polynomial f of degree 1 such that $f(v) \neq 0$;
- *geometrically reductive* if, for every linear action of G on k^n, and every invariant point v of k^n, $v \neq 0$, there exists a G-invariant homogeneous polynomial f of degree ≥ 1 such that $f(v) \neq 0$.

Warning. Although these terms are now widely accepted, many different terms have been used in the past. In particular, our definition of "linearly reductive" is synonymous with "every rational representation of G is completely reducible". (In fact, this has often been used as a definition of "reductive".) Here "rational representation" is the same as our "linear action on k^n" and "completely reducible" means that k^n is a direct sum of G-invariant subspaces which themselves have no invariant subspaces.

It is easy to see that the groups $\mathrm{GL}(p)$, $\mathrm{SL}(p)$ and $\mathrm{PGL}(p)$ are all reductive, also $(k^*)^m$ for any positive integer m, but **not** k as an additive group. It is not difficult to see that every geometrically reductive group is reductive; and, in characteristic 0, it is a theorem of Weyl that every reductive group is linearly reductive (so in this case all three conditions coincide). In finite characteristic, however, there are very few linearly reductive groups; certainly $\mathrm{GL}(p)$, $\mathrm{SL}(p)$ and $\mathrm{PGL}(p)$ are not linearly reductive for $p \geq 2$.

The link between reductivity and finite generation is provided by

Nagata's Theorem. Let G be a geometrically reductive group acting on a finitely generated k-algebra R by algebra automorphisms such that every element of R is contained in a finite-dimensional linear subspace of R on which G acts linearly. Then R^G is finitely generated.

In order to justify our assertions in Lecture 1, we need also

Mumford's Conjecture. Every reductive group is geometrically reductive.

This was conjectured by Mumford in the 1st. edition of [GIT] and proved by Haboush [H] in 1974. In fact, geometric reductivity is the key not only to finite generation but also to establishing the geometric properties in Theorem 1.1.

Remark 3.1. (i) It is possible for $A(X)^G$ to be finitely generated even when G is not reductive. Much work has been done on the extension of GIT to non-reductive groups (see [DK] and Kirwan's article in this volume).

(ii) It is possible for a (non-affine) categorical quotient to exist even if $A(X)^G$ is not finitely generated.

(iii) As already noted (Example 1.3), the finite generation of $A(X)^G$ does not imply that the corresponding affine variety is even a categorical quotient.

3.2 *Proof of Theorem 1.1*

The hypotheses of Theorem 1.1 imply, by Nagata's Theorem, that $A(X)^G$ is a finitely generated k-algebra; moreover it is a subalgebra of $A(X)$ and therefore has no nilpotent elements. Let Y be the corresponding affine variety, and $\phi : X \to Y$ the morphism corresponding to the inclusion $A(Y) = A(X)^G \subset A(X)$. We prove that Y and ϕ have the required properties.

(i) Suppose that there exist $g \in G$, $x \in X$ with $\phi(g.x) \neq \phi(x)$. Then, since Y is affine, there exists $f \in A(Y)$ such that

$$f(\phi(g.x)) \neq f(\phi(x)).$$

This contradicts the fact that $f \in A(X)^G$.

(ii) Let $y \in Y$ and let $f_1, \ldots, f_r$ generate the maximal ideal in $A(Y) = A(X)^G$ corresponding to the point y. It is not difficult to prove that

$$\sum f_i A(X) \neq A(X)$$

(see [Tata, Lemma 3.4.2]). Hence there exists a maximal ideal of $A(X)$ containing $\sum f_i A(X)$. If x is the point of X corresponding to this maximal ideal, then we have $f_i(x) = 0$ for all i; so $\phi(x) = y$.

(iii) First note that it is sufficient to prove (iii) for the open sets U of any fixed basis. Hence we can suppose that $U = Y_f$ for some non-zero $f \in A(Y) = A(X)^G$. In this case, $\phi^{-1}(U) = X_f$; so we require to prove that

$$\left(A(X)^G \right)_f = \left(A(X)_f \right)^G$$

for any $f \in A(X)^G$, where

$$A(X)_f = A(X_f) = \{h/f^r : r \geq 0, h \in A(X)\}.$$

This is easy to verify directly.

(iv) and (v). For any two disjoint closed G-invariant subsets W_1, W_2 of X, there exists [Tata, Lemma 3.3] an element $f \in A(X)^G$ such that

$$f(W_1) = 0, \quad f(W_2) = 1.$$

Regarding f as an element of $A(Y)$, we obtain from this

$$f\left(\phi(W_1)\right) = 0, \quad f\left(\phi(W_2)\right) = 1;$$

hence

$$\overline{\phi(W_1)} \cap \overline{\phi(W_2)} = \emptyset.$$

This proves (v); moreover, if W is a closed G-invariant subset of X and $y \in \overline{\phi(W)} \setminus \phi(W)$, we can apply the above result with $W_1 = W$, $W_2 = \phi^{-1}(y)$ to get a contradiction.

3.3 Proof of Corollaries 1.2, 1.3, 1.4, Proposition 1.6

The proofs of the corollaries use only the results of Theorem 1.1, hence also give proofs of Proposition 1.6. The only difficult proof is that of Corollary 1.2, the basic reason being that, even for $U = Y$, one has to prove the universal property for morphisms into any variety, not just into an affine variety. A detailed proof is contained in [Tata, pp64, 65]. The fact that the result extends to any open set U of Y requires the following lemma, which is a useful extension of Theorem 1.1(v).

Lemma 3.1. *Let X, G, Y, ϕ be as in Theorem 1.1, and let U be an open subset of Y. If W_1, W_2 are disjoint G-invariant subsets of $\phi^{-1}(U)$ and are closed in $\phi^{-1}(U)$, then*

$$\phi(W_1) \cap \phi(W_2) = \emptyset.$$

Proof Let $\overline{W_i}$ denote the closure of W_i in X, and suppose that $y \in \phi(W_1) \cap \phi(W_2)$. Then $\phi^{-1}(y) \cap \overline{W_1}$ and $\overline{W_2}$ are closed G-invariant subsets of X whose images under ϕ both contain y. Hence by Theorem 1.1(v)

$$\phi^{-1}(y) \cap \overline{W_1} \cap \overline{W_2} \neq \emptyset.$$

But W_1 and W_2 are closed in $\phi^{-1}(U)$; so

$$\phi^{-1}(U) \cap \overline{W_1} \cap \overline{W_2} = W_1 \cap W_2 = \emptyset.$$

This gives the required contradiction. $\qquad\square$

The other two proofs are straightforward.

Proof [Proof of Corollary 1.3] Take $W_i = \overline{G.x_i}$ in Theorem 1.1(v). $\quad\square$

Proof [Proof of Corollary 1.4] We need to show that, if $x_1, x_2 \in \phi^{-1}(U)$ do not belong to the same orbit, then $\phi(x_1) \neq \phi(x_2)$. This follows at once by taking $W_i = G.x_i$ in Lemma 3.1. $\qquad\square$

3.4 Proof of Theorem 1.7

(i) We begin by recalling some general facts about projective varieties and graded k-algebras.

Let $\hat{X}$ denote the affine variety in k^{n+1} lying over X; for each homogeneous element f of $A(\hat{X})$ of degree ≥ 1 we have an affine open subset $X_f = \{x \in X : f(x) \neq 0\}$ of X. Let R be a finitely generated homogeneous subalgebra of $A(\hat{X})$ and write X_R for the union of those X_f for which $f \in R$. Then there exists a projective variety Y (usually denoted by $\operatorname{Proj} R$) and a morphism $\phi : X_R \to Y$ such that

(a) Y is covered by affine open sets Y_f, one for each homogeneous element f of R of degree ≥ 1, and $A(Y_f)$ is isomorphic to the algebra

$$(R_f)_0 = \{h/f^r : r \geq 0, h \in R \text{ homogeneous of degree } r(\deg f)\};$$

(b) $\phi^{-1}(Y_f) = X_f$ and $\phi : X_f \to Y_f$ is the morphism corresponding to the inclusion of $(R_f)_0$ in $A(X_f) = \left(A(\hat{X})_f\right)_0$.

In our situation, note that the linear action of G on X induces an action on $A(\hat{X})$ which preserves the degree of every homogeneous element. Hence $R = A(\hat{X})^G$ is a homogeneous subalgebra of $A(\hat{X})$, and it is also finitely generated by Nagata's Theorem. Let Y and ϕ be as constructed above, and note that $X_R = X^{ss}$. By Theorem 1.1 and (b) above, together with the fact that $\left(\left(A(\hat{X})^G\right)_f\right)_0 = \left(\left(A(\hat{X})_f\right)_0\right)^G$, it follows that (Y_f, ϕ) is a good quotient of X_f for every homogeneous element f of $A(\hat{X})^G$ of degree ≥ 1. It follows easily that (Y, ϕ) is a good quotient of X^{ss}.

(ii) Put $Y^s = \phi(X^s)$ and let Y^0 be the union of those Y_f for which G acts on X_f with closed orbits. Clearly $X^s \subset \phi^{-1}(Y^0)$ and so $Y^s \subset Y^0$. Now by Proposition 1.6, (Y^0, ϕ) is a geometric quotient of $X^0 = \phi^{-1}(Y^0)$. It follows that $X^s = \phi^{-1}(Y^s)$ and

$$Y^0 \setminus Y^s = \phi(X^0 \setminus X^s).$$

Hence $Y^0 \setminus Y^s$ is closed in Y^0 by condition (iv) for a good quotient, and so Y^s is open in Y^0 and therefore also in Y. It follows that (Y^s, ϕ) is a good quotient of X^s and hence a geometric quotient.

(iii) follows at once from Proposition 1.6(ii).

(iv) If $x \in X^s$, then

$$\phi^{-1}\left(\phi(x)\right) \subset \phi^{-1}(Y^s) = X^s.$$

Since $\phi^{-1}\left(\phi(x)\right)$ is closed in X^{ss}, it follows that

$$\overline{G.x} \cap X^{ss} \subset X^s.$$

But G acts on X^s with closed orbits by (ii); hence $G.x$ is closed in X^s and therefore also in X^{ss}.

For the converse, note that, since $x \in X^{ss}$, there exists a G-invariant homogeneous polynomial f of degree ≥ 1 such that $x \in X_f$. If $G.x$ is closed in X^{ss}, then certainly $G.x$ is closed in X_f. It follows easily from this, and the fact that x has finite stabiliser, that x is stable (see [Tata, Lemma 3.15]).

3.5 Linearisation

The ideas of Lecture 1 can be slightly generalised and the definitions of stability and semistability placed in a more natural context. The point is that, if we are given an action of an algebraic group G on a variety X, there are many ways of embedding X in a projective space and linearising the action of G.

So let X be any variety, G an algebraic group acting on X, and L a line bundle over X; we denote by $p : L \to X$ the natural projection. A *linearisation* of the action of G *with respect to* L is an action of G on L such that

(i) for all $y \in L$, $g \in G$,

$$p(g.y) = g.p(y);$$

(ii) for all $x \in X$, $g \in G$, the map

$$L_x \to L_{g.x} : y \mapsto g.y$$

is linear.

An *L-linear action* of G on X is an action of G together with a linearisation of this action with respect to L.

This concept allows us to generalise the definitions of stable and semistable. Note first that an L-linear action of G on X determines an action on the space of sections of any tensor power L^r of L. Moreover, for any G-invariant section f of L^r,

$$X_f = \{x \in X : f(x) \neq 0\}$$

is an open set of X, invariant under G.

Definitions. Let X be a variety and L a line bundle over X. For any L-linear action of a reductive group G on X, a point x is called

- *semistable* if, for some positive integer r, there exists a G-invariant section f of L^r such that $f(x) \neq 0$ and X_f is affine;
- *stable* if x has finite stabiliser and, for some positive integer r, there exists a G-invariant section f of L^r such that $f(x) \neq 0$, X_f is affine and G acts on X_f with closed orbits.

We write $X^{ss}(L)$ ($X^s(L)$) for the set of semistable (stable) points, the dependence on the linearisation of the action of G being understood.

Example 3.1. Let X be a projective variety in $\mathbf{P}^n$ and L the restriction to X of the hyperplane bundle $\mathcal{O}(1)$. Any linear action of G on X then induces an L-linear action, so both sets of definitions make sense. Fortunately, in this situation,

$$X^{ss}(L) = X^{ss} \text{ and } X^s(L) = X^s.$$

This follows immediately from the standard fact that, for sufficiently large r, all sections of L^r are given by homogeneous polynomials of degree r.

Theorem 1.7 extends to our new situation and gives

Theorem 3.2. *Let X be a variety and L a line bundle over X. Then, for any L-linear action of a reductive group G on X,*

(i) *there exists a good quotient $\phi : X^{ss}(L) \to Y$ and $Y = X^{ss}(L)//G$ is quasi-projective;*

(ii) *there exists an open subset Y^s of Y such that $\phi^{-1}(Y^s) = X^s(L)$ and $Y^s = X^s(L)/G$ is a geometric quotient of $X^s(L)$;*

(iii) *for $x_1, x_2 \in X^{ss}(L)$, $\phi(x_1) = \phi(x_2) \Leftrightarrow \overline{G.x_1} \cap \overline{G.x_2} \cap X^{ss}(L) \neq \emptyset$;*

(iv) *for $x \in X^{ss}(L)$, x is stable if and only if x has finite stabiliser and $G.x$ is closed in $X^{ss}(L)$.*

Note that no assumption of projectivity (or quasi-projectivity) is made on X. The fact that Y is quasi-projective depends on the fact that $X^{ss}(L)$ is quasi-projective. This in turn depends on the concept of an ample line bundle (one for which some power L^r determines an embedding in a projective space) and the fact that the restriction of L to $X^{ss}(L)$ is always ample.

4 Lecture 4 – Examples

4.1 An elementary example

Example 4.1. Let k^* act on $\mathbf{P}^n$ $(n \geq 2)$ by

$$t.(x_0 : \ldots : x_n) = (tx_0 : \ldots, tx_{n-1} : t^{-n} x_n).$$

Writing $x = (x_0 : \ldots : x_n)$, we have

$$\mu(x) = \begin{cases} n & \text{if } x_n \neq 0 \\ -1 & \text{if } x_n = 0 \end{cases}$$

and

$$\mu^-(x) = \begin{cases} 1 & \text{if } x_0, \ldots, x_{n-1} \text{ are not all } 0 \\ -n & \text{if } x_0 = \ldots = x_{n-1} = 0. \end{cases}$$

So, by Proposition 2.2,

- x is stable if $x_n \neq 0$ and $x_0, \ldots, x_{n-1}$ are not all 0;
- x is not semistable if $x = (x_1 : \ldots : x_{n-1} : 0)$ or $x = (0 : \ldots : 0 : 1)$.
- there are no strictly semistable points.

Thus $(\mathbf{P}^n)^s = (\mathbf{P}^n)^{ss}$ can be identified with $k^n \setminus \{0\}$ and the action of k^* becomes

$$t.(x_0, \ldots, x_{n-1}) = (t^{n+1} x_0, \ldots, t^{n+1} x_{n-1}).$$

So

$$(\mathbf{P}^n)^s / k^* = (\mathbf{P}^n)^{ss} /\!/ k^* = \mathbf{P}^{n-1}.$$

(Of course the action on $k^n \setminus \{0\}$ is not quite the standard one but the quotient is the same. In the full theory the description above corresponds to constructing the quotient along with an ample line bundle, which in this case is $\mathcal{O}(n)$.)

Note that the orbits of stable points are not closed in $\mathbf{P}^n$. In fact, the closure of the orbit of $(x_0 : \ldots : x_n)$, where $x_0, \ldots, x_{n-1}$ are not all 0 and $x_n \neq 0$, contains the non-stable points $(x_0 : \cdots : x_{n-1} : 0)$ and $(0 : \ldots : 0 : 1)$. This is typical.

4.2 Binary forms

Example 4.2. We consider the vector space V_n of homogeneous polynomials in x_0, x_1 of degree n, i. e.

$$V_n = \left\{ f = \sum a_i x_0^{n-i} x_1^i : a_i \in k \right\}.$$

A *binary form* of degree n is an element of $\mathbf{P}(V_n)$.

Any binary form can be represented by a non-zero $f \in V_n$, which can be factorised into linear factors and so determines a set of n points (counted with multiplicities) in $\mathbf{P}^1$. Conversely a set of n points (with multiplicities) in $\mathbf{P}^1$ determines a binary form. It follows that the natural action of $\mathrm{SL}(2)$ on $\mathbf{P}^1$ induces an action on $\mathbf{P}(V_n)$. This action has a natural linearisation. We wish to determine the stable and semistable points using the Hilbert-Mumford criterion.

Any 1-PS of $\mathrm{SL}(2)$ is conjugate to one of the form λ_r for some $r \geq 1$, where

$$\lambda_r(t) = \begin{pmatrix} t^r & 0 \\ 0 & t^{-r} \end{pmatrix}.$$

By Proposition 2.4, a binary form is non-stable (non-semistable) if and only if it is equivalent under the action of $\mathrm{SL}(2)$ to one for which $\mu(f, \lambda_r) \leq 0 (< 0)$ for some r. Now we see by direct computation that

$$\lambda_r(t).\left(\sum a_i x_0^{n-i} x_1^i\right) = \sum t^{r(2i-n)} a_i x_0^{n-i} x_1^i.$$

So the action of λ_r is diagonal with respect to the obvious basis of V_n, and

$$\mu := \mu\left(\sum a_i x_0^{n-i} x_1^i, \lambda_r\right) = r(n - 2i_0),$$

where i_0 is the smallest value of i for which $a_i \neq 0$. Thus $\mu \leq 0 (< 0)$ if and only if $a_i = 0$ whenever $i < \frac{n}{2}$ $(i \leq \frac{n}{2})$, i.e. if and only if the point $(1 : 0)$ occurs as a point of multiplicity $\geq \frac{n}{2}$ $(> \frac{n}{2})$ for the given form. Taking account of the action of $\mathrm{SL}(2)$, we deduce that a binary form is

- stable if and only if no point of $\mathbf{P}^1$ occurs as a point of multiplicity $\geq \frac{n}{2}$;
- semistable if and only if no point of $\mathbf{P}^1$ occurs as a point of multiplicity $> \frac{n}{2}$.

Note that, when n is odd, the stable and semistable points coincide, and there is a projective geometric quotient. When n is even, there exist strictly semistable points and the geometric quotient of the stable points is not projective.

4.3 Cubic curves

Example 4.3. A cubic curve in $\mathbf{P}^2$ is defined by a non-zero polynomial

$$f = \sum a_{ij} x_0^{3-i-j} x_1^i x_2^j,$$

the summation being taken over all i, j with $i \geq 0$, $j \geq 0$, $i + j \leq 3$. The curve determines f up to a scalar multiple, so the set of cubic curves has a natural structure of a projective space $\mathbf{P}^9$ with coordinates given by the a_{ij}. The group $SL(3)$ acts naturally through its action on $\mathbf{P}^2$, which has a natural linearisation.

Before starting on the analysis of stability, we shall summarise information about singular points of the curve defined by f. We concentrate on a singular point at $(1 : 0 : 0)$, since we can translate this information to other points through the action of $SL(3)$.

- $(1 : 0 : 0)$ is singular if and only if $a_{00} = a_{10} = a_{01} = 0$;
- $(1 : 0 : 0)$ is a triple point if and only if $a_{00} = a_{10} = a_{01} = a_{20} = a_{11} = a_{02} = 0$;
- if $(1 : 0 : 0)$ is a double point, then the tangents at $(1 : 0 : 0)$ are the lines defined by the equation

$$a_{20}x_1^2 + a_{11}x_1x_2 + a_{02}x_2^2 = 0;$$

- in particular, $(1 : 0 : 0)$ is a double point with a unique tangent $x_2 = 0$ if and only if $a_{00} = a_{10} = a_{01} = a_{20} = a_{11} = 0$ and $a_{02} \neq 0$.

Any 1-PS of $SL(3)$ is conjugate to one of the form λ, where

$$\lambda(t) = \begin{pmatrix} t^{r_0} & 0 & 0 \\ 0 & t^{r_1} & 0 \\ 0 & 0 & t^{r_2} \end{pmatrix},$$

with r_i not all 0, $r_0 \geq r_1 \geq r_2$ and $r_0 + r_1 + r_2 = 0$. In a similar way to that of Example 4.2, we have

$$\lambda(t).f = \sum t^{(i+j-3)r_0 - ir_1 - jr_2} a_{ij} x_0^{3-i-j} x_1^i x_2^j$$

and hence

$$\mu(f, \lambda) = \max\{(3 - i - j)r_0 + ir_1 + jr_2 : a_{ij} \neq 0\}.$$

By Proposition 2.4, we know that a cubic curve is non-stable (non-semistable) if and only if it is equivalent under the action of $SL(3)$ to one for which $\mu(f, \lambda) \leq 0 (< 0)$ for some λ of the above form.

We consider first the condition for semistability. It is easy to check that, if $\mu(f, \lambda) < 0$, then

$$a_{00} = a_{10} = a_{01} = a_{20} = a_{11} = 0.$$

Conversely, if these conditions hold, and we take $r_0 = 3$, $r_1 = -1$, $r_2 = -2$, then $\mu(f, \lambda) < 0$. Hence $\mu(f, \lambda) < 0$ for some λ of the above form if and only if the cubic curve has either a triple point at $(1 : 0 : 0)$ or a double point at $(1 : 0 : 0)$ with unique tangent $x_2 = 0$. By Proposition 2.4, we deduce that a cubic curve is semistable if and only if it has no triple point and no double point with a unique tangent.

Next, suppose that f has a singular point; we can assume this to be the point $(1 : 0 : 0)$, so that $a_{00} = a_{10} = a_{01} = 0$. If we take $r_0 = 2$, $r_1 = -1$, $r_2 = -1$, we get $\mu(f, \lambda) \leq 0$; so f is non-stable.

Conversely, suppose that $\mu(f, \lambda) \leq 0$ for some λ. Then certainly $a_{00} = a_{10} = 0$. If $a_{01} \neq 0$, then

$$\mu(f, \lambda) \leq 0 \Rightarrow 2r_0 + r_2 \leq 0 \Rightarrow r_1 = r_0, \ r_2 = -2r_0.$$

For these values of the r_i, we have

$$\mu(f, \lambda) = \max\{(3 - 3j)r_0 : a_{ij} \neq 0\};$$

so $\mu(f, \lambda) \leq 0$ if and only if $a_{i0} = 0$ for all i. This implies that $f = x_2 f'$ for some f' of degree 2; hence f is singular at every point for which $x_2 = f = 0$. Thus in all cases f has a singular point, and we conclude that a cubic curve is stable if and only it is non-singular.

To summarise, the stable cubic curves are the non-singular ones, while the strictly semistable ones have one or more double points with distinct tangents. In fact, there are just three types of strictly semistable cubic curves, each forming a single orbit:

- nodal cubics, i. e. irreducible cubics possessing a double point with distinct tangents;
- cubics which degenerate into a conic and a line not tangent to it;
- cubics which degenerate into three non-concurrent lines.

We have $(\mathbf{P}^9)^{ss}//\mathrm{SL}(3) \cong \mathbf{P}^1$, with $(\mathbf{P}^9)^s/\mathrm{SL}(3) \cong k$ and the three strictly semistable orbits going to the "point at infinity". For further details, see [Tata, Chapter 4].

5 Appendix – Moduli of vector bundles on an algebraic curve

5.1 Moduli and quotients

Mumford's main motivation in introducing GIT was to use it to solve moduli problems in algebraic geometry, that is problems of classification of algebro-geometric objects (e. g. curves, vector bundles on a fixed projective variety, sets of points in $\mathbf{P}^n$) under some equivalence relation (e. g. isomorphism of curves, isomorphism of vector bundles, projective equivalence).

More precisely, a **moduli problem** concerns a set of objects A in algebraic geometry with some fixed **discrete invariants** (in the examples mentioned above, the discrete invariants are the genus of a curve, the Hilbert polynomial of a vector bundle, number of points) together with an equivalence relation $\sim$ on A. We need a concept of a **family** of objects parametrised by a variety (or scheme) S and an equivalence relation on families extending the given equivalence relation on A. Moreover, given a morphism $T \to S$, we must be able to pull back a family parametrised by S to one parametrised by T. We then define a contravariant functor

$$\mathbf{F} : \{\text{varieties}\} \to \{\text{sets}\}$$
$$\mathbf{F}(S) := \{\text{equivalence classes of families parametrised by } S\}.$$

We would like $\mathbf{F}$ to be **representable**, i.e. for there to exist a variety (or scheme) M which represents the functor; this means that there exists a **universal family** $\mathcal{U}$ parametrised by M such that any family parametrised by S is equivalent to the pullback of $\mathcal{U}$ by a unique morphism $S \to M$. In this case we call M a **fine moduli space**. However, we frequently have to settle for a lesser type of universality (**corepresentability**) in which case M is a **coarse moduli space**; this still allows the property that the (closed) points correspond bijectively to the equivalence classes of objects. Thus any family parametrised by S gives rise to a unique map $S \to M$ and this map is a morphism. For more details, see [Tata].

When attempting to solve a moduli problem by GIT, it is usually not possible to find a moduli space for the whole of A. We need to introduce concepts of **stability** and **semistability** for the objects of A (this is not a priori the same thing as (semi)stability in GIT). We can then proceed as follows.

- Find a family of objects in A parametrised by a variety (or scheme) R, together with an action of a reductive group G on R, such that all semistable objects of A are isomorphic to members of the family, the family has a local universal property and two objects are equivalent if and only if the corresponding points of R lie in the same orbit;
- linearise the action of G on R by embedding R in $\mathbf{P}^N$ as a quasi-projective variety such that the action of G extends to a linear action on $\mathbf{P}^N$;
- show that, for a suitable choice of R and the linearisation, (semi)stability in A corresponds to (semi)stability for the action of G on R.

It will then follow that R^s/G will be a quasi-projective coarse moduli space for the stable objects in A. If the closure Q of R in $\mathbf{P}^N$ has the property that $Q^{ss} = R^{ss}$, then $R^{ss}//G$ will be projective by Theorem 1.7 and is often referred to as the moduli space of semistable objects in A (strictly speaking, one should say "moduli space of S-equivalence classes of semistable objects", where S-equivalence may be defined by the property that the closures of the orbits in R^{ss} intersect).

5.2 Classification of coherent sheaves

One situation to which the methods of 5.1 have been applied is the classification of coherent sheaves on a projective variety (or scheme) X. This was first outlined by Mumford for vector bundles on a smooth curve, details being completed by Narasimhan and Seshadri [NS, Ses1, Ses2]. The method extends to singular curves (see [Tata]), but not to higher dimensions. Gieseker [G] introduced an alternative method which works for surfaces and was extended to arbitrary projective schemes by Maruyama [M]. Later Simpson [Sim] gave a more direct argument for projective schemes; it is his method that we shall present below for the case of smooth curves.

The basic result is that there exists a projective coarse moduli space for (S-equivalence classes of) semistable sheaves with fixed Hilbert polynomial on a projective scheme X. Subject to a coprimality condition on the coefficients of the Hilbert polynomial, stability and semistability are equivalent and we obtain a projective fine moduli space for the stable sheaves.

A variant of Simpson's method which is more direct and functorial has recently been introduced by Álvarez-Cónsul and King [AK]; an expository account is included in this volume.

5.3 Vector bundles on a curve

In this and the following sections, we describe the construction of the moduli spaces of stable bundles on an algebraic curve. Let C be an irreducible non-singular projective algebraic curve of genus g, defined over an algebraically closed field. We consider algebraic vector bundles E over C of rank r and degree d. (The *degree* of a vector bundle E on an algebraic curve may be defined as follows. Let s be a rational section of the line bundle $\det E$; then $\deg E = N - P$, where N and P

are the numbers of zeroes and poles of s, counted with multiplicities –
this number is independent of the choice of s.)

Definition.

(i) A *family* of vector bundles over C *parametrised by* a variety S is a
vector bundle V over $C \times S$; for any $s \in S$, we write $V_s = V|C \times \{s\}$ and refer to this as the *member* of the family corresponding
to s.

(ii) For any family V parametrised by S and any morphism $f : S' \to S$, we have an *induced family* $(id_C \times f)^*V$ parametrised by S'.

(iii) Two families V_1, V_2 parametrised by S are *equivalent* if $V_1 \cong V_2 \otimes p_S^* L$ for some line bundle L over S.

We can now define the contravariant functor **F** as indicated in 5.1.

Definition. A vector bundle E over C is *stable* (*semistable*) if, for every
proper subbundle F of E,

$$\frac{\deg F}{\operatorname{rk} F} < (\leq) \frac{\deg E}{\operatorname{rk} E}.$$

Remark 5.1. One can show that, in any family of bundles, the subsets
of stable and semistable members of the family correspond to open subsets of the parametrising variety. This is much harder than Lemma 1.5
and indeed was first proved by constructing moduli spaces as described
below and using an a posteriori argument.

5.4 Sheaves and cohomology

The correspondence between vector bundles and sheaves is fundamental
to the study of vector bundles. Let $\mathcal{O}$ denote the sheaf of local rings of
C. By considering local sections of a vector bundle E (that is, sections
defined over open subsets of C), we obtain the sheaf of sections of E.
This sheaf is a module over $\mathcal{O}$ and as such is **locally free**. Conversely
any locally free sheaf determines a vector bundle, unique up to isomorphism. It is conventional to identify the concepts of vector bundle and
locally free sheaf, and we shall do this.

It is sometimes necessary to use the bigger category of **coherent**
sheaves over C; unlike locally free sheaves, these form an abelian category. For our case, where C is a non-singular curve, a coherent sheaf
is locally free if and only if it is torsion-free; thus every coherent sheaf
is the direct sum of a locally free sheaf and a torsion sheaf – and the
latter is necessarily supported on a finite set of points of C. Note that

the concepts of rank and degree can be extended to coherent sheaves; in particular a sheaf of rank 0 is precisely a torsion sheaf.

The introduction of sheaves allows us to use sheaf cohomology. In our case we have just H^0 and H^1, where H^0 is the space of global sections. Since C is projective, H^0 and H^1 are finite-dimensional as vector spaces, and we write $h^i = \dim H^i$. We have the following fundamental results, valid for coherent sheaves over C, which as usual is an irreducible non-singular projective curve of genus g. We denote by K the **canonical** line bundle on C (that is, the dual of the tangent bundle).

Riemann-Roch Theorem. For any coherent sheaf E of rank r and degree d over C,

$$h^0(E) - h^1(E) = d + r(1 - g).$$

Serre Duality Theorem. Let E^* denote the dual sheaf $Hom(E, \mathcal{O})$. There is a natural duality of vector spaces between $H^1(E)$ and $H^0(E^* \otimes K)$. In particular

$$h^1(E) = h^0(E^* \otimes K).$$

Exercise. Using these theorems, show that

$$\deg K = 2g - 2, \quad h^0(K) = g.$$

5.5 Stable bundles as quotient sheaves

Lemma 5.1.   (i) *Every line bundle is stable.*

(ii) *If E is stable (semistable) and L is a line bundle, then $E \otimes L$ is stable (semistable).*

Exercise. Prove Lemma 5.1.

Definition. A bundle E is *generated* if the evaluation map $H^0(E) \to E_x$ is surjective for all $x \in C$, or equivalently if there is an exact sequence

$$0 \to F \to V_0 \otimes \mathcal{O} \to E \to 0,$$

where V_0 is a vector space.

Lemma 5.2. *Let E be a semistable bundle of rank r and degree d with $d > r(2g - 1)$. Then*

- *E is generated;*
- *$h^1(E) = 0$.*

Exercise. Prove Lemma 5.2.

Now fix r, d with $d > r(2g - 1)$. Assuming E is semistable, we have by Lemma 5.2 an exact sequence

$$0 \to F \to V_0 \otimes \mathcal{O} \to E \to 0,$$

where the induced linear map $V_0 \to H^0(E)$ is an isomorphism and $h^1(E) = 0$. So by Riemann-Roch

$$p := \dim V_0 = h^0(E) = d + r(1 - g).$$

The set of all sheaf quotients of $V_0 \otimes \mathcal{O}$ of rank r and degree d gives Grothendieck's Quot scheme Q [Gro2], which is a projective scheme. Moreover there exists an exact sequence

$$0 \to \mathcal{F} \to V_0 \otimes \mathcal{O}_{C \times Q} \to \mathcal{E} \to 0$$

on $C \times Q$, flat over Q, whose restriction to $C \times \{q\}$ is the sequence

$$0 \to \mathcal{F}_q \to V_0 \otimes \mathcal{O} \to \mathcal{E}_q \to 0$$

corresponding to the point $q \in Q$ (we refer to this as the **universal quotient sequence**). (Flatness here is a technical condition which ensures that pull-backs behave well.) Note that, if we identify $\mathrm{GL}(p)$ with the automorphism group of V_0, then $\mathrm{GL}(p)$ acts on Q; moreover scalar matrices act trivially, so the action goes down to an action of $\mathrm{PGL}(p)$ on Q.

We call a point $q \in Q$ **good** if the induced map $V_0 \to H^0(\mathcal{E}_q)$ is an isomorphism.

Lemma 5.3. *Let q_1, q_2 be good points of Q. Then $\mathcal{E}_{q_1} \cong \mathcal{E}_{q_2}$ if and only if q_1, q_2 lie in the same orbit for the action of $\mathrm{PGL}(p)$ on Q.*

Exercise. Prove Lemma 5.3.

Now define

$$R := \{q \in Q : \mathcal{E}_q \text{ is locally free and } q \text{ is good}\}.$$

It is easy to see that R is an open $\mathrm{PGL}(p)$-invariant subset of Q. This completes the first part of the construction of the moduli space.

5.6 Linearisation of Q and computation of stability

The next step is to embed R in $\mathbf{P}^N$. We start by fixing a line bundle $\mathcal{O}(1)$ of degree 1 on C and defining $E(n) := E \otimes \mathcal{O}(1)^n$ for every integer n. We require the following lemma.

Lemma 5.4. _For any family V of coherent sheaves parametrised by S, there exists an integer n_0 such that, for all $n \geq n_0$, $h^1(V_s(n)) = 0$ for all $s \in S$._

Applying Lemma 5.4 to the universal quotient sequence, we can choose an integer n_0 such that, for $n \geq n_0$, we have $h^1(\mathcal{F}_q(n)) = h^1(\mathcal{E}_q(n)) = 0$ for all $q \in Q$. We therefore have surjections

$$V_0 \otimes H^0(\mathcal{O}(n)) \to H^0(\mathcal{E}_q(n))$$

with $\dim H^0(\mathcal{E}_q(n)) = p + rn$ for all $q \in Q$. From this, we obtain a morphism

$$\phi_n : Q \to \mathrm{Gr},$$

where $\mathrm{Gr} := \mathrm{Gr}(p+rn, V_0 \otimes H^0(\mathcal{O}(n)))$ is the Grassmannian of $(p+rn)$-dimensional quotient spaces of $V_0 \otimes H^0(\mathcal{O}(n))$. The group $\mathrm{PGL}(p) = \mathrm{PGL}(V_0)$ acts naturally on Q and Gr and ϕ_n is equivariant. If we embed Gr in a projective space $\mathbf{P}^N$ by means of the Plücker coordinates, the action of $\mathrm{PGL}(p)$ extends to an action on $\mathbf{P}^N$ and the corresponding action of $\mathrm{SL}(p)$ is linear.

Proposition 5.5. _(Grothendieck) There exists an integer $n_1 \geq n_0$ (depending on r and d) such that, for $n \geq n_1$, ϕ_n is an embedding._

Thus Theorem 1.7 allows us to construct $Q^{ss}//\mathrm{SL}(p)$ and $Q^s/\mathrm{SL}(p)$.

It remains to match the stability condition for bundles with the GIT definition for the action of $\mathrm{SL}(p)$ on Q.

Proposition 5.6. _There exists an integer d_0 such that, for all $d \geq d_0$ and sufficiently large n (depending on r and d),_

(i) _if $\mathcal{E}_q$ is stable (semistable), then $\phi_n(q)$ is stable (semistable);_
(ii) _if $\phi_n(q)$ is semistable, then $q \in R$;_
(iii) _if $\phi_n(q)$ is stable (semistable), then $\mathcal{E}_q$ is stable (semistable)._

The proof may be found in [LeP] and involves a lengthy computation with the Hilbert-Mumford criterion.

Proposition 5.6 allows us to define the moduli spaces of stable (semistable) bundles on C for $d \geq d_0$ by

$$M(r,d) := Q^s/\mathrm{SL}(p) \quad (\widetilde{M}(r,d) := Q^{ss}//\mathrm{SL}(p)).$$

This definition can be extended to arbitrary d using Lemma 5.1(ii).

5.7 Consequences

It follows from the infinitesimal theory of Quot schemes that

- $M(r,d)$ is non-singular;
- if $M(r,d)$ is non-empty, then $\dim M(r,d) = n^2(g-1) + 1$.

Moreover

- $\widetilde{M}(r,d)$ is a normal projective variety;
- if r and d are coprime, then $M(r,d) = \widetilde{M}(r,d)$ (this is a purely numerical deduction); hence $M(r,d)$ is a non-singular projective variety.

We can describe in bundle theoretic terms when two semistable bundles give rise to the same point of $\widetilde{M}(r,d)$. In fact, any semistable bundle E possesses a *Jordan-Hölder filtration*

$$0 = E_0 \subset E_1 \subset E_2 \subset \ldots \subset E_m = E$$

by subbundles E_i, where E_i/E_{i-1} is a stable bundle of rank r_i and degree d_i with $d_i/r_i = d/r$ for $1 \le i \le m$. This filtration is not unique, but the associated graded object

$$\mathrm{gr}(E) := \bigoplus_{i=1}^{m} E_i/E_{i-1}$$

is determined by E.

Definition. Two semistable bundles E_1, E_2 are *S-equivalent* if $\mathrm{gr}(E_1) = \mathrm{gr}(E_2)$.

The following proposition is not hard to prove.

Proposition 5.7. *For sufficiently large d and n (depending on d), two semistable bundles of rank r and degree d map to the same point of $\widetilde{M}(r,d)$ if and only if they are S-equivalent.*

From the general theory, $M(r,d)$ is a coarse moduli space for stable bundles of rank r and degree d on C. We can ask whether it is a fine moduli space, that is whether there exists a universal bundle.

Proposition 5.8. *$M(r,d)$ is a fine moduli space if and only if $(r,d) = 1$.*

The problem with constructing a universal bundle and hence proving this proposition is that the action of $\mathrm{PGL}(p)$ on R does not lift to the universal quotient $\mathcal{E}$. However the action of $\mathrm{GL}(p)$ does lift in such a way that the scalar matrix λI acts by multiplication by λ. It turns out that, when $(r,d) = 1$, one can use $\mathcal{E}$ to construct a line bundle L on R

to which the action of GL(p) lifts in such a way that λI acts by λ^{-1}. We can now replace $\mathcal{E}$ by $\mathcal{E} \otimes p_R^* L$, which as a family is equivalent to $\mathcal{E}$ and on which PGL(p) acts. A descent argument now gives a universal bundle on $C \times M(r, d)$. This argument fails if $(r, d) \neq 1$, and Ramanan [Ram] showed that in this case the universal bundle does not exist.

Finally, none of this proves the most basic property of all, namely non-emptiness. In fact,

- if $g \geq 2$, $M(r, d)$ is non-empty for all r, d;
- if $g = 1$, $M(r, d)$ is non-empty if and only if $(r, d) = 1$ [A];
- if $g = 0$, $M(r, d)$ is non-empty if and only if $r = 1$; in fact in this case every vector bundle is a direct sum of line bundles [Gro1].

Bibliography

[Tata] P. E. Newstead, Introduction to moduli problems and orbit spaces, TIFR, Bombay, 1978 (out of print but many people have copies).

[GIT] D. Mumford, J. Fogarty and F. Kirwan, Geometric invariant theory, 3rd. edition, Ergebnisse Math. 34, Springer-Verlag, Berlin, 1994 (the 1st. edition, published in 1965, is the "bible" on the subject).

[Dol] I. V. Dolgachev, Lectures on invariant theory, LMS Lecture Notes Series 296, CUP 2003.

[Tho] R. P. Thomas, Notes on GIT and symplectic reduction for bundles and varieties, arXiv:math.AG/0512411 v3, 2006, to appear in Surveys in Differential Geometry 10 (2006): A Tribute to Professor S.-S. Chern (this is mainly useful for explaining the link between GIT and symplectic reduction, a topic not covered here).

[Pol] P. E. Newstead, Vector bundles on algebraic curves (my "Polish" notes, available from http://www.mimuw.edu.pl/~jarekw/EAGER/Lukecin02. html) (Lecture 1 in these notes is similar to Lecture 1 in the Polish notes and to a lecture given in the Boston/Tufts Algebraic Geometry Seminar, October 2006; the appendix is an adapted version of Lecture 2 in the Polish notes and was also presented in the Boston/Tufts seminar).

[Muk] S. Mukai, An introduction to invariants and moduli (translated by W. M. Oxbury), Cambridge Studies in Advanced Mathematics 81, CUP 2003 (includes classical invariant theory as well as GIT and moduli of vector bundles on curves).

[LeP] J. Le Potier, Lectures on vector bundles (translated by A. Maciocia), Cambridge Studies in Advanced Mathematics 54, CUP 1997 (includes a chapter on GIT).

[A] M. F. Atiyah, *Vector bundles over an elliptic curve,* Proc. London Math. Soc. **7** (1957), 414–452.

[AK] L. Álvarez-Cónsul and A. King, *A functorial construction of moduli of sheaves,* Invent. Math. **168** (2007), 613–666.

[DK] B. Doran and F. Kirwan, *Towards non-reductive geometric invariant theory,* Pure Appl. Math. Q. **3** (2007), 61–105.

[G] D. Gieseker, *On the moduli of vector bundles on an algebraic surface,* Ann. of Math. **106** (1977), 45–60.

[Gro1] A. Grothendieck, *Sur la classification des fibrés holomorphes sur la sphère de Riemann,* Amer. J. Math. **79** (1957), 121–138.

[Gro2] A. Grothendieck, *Techniques de construction et théorèmes d'existence en géométrie algébrique IV. Les schémas de Hilbert* (Séminaire Bourbaki 1960/61, Exp. 221), Séminaire Bourbaki Vol. 6, 249–276, Soc. Math. France, Paris, 1995.

[H] W. J. Haboush, *Reductive groups are geometrically reductive,* Ann. of Math. **102** (1975), 67–83.

[M] M. Maruyama, *Moduli of stable sheaves I, II,* J. Math. Kyoto Univ. **17** (1977), 91–126, **18** (1978), 557–614.

[N1] M. Nagata, *On the fourteenth problem of Hilbert,* Proc. Internat. Congress of Math. (Edinburgh 1958), 459–462.

[N2] M. Nagata, *On the fourteenth problem of Hilbert,* Amer. J. Math. **81** (1959), 766–772.

[NS] M. S. Narasimhan and C. S. Seshadri, *Stable and unitary vector bundles on a compact Riemann surface,* Ann. of Math. **82** (1965), 540–567.

[Ram] S. Ramanan, *The moduli spaces of vector bundles over an algebraic curve,* Math. Ann. **200** (1973), 69–84.

[Ros] M. Rosenlicht, *A remark on quotient spaces,* Annaes Academia Brasileira de Ciencias **35** (1963), 487–489.

[Sch] A. H. W. Schmitt, *Geometric invariant theory relative to a base curve,* in *Algebraic Cycles,* Sheaves, Shtukas and Moduli, Trends in Mathematics (2007), 131–183.

[Ses1] C. S. Seshadri, *Space of unitary vector bundles on a compact Riemann surface,* Ann. of Math. **85** (1967), 303–336.

[Ses2] C. S. Seshadri, *Fibrés vectoriels sur les courbes algébriques (rédigé par J.-M. Drezet),* Astérisque **96** (1982).

[Sim] C. Simpson, *Moduli of representations of the fundamental group of a smooth projective variety I,* Inst. Hautes Études Sci. Publ. Math. **79** (1994), 47–129.

5
Deformation Theory for Vector Bundles

Nitin Nitsure

*School of Mathematics, Tata Institute of Fundamental Research,
Homi Bhabha Road, Mumbai 400 005, India.
e-mail: nitsure@math.tifr.res.in*

Abstract

These expository notes give an introduction to the elements of deformation theory which is meant for graduate students interested in the theory of vector bundles and their moduli.

1 Introduction : Basic examples

For simplicity, we will work over a fixed base field k which may be assumed to be algebraically closed. All schemes and all morphisms between them will be assumed to be over the base k, unless otherwise indicated. In this section we introduce four examples which are of basic importance in deformation theory, with special emphasis on vector bundles.

Basic example 1: Deformations of a point on a scheme

We begin by setting up some notation. Let $\mathbf{Art}_k$ be the category of all artin local k-algebras, with residue field k. In other words, the objects of $\mathbf{Art}_k$ are local k-algebras with residue field k which are finite-dimensional as k-vector spaces, and morphisms are all k-algebra homomorphisms. Note that k is both an initial and a final object of $\mathbf{Art}_k$. By a **deformation functor** we will mean a covariant functor $F : \mathbf{Art}_k \to \mathbf{Sets}$ for which $F(k)$ is a singleton point. As k is an initial object of $\mathbf{Art}_k$, this condition means that we can as well regard F to be a functor to the category of pointed sets.

For any A in $\mathbf{Art}_k$, let $h_A : \mathbf{Art}_k \to \mathbf{Sets}$ be the deformation functor defined by taking $h_A(B) = Hom_{k\text{-alg}}(A, B)$. Recall the well-known Yoneda lemma, which asserts that there is a natural bijection $Hom(h_A, F) \to F(A)$ under which a natural transformation $\alpha : h_A \to F$ is identified with the element $\alpha(id_A) \in F(A)$. To simplify notation,

128

given any natural transformation $\alpha : h_A \to F$, we denote again by $\alpha \in F(A)$ the element $\alpha(id_A) \in F(A)$. Any element $\alpha \in F(A)$ will be called a **family** parametrised by A (the reason for this nomenclature will be clear from the examples). Given $f : B \to A$ and $\beta \in F(B)$, we denote the family $F(f)\beta \in F(A)$ simply by $\beta|_A$, when f is understood.

Let $\widehat{\mathbf{Art}}_k$ denote the category of complete local noetherian k-algebras with residue field k as objects and all k-algebra homomorphisms as morphisms. Given any R in $\widehat{\mathbf{Art}}_k$, we denote by $h_R : \mathbf{Art}_k \to \mathbf{Sets}$ the deformation functor defined by taking $h_R(A) = Hom_{k\text{-alg}}(R, A)$. A deformation functor F will be called **pro-representable** if there exists a natural isomorphism $r : h_R \to F$ where R is in $\widehat{\mathbf{Art}}_k$. The pair (R, r) will be called a **universal pro-family** for F. (To understand this name, see Lemma 2.7.)

If X is any scheme over k and $x \in X$ a k-rational point, we define a deformation functor $h_{X,x}$ by taking for any A in $\mathbf{Art}_k$ the set $h_{X,x}(A)$ to be the set of all morphisms $\operatorname{Spec} A \to X$ over k, for which the closed point of $\operatorname{Spec} A$ maps to x. Any such morphism is the same as a k-algebra homomorphism $\mathcal{O}_{X,x} \to A$. As the maximal ideal of A is nilpotent, such homomorphisms are in a natural bijection with k-algebra homomorphisms $R \to A$ where R denotes the completion of the local ring $\mathcal{O}_{X,x}$ at its maximal ideal. This shows that the above functor $h_{X,x}$ is naturally isomorphic to h_R, hence is pro-representable.

The functor $h_{X,x}$ is the ultimate example of a deformation functor in the sense that it is the simplest and has the best possible properties. One of the aims of the general theory is to determine when a given deformation functor is of this kind, at least, to determine whether it shares some nice properties with $h_{X,x}$.

Basic example 2: Deformations of a coherent sheaf

Let X be a proper scheme over a field k, and let E be a coherent sheaf of $\mathcal{O}_X$-modules. The **deformation functor $\mathcal{D}_E$ of E** is defined as follows. For any A in $\mathbf{Art}_k$, we take $\mathcal{D}_E(A)$ to be the set of all equivalence classes of pairs $(\mathcal{F}, \theta)$ where $\mathcal{F}$ is a coherent sheaf on $X_A = X \otimes_k A$ which is flat over A, and $\theta : i^*\mathcal{F} \to E$ is an isomorphism where $i : X \hookrightarrow X_A$ is the closed embedding induced by the residue map $A \to k$, with $(\mathcal{F}, \theta)$ and $(\mathcal{F}', \theta')$ to be regarded as equivalent when there exists some isomorphism $\eta : \mathcal{F} \to \mathcal{F}'$ such that $\theta' \circ (\eta|_X) = \theta$. It can be seen that $\mathcal{D}_E(A)$ is indeed a set. Given any homomorphism $f : B \to A$ in $\mathbf{Art}_k$ and an equivalence class $(\mathcal{F}, \theta)$ in $\mathcal{D}_E(B)$, we define $f(\mathcal{F}, \theta)$ in

$\mathcal{D}_E(A)$ to be its pull-back under the induced morphism $X_A \to X_B$ (by applying $- \otimes_B A$). This preserves equivalences, and so we get a functor $\mathcal{D}_E : \mathbf{Art}_k \to \mathbf{Sets}$.

If we assume that E is a vector bundle (that is, locally free), then flatness of $\mathcal{F}$ over A just amounts to assuming that $\mathcal{F}$ is a vector bundle on X_A.

Basic example 3: Deformations of a quotient

Let X be a proper scheme over k, E be a coherent $\mathcal{O}_X$-module over X, and $q_0 : E \to F_0$ be a coherent quotient $\mathcal{O}_X$-module. For any A in $\mathbf{Art}_k$, let E_A denote the pull-back of E to $X_A = X \otimes_k A$. Let $i : X \hookrightarrow X_A$ be the inclusion of the special fiber of X_A. We consider all $\mathcal{O}_{X_A}$-linear surjections $q : E_A \to \mathcal{F}$ such that $\mathcal{F}$ is flat over A and the kernel of $q|_X : E \to \mathcal{F}|_X$ equals $\ker(q_0)$. For any such, there exists a unique isomorphism $\theta : i^*\mathcal{F} \to F_0$ such that the following square commutes.

$$
\begin{array}{ccc}
i^*E_A & = & E \\
{\scriptstyle i^*q} \downarrow & & \downarrow {\scriptstyle q_0} \\
i^*\mathcal{F} & \xrightarrow{\theta} & F_0
\end{array}
$$

Two such surjections $q : E_A \to \mathcal{F}$ and $q' : E_A \to \mathcal{F}'$ will be called equivalent if $\ker(q) = \ker(q')$. For any object A of $\mathbf{Art}_k$, let $Q(A)$ be the set of all equivalence classes of such $q : E_A \to \mathcal{F}$ (it can be seen that $Q(A)$ is indeed a set). For any morphism $B \to A$ in $\mathbf{Art}_k$, we get by pull-back (by applying $- \otimes_B A$) a well-defined set map $Q(B) \to Q(A)$, so we have a deformation functor $Q : \mathbf{Art}_k \to \mathbf{Sets}$ (note that $Q(k)$ is clearly a singleton). Sending $(q : E_A \to \mathcal{F}) \mapsto (\mathcal{F}, \theta)$, where $\theta : i^*\mathcal{F} \to F_0$ is defined as above, defines a natural transformation $Q \to \mathcal{D}_{F_0}$.

In the special case when $E = \mathcal{O}_X$, a coherent quotient $q_0 : E \to F_0$ is the same as a closed subscheme $Y_0 \subset X$, and the functor Q becomes the functor of its proper flat deformations of Y_0 inside X.

If X is projective over X then by a fundamental theorem of Grothendieck there exists a k-scheme $Z = Quot_{E/X}$ (called the quot scheme of E or the Hilbert scheme of X when $E = \mathcal{O}_X$), and q_0 corresponds to a k-valued point z on Z. The functor Q in this case becomes just $h_{Z,z}$, the deformation functor of a point on a scheme which was our basic example 1 introduced above. However, even in the projective case, it is useful to study the deformation theory of Q from a general functorial point of view, as we will do later.

Basic example 4: Deformations of a scheme

Given a scheme X of finite type over a field k, let the deformation functor $\mathbf{Def}_X : \mathbf{Art}_k \to \mathbf{Sets}$ be defined as follows. For any $A \in \mathbf{Art}_k$, consider pairs $(p : \mathfrak{X} \to \operatorname{Spec} A, i : X \to X_0)$ where p is a flat morphism of k-schemes, $X_0 = p^{-1}(\operatorname{Spec} k) = \mathfrak{X}|_{\operatorname{Spec} k}$ is the schematic special fibre of p, and i is an isomorphism. Denoting again by i the composite $X \to X_0 \hookrightarrow \mathfrak{X}$, this means that the following square is cartesian.

$$
\begin{array}{ccc}
X & \xrightarrow{i} & \mathfrak{X} \\
\downarrow & & \downarrow p \\
\operatorname{Spec} k & \to & \operatorname{Spec} A
\end{array}
$$

We say that two such pairs (p, i) and (p', i') are equivalent if there exists an A-isomorphism between $\mathfrak{X}$ and $\mathfrak{X}'$ which takes i to i'. We take $\mathbf{Def}_X(A)$ to be the set of all equivalence classes of pairs (p, i). It can be seen that this is indeed a set, and moreover it is clear that a morphism $A \to B$ in $\mathbf{Art}_k$ gives by pull-back a well-defined set map $\mathbf{Def}_X(A) \to \mathbf{Def}_X(B)$ which indeed gives a functor $\mathbf{Def}_X : \mathbf{Art}_k \to \mathbf{Sets}$.

The above example and its variants and special cases quickly bring out all the possible complications in deformation theory, and have historically led to its major developments. In these notes, which are designed to be a short introduction aimed at graduate students interested in vector bundles, we will not treat this example in any detail, but will just mention some basic results.

Relation with moduli functors

The theory of moduli may suggest that rather than the deformation functor $\mathcal{D}_E$ of basic example 2, we should consider the functor $\mathcal{M}_E$ defined as follows. For any A in $\mathbf{Art}_k$, we take $\mathcal{M}_E(A)$ to be the set of isomorphism classes $[\mathcal{F}]$ of coherent sheaves $\mathcal{F}$ on X_A that are flat over A, such that the restriction $\mathcal{F}|_X$ to the special fiber of $\operatorname{Spec} A$ is *isomorphic* to E. Note that this does not involve a choice of a specific isomorphism $\theta : \mathcal{F}|_X \to E$, so this functor differs from the deformation functor $\mathcal{D}_E$.

Clearly, there is a natural transformation $\mathcal{D}_E \to \mathcal{M}_E$, which forgets the choice of θ. If the sheaf E has the special property that for each A and each $(\mathcal{F}, \theta)$ in $\mathcal{D}_E(A)$ every automorphism of E is the restriction of an automorphism of $\mathcal{F}$, then the natural transformation $\mathcal{D}_E \to \mathcal{M}_E$ is an isomorphism. For example, this may happen when E is stable in a certain sense, where the stability condition ensures that all automorphisms of E are just scalars. If moreover a fine moduli scheme M for stable sheaves exists, with E defining a point $[E] \in M$, then $\mathcal{M}_E$ is the same as the corresponding local moduli functor $h_{M,[E]}$ which is a case of the deformation functor of our basic example 1. Hence in this case, $\mathcal{D}_E$ will just be $h_{M,[E]}$, and the study of $\mathcal{D}_E$ will shed light on the local structure of M around $[E]$.

But even when the above condition (that automorphisms of E must be extendable to any infinitesimal family $\mathcal{F}$ around it) is not fulfilled, the study of the deformation functor $\mathcal{D}_E$ continues to be of importance, for it sheds light on the local structure of the corresponding moduli stacks. On the other hand, the functor $\mathcal{M}_E$, which may at first sight look more natural than $\mathcal{D}_E$, does not have good properties in general. The Example 3.9 illustrates this point.

Similar remarks apply to the functor $\mathbf{Def}_X$ of basic example 4 vis à vis the corresponding local moduli functor.

2 General theory

Tangent space to a functor

2.1 Let $\mathbf{Vect}_k$ be the category of all vector spaces over k, and let $\mathbf{FinVect}_k$ be its full subcategory consisting of all finite dimensional vector spaces. Let $\varphi : \mathbf{FinVect}_k \to \mathbf{Sets}$ be a functor into the category of sets which satisfies the following:

(T0) For the zero vector space 0, the set $\varphi(0)$ is a singleton set.

(T1) The natural map $\beta_{V,W} : \varphi(V \times W) \to \varphi(V) \times \varphi(W)$ induced by applying φ to the projections $V \times W \to V$ and $V \times W \to W$ is bijective.

Then for each V in $\mathbf{FinVect}_k$, there exists a unique structure of a k vector space on the set $\varphi(V)$ which gives a lift of φ to a k-linear functor which we again denote by $\varphi : \mathbf{FinVect}_k \to \mathbf{Vect}_k$. The addition map $\varphi(V) \times \varphi(V) \to \varphi(V)$ is the composite

$$\varphi(V) \times \varphi(V) \stackrel{\beta_{V,V}^{-1}}{\to} \varphi(V \times V) \stackrel{\varphi(+)}{\to} \varphi(V)$$

where $\beta_{V,V}^{-1}$ is the inverse of the natural isomorphism given by the assumption on φ, and $+ : V \times V \to V$ is the addition map of V. Also, for any $\lambda \in k$, the scalar multiplication map $\lambda_{\varphi(V)} : \varphi(V) \to \varphi(V)$ is just $\varphi(\lambda_V)$.

2.2 The **tangent space** T_φ to a functor $\varphi : \mathbf{FinVect}_k \to \mathbf{Sets}$ which satisfies **(T0)** and **(T1)** is defined to be the vector space $\varphi(k)$. This may not necessarily be finite dimensional. We now show that there exists a linear isomorphism

$$\Psi_{\varphi,V} : \varphi(V) \to T_\varphi \otimes_k V$$

which is functorial in both V and φ. For this, choose an isomorphism $g : V \to k^n$, and apply **(T1)** repeatedly to get a composite isomorphism

$$\varphi(V) \overset{\varphi(g)}{\to} \varphi(k^n) \overset{\beta}{\to} \varphi(k)^n = \varphi(k) \otimes k^n \overset{id \otimes \varphi(g)^{-1}}{\to} \varphi(k) \otimes V = T_\varphi \otimes V$$

which can be verified to be independent of the choice of g.

Thus, a functor $\varphi : \mathbf{FinVect}_k \to \mathbf{Sets}$ satisfying **(T0)** and **(T1)** is completely described by its tangent space T_φ. Conversely, given any vector space T, the association $V \mapsto T \otimes V$ defines a functor φ that satisfies **(T0)** and **(T1)**, for which T_φ is just T.

Artin local algebras

If A_1 and A_2 are local k-algebras with residue field k, that is, the composite $k \to A_i \to A_i/\mathfrak{m}_i$ is an isomorphism where $\mathfrak{m}_i \subset A_i$ is its maximal ideal, then any k-algebra homomorphism $f : A_1 \to A_2$ is necessarily *local*, that is, $f^{-1}(\mathfrak{m}_2) = \mathfrak{m}_1$.

If $f : B \to A$ and $g : C \to A$ are homomorphisms in $\mathbf{Art}_k$, the fibred product

$$B \times_A C = \{(b,c)|f(b) = g(c) \in A\}$$

with component-wise operations is again an object in $\mathbf{Art}_k$ (Exercise). Also, for homomorphisms $A \to B$ and $A \to C$ in $\mathbf{Art}_k$, the tensor product $B \otimes_A C$ is again an object in $\mathbf{Art}_k$ (Exercise). Thus, $\mathbf{Art}_k$ admits both fibred products (pull-backs) $B \times_A C$ and tensor products (push-outs) $B \otimes_A C$.

As k is the final object in $\mathbf{Art}_k$, the fibred product $A \times_k B$ serves as the direct product in the category $\mathbf{Art}_k$, and as k is the initial object in $\mathbf{Art}_k$, the tensor product $B \otimes_A C$ serves as the Co-product in the category $\mathbf{Art}_k$.

For a k-vector space V, let $k\langle V \rangle = k \oplus V$ with ring multiplication defined by putting $(a,v)(b,w) = (ab, aw + bv)$, and obvious k-algebra structure. Note that $k\langle V \rangle$ is artinian if and only if V is finite dimensional. It can be seen that $V \mapsto k\langle V \rangle$ defines a fully faithful functor $\mathbf{FinVect}_k \to \mathbf{Art}_k$, and its image consists of all A in $\mathbf{Art}_k$ with $\mathfrak{m}_A^2 = 0$, as such an A is naturally isomorphic to $k\langle \mathfrak{m}_A \rangle$. The functor $V \mapsto k\langle V \rangle$ takes the zero vector space (which is both an initial and final object of $\mathbf{FinVect}_k$) to the algebra k (which is both an initial and final object of

$\mathbf{Art}_k$). If $V \to U$ and $W \to U$ are morphisms in $\mathbf{FinVect}_k$, then it can be seen that the natural map

$$k\langle V \times_U W \rangle \to k\langle V \rangle \times_{k\langle U \rangle} k\langle W \rangle$$

(which is induced by the projections from $V \times_U W$ to V and W) is an isomorphism. Therefore the functor $\mathbf{FinVect}_k \to \mathbf{Art}_k : V \mapsto k\langle V \rangle$ preserves all finite inverse limits, in particular, it preserves equalisers.

For any deformation functor $F : \mathbf{Art}_k \to \mathbf{Sets}$, let the composite $\mathbf{FinVect}_k \to \mathbf{Art}_k \to \mathbf{Sets}$ be denoted by φ. As $F(k)$ is a singleton, $\varphi(0) = F(k\langle 0 \rangle) = F(k) = 0$, so φ satisfies the condition $(\mathbf{T0})$ on functors $\mathbf{FinVect}_k \to \mathbf{Sets}$. We now introduce another condition $(\mathbf{H}\epsilon)$ on a deformation functor F, which just amounts to demanding that φ should satisfy $(\mathbf{T1})$.

2.3 Deformation condition $(\mathbf{H}\epsilon)$ For any two A and B in $\mathbf{Art}_k$ with $\mathfrak{m}_A^2 = 0$ and $\mathfrak{m}_B^2 = 0$, the map $F(A \times_k B) \to F(A) \times F(B)$ that is induced by applying F to the projections of $A \times_k B$ on A and B is a bijection.

Note that $k\langle k \rangle$ is just the ring $k[\epsilon]/(\epsilon^2)$ of dual numbers over k.

2.4 The **tangent space T_F to a deformation functor F** that satisfies $(\mathbf{H}\epsilon)$ is defined to be the resulting k-vector space

$$T_F = F(k[\epsilon]/(\epsilon^2)) = \varphi(k) = T_\varphi.$$

Exercise 2.5 Let X be a k-scheme, $x \in X$ a k-valued point, and let F be the deformation functor $h_{X,x}$ of basic example 1. Then T_F equals the tangent space $T_x X$, which is the dual to $\mathfrak{m}_x/\mathfrak{m}_x^2$ where $\mathfrak{m}_x \subset \mathcal{O}_{X,x}$ is the maximal ideal of x. This is the motivation for defining T_F for deformation functors.

Exercise 2.6 Universal first-order family Let a deformation functor F satisfy $(\mathbf{H}\epsilon)$, and moreover let the resulting tangent vector space T_F be finite dimensional. Let T_F^* be the dual vector space of T_F, and let $A = k\langle T_F^* \rangle \in \mathbf{Art}_k$. Note that $T_A = T_F$. The identity endomorphism $\theta \in End(T_F) = T_F \otimes T_F^* = F(A)$ defines a family (A, θ). Show that this family has the following properties.

(i) The map $\theta : h_A \to F$ induces the identity isomorphism $T_F \to T_F$.

(ii) Let (R, r) be any pro-family for F parametrised by $R \in \widehat{\mathbf{Art}_k}$.

Let $R_1 = R/\mathfrak{m}_R^2$ and let $r_1 = r|_{R_1}$. Then there exists a unique k-homomorphism $A \to R_1$ such that $r_1 \in F(R_1)$ is the image of $\theta \in F(A)$.

In view of property (ii), the family (A, θ) is called as the universal first-order family for F.

Pro-families and limit Yoneda lemma

Recall that the Yoneda lemma asserts the following. If $\mathcal{C}$ is any category, A is any object of $\mathcal{C}$ and $h_A = Hom(A, -)$ the corresponding representable functor, and $F : \mathcal{C} \to \mathbf{Sets}$ another functor, then there is a natural bijection $Hom(h_A, F) \to F(A)$, under which a natural transformation $\alpha : h_A \to F$ is mapped to the element $\alpha(\mathrm{id}_A) \in F(A)$. By abuse of notation, we denote $\alpha(\mathrm{id}_A)$ just by $\alpha \in F(A)$.

A **pro-family** for a deformation functor $F : \mathbf{Art}_k \to \mathbf{Sets}$ is a pair (R, r) where R is in $\widehat{\mathbf{Art}_k}$ and $r \in \widehat{F}(R)$ where by definition

$$\widehat{F}(R) = \varprojlim F(R/\mathfrak{m}^n)$$

where $\mathfrak{m} \subset R$ is the maximal ideal. By the following lemma, r is same as a morphism of functors $h_R \to F$.

Lemma 2.7. (Limit Yoneda Lemma)
Let $F : \mathbf{Art}_k \to \mathbf{Sets}$ be a deformation functor, and let $\widehat{F} : \widehat{\mathbf{Art}_k} \to \mathbf{Sets}$ be its prolongation as constructed above. Let $\alpha_R : Hom(h_R, F) \to \widehat{F}(R)$ be the map defined as follows. Given $f \in Hom(h_R, F)$, and $n \geq 1$, let $f_{R/\mathfrak{m}^n}(q_n) \in F(R/\mathfrak{m}^n)$ denote the image of the quotient $q_n \in Hom_{k\text{-}alg}(R, R/\mathfrak{m}^n)$. This defines an inverse system as n varies, so gives an element $\alpha_R(f) = (f(R/\mathfrak{m}^n)(q_n))_{n \in \mathbf{N}} \in \widehat{F}(R)$. The map $\alpha_R : Hom(h_R, F) \to \widehat{F}(R)$ so defined is a bijection, functorial in both R and F. ☐

We leave the proof of this lemma (which is a straight-forward generalisation of the usual Yoneda lemma) as an exercise.

Versal, miniversal, universal families

For a quick review of basic notions about smoothness and formal smoothness, see for example Milne [Mi].

Let $F : \mathbf{Art}_k \to \mathbf{Sets}$ and $G : \mathbf{Art}_k \to \mathbf{Sets}$ be functors. Recall that a morphism of functors $\phi : F \to G$ is called **formally smooth** if given any surjection $q : B \to A$ in $\mathbf{Art}_k$ and any elements $\alpha \in F(A)$ and $\beta \in G(B)$ such that

$$\phi_A(\alpha) = G(q)(\beta) \in F(A),$$

there exists an element $\gamma \in F(B)$ such that

$$\phi_B(\gamma) = \beta \in G(B) \text{ and } F(q)(\gamma) = \alpha \in F(A)$$

In other words, the following diagram of functors commutes, where the diagonal arrow $h_B \to F$ is defined by γ.

$$
\begin{array}{ccc}
h_A & \xrightarrow{\alpha} & F \\
q\downarrow & \nearrow & \downarrow \phi \\
h_B & \xrightarrow{\beta} & G
\end{array}
$$

The morphism $\phi : F \to G$ is called **formally étale** if it is formally smooth, and moreover the element $\gamma \in F(B)$ is unique.

Caution If the functors F and G are of the form h_R and h_S for rings R and S, then ϕ is formally étale if and only if it is formally smooth and the tangent map $T_R \to T_S$ is an isomorphism. However, if F and G are not both of the above form, then a functor ϕ can be formally smooth, and moreover the map $T_F \to T_G$ can be an isomorphism, yet ϕ need not be formally étale. It is because of this subtle difference that a miniversal family can fail to be universal, as we will see in examples later.

2.8 A versal family for a deformation functor $F : \mathbf{Art}_k \to \mathbf{Sets}$ is a pro-family (R, r) (where R is a complete local noetherian k-algebra with residue field k, and $r \in \widehat{F}(R)$) such that the morphism of functors $r : h_R \to F$ is formally smooth.

If (R, r) is a versal family, then for any A in $\mathbf{Art}_k$, the induced set map $r(A) : h_R(A) \to F(A)$ is surjective. For, given any $v \in F(A)$, we can regard it as a morphism $v : h_A \to F$. Now consider the following commutative square.

$$
\begin{array}{ccc}
h_k & \longrightarrow & h_R \\
\downarrow & & \downarrow \\
h_A & \xrightarrow{v} & F
\end{array}
$$

By formal smoothness of $h_R \to F$, there exists a morphism $u : h_A \to h_R$ which makes the above diagram commute. But such a morphism is just an element of $h_R(A)$ which maps to $v \in F(A)$, which proves that $r(A) : h_R(A) \to F(A)$ is surjective. In other words, every family over A is a pull-back of the versal family over R, under a morphism $u : \operatorname{Spec} A \to \operatorname{Spec} R$. However, the morphism u need not be unique.

For any deformation functor $F : \mathbf{Art}_k \to \mathbf{Sets}$, the pointed set

$$T_F = F(k[\epsilon]/(\epsilon^2))$$

is called the **tangent set** to F, or the set of **first order deformations** under F.

2.9 A minimal versal ('miniversal') family (also called as a **hull**) for a deformation functor $F : \mathbf{Art}_k \to \mathbf{Sets}$ is a versal family for which the set map

$$dr : T_R = h_R(k[\epsilon]/(\epsilon^2)) \to F(k[\epsilon]/(\epsilon^2)) = T_F$$

is a bijection.

Exercise 2.10 If $r : h_R \to F$ is a hull for a deformation functor F, then show that F satisfies the deformation condition $(\mathbf{H}\epsilon)$, and the bijection of sets $dr : T_R \to T_F$ is in fact a linear isomorphism.

2.11 A universal family for a deformation functor $F : \mathbf{Art}_k \to \mathbf{Sets}$ is a pro-family (R, r) such that $r : h_R \to F$ is a natural bijection. If a universal family exists, it is clearly unique up to a unique isomorphism. A deformation functor $F : \mathbf{Art}_k \to \mathbf{Sets}$ is called **pro-representable** if a universal family exists. (The reason for the prefix 'pro-' is that R need not be in the subcategory $\mathbf{Art}_k$ of $\widehat{\mathbf{Art}_k}$.)

Exercise 2.12 Show the following.

(i) A pro-family (R, r) is universal if and only if the morphism of functors $r : h_R \to F$ is formally étale.

(ii) If F is pro-representable, then each hull pro-represents it.

(iii) **A miniversal family that is not universal.** Let $F : \mathbf{Art}_k \to \mathbf{Sets}$ be the functor $A \mapsto \mathfrak{m}_A/\mathfrak{m}_A^2$. Show that a hull (R, r) for F is given by $R = k[[t]]$ with $r = dt \in \mathfrak{m}_R/\mathfrak{m}_R^2 = \widehat{F}(R)$, but F is not pro-representable.

(iv) If a deformation functor F admits a hull and moreover if $T_F = 0$ then $F(A)$ is the singleton set $F(k)$ for all A in $\mathbf{Art}_k$.

(v) Let a deformation functor F have a versal family $(R, r : h_R \to \varphi)$, such that h_R is formally smooth. Then F is formally smooth. Conversely, if F is formally smooth, then each versal family is formally smooth.

Grothendieck's pro-representability theorem

The following condition on a deformation functor F is obviously satisfied by any pro-representable functor h_R.

2.13 Deformation condition (Lim) The functor F preserves fibred products: the induced map $F(B \times_A C) \to F(B) \times_{F(A)} F(C)$ is bijective for any pair of homomorphisms $B \to A$ and $C \to A$ in $\mathbf{Art}_k$.

As $\mathbf{Art}_k$ has a final object and admits fibred products, this is equivalent to the condition that F preserves all finite inverse limits in $\mathbf{Art}_k$, hence the name **(Lim)**.

As **(Lim)** implies the deformation condition **(Hϵ)** (see 2.3 above), the set $T_F = F(k[\epsilon]/(\epsilon^2))$ is naturally a k-vector space whenever **(Lim)** is satisfied.

Theorem 2.14. (Grothendieck) *A deformation functor F is pro-representable if and only if the following two conditions* **(Lim)** *and* **(H3)** *are satisfied.*

(Lim) *The deformation functor F preserves fibred products.*

(H3) *The k-vector space T_F is finite dimensional.*

Obviously, a pro-representable functor satisfies the above conditions **(Lim)** and **(H3)**. The sufficiency of these conditions follows from Schlessinger's theorem (Theorem 2.19), in which the conditions **(Lim)** and **(H3)** are weakened to the conditions **(H1)**, **(H2)**, **(H3)** and **(H4)**. In practice, the conditions **(Hϵ)** and **(H3)** are the easiest to verify, the conditions **(H1)** and **(H2)** are of intermediate difficulty, while the condition **(Lim)** is quite difficult to check in most examples. Hence Schlessinger's theorem is more useful in actual practice than Theorem 2.14.

The interested reader can take the proof of Schlessinger's theorem, and shorten and simplify it using the stronger hypothesis **(Lim)** to get a proof of Theorem 2.14. Though this exercise makes a reversal of the actual history, it helps us understand the Schlessinger theorem better.

Schlessinger's conditions and the resulting group action

2.15 A small extension e in $\mathbf{Art}_k$ is a surjective homomorphism $B \to A$ whose kernel I satisfies $\mathfrak{m}_B I = 0$. The small extension e is called a **principal small extension** if moreover I is principal. We often use the notation $e = (0 \to I \to B \to A \to 0)$ for a small extension.

2.16 We now state the famous Schlessinger conditions **(H1)**, **(H2)**, **(H3)** and **(H4)** on a deformation functor, and their variants **(H1')** and **(H2')**.

(H1) For any homomorphisms $B \to A$ and $C \to A$ in $\mathbf{Art}_k$ such that $C \to A$ is a principal small extension, the induced map $F(B \times_A C) \to F(B) \times_{F(A)} F(C)$ is surjective.

The condition **(H1)** is equivalent to the following seemingly stronger condition:

(H1') For any homomorphisms $B \to A$ and $C \to A$ in $\mathbf{Art}_k$ such that $C \to A$ is surjective, the induced map $F(B \times_A C) \to F(B) \times_{F(A)} F(C)$ is surjective.

To see this, first note that a surjective homomorphism $p : C \to A$ can be factored in $\mathbf{Art}_k$ as the composite of a finite sequence of surjections $C = C_n \to C_{n-1} \to \ldots \to C_1 \to C_0 = A$ where $n \geq 1$ is an integer such that $\mathfrak{m}_C^n = 0$, and $C_j = C/\mathfrak{m}^j I$ where I is the kernel of $C \to A$. Then each $C_i \to C_{i-1}$ is a small extension. Moreover, a small extension $0 \to I \to C \to A \to 0$ can be factored as $C = C_n \to C_{n-1} \to \ldots \to C_1 \to C_0 = A$ where $n = \dim_k I$, and each $C_i \to C_{i-1}$ is a principal small extension. Hence **(H1')** follows by applying **(H1)** successively to a finite sequence of principal small extensions.

(H2) For any B in $\mathbf{Art}_k$, the induced map $F(B \times_k k[\epsilon]/(\epsilon^2)) \to F(B) \times F(k[\epsilon]/(\epsilon^2))$ is bijective.

Similarly, the condition **(H2)** is equivalent to the following:

(H2') Let B be any object in $\mathbf{Art}_k$, and let $C = k\langle V \rangle$ where V is a finite dimensional k-vector space. Then the induced map $F(B \times_k C) \to F(B) \times F(C)$ is bijective.

Note As **(H2')** implies **(Hϵ)**, the set $T_F = F(k[\epsilon]/(\epsilon^2))$ gets a natural k-vector space structure whenever **(H2)** is satisfied. Hence the condition **(H3)** below makes sense whenever **(H2)** is satisfied.

(H3) The k-vector space T_F is finite dimensional.

(H4) If $B \to A$ is a principal small extension, then the induced map $F(B \times_A B) \to F(B) \times_{F(A)} F(B)$ is a bijection.

2.17 (Definition of an action) Let $F : \mathbf{Art}_k \to \mathbf{Sets}$ be a functor with $F(k)$ a singleton, which satisfies the Schlessinger conditions **(H1)** and **(H2)**. Let $0 \to I \to B \to A \to 0$ be a small extension in $\mathbf{Art}_k$. We now define an action of the group $T_F \otimes I$ on the set $F(B)$. Note that we have a k-algebra isomorphism $f : B \times_k k\langle I \rangle \to B \times_A B$ defined by $(b, \overline{b} + u) \mapsto (b, b + u)$ where $\overline{b} \in k$ is the residue class of b. Applying F and using **(H1')** and **(H2')**, the following composite map is a surjection:

$$F(B) \times T_F \otimes I = F(B) \times F(k\langle I \rangle) \xrightarrow{F(f)} F(B \times_A B) \to F(B) \times_{F(A)} F(B).$$

From its definition, the above map sends any pair (β, x) to a pair of the form (β, γ). We define a map $F(B) \times T_F \otimes I \to F(B)$ by $(\beta, x) \mapsto \gamma$, and this can be verified to define an action of the abelian group $T_F \otimes I$ on the set $F(B)$.

Proposition 2.18. *Let F be a deformation functor which satisfies the Schlessinger conditions* **(H1)** *and* **(H2)**. *Then for any small extension $0 \to I \to B \to A \to 0$ in* **Art**$_k$, *the induced action of the abelian group $T_F \otimes I$ on the set $F(B)$ has the following properties.*

(i) The orbits of $T_F \otimes I$ in $F(B)$ are exactly the fibers of the map $F(B) \to F(A)$.

(ii) When $A = k$, the action is free.

(iii) The action is functorial in small extensions: given a commutative diagram

$$
\begin{array}{ccccccccc}
0 \to & I & \to B \to & A & \to 0 \\
& \downarrow & \downarrow & \downarrow & \\
0 \to & I' & \to B' \to & A' & \to 0
\end{array}
$$

where the rows are small extensions, the induced map $F(B) \to F(B')$ is equivariant w.r.t. the induced group homomorphism $T_F \otimes I \to T_F \otimes I'$ and the actions of $T_F \otimes I$ and $T_F \otimes I'$ on $F(B)$ and $F(B')$.

(iv) The action is functorial in F: if $F \to G$ is a natural transformation of deformation functors which satisfy **(H1)** *and* **(H2)**, *then the induced set-map $F(B) \to G(B)$ is equivariant w.r.t. the induced group homomorphism $T_F \otimes I \to T_G \otimes I$.*

Proof Assertion (i) is just the surjectivity of

$$
F(B) \times T_F \otimes I \to F(B) \times_{F(A)} F(B).
$$

(ii) amounts to the bijectivity of $F(B) \times T_F \otimes I \to F(B) \times_{F(A)} F(B)$ when $A = k$, and so follows from **(H2)**. (iii) and (iv) can be verified in a straight-forward manner from the definition of the action. $\square$

Schlessinger's theorem

Theorem 2.19. (Schlessinger) *A deformation functor F admits a hull if and only if the conditions* **(H1)**, **(H2)**, **(H3)** *are satisfied. Moreover, F is pro-representable if and only if the conditions* **(H1)**, **(H2)**, **(H3)** *and* **(H4)** *are satisfied.*

Proof It is a simple exercise to show that if R is in $\widehat{\mathbf{Art}_k}$ then h_R satisfies **(H1)**, **(H2)**, **(H3)** and **(H4)**, and if $r : h_R \to F$ is a hull for F then F satisfies **(H1)**, **(H2)** and **(H3)**. We now prove the reverse implications.

Existence of hull together with (H4) implies pro-representability : We will show that if **(H4)** is satisfied then any hull (R, r) is in fact a universal family. Let $0 \to I \to B \to A \to 0$ be a small extension in $\mathbf{Art}_k$, and consider the action of $T_F \otimes I$ on $F(B)$, which satisfies the properties given by Proposition 2.18. If **(H4)** holds, then $F(B \times_A B) \to F(B) \times_{F(A)} F(B)$ is bijective, so the surjective map $F(B) \times (T_F \otimes I) \to F(B) \times_{F(A)} F(B)$ is actually a bijection, which means that each fibre of $F(B) \to F(A)$ is a principal set (possibly empty) under the group $T_F \otimes I$.

To show that a miniversal family (R, r) is universal, we must show that the map $r(B) : h_R(B) \to F(B)$ is a bijection for each object B of $\mathbf{Art}_k$. This is clear for $B = k$. So now we proceed by induction on the smallest positive integer $n(B)$ for which $\mathfrak{m}_B^{n(B)} = 0$ (for $B = k$ we have $n = 1$). For a given B, suppose $n(B) \geq 2$. Let $I = \mathfrak{m}_B^{n(B)-1}$ so that $\mathfrak{m}_B I = 0$. Let $A = B/I$, so that $n(A) = n(B) - 1$, which by induction gives a bijection $r(A) : h_R(A) \to F(A)$. Consider the commutative square

$$
\begin{array}{ccc}
h_R(B) & \to & F(B) \\
\downarrow & & \downarrow \\
h_R(A) & = & F(A)
\end{array}
$$

Note that each fiber of $h_R(B) \to h_R(A)$ is a principal $T_R \otimes I$-set over $h_R(A)$ (possibly empty) and the map $r(B) : h_R(B) \to F(B)$ is $T_F \otimes I$-equivariant, where we identify T_R with T_F via $r : h_R \to F$. It follows that $r(B) : h_R(B) \to F(B)$ is injective. As $r(B) : h_R(B) \to F(B)$ is already known to be surjective by versality, this shows that $r(B)$ is bijective, thus (R, r) pro-represents F.

(H1), (H2), (H3) imply the existence of a hull : The proof will go in two stages. First, we will construct a family (R, r), which will be our candidate for a hull. Next, we prove that the family (R, r) is indeed a hull.

Construction of a family (R, r) : Let S be the completion of the local ring at the origin of the affine space $\operatorname{Spec} \operatorname{Sym}_k(T_F^*)$. If $x_1, \ldots, x_d$ is a linear basis for T_F^*, then $S = k[[x_1, \ldots, x_d]]$. Let $\mathfrak{n} = (x_1, \ldots, x_d) \subset S$ denote the maximal ideal of S. We will construct a versal family (R, r) where $R = S/J$ for some ideal J. The ideal J will be constructed as the intersection of a decreasing chain of ideals

$$
\mathfrak{n}^2 = J_2 \supset J_3 \supset J_4 \supset \ldots \supset \cap_{q=2}^\infty J_q = J
$$

such that at each stage we will have $J_q \supset J_{q+1} \supset \mathfrak{n} J_q$. Consequently, we will have $J_q \supset \mathfrak{n}^q$ which in particular means $R/J_q \in \mathbf{Art}_k$, and J_q/J is a fundamental system of open neighbourhoods in $R = S/J$ for the $\mathfrak{m}$-adic topology on R, where $\mathfrak{m} = \mathfrak{n}/J$ is the maximal ideal of R. Note that R will be automatically complete for the $\mathfrak{m}$-adic topology.

Starting with $q = 2$, we will define for each q an ideal J_q and a family (R_q, r_q) parametrised by $R_q = S/J_q$, such that $r_{q+1}|_{R_q} = r_q$. We take $J_2 = \mathfrak{n}^2$. On $R_2 = S/\mathfrak{n}^2 = k\langle T_F^* \rangle$ we take q_2 to be the 'universal first order family' θ (see Exercise 2.6 above). Having already constructed (R_q, r_q), we next take J_{q+1} to be a minimal ideal in the non-empty set Ψ of all ideals $I \subset S$ which satisfy the following two conditions:

(1) We have inclusions $J_q \supset I \supset \mathfrak{n}J_q$.

(2) There exists a family α (need not be unique) parametrised by R/I which prolongs r_q, that is, $\alpha|_{R_q} = r_q$.

Note that Ψ is non-empty as $J_q \in \Psi$, and Ψ has at least one minimal element as $S/\mathfrak{n}J_q$ is artinian being a quotient of $S/\mathfrak{n}^{q+1}$. (In fact, Ψ can be shown to be closed under finite intersections, so has a unique minimal element, but we will not use this fact here.)

We now choose $r_{q+1} \in F(S/J_{q+1})$ to be an arbitrary prolongation of r_q (not claimed to be unique).

Let J be the intersection of all the J_n, and let $R = S/J$. We want to define an element $r \in \widehat{F}(R)$ which restricts to r_q on S/J_q for each q. This makes sense and is indeed possible, as we will show using the following lemma, whose proof we leave as an exercise in the application of the familiar Mittag-Leffler condition for exactness of inverse limits.

Lemma 2.20. *Let R be a complete noetherian local ring with with maximal ideal $\mathfrak{m}$. Let $I_1 \supset I_2 \supset \ldots$ be a decreasing sequence of ideals such that (i) the intersection $\cap_{n \geq 1} I_n$ is 0, and (ii) for each $n \geq 1$, we have $I_n \supset \mathfrak{m}^n$. Then the natural map $f : R \to \lim_{\leftarrow} R/I_n$ is an isomorphism. Moreover, for any $n \geq 1$ there exists an $q \geq n$ such that $\mathfrak{m}^n \supset I_q$.*

Let $R = S/J$ as before, which is a complete noetherian local ring with maximal ideal $\mathfrak{m} = \mathfrak{n}/J$, and let $I_q = J_q/J$ for $q \geq 2$. By construction, we have $J_q \supset J_{q+1} \supset \mathfrak{n}J_q$, which means $I_q \supset I_{q+1} \supset \mathfrak{m}I_q$. In particular, this means $I_q \supset \mathfrak{m}^q$. As $J = \cap J_q$, we get $\cap I_q = 0$. Therefore by Lemma 2.20, for each $n \geq 1$ there exists a $q \geq n$ with $I_n \supset \mathfrak{m}^n \supset I_q$, and in particular the natural map $R \to \lim_{\leftarrow} R/I_n$ is an isomorphism.

Recall that we have already chosen an inverse system of elements $r_q \in F(R/I_q)$, as $R/I_q = S/J_q$. For each n choose the smallest $q_n \geq n$ such that $\mathfrak{m}^n \supset I_{q_n}$. We have a natural surjection $R/I_{q_n} \to R/\mathfrak{m}^n$. Let $\theta_n = r_{q_n}|_{R/\mathfrak{m}^n}$. Then from its definition it follows that under $R/\mathfrak{m}^{n+1} \to$

$R/\mathfrak{m}^n$, we have $\theta_n = \theta_{n+1}|_{R/\mathfrak{m}^n}$. Therefore (θ_n) defines an element

$$r = (\theta_n) \in \varprojlim F(R/\mathfrak{m}^n) = \widehat{F}(R)$$

Verification that (R, r) is a hull for F : By its construction, the map $T_R \to T_F$ is an isomorphism. So all that remains is to show that $h_R \to F$ is formally smooth. This means given any surjection $p : B \to A$ in $\mathbf{Art}_k$ and a commutative square

$$
\begin{array}{ccc}
h_A & \overset{h_u}{\to} & h_R \\
{\scriptstyle h_p} \downarrow & & \downarrow {\scriptstyle r} \\
h_B & \overset{b}{\to} & F
\end{array}
$$

there exists a diagonal morphism $h_v : h_B \to h_R$ (that is, a homomorphism $v : R \to B$) which makes the resulting diagram (the above square together with a diagonal) commute. If $\dim_k(B) = \dim_k(A)$ as k-vector space, then the surjection $B \to A$ is an isomorphism, and we are done. Otherwise, we can reduce to the case where $\dim_k(B) = \dim_k(A) + 1$ (the case of a small extension) by factoring $p : B \to A$ as the composite of a finite sequence of surjections $B = B_n \to B_{n-1} \to \ldots \to B_1 \to B_0 = A$ where each $B_i \to B_{i-1}$ is a principal small extension, and lifting step-by-step, making the analogue of the above square commute at each step.

Suppose there exists a homomorphism $w : R \to B$ such that $u = p \circ w : R \to B \to A$. Using such a w, we will construct a homomorphism $v : R \to B$ as needed in the proof of formal smoothness of $h_R \to F$, which satisfies both $u = p \circ v : R \to B \to A$ and $r \circ h_v = b : h_B \to F(B) \to F(A)$.

Consider the following commutative square:

$$
\begin{array}{ccc}
h_R(B) & \overset{r(B)}{\to} & F(B) \\
{\scriptstyle h_R(p)} \downarrow & & \downarrow {\scriptstyle F(p)} \\
h_R(A) & \overset{r(A)}{\to} & F(A)
\end{array}
$$

As $B \to A$ is a small extension, and as both h_R and F satisfy **(H1)** and **(H2)**, and as $T_R = T_F$, there is a natural transitive action of the additive group $T_F \otimes I$ on each fibre of the set maps $h_R(B) \to h_R(A)$ and $F(B) \to F(A)$. By Proposition 2.18.(iv), the top map $r(B) : h_R(B) \to F(B)$ in the above square is $T_F \otimes I$-equivariant. As $u = p \circ w$, the elements $r(B)w$ and b both lie in the same fibre of $F(B) \to F(A)$, over $r(A)u \in F(A)$. Therefore, there exists some $\alpha \in G$ (not necessarily unique) such that $b = r(B)w + \alpha$. Let $v = w + \alpha \in h_R(B)$. By G-equivariance of $r(B)$, we get $r(B)v = r(B)(w + \alpha) = r(B)w + \alpha = b$. Also, as the action of G preserves the fibers of $h_R(B) \to h_R(A)$, we have $p \circ v = p \circ (w + \alpha) = p \circ w = u$. Therefore v has the desired property. It therefore just remains to show the existence of $w : R \to B$ with $p \circ w = u$. For this we first make the following elementary observation.

Remark 2.21 Let $B \to A$ be a surjection in $\mathbf{Art}_k$ such that $\dim_k(B) = \dim_k(A) + 1$ (equivalently, the kernel I of the surjection satisfies $\mathfrak{m}_B I = 0$ and $\dim_k(I) = 1$). Suppose that $B \to A$ does not admit a section $A \to B$. Then for any k-algebra homomorphism $C \to B$, the composite $C \to B \to A$ is surjective (if and) only if $C \to B$ is surjective.

We now show the existence of $w : R \to B$ with $p \circ w = u$. As A is artinian, the homomorphism $u : R \to A$ must factor via $R_q = R/\mathfrak{m}^q$ for some $q \geq 1$, giving a homomorphism $u_q : R_q \to A$. We are given a diagram

$$\begin{array}{ccccc}
\operatorname{Spec} A & \overset{u_q^*}{\to} & \operatorname{Spec} R_q \to \operatorname{Spec} R & \hookrightarrow & \operatorname{Spec} S \\
\downarrow & & & & \downarrow \\
\operatorname{Spec} B & & \to & & \operatorname{Spec} k
\end{array}$$

The morphism $\operatorname{Spec} S \to \operatorname{Spec} k$ is formally smooth, therefore, there exists a diagonal homomorphism $f^* : \operatorname{Spec} B \to \operatorname{Spec} S$ which makes the resulting diagram commute. Equivalently, there exists a k-algebra homomorphism $f : S \to B$ such that $p \circ f = u \circ \pi : S \to A$ where $\pi : S \to R = S/J$ is the quotient map. Therefore, we get a commutative square

$$\begin{array}{ccc}
S & \overset{f}{\to} & B \\
\downarrow & & \downarrow \\
R_q & \overset{u_q}{\to} & A
\end{array}$$

where the vertical maps are the quotient maps $\pi_q : S \to S/J_q = R_q$, and $p : B \to A$. This defines a k-homomorphism

$$\varphi = (\pi_q, f) : S \to R_q \times_A B$$

The composite $S \to R_q \times_A B \to R_q$ is π_q which is surjective. As by assumption $\dim_k(B) = \dim_k(A) + 1$, it follows that

$$\dim_k(R_q \times_A B) = \dim_k(R_q) + 1$$

Therefore by Remark 2.21, at least one of the following holds:

(1) The projection $R_q \times_A B \to R_q$ admits a section $(\mathrm{id}, s) : R_q \to R_q \times_A B$, in other words, there exists some $s : R_q \to B$ such that $p \circ s = u_q : R_q \to A$.

(2) The homomorphism $\varphi : S \to R_q \times_A B$ is surjective.

If (1) holds, then we immediately get a lift

$$v : R \to R_q \xrightarrow{s} B$$

of $u : R \to A$, completing the proof.

If (2) holds, then we claim that $\varphi : S \to R_q \times_A B$ factors through $S \to S/J_{q+1} = R_{q+1}$, thereby giving a homomorphism $s' : R_{q+1} \to B$ such that $p \circ s' = u_{q+1} : R_{q+1} \to A$. This would immediately give a lift

$$v : R \to R_{q+1} \xrightarrow{s'} B$$

of $u : R \to A$, again completing the proof.

Therefore, all that remains is to show that if $\varphi : S \to R_q \times_A B$ is surjective, then it must factor through $S \to S/J_{q+1} = R_{q+1}$. To see this, let $K = ker(\varphi) \subset S$, so that $R_q \times_A B$ gets identified with S/K by surjectivity of φ. We have the families $r_q \in F(R_q)$, $a \in F(A)$ and $b \in F(B)$ such that both r_q and b map to a under $R_q \to A$ and $B \to A$. By **(H1)** the map $F(R_q \times_A B) \to F(R_q) \times_{F(A)} F(B)$ is surjective, so there exists a family $\mu \in F(R_q \times_A B) = F(S/K)$ which restricts to $r_q \in F(R_q)$. This means the ideal K is in the set of ideals Ψ defined earlier while constructing the nested sequence $J_2 \supset J_3 \supset \ldots$ of ideals in S. By minimality of J_{q+1}, we have $K \supset J_{q+1}$. Therefore $\varphi : S \to R_q \times_A B = S/K$ factors through $S \to S/J_{q+1} = R_{q+1}$ as desired.

This completes the proof of Schlessinger's theorem.   □

Obstruction theory

A deformation functor F is called **formally smooth** or **unobstructed** if for each surjection $B \to A$ in $\mathbf{Art}_k$ the map $F(B) \to F(A)$ is surjective.

Note in particular that the deformation functor $h_{X,x}$ of basic example 1 is unobstructed if and only X is smooth at x.

The above notion is generalised by the notion of an obstruction theory, defined below. In these terms, F will be formally smooth if it admits an obstruction theory $(O_F, (o_e))$ where $O_F = 0$ (or more generally where each $o_e = 0$).

2.22 An **obstruction theory** $(O_F, (o_e))$ for a deformation functor F is a k-vector space O_F together with additional data (o_e) consisting of a set-map $o_e : F(A) \to O_F \otimes_k I$ associated to each small extension $e = (0 \to I \to B \to A \to 0)$ in $\mathbf{Art}_k$ such that the following conditions **(O1)** and **(O2)** are satisfied.

(O1) An element $\alpha \in F(A)$ lifts to an element of $F(B)$ if and only if $o_e(\alpha) = 0$.

(O2) The map o_e is functorial in e in the sense that given any commutative diagram

$$
\begin{array}{ccccccccc}
0 \to & I & \to B \to & A & \to 0 \\
& \downarrow & \downarrow & \downarrow & \\
0 \to & I' & \to B' \to & A' & \to 0
\end{array}
$$

with rows e and e' small extensions in $\mathbf{Art}_k$, the following induced square commutes:

$$
\begin{array}{ccc}
F(A) & \xrightarrow{o_e} & O_F \otimes I \\
\downarrow & & \downarrow \\
F(A') & \xrightarrow{o_{e'}} & O_F \otimes I'
\end{array}
$$

Let R be a complete noetherian local k-algebra with residue field k, so that R can be expressed as a quotient S/J where $S = k[[t_1, \ldots, t_n]]$ is the power-series ring in n variables where $n = \dim_k \mathfrak{m}/\mathfrak{m}^2$, $\mathfrak{m} \subset R$ is its maximal ideal, and $J \subset S$ is an ideal with $J \subset \mathfrak{n}^2$ where $\mathfrak{n} = (t_1, \ldots, t_n)$ is the maximal ideal of S. Given any small extension $e = (0 \to I \to B \to A \to 0)$ in $\mathbf{Art}_k$ and a homomorphism $\alpha : R \to A$, by arbitrarily lifting the generators t_i we get a homomorphism $\alpha' : S \to B$. This induces a homomorphism $\alpha'' : J \to I$, such that the following diagram commutes.

$$
\begin{array}{ccccccccc}
0 \to & J & \to S \to & R & \to 0 \\
& \alpha'' \downarrow & \alpha' \downarrow & \alpha \downarrow & \\
0 \to & I & \to B \to & A & \to 0
\end{array}
$$

As $\alpha'(\mathfrak{n}) \subset \mathfrak{m}_B$ for the maximal ideal $\mathfrak{n} \subset S$, it follows that $\alpha''(\mathfrak{n}J) \subset \mathfrak{m}_B I = 0$, hence α'' induces a map $\overline{\alpha} : J/\mathfrak{n}J \to I$. We can regard this as an element $\overline{\alpha} \in (J/\mathfrak{n}J)^* \otimes_k I$. This defines a linear map

$$
o_e : F(A) \to (J/\mathfrak{n}J)^* \otimes_k I : \alpha \mapsto \overline{\alpha}
$$

which can be seen to be well-defined (independent of the intermediate choice of α') and functorial in e. As the map $\alpha : R \to A$ admits a lift to a map $\beta : R \to B$ if and only if $\alpha'' = 0$ (equivalently, $\overline{\alpha} = 0$), the following proposition is proved.

Proposition 2.23. *For any complete noetherian local k-algebra R with residue field k, the above data $((J/\mathfrak{n}J)^*, (o_e))$ is an obstruction theory for h_R.*

Remark 2.24 The above obstruction theory for h_R is *minimal* in the sense that given any other obstruction theory (V, v_e) there exists a unique linear injection $(J/\mathfrak{n}J)^* \to V$ making the obvious diagrams commute. Consequently, if the deformation functor $F = h_{X,x}$ has an obstruction theory $(O_F, (o_e))$ with $\dim_x X = \dim T_x X - \dim O_F$, then X is a local complete intersection at x. (See for example [F-G] Theorem 6.2.4 and its corollaries.)

2.25 A tangent-obstruction theory $(T^1, T^2, (\phi_e), (o_e))$ for a deformation functor F consists of finite-dimensional k-vector spaces T^1 and T^2 together with the following additional data. For each small extension $e = (0 \to I \to B \to A \to 0)$ in $\mathbf{Art}_k$, we are given an action $\phi_e : F(B) \times T^1 \otimes I \to F(B)$ satisfying the conclusions (i)-(iii) of Proposition 2.18, and a set map $o_e : F(A) \to T^2 \otimes I$ such that $(T^2, (o_e))$ is an obstruction theory for F. Then such an F automatically satisfies **(H1)**, **(H2)** and **(H3)**, and $(T^1, (\phi_e))$ is isomorphic to T_F together with its natural action (exercise). Moreover, the action of $T^1 \otimes I$ on $F(B)$ is free if an only if F also satisfies **(H4)**. Hence the following chain of implications holds for any deformation functor F.

Pro-representability $\Rightarrow$ Existence of a tangent-obstruction theory

$\Rightarrow$ Existence of a hull.

3 Calculations for basic examples

Preliminaries on flatness and base-change

All our examples involve flat families over base A, where $A \in \mathbf{Art}_k$, and so tools for verification of flatness are important. Here we have gathered together all the flatness statements we need. The reader in a hurry can skip this part and return to it as needed.

Exercise 3.1 (i) (Nilpotent Nakayama) Let A be a ring and $J \subset A$ a nilpotent ideal (means $J^n = 0$ for $n \gg 0$). If M is any A-module (not necessarily finitely generated) with $M = JM$, then show that $M = 0$.

(ii) (Schlessinger [S] Lemma 3.3) Apply (i) to show the following. Let A be a ring and $J \subset A$ a nilpotent ideal. Let $u : M \to N$ be a homomorphism of A-modules where N is flat over A. If $\bar{u} : M/JM \to N/JN$ is an isomorphism, then u is an isomorphism.

(iii) (Flat equivalent to free) Deduce from (ii) that flatness is equivalent to freeness for any module over an artin local ring.

(iv) (Tor vanishing implies flatness) Let A be an artin local ring, and M an A-module (not necessarily finitely generated). Then M is flat if and only if $Tor_1^A(A/\mathfrak{m}, M) = 0$.

(v) (Flatness over $k\langle V \rangle$) Apply the above Tor_1-vanishing criterion to prove the following. Let k be a field and V a finite dimensional k-vector space. Let M a module over $k\langle V \rangle$, not necessarily finitely generated. Then M is flat over $k\langle V \rangle$ if and only if the map $V \otimes_k (M/VM) \to VM$, induced by scalar multiplication, is an isomorphism.

The following lemma is an example of **non-flat descent**: even though $\mathrm{Spec}(A') \to \mathrm{Spec}(B)$ and $\mathrm{Spec}(A'') \to \mathrm{Spec}(B)$ is not necessarily a flat cover of $\mathrm{Spec}(B)$, we get a flat module N on B from flat modules M' and M'' on A' and A''.

Lemma 3.2. (Schlessinger [S] Lemma 3.4) *Let $A' \to A$ and $A'' \to A$ be ring homomorphisms, such that $A'' \to A$ is surjective with its kernel a nilpotent ideal $J \subset A''$. Let $B = A' \times_A A''$, with $B \to A'$ and $B \to A''$ the projections. Let M, M' and M'' be modules over A, A', A'', together with A'-linear homomorphism $u' : M' \to M$ and A''-linear homomorphism $u'' : M'' \to M$ which give isomorphisms $M' \otimes_{A'} A \to M$ and $M'' \otimes_{A''} A \to M$. Let N be the B-module $N = M' \times_M M'' = \{(x', x'') \in M' \times M'' \,|\, u'(x') = u''(x'') \in M\}$, where scalar multiplication by elements $(a', a'') \in B$ is defined by $(a', a'') \cdot (x', x'') = (a'x', a''x'')$. If M' and M'' are flat modules over A' and A'' respectively, then N is flat over B. Moreover, the projection maps $N \to M'$ and $N \to M''$ induce isomorphisms $N \otimes_B A' \xrightarrow{\sim} M'$ and $N \otimes_B A'' \xrightarrow{\sim} M''$.*

Proof We will prove this only in the case where M' is a free A'-module. Note that if A' is artin local, then we are automatically in this case by Exercise 3.1.(iii). This is therefore the only case which we need in these notes.

Let $(x_i')_{i \in I}$ be a free basis for M' over A'. As $M' \otimes_{A'} A \to M$ is an isomorphism, this gives a free basis $u'(x_i')$ of M over A. As $A'' \to A$ is surjective, any element $\sum y_j'' \otimes a_j''$ of $M'' \otimes_{A''} A$ equals $x'' \otimes 1$ for some (not necessarily uniquely determined) element $x'' \in M''$. Therefore the assumption of surjectivity of $M'' \otimes_{A''} A \to M$ tells us that $u'' : M'' \to M$ must be surjective. Hence we can choose elements $x_i'' \in M''$ such that $u''(x_i'') = u'(x_i')$. Let $N = \oplus_I A''$ be the free A''-module on the set I, with standard basis denoted by $(e_i)_{i \in I}$, and let $u : N \to M''$ be defined by $e_i \mapsto x_i''$. Then $\bar{u} : N/JN \to M''/JM'' = M$ is an isomorphism. Therefore by Exercise 3.1.(ii), u is an isomorphism, which shows M'' is free with basis $(x_i'')_{i \in I}$. It follows that N is free over B, with basis $(x_i', x_i'')_{i \in I}$. It is now immediate that the projections $N \to M'$ and $N \to M''$ induce isomorphisms $N \otimes_B A' \xrightarrow{\sim} M'$ and $N \otimes_B A'' \xrightarrow{\sim} M''$. $\qquad\square$

Corollary 3.3. (Schlessinger [S] Corollary 3.6) *With hypothesis and notation as in the above lemma, let L be a B-module, and $q' : L \to M'$ and $q'' : L \to M''$ be B-linear homomorphisms, such that the following diagram commutes:*

$$
\begin{array}{ccc}
L & \xrightarrow{q''} & M'' \\
q' \downarrow & & \downarrow u'' \\
M' & \xrightarrow{u'} & M
\end{array}
$$

Suppose that q' induces an isomorphism $L \otimes_B A' \to M'$. Then the map $(q', q'') : L \to N = M' \times_M M''$ is an isomorphism of B-modules.

Proof The kernel of the projection $B \to A'$ is the ideal $I = 0 \times J \subset A' \times_A A'' = B$. The ideal I is nilpotent as by assumption J is nilpotent. The desired result follows by applying Exercise 3.1.(ii) to the B-homomorphism $u = (q', q'') : L \to N$, which becomes the given isomorphism $L \otimes_B A' \to M'$ on going modulo the nilpotent ideal $I \subset B$. $\qquad\square$

We will need the following well-known base change result of Grothendieck [EGA] (for an exposition also see Hartshorne [H] Theorem 12.11).

Theorem 3.4. *Let $S = \mathrm{Spec}(A)$ where A is a noetherian local ring. Let $\pi : \mathfrak{X} \to S$ be a proper morphism and $\mathcal{F}$ a coherent $\mathcal{O}_{\mathfrak{X}}$-module which is flat over S. Let $s \in S$ be the closed point with residue field denoted by k. Let $\mathfrak{X}_s$ be the fiber over s and let $\mathcal{F}_s = \mathcal{F}|_{\mathfrak{X}_s}$ denote the restriction of $\mathcal{F}$ to $\mathfrak{X}_s$. Let i be an integer, such that the natural map $H^i(\mathfrak{X}, \mathcal{F}) \otimes_A k \to H^i(\mathfrak{X}_s, \mathcal{F}_s)$ is surjective. Then for any A-module M, the induced map $H^i(\mathfrak{X}, \mathcal{F}) \otimes_A M \to H^i(\mathfrak{X}, \mathcal{F} \otimes_{\mathcal{O}_{\mathfrak{X}}} \pi^* M)$ is an isomorphism. In particular if $H^i(\mathfrak{X}_s, \mathcal{F}_s) = 0$ then $H^i(\mathfrak{X}, \mathcal{F}) = 0$.*

Both [EGA] and [H] give rather complicated proofs of the above, involving inverse limits over modules of finite length (which in [H] is done by invoking the theorem on formal functions). These can be replaced by a simple application of Nakayama lemma to the semi-continuity complex.

The following lemma will be used in the deformation theory for a coherent sheaf E which is simple (that is, $End(E) = k$), to prove the theorem that the deformation functor $\mathcal{D}_E$ of such a sheaf is pro-representable.

Lemma 3.5. *Let A be a noetherian local ring, let $S = \operatorname{Spec} A$, and let $\pi : \mathfrak{X} \to S$ be a proper morphism. Let X denote the schematic fiber of π over the closed point $\operatorname{Spec} k$, where k is the residue field of A. Let $\mathcal{E}$ be a coherent sheaf on $\mathfrak{X}$ such that $\mathcal{E}$ is flat over S. Assume that there exists an exact sequence $\mathcal{F}_1 \to \mathcal{F}_0 \to \mathcal{E} \to 0$ of $\mathcal{O}_{\mathfrak{X}}$-modules, where $\mathcal{F}_1$ and $\mathcal{F}_0$ are locally free (note that this condition is automatically satisfied when $\mathcal{E}$ itself is locally free, or when $\pi : \mathfrak{X} \to S$ is a projective morphism). Let $E = \mathcal{E}|_X$ be the restriction of $\mathcal{E}$ to X. If the ring homomorphism $k \to End_X(E)$ (under which k acts on E by scalar multiplication) is an isomorphism, then for any morphism $f : T \to S$, the natural ring homomorphism*

$$H^0(T, \mathcal{O}_T) \to End_{\mathfrak{X}_T}((\operatorname{id} \times f)^*\mathcal{E})$$

(under which $H^0(T, \mathcal{O}_T)$ acts on $(\operatorname{id} \times f)^*\mathcal{E}$ by scalar multiplication) is an isomorphism.

Proof Consider the contravariant functor $End(\mathcal{E})$ from S-schemes to sets, which associates to any S-scheme $f : T \to S$ the set $End(\mathcal{E})(T) = End_{\mathfrak{X}_T}((\operatorname{id} \times f)^*\mathcal{E})$. Then by a fundamental theorem of Grothendieck (EGA III 7.7.8, 7.7.9), there exists a coherent sheaf $\mathcal{Q}$ on S and a functorial $H^0(T, \mathcal{O}_T)$-module isomorphism $\alpha_T : End_{\mathfrak{X}_T}((\operatorname{id} \times f)^*\mathcal{E}) \to Hom_T(f^*\mathcal{Q}, \mathcal{O}_T)$. As $S = \operatorname{Spec} A$, the coherent sheaf $\mathcal{Q}$ corresponds to the finite A-module $Q = H^0(S, \mathcal{Q})$. Consider the isomorphism $\alpha_S : End_{\mathfrak{X}}(\mathcal{E}) \to Hom_S(\mathcal{Q}, \mathcal{O}_S) = Hom_A(Q, A)$. Let $\theta : Q \to A$ be the image of $1_\mathcal{E}$ under α_S. By functoriality, the restriction $\theta_k : Q \otimes_A k \to k$ of θ to $\operatorname{Spec} k$ is the image of 1_E under the isomorphism $\alpha_k : End_X(E) \to Hom_k(Q \otimes_A k, k) = Hom_A(Q, A)$.

As by assumption $k \to End_X(E)$ is an isomorphism, by composing with α_k we get an isomorphism $k \mapsto Hom_k(Q \otimes_A k, k)$ under which $1 \mapsto \theta_k$. Hence $Hom_k(Q \otimes_A k, k)$ is 1-dimensional as a k-vector space with basis θ_k. Therefore θ_k is surjective, and so by Nakayama it follows that $\theta : Q \to A$ is surjective. Hence we have a splitting $Q = A \oplus N$ where

$N = \ker(\theta)$, under which the map $\theta : Q \to A$ becomes the projection $p_1 : A \oplus N \to A$ on the first factor. But as θ_k is an isomorphism, it again follows by Nakayama that $N = 0$. This shows that $\theta : Q \to A$ is an isomorphism.

Identifying $\mathcal{Q}$ with $\mathcal{O}_S$ under θ, for any $f : T \to S$ we have

$$Hom_T(f^*\mathcal{Q}, \mathcal{O}_T) = H^0(T, \mathcal{O}_T),$$

and so we get a functorial $H^0(T, \mathcal{O}_T)$-module isomorphism

$$\alpha_T : End_{\mathfrak{X}_T}((\mathrm{id} \times f)^*\mathcal{E}) \to H^0(T, \mathcal{O}_T)$$

which maps $1 \mapsto 1$. The composite map

$$H^0(T, \mathcal{O}_T) \to End_{\mathfrak{X}_T}((\mathrm{id} \times f)^*\mathcal{E}) \to H^0(T, \mathcal{O}_T)$$

is identity, so it follows that $H^0(T, \mathcal{O}_T) \to End_{\mathfrak{X}_T}((\mathrm{id} \times f)^*\mathcal{E})$ is an isomorphism. $\qquad\square$

Deformations of a coherent sheaf

Let X be a k-scheme of finite type, and E a coherent sheaf of $\mathcal{O}_X$-modules. We now return to the deformation functor $\mathcal{D}_E$ introduced earlier as our basic example 2.

The calculation of the tangent space to $\mathcal{D}_E$ in the special case where E is locally free is the first exposure many of us have had to deformation theory. So let us begin with this illuminating case. Let the vector bundle E be described by transition functions $g_{i,j}$ w.r.t an open cover U_i of X. If $(\mathcal{F}, \theta)$ is a deformation of E over $k[\epsilon]/(\epsilon^2)$, then $\mathcal{F}$ is again a vector bundle, which is trivial over each $U_i[\epsilon] = U_i \times \mathrm{Spec}\, k[\epsilon]/(\epsilon^2)$. Choose a trivialization for $\mathcal{F}|_{U_i[\epsilon]}$ which restricts under θ to the chosen trivialization for $E|_{U_i}$, so that $\mathcal{F}$ is described by transition functions of the form $g_{i,j} + \epsilon h_{i,j}$. The cocycle condition on the transition function now reads

$$h_{i,k} = g_{i,j}h_{j,k} + h_{i,j}g_{j,k}$$

over $U_{i,j,k}$, which means the $h_{i,j}$ regarded as an endomorphism of E over $U_{i,j}$ – going from the basis restricted from U_j to the basis restricted from U_i – define a Cech 1-cocycle $(h_{i,j}) \in Z^1((U_i), \underline{End}(E))$. The corresponding cohomology class $h \in H^1(X, \underline{End}(E))$ can be seen to be independent of the choice of open cover and local trivializations, and so we get a map $\mathcal{D}_E(k[\epsilon]/(\epsilon^2)) \to H^1(X, \underline{End}(E))$. We can define an inverse to this map by sending $(h_{i,j})$ to the pair $(\mathcal{F}, \theta)$ consisting of the bundle $\mathcal{F}$ on $X[\epsilon]$ defined by transition functions $g_{i,j} + \epsilon h_{i,j}$, with θ induced by identity. Hence the map $\mathcal{D}_E(k[\epsilon]/(\epsilon^2)) \to H^1(X, \underline{End}(E))$

is a bijection. An exact analogue of the above argument, where we replace $k[\epsilon]/(\epsilon^2)$ by $k\langle V\rangle$ for a finite-dimensional k-vector space V, gives a bijection $\mathcal{D}_E(k\langle V\rangle) \to H^1(X, \underline{End}(E)) \otimes V$. If $\phi : V \to W$ is a linear map, then the following square commutes.

$$
\begin{array}{ccc}
\mathcal{D}_E(k\langle V\rangle) & \to & H^1(X, \underline{End}(E)) \otimes V \\
{\scriptstyle k\langle \phi \rangle} \downarrow & & \downarrow {\scriptstyle id \otimes \phi} \\
\mathcal{D}_E(k\langle W\rangle) & \to & H^1(X, \underline{End}(E)) \otimes W
\end{array}
$$

Hence $\mathcal{D}_E$ satisfies **(Hϵ)**, and its tangent space is $H^1(X, \underline{End}(E))$. If X is proper over k, this is finite dimensional, so $\mathcal{D}_E$ satisfies **(H3)** in that case.

Theorem 3.6. *Let X be a proper scheme over a field k. Let E be a coherent sheaf on X. Then the deformation functor $\mathcal{D}_E$ admits a hull, with tangent space $Ext^1(E, E)$.*

Proof We will show that the conditions **(H1)**, **(H2)**, **(H3)** in the Schlessinger Theorem 2.19 are satisfied by our functor $\mathcal{D}_E$.

Verification of (H1): An element of $\mathcal{D}_E(A') \times_{\mathcal{D}_E(A)} \mathcal{D}_E(A'')$ is an ordered tuple $(\mathcal{F}', \theta', \mathcal{F}'', \theta'')$ where $(\mathcal{F}', \theta') \in \mathcal{D}_E(A')$ and $(\mathcal{F}'', \theta'') \in \mathcal{D}_E(A'')$, such that there exists an isomorphism $\eta : \mathcal{F}'|_A \to \mathcal{F}''|_A$ which makes the following diagram commute:

$$
\begin{array}{ccc}
\mathcal{F}'|_X & \overset{i^*(\eta)}{\to} & \mathcal{F}''|_X \\
{\scriptstyle \theta'} \downarrow & & \downarrow {\scriptstyle \theta''} \\
E & = & E
\end{array}
$$

We fix one such η. Let $\mathcal{F} = \mathcal{F}''|_A$, let $u'' : \mathcal{F}'' \to \mathcal{F}$ be the quotient and let $u' : \mathcal{F}' \to \mathcal{F}$ be induced by η. Let $B = A' \times_A A''$, and let $\mathcal{G}$ be the sheaf of $\mathcal{O}_{X_B}$-modules defined by

$$
\mathcal{G} = \mathcal{F}' \times_{u', \mathcal{F}, u''} \mathcal{F}''
$$

This is clearly coherent, as the construction can be done on each affine open and glued. By Lemma 3.2 applied stalk-wise, the sheaf $\mathcal{G}$ is flat over B. By Lemma 3.3 applied stalk-wise, this is up to isomorphism the only coherent sheaf on X_B, flat over B, which comes with homomorphisms $p' : \mathcal{G} \to \mathcal{F}'$ and $p'' : \mathcal{G} \to \mathcal{F}''$ which make the following square commute:

$$
\begin{array}{ccc}
\mathcal{G} & \overset{p''}{\to} & \mathcal{F}'' \\
{\scriptstyle p'} \downarrow & & \downarrow {\scriptstyle u''} \\
\mathcal{F}' & \overset{u'}{\to} & \mathcal{F}
\end{array}
$$

This shows that $\mathcal{D}_E(B) \to \mathcal{D}_E(A') \times_{\mathcal{D}_E(A)} \mathcal{D}_E(A'')$ is surjective, as desired. Thus, Schlessinger condition **(H1)** is satisfied.

Caution: If we choose another η, we might get a different $\mathcal{G}$, and so the map $\mathcal{D}_E(B) \to \mathcal{D}_E(A') \times_{\mathcal{D}_E(A)} \mathcal{D}_E(A'')$ may not be injective.

Verification of (H2): If we take A to be k in the above verification of the condition **(H1)**, then η would be unique, and so we will get a bijection $\mathcal{D}_E(A' \times_k A'') \to \mathcal{D}_E(A') \times_{\mathcal{D}_E(k)} \mathcal{D}_E(A'')$. In particular, this implies that **(H2)** is satisfied.

Verification of (H3): We have already seen in the special case when E is locally free that the finite dimensional vector space $H^1(X, \underline{End}(E)) = Ext^1(E, E)$ is the tangent space to $\mathcal{D}_E$. Now we give a proof that for a general coherent E, the tangent space is $Ext^1(E, E)$. This proof is very different in spirit, and in particular it gives another proof in the vector bundle case. For any finite dimensional vector space V over k, we define a map

$$f_V : V \otimes_k Ext^1(E, E) = Ext^1(V \otimes_k E, E) \to \mathcal{D}_E(k\langle V \rangle)$$

as follows. An element of $Ext^1(V \otimes_k E, E)$ is represented by a short exact sequence $S = (0 \to V \otimes_k E \xrightarrow{i} F \xrightarrow{j} E \to 0)$ $\mathcal{O}_X$-modules. We give F the structure of an $\mathcal{O}_{X[V]}$-module (where $X[V] = X \otimes_k k\langle V \rangle$) by defining the scalar-multiplication map $V \otimes_k F \to F$ as the composite $V \otimes_k F \xrightarrow{(\mathrm{id}_V, j)} V \otimes_k E \xrightarrow{i} F$. We denote the resulting $\mathcal{O}_{X[V]}$-module by $\mathcal{F}_S$. Note that the induced homomorphism $V \otimes_k (\mathcal{F}_S / V \mathcal{F}_S) \to V \mathcal{F}_S$ is an isomorphism, as it is just the identity map on $V \otimes_k E$. Hence by Exercise 3.1.(v), it follows that $\mathcal{F}_S$ is flat over $k\langle V \rangle$. Hence we indeed get an element of $\mathcal{D}_E(k\langle V \rangle)$, which completes the definition of the map $f_V : V \otimes_k Ext^1(E, E) \to \mathcal{D}_E(k\langle V \rangle)$.

It can be seen from its definition that f is functorial in V, that is, if $\phi : V \to W$ is a linear map, then the following square commutes.

$$\begin{array}{ccc} Ext^1(E, E) \otimes V & \to & \mathcal{D}_E(k\langle V \rangle) \\ {\scriptstyle id \otimes \phi} \downarrow & & \downarrow {\scriptstyle k\langle \phi \rangle} \\ Ext^1(E, E) \otimes W & \to & \mathcal{D}_E(k\langle W \rangle) \end{array}$$

Next, we give an inverse $g_V : \mathcal{D}_E(k\langle V \rangle) \to V \otimes_k Ext^1(E, E)$ to f_V as follows. Given any $(\mathcal{F}, \theta) \in \mathcal{D}_E(k\langle V \rangle)$, let $F = \pi_*(\mathcal{F})$ where $\pi : X[V] \to X$ is the projection induced by the ring homomorphism $k \hookrightarrow k\langle V \rangle$. Let $j : F \to E$ be the $\mathcal{O}_X$-linear map which is obtained from the $\mathcal{O}_X[V]$-linear map $\mathcal{F} \to \mathcal{F}|_X \xrightarrow{\theta} E$ by forgetting scalar multiplication by V. By flatness of $\mathcal{F}$ over $k\langle V \rangle$, the sequence $0 \to V \otimes_{k\langle V \rangle} \mathcal{F} \to \mathcal{F} \to \mathcal{F}|_X \to 0$ obtained by applying $- \otimes_{k\langle V \rangle} \mathcal{F}$ to $0 \to V \to k\langle V \rangle \to k \to 0$ is again exact. As $V \otimes_{k\langle V \rangle} \mathcal{F} = V \otimes_k (\mathcal{F}/V\mathcal{F})$, by composing with θ (and its inverse) this gives an exact sequence $S = (0 \to V \otimes_k E \xrightarrow{i} \mathcal{F} \xrightarrow{j} E \to 0)$. We define $g_V : \mathcal{D}_E(k\langle V \rangle) \to V \otimes_k Ext^1(E, E)$ by putting $g_V(\mathcal{F}, \theta) = S$.

Hence $\mathcal{D}_E$ satisfies **(Hε)**, and its tangent space is $Ext^1(E, E)$. If X is proper over k, this is finite dimensional, so $\mathcal{D}_E$ satisfies **(H3)** in that case. This completes the proof of the Theorem 3.6 in the general case of coherent sheaves.   □

Pro-Representability for a simple sheaf

Theorem 3.7. *Let X be a proper scheme over a field k, and let F be a coherent sheaf on X. Assume that there exists an exact sequence $E_1 \to E_0 \to F \to 0$ of $\mathcal{O}_X$-modules, where E_1 and E_0 are locally free (note that this condition is automatically satisfied when F itself is locally free, or when X is projective over k). If the ring homomorphism $k \to End(F)$ (under which k acts on F by scalar multiplication) is an isomorphism, then the deformation functor $\mathcal{D}_F$ is pro-representable.*

Proof Let A be artin local, and let I be a proper ideal. Let $(\mathcal{F}, \theta) \in \mathcal{D}_F(A)$, and let $(\mathcal{F}', \theta')$ denote its restriction to A/I. By Lemma 3.5, the natural ring homomorphisms $A \to End_{X \otimes A/I}(\mathcal{F})$ and $A/I \to End_{X \otimes A/I}(\mathcal{F}')$, under which A and A/I act respectively on $\mathcal{F}$ and $\mathcal{F}'$ by scalar multiplication, are isomorphisms. In particular, we get induced group isomorphisms $A^\times \to Aut(\mathcal{F})$ and $(A/I)^\times \to Aut(\mathcal{F}')$. The subgroups $1 + \mathfrak{m}_A \subset A^\times$ and $1 + \mathfrak{m}_{A/I} \subset (A/I)^\times$ therefore map isomorphically onto $Aut(\mathcal{F}, \theta)$ and $Aut(\mathcal{F}', \theta')$ respectively. As the homomorphism $1 + \mathfrak{m}_A \to 1 + \mathfrak{m}_{A/I}$ is surjective, the restriction map $Aut(\mathcal{F}, \theta) \to Aut(\mathcal{F}', \theta')$ is again surjective. From this, it follows that the Schlessinger condition **(H4)** is satisfied, and so the functor $\mathcal{D}_F$ is pro-representable by Theorem 2.19.   □

Obstruction theory for deformations of a coherent sheaf

Theorem 3.8. *Let X be a projective scheme over a field k, and let F be a coherent sheaf on X. Then the deformation functor $\mathcal{D}_F$ admits an obstruction theory $(Ext^2(F, F), (o_e))$. In particular when $Ext^2(F, F) = 0$*

the functor $\mathcal{D}_F$ is smooth, that is, for any surjection $B \to A$ in $\mathbf{Art}_k$, the induced map $\mathcal{D}_F(B) \to \mathcal{D}_F(A)$ is surjective.

Proof Let $\mathcal{O}_X(1)$ be a chosen very ample line bundle on X. Then for for all n sufficiently large, the higher cohomologies $H^i(X, F(n))$ vanish, and by evaluating global sections we get a surjection $q_0 : H^0(X, F(n)) \otimes_k \mathcal{O}_X(-n) \to F$. Choose a large enough n, and let E be the corresponding vector bundle $E = H^0(X, F(n)) \otimes_k \mathcal{O}_X(-n)$. Let Q be the deformation functor of basic example 3, which keeps E fixed and deforms the quotient q_0. For A in $\mathbf{Art}_k$, given any element $q : E \otimes_k A \to \mathcal{F}$ of $Q(A)$, the sheaf $\mathcal{F}$ together with the unique isomorphism $\theta : F \to \mathcal{F} \otimes_A k$ which takes q_0 to $q \otimes_A k$ defines an element $(\mathcal{F}, \theta)$ of $\mathcal{D}_F(A)$. This association is functorial, and so we have a forgetful functor $f : Q \to \mathcal{D}_F$. Let A be in $\mathbf{Art}_k$, and let $(\mathcal{F}, \theta) \in \mathcal{D}_F(A)$. The surjectivity of $q_0 : H^0(X, F(n)) \otimes_k \mathcal{O}_X(-n) \to F$ implies the surjectivity of the evaluation map $p : H^0(X_A, \mathcal{F}(n)) \otimes_A \mathcal{O}_{X_A}(-n) \to \mathcal{F}$. As $\mathcal{F}$ is flat over A and as higher cohomologies of $F(n)$ are zero, it follows that $H^0(X_A, \mathcal{F}(n))$ is a free A-module, of the same rank as $\dim_k H^0(X, F(n))$. Hence we can choose an isomorphism $\phi : H^0(X, F(n)) \otimes_k A \to H^0(X_A, \mathcal{F}(n))$ which restricts to identity modulo $\mathfrak{m}_A$. Consider the composite surjection $q = p \circ \phi : E \otimes_k A \to \mathcal{F}$. Then $q \in Q(A)$ maps to $(\mathcal{F}, \theta) \in \mathcal{D}_F(A)$ under the forgetful functor $f : Q \to \mathcal{D}_F$. This shows that the forgetful functor $f : Q \to \mathcal{D}_F$ is formally smooth. We have shown later (Theorem 3.11) that Q has an obstruction theory taking values in $Ext_X^1(G, F)$ where G is the kernel of $q : E \to F$. We have $Ext_X^i(E, F) = H^i(X, \underline{Hom}(E, F)) = H^i(X, F(n)) \otimes H^0(X, F(n))^* = 0$ for all $i \geq 1$. Applying $Hom_X(-, F)$ to the short-exact sequence $0 \to G \to E \to F \to 0$ therefore gives an isomorphism $\partial : Ext_X^1(G, F) \to Ext_X^2(F, F)$. If $e = (0 \to I \to B \to A \to 0)$ is a small extension in $\mathbf{Art}_k$, we have an obstruction map $o_e : Q(A) \to Ext_X^1(G, F) \otimes I$.

Let $q' : E_A \to \mathcal{F}$ be another homomorphism with $q'|_X = q_0 = q|_X$. Then q' is necessarily surjective, and all such q' form the fiber of $Q(A) \to \mathcal{D}_F(A)$ containing q. Let $\mathcal{G} = \ker(q)$ and $\mathcal{G}' = \ker(q')$. These are flat over A, with $\mathcal{G}|_X = G = \mathcal{G}'_X$. The corresponding obstruction classes ω_A and ω'_A for lifting these to $Q(B)$ (see the proof of Theorem 3.11) respectively lie in the vector spaces $Ext_{X_A}^1(I \otimes_k F, \mathcal{G})$ and $Ext_{X_A}^1(I \otimes_k F, \mathcal{G}')$. But under the isomorphisms of these spaces with $Ext_X^1(G, F) \otimes I$, it follows from $q'|_X = q_0 = q|_X$ and $\mathcal{G}|_X = G = \mathcal{G}'_X$ that ω_A and ω'_A map to the same element of $Ext_X^1(G, F) \otimes I$, that is, $o_e(q) = o_e(q') \in Ext_X^1(G, F) \otimes I$. Therefore o_e is constant on fibers of $Q(A) \to \mathcal{D}_F(A)$, so we get a unique map $o'_e : \mathcal{D}_F(A) \to Ext^2_X(F, F) \otimes I$ such that the following diagram commutes:

$$
\begin{array}{ccccc}
Q(B) & \to & Q(A) & \xrightarrow{o_e} & Ext^1_X(G, F) \otimes I \\
f_B \downarrow & & f_A \downarrow & & \downarrow \partial \otimes id_I \\
\mathcal{D}_F(B) & \to & \mathcal{D}_F(A) & \xrightarrow{o'_e} & Ext^2_X(F, F) \otimes I
\end{array}
$$

From the commutativity of the above diagram, the surjectivity of the first two vertical maps and the fact that the last vertical map is an isomorphism, it follows that the lower row is exact. We leave the verification of the functoriality condition on o'_e (using the functoriality of o_e) as an exercise to the reader. Thus, $(Ext^2_X(F, F), (o'_e))$ is an obstruction theory for $\mathcal{D}_F$. $\square$

Example 3.9 Consider the projective line $\mathbf{P}^1$ over k with standard open cover $U_0 = \operatorname{Spec} k[z]$ and $U_\infty = \operatorname{Spec} k[z^{-1}]$. Let $E = \mathcal{O}(-1) \oplus \mathcal{O}(1)$. Then the tangent space $Ext^1(E, E)$ to $\mathcal{D}_E$ is 1-dimensional, and $\mathcal{D}_E$ is formally smooth as $Ext^2(E, E) = 0$. Hence E admits a hull parametrised by the formal power series ring $k[[t]]$. The transition function

$$
g_{0,\infty}(t) = \begin{pmatrix} z^{-1} & t \\ 0 & z \end{pmatrix}
$$

over the open cover of $\mathbf{P}^1 \times \mathbf{A}^1$ given by $U_0 \times \mathbf{A}^1$ and $U_\infty \times \mathbf{A}^1$ defines a vector bundle $\mathcal{E}(t)$ on $\mathbf{P}^1 \times \mathbf{A}^1$. As $g_{0,\infty}(0) = \operatorname{diag}(z^{-1}, z)$, this comes with an isomorphism $\theta : \mathcal{E}(t)|_{\mathbf{P}^1} \to E$. (This is actually the universal family of extensions of $\mathcal{O}(1)$ by $\mathcal{O}(-1)$.) Going modulo (t^2), the restriction $(\mathcal{E}(\epsilon), \theta)$ gives a universal first order family for $\mathcal{D}_E$ parametrised by $k[\epsilon]/(\epsilon^2)$. Hence by inverse function theorem for $k[[t]]$, the pro-family $(\mathcal{E}(t), \theta)$ (obtained by restrictions to each $k[t]/(t^n)$) is miniversal. The pro-family $(\mathcal{E}(t + t^2), \theta)$ defined by transition matrix $g(t+t^2)$ is isomorphic to the original family, as $g(t+t^2) = h(t)g(t)h(t)^{-1}$ where $h(t) = diag(1 + t, 1)$ is invertible over $k[[t]]$, with $h(0) = I$. Hence the non-trivial automorphism $k[[t]] \to k[[t]]$ defined $t \mapsto t + t^2$ pulls back the miniversal family to another such. Hence the family is not universal. Hence the condition of simplicity in Theorem 3.7 is not superfluous.

Finally, for the moduli functor $\mathcal{M}_E$, the set $\mathcal{M}_E(k[\epsilon]/(\epsilon^2))$ has exactly two elements, namely $[\mathcal{E}(\epsilon)]$ and $[\mathcal{E}(0)]$. This cannot be a vector space when k is larger than $\mathbf{Z}/(2)$. In particular the functor $\mathcal{M}_E$ does not have the good properties of $\mathcal{D}_E$.

Homological preliminaries for the Quot functor

Consider a short-exact sequence $s = (0 \to M' \to M \to M'' \to 0)$ in an abelian category, together with monomorphisms $u' : N' \to M'$ and $u'' : N'' \to M''$. An **exact filler** for (s, u', u'') will mean a monomorphism $u : N \hookrightarrow M$ such that we have morphisms $N' \to N$ and $N \to N''$ (necessarily unique) which give the following commutative diagram with short-exact rows.

$$
\begin{array}{ccccccccc}
0 \to & N' & \to & N & \to & N'' & \to 0 \\
 & \downarrow & & \downarrow & & \downarrow & \\
0 \to & M' & \to & M & \to & M'' & \to 0
\end{array}
$$

Lemma 3.10. *Let $\mathcal{C}$ be an abelian category and let $s = (0 \to M' \xrightarrow{i} M \xrightarrow{j} M'' \to 0)$ be a short exact sequence in $\mathcal{C}$. Let $u' : N' \hookrightarrow M'$ and $u'' : N'' \hookrightarrow M''$ be given sub-objects. Then the following holds.*

(1) Under the natural map $Ext^1_{\mathcal{C}}(M'', M') \to Ext^1_{\mathcal{C}}(N'', M'/N')$, the image of the class s is the class ω of the induced short exact sequence

$$\omega = (0 \to M'/N' \to j^{-1}N''/N' \to N'' \to 0).$$

There exists an exact filler for (s, u', u'') if and only if $\omega = 0$.

(2) The set $\mathbf{S}$ of all isomorphism classes of exact fillers for the above diagram is in a natural bijection φ (described within the proof) with the set $\mathbf{L}$ of all lifts $h : N'' \to M/N'$ of $N'' \hookrightarrow M''$. The set $\mathbf{L}$ admits a natural action of the abelian group $Hom_{\mathcal{C}}(N'', M'/N')$, under which an element $\alpha \in Hom_{\mathcal{C}}(N'', M'/N')$ acts by $h \mapsto h + \alpha$. This action makes $\mathbf{L}$ (and hence also $\mathbf{S}$ via φ) a principal $Hom_{\mathcal{C}}(N'', M'/N')$-set (which by (1) is non-empty if and only if $\omega = 0$).

*(3) (**Equivariance**)   With notation as before, suppose we have a commutative square*

$$
\begin{array}{ccc}
N' & \to & L' \\
u' \downarrow & & \downarrow v' \\
M' & \to & K'
\end{array}
$$

where $v' : L' \to K'$ is monic. Let

$$f : Hom_{\mathcal{C}}(N'', M'/N') \to Hom_{\mathcal{C}}(N'', K'/L')$$

denote the homomorphism induced by the above commutative square. Let $K = K' \coprod_{M'} M$ be the push-out, so that we have the following commutative diagram with short-exact rows.

$$
\begin{array}{ccccccccc}
0 \to & M' & \to & M & \to & M'' & \to 0 \\
 & \downarrow & & \downarrow & & \| & \\
0 \to & K' & \to & K & \to & M'' & \to 0
\end{array}
$$

Let s' denote the bottom row of the above diagram, and let $\mathbf{S}'$ be the set of isomorphism classes of exact fillers for (s', v', u'') which by (2) is in natural bijection with the set $\mathbf{L}'$ of section over N'' of $K/L' \to M''$. Given any $h : N'' \to M/N'$ in $\mathbf{L}$, we get an element $h' : N'' \to K/L'$ in $\mathbf{L}'$ by composing with the homomorphism $M/N' \to K/L'$. This defines a natural map $\mathbf{S} \to \mathbf{S}'$, which is equivariant under the homomorphism $f : Hom_{\mathcal{C}}(N'', M'/N') \to Hom_{\mathcal{C}}(N'', K'/L')$.

Proof Let $j^{-1}N'' \subset M$ denote the pull-back of $N'' \subset M''$ under $j : M \to M''$. By definition, the image ω of e in $Ext_{\mathcal{C}}^1(N'', M'/N')$ is the extension class of the short exact sequence $0 \to M'/N' \to j^{-1}N''/N' \to N'' \to 0$. Therefore $\omega = 0$ if and only if there exists a 'lift' $h : N'' \hookrightarrow M/N'$ with $\bar{j} \circ h = u'' : N'' \hookrightarrow M''$ where $\bar{j}$ is induced by j.

An exact filler $u : N \hookrightarrow M$ induces a sub-object $\bar{u} : N/N' \hookrightarrow M/N'$. As we have an isomorphism $N/N' \to N''$, this gives a lift $h : N'' \to M/N'$ of $u'' : N'' \hookrightarrow M''$. We define $\varphi : \mathbf{S} \to \mathbf{L}$ by $u \mapsto h$. Conversely, the pull-back of the sub-object (N'', h) of M/N' under the quotient morphism $M \to M/N'$ is a sub-object (N, u) of M, which defines an inverse for φ, showing it is a bijection.

It is clear that $\mathbf{L}$ is a principal $Hom_{\mathcal{C}}(N'', M'/N')$-set under the given action. The rest of the lemma is now a simple exercise. $\square$

Pro-representability and tangent space for the Quot functor

Let X be a proper scheme over k. Let E be a coherent $\mathcal{O}_X$-module over X, and let $q_0 : E \to F_0$ be a coherent quotient $\mathcal{O}_X$-module. Let Q be the deformation functor for the above quotient, which was introduced as our basic example 3.

The following result is essentially due to Grothendieck, though re-cast in the language of Schlessinger.

Theorem 3.11. *Let k be any field, X proper over k, and let $q_0 : E \to F_0$ be a surjective morphism of coherent $\mathcal{O}_X$-modules. Let Q denote the corresponding deformation functor. Then we have the following.*

(1) The functor Q is pro-representable.

(2) It has tangent space $T_Q = Hom_X(G_0, F_0)$ where $G_0 = \ker(q_0)$.

(3) There exists a deformation theory for Q taking values in $Ext_X^1(G_0, F_0)$. In particular if $Ext_X^1(G_0, F_0) = 0$, then the functor Q is formally smooth.

We begin with the proof of pro-representability.

If $X \to \operatorname{Spec} k$ is projective, then as proved by Grothendieck (see for example [Ni]), there exists a scheme $Quot_{E/X}$ (the **quot scheme**) of locally finite type over k, whose S-valued points for any k-scheme S are equivalence classes of S-flat families of coherent quotients of E over $X \times_k S$. The given quotient $E \to F_0$ defines a k-rational point $q_0 \in Quot_{E/X}$. Hence the functor Q is just the functor of deformations of the point q_0 in $Quot_{E/X}$, so this is a case of basic example 1, hence is representable (which is more than being pro-representable).

The general case, where the proper morphism $X \to \operatorname{Spec} k$ need not be projective, is treated via Schlessinger's theorem for pro-representability, which has the following obvious corollary.

Theorem 3.12. *A deformation functor φ is pro-representable if and only if the following two conditions are satisfied.*

(1) For any morphisms $A' \to A$ and $A'' \to A$ in $\mathbf{Art}_k$ such that $A'' \to A$ is surjective, the induced map $\varphi(A' \times_A A'') \to \varphi(A') \times_{\varphi(A)} \varphi(A'')$ is a bijection of sets.

(2) The tangent vector space $T_\varphi = \varphi(k[\epsilon]/(\epsilon)^2)$ is finite dimensional (this is indeed a vector space when (1) is satisfied).

We now show that the condition (1) in the Theorem 3.12 is satisfied by our functor Q. An element of $Q(A') \times_{Q(A)} Q(A'')$ has the form (q', q''), where $q' : E_{A'} \to \mathcal{F}'$ and $q'' : E_{A''} \to \mathcal{F}''$ are coherent quotients over $X_{A'}$ and $X_{A''}$ respectively, with $\mathcal{F}'$ flat over A' and $\mathcal{F}''$ flat over A'', such that there exists an isomorphism $\eta : \mathcal{F}'|_A \to \mathcal{F}''|_A$ such that the following diagram commutes:

$$
\begin{array}{ccc}
E_A & = & E_A \\
{\scriptstyle q'|_A}\downarrow & & \downarrow{\scriptstyle q''|_A} \\
\mathcal{F}'|_A & \xrightarrow{\eta} & \mathcal{F}''|_A
\end{array}
$$

Note that if a η exists as above, it is necessarily unique by surjectivity of the vertical maps.

Note The above uniqueness of η is the reason why the functor Q is pro-representable, in contrast with the functor of deformations of a coherent sheaf, where we only had a hull. Recall that the corresponding isomorphism η was not unique in the case of the functor of deformations of a coherent sheaf.

Let $\mathcal{F} = \mathcal{F}''|_A$, let $u'' : \mathcal{F}'' \to \mathcal{F}$ be the quotient and let $u' : \mathcal{F}' \to \mathcal{F}$ be induced by η. Let $B = A' \times_A A''$, and let $\mathcal{G}$ be the sheaf of $\mathcal{O}_{X_B}$-modules defined by

$$
\mathcal{G} = \mathcal{F}' \times_{u', \mathcal{F}, u''} \mathcal{F}''
$$

This is clearly coherent, as the construction can be done on each affine open and glued. By Lemma 3.2 applied stalk-wise, the sheaf $\mathcal{G}$ is flat over

B. By Lemma 3.3 applied stalk-wise, this is up to isomorphism the only coherent sheaf on X_B, flat over B, which comes with homomorphisms $p' : \mathcal{G} \to \mathcal{F}'$ and $p'' : \mathcal{G} \to \mathcal{F}''$ which make the following square commute:

$$
\begin{array}{ccc}
\mathcal{G} & \overset{p''}{\to} & \mathcal{F}'' \\
{\scriptstyle p'}\downarrow & & \downarrow{\scriptstyle u''} \\
\mathcal{F}' & \overset{u'}{\to} & \mathcal{F}
\end{array}
$$

Next, let $p : E_B \to \mathcal{G}$ be the $\mathcal{O}_{X_B}$-linear homomorphism induced by (q', q''). This is clearly surjective, and is the only $\mathcal{O}_{X_B}$-linear homomorphism which restricts to q' and q'' over A' and A''. This shows that $Q(B) \to Q(A') \times_{Q(A)} Q(A'')$ is bijective, as desired.

Therefore, to complete the proof of pro-representability, it only remains to verify the condition (2) of Theorem 3.12. This we do next, when we determine the space T_Q.

Remark 3.13 Let $B \to A$ be surjection of rings with kernel I, such that $I^2 = 0$, so that I is naturally an A-module. Then the natural map $I \otimes_B M \to I \otimes_A (M/IM) : \sum b_i \otimes_B x_i \mapsto \sum b_i \otimes_A \overline{x_i}$ is an isomorphism of B-modules for any B-module M.

Lemma 3.14. *Let $B \to A$ be surjection of rings with kernel I, such that $I^2 = 0$. Let M be a flat B-module (not necessarily finitely generated). Let $u'' : \mathcal{G} \hookrightarrow A \otimes_B M = M/IM$ be an A-submodule, such that the quotient $\mathcal{F} = (A \otimes_B M)/\mathcal{G}$ is a flat A-module. In particular, the induced map $u' : I \otimes_A \mathcal{G} \to I \otimes_A (A \otimes_B M) = I \otimes_B M$ is monic, where the last equality is by Remark 3.13. As M is B-flat, the sequence $s = (0 \to I \otimes_B M \to M \to A \otimes_B M \to 0)$ is exact. For any exact filler $u : N \hookrightarrow M$ of (s, u', u''), consider the resulting commutative diagram of B-modules*

$$
\begin{array}{ccccccccc}
0 \to & I \otimes_A \mathcal{G} & \to & N & \to & \mathcal{G} & \to 0 \\
 & {\scriptstyle u'}\downarrow & & {\scriptstyle u}\downarrow & & {\scriptstyle u''}\downarrow & \\
0 \to & I \otimes_B M & \to & M & \to & A \otimes_B M & \to 0
\end{array}
$$

where the rows are exact. Then we have the following:

(1) The submodule $I \otimes_A \mathcal{G} \subset N$ from the top row is the submodule $IN \subset N$. Consequently, the quotient map $N \to \mathcal{G}$ induces an isomorphism $A \otimes_B N \to \mathcal{G}$.

(2) If moreover $B \to A$ is a local homomorphism of artin local rings, then the quotient module M/N is flat over B.

Proof (1) As I annihilates $\mathcal{G}$, from the surjection $N \to \mathcal{G}$ it follows that $IN \subset \ker(N \to \mathcal{G}) = I \otimes_A \mathcal{G}$. For the reverse inclusion, consider an element $b \otimes_A g$ of $I \otimes_A \mathcal{G}$. Under the inclusion $\mathcal{G} \subset A \otimes_B M$, it follows that g can be written as $1 \otimes_B x$ for some $x \in M$. Then $b \otimes_A g$ maps to the element $b \otimes_A (1 \otimes_B x) = bx \in M$ under the composite $I \otimes_A \mathcal{G} \hookrightarrow I \otimes_B M \hookrightarrow M$. As $N \to \mathcal{G}$ is surjective, there exists some element $y \in N \subset M$ which maps to $g \in \mathcal{G}$, that is, $1 \otimes_B y = g \in A \otimes_B M$. This means $1 \otimes_B x = g = 1 \otimes_B y$, so $1 \otimes_B (x - y) = 0 \in A \otimes_B M$, which means $x - y \in IM$. As by assumption $I^2 = 0$, it follows that $bx = by$. This shows $bx \in IN$, proving the desired inclusion $I \otimes_A \mathcal{G} \subset IN$.

(2) As $N \to \mathcal{G}$ is the quotient $N \to N/IN$, it follows that $M/N \to \mathcal{F}$ is the quotient $(M/N)/I(M/N)$. In other words, applying $A \otimes_B -$ to $0 \to N \to M \to M/N \to 0$ produces the exact sequence $0 \to \mathcal{G} \to M/IM \to \mathcal{F} \to 0$ of A-modules. If $\mathfrak{m}_A \subset A$ and $\mathfrak{m}_B \subset B$ are the maximal ideals, then we have $B/\mathfrak{m}_B = A/\mathfrak{m}_A = k$ say. Therefore, applying $k \otimes_B -$ to $0 \to N \to M \to M/N \to 0$ is the same as applying $k \otimes_A -$ to $0 \to \mathcal{G} \to M/IM \to \mathcal{F} \to 0$, and as $\mathcal{F}$ is A-flat, this gives an injection $k \otimes_A \mathcal{G} \to k \otimes_A (M/IM)$, which is just the map $k \otimes_B N \to k \otimes_B M$. As M is assumed to be B-flat, the injectivity of the above map shows that $Tor_1^B(k, M/N) = 0$. Hence by Exercise 3.1.(v), it follows that M/N is B-flat as desired. $\qquad\square$

Let $B \to A$ be a small extension with kernel I, let $q : E_A \to \mathcal{F}$ be in $Q(A)$, and let $\mathcal{G} \hookrightarrow E_A$ be the kernel of $q : E_A \to \mathcal{F}$. Note that $I \otimes_A \mathcal{G} = I \otimes_k G_0$ as $\mathfrak{m}_B I = 0$. Hence the above lemma has the following immediate corollary.

Lemma 3.15. *If $B \to A$ is a small extension with kernel I, then the fiber of $Q(B) \to Q(A)$ over $(q : E_A \to \mathcal{F}) \in Q(A)$ is in a natural bijection with the set of all exact fillers of the diagram*

$$
\begin{array}{ccccccccc}
& & I \otimes_k G_0 & & & & \mathcal{G} & & \\
& & \downarrow & & & & \downarrow & & \\
0 \to & & I \otimes_k E & \to & B \otimes_k E & \to & A \otimes_k E & \to 0
\end{array}
$$

To determine the tangent space T_Q, we apply the above description of the fibers of $Q(B) \to Q(A)$ in the case where $A = k$ and $B = k\langle V \rangle$ for any finite dimensional k-vector space V. As $Q(k)$ is a singleton set, this shows that $Q(k\langle V \rangle)$ is in a natural bijection with the set $\mathbf{S}_V$ of all exact fillers of the diagram

$$
\begin{array}{ccccccccc}
& & V \otimes_k G_0 & & & & G_0 & & \\
& & \downarrow & & & & \downarrow & & \\
0 \to & & V \otimes_k E & \to & k\langle V \rangle \otimes_k E & \to & E & \to 0
\end{array}
$$

The set $\mathbf{S}_V$ has a natural base-point $*_V$, given by the filler $k\langle V \rangle \otimes_k G_0 \hookrightarrow k\langle V \rangle \otimes_k E$. By Lemma 3.10, the set $\mathbf{S}_V$ is naturally a principal set under $Hom_X(G_0, V \otimes_k F_0)$. Therefore the base point gives a bijection $Hom_X(G_0, V \otimes_k F_0) \to \mathbf{S}_V$. Given any linear map $V \to W$ of finite dimensional k-vector spaces, by Lemma 3.10, we get an induces map $\mathbf{S}_V \to \mathbf{S}_W$, which is equivariant under the induces group homomorphism $Hom_X(G_0, V \otimes_k F_0) \to Hom_X(G_0, W \otimes_k F_0)$. Also, it maps the base point $*_V$ to the base point $*_W$. Hence the bijection $Hom_X(G_0, V \otimes_k F_0) \to \mathbf{S}_V$ is functorial on the category of finite dimensional k-vector spaces. Composing with the natural bijection $\mathbf{S}_V \to Q(k\langle V \rangle)$, we get a natural bijection $Hom_X(G_0, V \otimes_k F_0) \to Q(k\langle V \rangle)$ in the category of finite dimensional k-vector spaces.

This shows that the tangent space to Q is $Hom_X(G_0, F_0)$, proving Theorem 3.11.(2). As X is proper over k, the vector space $Hom_X(G_0, F_0)$ is finite dimensional, which completes the proof of Theorem 3.11.(1) via the Schlessinger criterion (**H3**).

Obstruction theory for Q

We now prove Theorem 3.11.(3). Given a small extension $e = (0 \to I \to B \to A \to 0)$, we define a map $o_e : Q(A) \to Ext^1_X(G_0, F_0) \otimes_k I$ as follows. By Lemma 3.15, the fiber of $Q(B) \to Q(A)$ over an element $(E_A \to \mathcal{F})$ is the set of exact fillers of the diagram

$$
\begin{array}{ccccccc}
 & I \otimes_A \mathcal{G} & & & & \mathcal{G} & \\
 & \downarrow & & & & \downarrow & \\
0 \to & I \otimes_A E_A & \to & E_B & \to & E_A & \to 0
\end{array}
$$

where $\mathcal{G} = \ker(E_A \to \mathcal{F})$. Let $s \in Ext^1_{X_B}(E_A, I \otimes_A E_A)$ be the extension class of $0 \to I \otimes_A E_A \xrightarrow{i} E_B \xrightarrow{j} E_A \to 0$. By Lemma 3.10(1), an exact filler exists for the above diagram if and only if its image $\omega = 0$, where $\omega \in Ext^1_{X_B}(\mathcal{G}, I \otimes_A \mathcal{F})$ is the extension class of

$$
0 \to \frac{I \otimes_A E_A}{I \otimes_A \mathcal{G}} \to \frac{j^{-1}(\mathcal{G})}{i(I \otimes_A \mathcal{G})} \to \mathcal{G} \to 0
$$

As $(I \otimes_A E_A)/(I \otimes_A \mathcal{G}) = I \otimes_A \mathcal{F} = I \otimes_k F_0$, we regard ω as an element of $Ext^1_{X_B}(\mathcal{G}, I \otimes_k F_0)$. As the module $j^{-1}(\mathcal{G})/i(I \otimes_A \mathcal{G})$ is annihilated by I, the above short exact sequence is a sequence of $\mathcal{O}_{X_A}$-modules, hence ω corresponds to an element

$$
\omega_A \in Ext^1_{X_A}(\mathcal{G}, I \otimes_k F_0).
$$

As X_A is projective over A, the module $\mathcal{G}$ admits a locally free resolution $\ldots \to \mathcal{L}_1 \to \mathcal{L}_0 \to \mathcal{G} \to 0$. As $\mathcal{G}$ is flat over A, the above restricted to X gives a resolution $\ldots \to L_1 \to L_0 \to G_0 \to 0$. The functor Ext can be calculated by using a locally free resolution (see for example Hartshorne [H] Chapter III Proposition 6.5). We therefore have

$$
\begin{aligned}
Ext^1_{X_A}(\mathcal{G}, I \otimes_k F_0) &= h^1(\underline{Hom}_{\mathcal{O}_{X_A}}(\mathcal{L}_\bullet, I \otimes_k F_0)) \\
&\quad \text{by [H] Chapter III Proposition 6.5} \\
&= h^1(\underline{Hom}_{\mathcal{O}_X}(\mathcal{L}_\bullet, I \otimes_k F_0)) \\
&\quad \text{as } I \otimes_k F_0 \text{ is annihilated by } \mathfrak{m}_A \\
&= I \otimes_k h^1(\underline{Hom}_{\mathcal{O}_X}(L_\bullet, F_0)) \\
&= I \otimes_k Ext^1_X(G_0, F_0) \\
&\quad \text{again by [H] Chapter III Proposition 6.5.}
\end{aligned}
$$

Hence the element ω associated to a given element of $Q(A)$ and a small extension e defines an element $\omega_k \in I \otimes_k Ext^1_X(G_0, F_0)$. We now define $o_e : Q(A) \to Ext^1_X(G_0, F_0) \otimes I$ by sending $q \mapsto \omega_A$. By its definition, this gives an exact sequence $Q(B) \to Q(A) \xrightarrow{o_e} Ext^1_X(G_0, F_0) \otimes I$. The reader may verify from its definition that o_e is functorial in e.

This completes the proof of the theorem. □

Deformations of a variety

Though our focus in these notes has been on vector bundles, historically the main source of motivation and ideas in deformation theory has been the study of deformations a variety. We have room here only to mention some basic facts.

Given a smooth variety X over a field k, consider the deformation functor $\mathbf{Def}_X$ of our basic example 4. Then $\mathbf{Def}_X$ admits a hull, with tangent space $H^1(X, T_X)$, where T_X is the tangent bundle of X. In particular if X is affine then all its infinitesimal deformations are trivial. When C is a smooth projective curve of genus $g \geq 2$, the tangent space is of dimension $3g - 3$. This is the so called 'Riemann's count', which is the historical beginning of deformation theory. When X is smooth, the functor $\mathbf{Def}_X$ admits an obstruction theory taking values in $H^2(X, T_X)$. For more on this subject, the reader can see the book by Sernesi [Se] and the notes by Vistoli [V].

Some suggestions for further reading

The above brief notes give just the beginning of the algebraic approach to deformation theory. For a graduate student wishing to study further, the next basic topic I would like to suggest is the use of the Grothendieck existence theorem to convert pro-families into 'actual' families. For this and further theoretical development of algebraic deformation theory (including the cotangent complex) with some of its important applications, a good introductory source is the lecture notes of Luc Illusie [I]. To see some examples of the application of basic deformation theory to vector bundles and moduli, the reader can consult the textbook of Huybrecht and Lehn [L-H].

Finally, I take this opportunity to express my gratitude to Peter Newstead. His lecture notes [Ne] and other writings have helped me (and in fact an entire generation of algebraic geometers) learn the foundations of vector bundles and moduli theory.

Bibliography

[EGA] A. Grothendieck : *Eléments de Géométrie Algébrique I - IV*, Pub. Math. IHES, 1960–1964.

[FGA] A. Grothendieck : *Fondements de Géométrie Algébrique*, Bourbaki Seminar talks, 1957–62.

[FGA Explained] B. Fantechi et al : *Fundamental Algebraic Geometry - Grothen-dieck's FGA Explained*, Math. Surveys and Monographs 123, Amer. Math. Soc., 2005.

[F-G] B. Fantechi and L. Göttsche : *Elementary deformation theory*. Chapter 5 of [FGA Explained].

[H] R. Hartshorne : *Algebraic Geometry*, Springer Verlag, 1977.

[I] L. Illusie : *Grothendieck's existence theorem in formal geometry*. Chapter 8 of [FGA Explained].

[H-L] D. Huybrecht and and M. Lehn : *The Geometry of Moduli Spaces of Sheaves*, Aspects of Mathematics 31, Vieweg Verlag, 1997.

[Mi] J.S. Milne : *Etale cohomology*, Princeton Univ. Press, 1980.

[Ne] P.E. Newstead : *Introduction to moduli problems and orbit spaces*, TIFR lect. notes, Springer 1978.

[Ni] N. Nitsure : *Construction of Hilbert and Quot schemes*. Chapter 2 of [FGA Explained].

[S] M. Schlessinger : *Functors of Artin rings*. Trans. Amer. Math. Soc. 130 (1968), 208–222.

[Se] E. Sernesi : *Deformations of Algebraic Schemes*, Springer Verlag, 2006.

[V] A. Vistoli : *The deformation theory of local complete intersections*. e-print arXiv:alg-geom/9703008v2.

6

The Theory of Vector Bundles on Algebraic Curves with some applications

S. Ramanan

Chennai Mathematical Institute
H1, SIPCOT Information Technology Park
Padur Post, Siruseri-603103
Tamil Nadu, India.
e-mail: sramanan@cmi.ac.in

Peter Newstead and I started working on vector bundles at about the same time. We have shared many ideas for over four decades although we never actually collaborated. I am happy to contribute this summary of some of these areas of common interest to this Festschrift on the occasion of his turning sixty-five.

1 Introduction

The theory of vector bundles has many ramifications. One can study it from number theoretic, algebraic geometric and differential geometric points of view. It has also proved useful to mathematical physicists interested in Conformal Field theory, String theory, etc. In this account, I will mainly deal with the geometric aspects, both algebraic and differential, and will confine myself to just a few remarks on the number theoretic point of view.

The classical theory of abelian class fields seeks to understand Galois extensions of a number field in terms of the number theoretic behaviour of the corresponding integral extensions of the ring of integers in the number field. This has a geometric analogy. Consider any compact Riemann surface. Any finite covering of the surface gives a (finite) extension of the field of meromorphic functions on it. The attempt to try and understand abelian extensions of this field in terms of geometric data on the Riemann surface leads to the theory of *Jacobians* of Riemann surfaces.

A. Weil initiated the theory of vector bundles over an algebraic curve motivated by the desire to develop a 'non-abelian class field theory' for function fields. The number theoretic analogues are more pertinent when the function field have finite field of constants, but the study makes sense and is interesting when the function field in question is the field of meromorphic functions on a compact Riemann surface in the traditional sense.

The study of Jacobians is complex analytic or algebraic and can be understood purely geometrically. Jacobi's work in this respect may be interpreted, as establishing in particular a correspondence between the fundamental group and the Jacobian, that is to say, the variety of divisor classes.

We will start with a brief survey of basic concepts centering around divisors and explain the origins of this aspect and later lead up to the theme of vector bundles.

2 Line bundles on a compact Riemann surface

2.1 Periods of holomorphic differentials

Let X be a compact Riemann surface, that is to say, a compact, connected, complex manifold of dimension 1. Its topological type is determined by a non-negative integer g, called its *genus*. The genus of the Riemann sphere is 0. The quotient of $\mathbf{C}$ by a discrete subgroup of rank 2, is topologically a two-dimensional torus, but it also inherits a complex structure from $\mathbf{C}$. Riemann surfaces so obtained are called *elliptic curves* and have genus 1. Since such a surface is homeomorphic to $S^1 \times S^1$, its fundamental group is free abelian of rank 2. In general, the first homology group $H_1(X)$ of any compact Riemann surface is free of even rank and one way to define g is to say that this rank is $2g$. One can actually write out explicitly its fundamental group $\pi(X)$ in terms of the genus. It is isomorphic to a group on $2g$ generators a_i, b_i, $i = 1, \cdots, g$ with a single relation $\prod a_i b_i a_i^{-1} b_i^{-1} = 1$. Of course this implies that its abelianisation, namely $H_1(X, \mathbf{Z})$, is free of rank $2g$.

One has also an *analytic* interpretation of the invariant g. Indeed, if g is the genus of X, the space Ω of all holomorphic differentials on X is a vector space of dimension g. The origin of the theory of line bundles from Jacobi's point of view was the attempt to integrate a holomorphic differential on X. Let ω be a holomorphic differential on X. We fix a point x_0 and seek to compute the *indefinite* integral $\int_{x_0}^x \omega$ as a function of x. In order for this to make sense, we have to integrate the differential along a path c connecting x_0 and x. (For purposes of integration we will assume the path to be a smooth, or piecewise smooth, map $[0,1] \to X$.) One may think of the integral over c as a linear form on Ω, by varying ω. The integral of course depends on the path c, but the monodromy theorem says that the integral is the same if we replace c by a homotopically equivalent path. In other words, the integral depends only on the *homotopy class* of the path c. In general, if c and c' are any two paths connecting x_0 and x, the two integrals differ by the integral of ω along the loop $c'.c^{-1}$. Therefore we are led to consider the linear forms on Ω obtained by integrating holomorphic forms on loops based at x_0. These special linear forms are called *periods* of holomorphic differentials. We have thus given a *homomorphism* of the fundamental group $\pi(X, x_0)$ into Ω^*. This obviously goes down to a homomorphism of the abelianised fundamental group $H_1(X, \mathbf{Z})$ into Ω^*. It also follows that this homomorphism does not depend on the choice of x_0. The first important result is the following fact.

Theorem 2.1. *The above homomorphism of $H_1(X)$ into Ω^* is injective and the image is a discrete subgroup of maximal rank.*

We will call the image the *period group*. Our remarks above amount to saying that integration of holomorphic differentials leads to a map of X into the *quotient* of Ω^* by the period group. This map is called the *period map*. Here we will denote it by σ.

2.2 The Albanese variety

Consider now the following situation. Let V be a complex vector space of dimension g and Γ a discrete subgroup of rank $2g$. The quotient $A = V/\Gamma$ has a lot of structure. First of all, it is compact and connected. Topologically it is actually a torus of dimension $2g$. Secondly the natural map $V \to A$ is a local homeomorphism. Using this we can equip A with a complex structure, making it a compact complex manifold of dimension g. On the other hand, it is also an abelian group. The complex structure and the group structure are obviously compatible in the sense that the group operations are holomorphic maps. In other words, it is a *complex Lie group*. Now we have the following fact.

Theorem 2.2. *Any compact connected complex Lie group is isomorphic to the quotient of $\mathbf{C}^g$ by a discrete subgroup of maximal rank.*

Proof: Consider the adjoint representation of the group G in question in its Lie algebra. It is a holomorphic map of the group into the group of linear automorphisms of the Lie algebra. Since any holomorphic function on a compact, connected, complex manifold is a constant, the adjoint representation is the trivial one. It follows that the group G is abelian. The exponential map of the Lie algebra into the Lie group is then a surjective homomorphism of the additive group underlying the Lie algebra into G, and the kernel is a subgroup Γ of $\mathbf{C}^g$ such that $\mathbf{C}^g/\Gamma$ is compact. It is easily seen that this means that Γ is a lattice in $\mathbf{C}^g$. $\square$

The complex Lie groups obtained as above are called *complex tori*.

In our case, V is Ω^* and Γ is the period subgroup. The map $X \to A$ that we have defined above is easily seen to be holomorphic, and we have actually the following universal property.

Theorem 2.3. *Any holomorphic map of X into a complex torus T taking x_0 to 0 factors through a unique holomorphic homomorphism of A into T.*

Proof: Note first that the space of holomorphic 1-differentials on T is canonically identified with $\mathfrak{t}^*$, where $\mathfrak{t}$ is its Lie algebra. The given map of X into T gives rise, via pull back, to a linear map $\mathfrak{t}^* \to \Omega$, by simply pulling back holomorphic differentials. Its transpose is a linear map $\Omega^* \to \mathfrak{t}$. This takes Γ into $\mathcal{L}$ and induces a map of $\Omega^*/\Gamma \to A/\mathcal{L}$ where $\mathcal{L}$ is the discrete subgroup of $\mathfrak{t}$ which is the kernel of the exponential map $\mathfrak{t} \to T$. $\square$

This is called the *Albanese property* of the period map. Using the group structure of A, we may analyse the period map a little further. As we observed above, the period group does not depend on the base point x_0 that we chose. But the map σ does depend on the choice. If we wish to do away with this dependence, we may take, instead of a single point, any finite set of points x_i. In order to allow repetitions, we will assign multiplicities m_i to each x_i. We do not insist that these integers m_i be positive. Such a datum is called a *divisor* in X. (The name has its origin in Number theory. Any nonzero rational number can be thought of as assigning multiplicities to finitely many primes). If all the m_i are non-negative, we call it an *effective divisor*. A concise definition of a divisor is that it is an element of the free abelian group $Div(X)$ with the underlying set X as basis. Given a divisor $D = \sum m_i x_i$, we can define an element of A as follows. Take $\sum m_i \sigma(x_i)$ in the sense of the group addition in A. The point is that this map is *independent of x_0 if $\sum m_i = 0$*. The integer $\sum m_i$ is called *the degree* of the divisor

$\sum m_i x_i$. Degree is then obviously a homomorphism of $Div(X)$ into **Z**. What we asserted above is that we have a canonical homomorphism of the group of divisors of degree 0 into A. We will denote this map by α. Actually this homomorphism is surjective. One would like to understand the kernel of this map.

Theorem 2.4. *The kernel of α consists of those divisors $\sum m_x x$ which have the property that there exists a nonzero meromorphic function f on X with $ord_x(f) = m_x$ for all $x \in X$.*

See below for an explicit definition of $ord_x(f)$. We do not prove this theorem here. See [Mu2].

2.3 Divisor classes

Any nonzero meromorphic function f gives rise to a divisor as follows. For any $x \in X$ associate the integer $ord_x(f)$, namely the integer i such that $f = t^i.g$, where g is a nonzero holomorphic function in a neigbourhood of x and t is a local coordinate. Since the zeros and poles of f are finite in number, $\sum_{x \in X} ord_x(f)x$ is actually a divisor $div(f)$. This is called *the divisor associated to f*. Divisors of nonzero meromorphic functions are called *principal divisors*. This in fact gives a homomorphism of the multiplicative group of nonzero meromorphic functions into the group of divisors. It is an easy consequence of the integral formula that the degree of any principal divisor is zero. If two divisors are considered *equivalent* whenever their difference is a principal divisor, then the equivalence classes are called *divisor classes*. Since principal divisors obviously form a subgroup of the divisor group, divisor classes form a group. What the above theorem asserts therefore is that the the map α induces an isomorphism of the group of divisor classes of degree 0, onto the Albanese variety Ω^*/Γ.

2.4 Invertible sheaves and line bundles

Suppose $D = \sum m_x x$ is a divisor. To every open set U in X, we associate the vector space of all meromorphic functions f in U such that $div(f)+D$ is effective. In long hand, this means that the order of the function at any point x is at least $-m_x$. For example, if the divisor is simply a for some point $a \in X$, then the above space consists of all meromorphic functions in U, which have at most a simple pole at a. This assignment gives a sheaf on X. Indeed it is a sheaf of $\mathcal{O}$-Modules, since multiplication by a holomorphic function preserves the above property. We denote this sheaf by $\mathcal{O}(D)$. Note that if U is a coordinate neighbourhood of a point x, with the coordinate t, then any function with the above property has the Laurent expansion at x of the form $t^{-m_x}.g$ where g is a holomorphic function. Thus locally this sheaf is isomorphic, as an $\mathcal{O}$-Module, to $\mathcal{O}$. However globally it is not in general isomorphic to $\mathcal{O}$.

Definition 2.5. A sheaf of $\mathcal{O}$-Modules which is locally isomorphic to $\mathcal{O}$ is called *an invertible sheaf* on X. The invertible sheaf $\mathcal{O}$ on X is said to be *trivial*.

The set of all invertible sheaves on X is an abelian group under tensor product as group operation, the trivial sheaf $\mathcal{O}$ being the identity element. We have observed above that every divisor gives rise to an invertible sheaf. This yields an isomorphism of the group of divisor classes onto the group of invertible sheaves. An important invertible sheaf is the sheaf which associates to any open set U of X the space of holomorphic differentials on U. We will generally denote this by K_X or ω_X.

The notion of invertible sheaves is entirely equivalent to that of *line bundles*. A line bundle consists of a variety L and a holomorphic map π of L onto X, with fibres equipped with the structure of a vector space of rank 1. It is supposed to be *locally trivial* in the sense that every point of X has an open neighbourhood U such that there is an isomorphism of $\pi^{-1}(U)$ with the product $U \times \mathbf{C}$ which is linear on fibres. The sheaf of holomorphic sections of L is an invertible sheaf, and this assignment of invertible sheaves to line bundles makes the two notions interchangeable.

2.5 Cohomological interpretation

If we trivialise a line bundle on an open set U in two different ways, they 'differ' by an automorphism of the trivial bundle on this open set. Since an automorphism of a one-dimensional vector space consists only of nonzero scalars, we conclude that the two trivialisations differ (multiplicatively) by a function on U with values in $\mathbf{C}^\times$. If we cover the manifold with open sets U_i with trivialisations of the line bundle over each of these, we obtain on pairs of intersections $U_i \cap U_j$, functions $m_{i,j} : U_i \cap U_j \to \mathbf{C}^\times$ as above. These satisfy the compatibility equation:

$$m_{i,j} m_{j,k} = m_{i,k}$$

for all triples i, j, k. These can be interpreted as co-cycles in the Čech sense with respect to the given covering. Hence one gets an element of $H^1(X, \mathcal{O}^\times)$, where $\mathcal{O}^\times$ is the sheaf of nonzero holomorphic functions.

Assuming the theory of cohomology of sheaves, one can check that this gives an isomorphism of the divisor class group with $H^1(X, \mathcal{O}^\times)$.

2.6 Linear systems

To an invertible sheaf L (and indeed to any sheaf of abelian groups) is associated *cohomology groups* $H^i(X, L)$. If L is the invertible sheaf associated to a divisor D, the vector space $H^0(X, L)$ is simply the space of (global) holomorphic functions f with $div(f) + D$ effective. When X is a curve, these groups vanish for $i \geq 2$. Towards the computation of H^0 and H^1, which are finite dimensional vector spaces, we have the following famous result, called the *Riemann-Roch theorem*.

Theorem 2.6. $dim H^0(X, L) - dim H^1(X, L) = deg(L) + 1 - g$.

Ideally we would have liked a formula for computing $dim\ H^0(X, L)$, but this number can vary when L is perturbed a little. So one *cannot* have a formula which computes $dim H^0(X, L)$ in terms of discrete invariants of L and X. On the other hand, the left side of the above formula is indeed a *deformation invariant*. It is called the *Euler characteristic* of L and is usually denoted by $\chi(X, L)$.

The Riemann-Roch theorem and the following duality theorem are important tools in the study of curves.

Theorem 2.7. $H^0(X, L)$ *and* $H^1(X, K \otimes L^{-1})$ *are canonically dual, where k is the cotangent bundle.*

If a line bundle admits a nonzero section, then its degree is non-negative. For if L is represented by a divisor D and f a meromorphic function on X with $div(f) + D$ effective, then since degree of $div(f)$ is zero, we ought to have $deg(L) \geq 0$. Thus if $deg\ (L) < 0$, then $H^0(X, L) = 0$. By duality we have, as a corollary, $H^1(X, L) = 0$ if $deg(L) > deg(K)$.

If we take L to be the trivial bundle, then $H^0(X, L)$ is the space of holomorphic functions on X and is therefore one dimensional. Hence $H^1(X, K)$ is also one dimensional. On the other hand, if we take $L = K$ and use the fact that $\Omega = H^0(X, K)$ has rank g, we get $deg(K) = 2g - 2$. In other words, the divisor of zeros of any nonzero holomorphic differential is of degree $2g - 2$. Also we can restate the vanishing theorem as follows.

If $deg(L) > 2g - 2$, then $H^1(X, L) = 0$.

Thus the Riemann Roch theorem computes $dim(H^0(X,L))$ if the degree is greater than $2g - 2$.

Using these facts, one can show for example that the complex manifold X is indeed isomorphic to a submanifold of some complex projective space $\mathbf{CP}^n$. This is accomplished as follows. Let L be a line bundle which has 'lots of sections'. Let $s_0, \cdots, s_n$ be a basis for $H^0(X,L)$. Locally these can be regarded as functions. Then the assignment $x \mapsto (s_0(x), \cdots, s_n(x))$ gives a holomorphic map of X into $\mathbf{CP}^n$ locally. It is easy to see that indeed this map is global, and gives an imbedding under our assumption that there are sufficiently many sections. Canonically speaking, the projective space $\mathbf{CP}^n$ is the projective space associated to the vector space of linear forms on $H^0(X,L)$, and to each x we associate the one dimensional subspace of sections vanishing at x.

From the Riemann Roch formula we deduce that any line bundle L of large degree (for example $> 2g$), serves this purpose. This is because for any two $x, y \in X$, we can compute the dimension of $H^0(L)$, $H^0(X, L \otimes \mathcal{O}(-x))$ and $H^0(L \otimes \mathcal{O}(-x - y))$ by the Riemann Roch theorem to be $deg(L) + 1 - g$, $(deg(L) - 1) + 1 - g$ and $(deg(L) - 2) + 1 - g$, respectively. Since these are respectively the space of sections of L, the space of sections of L vanishing at x, and the space of sections of L vanishing at both x and y, it follows that for any two points x and y, there is a section of L vanishing at x but not at y. This proves that the map we defined above is injective. Taking y to be x in the above computation, one can show that the differential of the map at x is also injective. In other words, X can be imbedded in a complex projective space as a submanifold. Indeed, it can be shown that any such submanifold can actually be defined by the vanishing of homogeneous polynomials, thus showing that X is a *projective algebraic curve*.

What we explained above is a general procedure, applicable to higher dimensional manifolds as well. If a compact manifold admits a line bundle with a lot of sections, then the manifold is actually a projective algebraic manifold. This statement is almost tautologous, but it concentrates the effort of imbedding a manifold in a projective space to finding such line bundles, which we accomplished above in the case of compact Riemann surfaces.

2.7 Polarisation

Let us now go back to our starting point. We gave an isomorphism of group of divisors classes of degree 0 onto Ω^*/Γ. We noticed that the divisor class group can be identified with the group of line bundles. We have also seen that the latter group can be identified in turn with the cohomology group $H^1(X, \mathcal{O}^\times)$. Consider the exact sequence of sheaves:

$$0 \to \mathbf{Z} \to \mathcal{O} \to \mathcal{O}^\times \to 0$$

Here $\mathbf{Z}$ is the constant sheaf of integers and the right map is the exponential map $f \mapsto exp(2\pi i f)$.

The connecting homomorphism $H^1(X, \mathcal{O}^\times) \to H^2(X, \mathbf{Z})$ associates to each line bundle its degree. The image is the *Chern class* of the line bundle. So the kernel, namely the group of divisor classes of degree 0, is isomorphic to $H^1(X, \mathcal{O})/H^1(X, \mathbf{Z})$. This is thus again the quotient of a complex vector space by a lattice. As a matter of fact, there is a duality between $H^1(X, \mathcal{O})$ and $H^0(X, K) = \Omega$. Thus the two spaces $H^1(X, \mathcal{O})$ and Ω^* are canonically isomorphic. The lattice $H^1(X, \mathbf{Z})$ goes into the period subgroup under this isomorphism, thereby yielding an isomorphism of these complex tori.

Now Hodge theory gives a decomposition

$$H^1(X, \mathbf{C}) = H^1(X, \mathcal{O}) \oplus \Omega$$

and an anti-isomorphism of $H^1(X, \mathcal{O})$ with Ω given by complex conjugation on 1-forms. Using these, we get a positive definite hermitian form on the vector space $H^1(X, \mathcal{O})$. Its imaginary part restricts to $H^1(X, \mathbf{Z})$ as the Poincaré pairing $H^1(X, \mathbf{Z}) \times H^1(X, \mathbf{Z}) \to H^2(X, \mathbf{Z}) = \mathbf{Z}$.

2.8 Abelian varieties

Let us turn to the abstract situation of the quotient of a complex vector space V by a lattice Γ. We have already remarked that it is a complex torus. Suppose in addition that there is a positive definite hermitian form on V with imaginary part α. Assume that the restriction e of α to Γ is integral and nondegenerate (in the sense that $e(x, y) = 0$ for all $y \in \Gamma$ if and only if $x = 0$). Then one can actually show that $A = V/\Gamma$ is a projective variety, i.e. a submanifold of $\mathbf{CP}^n$ for some n, given by the vanishing of some homogeneous polynomials. In such a case, we call $A = V/\Gamma$ an *Abelian variety*.

We will now elaborate on this a little bit. The Chern class of any line bundle is an element of $H^2(A, \mathbf{Z})$. For a torus A this can be identified with the space of alternating 2-forms on $H_1(A, \mathbf{Z}) = \Gamma$. Thus the imaginary part of the hermitian form we hypothesized above, can be interpreted to be an element of $H^2(A, \mathbf{Z})$. One can show that it is actually the Chern class of a holomorphic line bundle on A. Some positive tensor power of this line bundle possesses enough sections to imbed A in a suitable projective space. We will collect all these remarks in the following statement.

Theorem 2.8. *The complex torus V/Γ is isomorphic to a projective variety if and only if there exists a positive definite hermitian form on V whose imaginary part is integral on Γ.*

Definition 2.9. When V/Γ is a projective variety, it is called *an Abelian variety*. The Abelian variety Ω^*/Γ associated to X is called the *Jacobian* of X.

2.9 Theta functions

We have remarked above that there is a line bundle L whose Chern class is the alternating form e, namely, the imaginary part of the given positive definite hermitian form h. This line bundle is not unique, but is determined up to a translation. The classical proof of the theorem quoted above is via the theory of *theta functions*.

In fact, the line bundle when pulled up to V itself is trivial. Let us choose a trivialization. Sections of L when pulled up to V are therefore holomorphic functions. The fact that the line bundle came from A would impose some conditions on these functions, under translation by elements of Γ. Explicitly, these are functions that satisfy the functional equation:

$$f(v + \gamma) = \alpha(\gamma)exp(\pi h(v, \gamma) + \frac{\pi}{2}h(\gamma, \gamma))f(v)$$

for all $v \in V$ and $\gamma \in \Gamma$. Here, α is a fixed map of Γ into $S^1 = \{z \in \mathbf{C} : |z| = 1\}$, satisfying

$$\alpha(\gamma_1 + \gamma_2) = exp(i\pi e(\gamma_1, \gamma_2))\alpha(\gamma_1)\alpha(\gamma_2)$$

for all $\gamma_1.\gamma_2 \in \Gamma$. Note that α is not uniquely determined by e, but can only be altered multiplicatively by a unitary character on Γ. This of course depends on the particular line bundle L we take with e as its Chern class.

By explicit analysis of these equations, one can actually write down a basis for these. These are called *theta functions*. The choice of the hermitian form (and therefore e) is referred to as a *polarisation*. It can be shown that some positive power of L – indeed Lefschetz showed that L^3 would do – has enough sections to imbed A in a projective space.

3 Poincaré bundle

3.1 Families of line bundles

We have seen above that the group of divisor classes of degree 0 on X is naturally bijective with points of $H^1(X, \mathcal{O})$ modulo the period group. In particular this group has been provided with the structure of an Abelian variety. This is called the *Jacobian* of X and denoted by $J(X)$ or simply J. In what sense is this structure natural? In order to answer this, one needs to understand what is meant by *classification* of line bundles.

Let T be any (parameter) variety. A *family* of line bundles on X, parametrised by T, should be a line bundle L on $T \times X$. For each point $t \in T$, we may restrict L to $\{t\} \times X$ and obtain a line bundle L_t on X. We would like to think of the family to be the collection $\{L_t\}_{t \in T}$. But then two non-isomorphic line bundles L and L' on $X \times T$ can give rise to the same family in this sense. For example, one can take a line bundle ξ on T and take $L' = L \otimes p_T^* \xi$. Fortunately, this is is only thing that can go wrong! In any case, we will consider such families to be *equivalent*. Given any such family L of line bundles of degree 0, one gets a map of T into $J(X)$ which associates to each $t \in T$, the isomorphism class of L_t considered as a point of J. We call it the *classifying map* $\varphi_L : T \to J$. We require that this map be holomorphic. In fact, this nails down the structure of J as a complex manifold. Since both X and J are algebraic varieties we may equally well work with algebraic varieties, or even schemes.

We may actually wish to construct a (universal) family of line bundles on X parametrised by J. In other words, we seek to construct a line bundle P on $X \times J$ such that for any $\xi \in J$, the equivalence class of the line bundle P_ξ, considered as a point of J, is ξ. Such a family does exist and is called the *Poincaré* bundle P. Given any family L parametrised by T we have the classifying map $\varphi_L : T \to J$. We can thus use the map $\varphi_L \times Id_X : T \times X \to J \times X$ to pull back P. This and L are obviously equivalent families.

We have explained the construction of a variety structure on the set of divisor classes of degree 0. We could also do the same for the set of divisor classes of any fixed degree d. This is of course not a group, but a coset of J in the divisor class group. As such, the group J acts simply transitively on it and so we can transfer the strucure of a projective variety on it. An appropriate notation for this variety would be $J^d(X)$.

3.2 Algebraic geometric point of view

Any compact Riemann surface is a projective variety, and its Jacobian also turned out to be a projective variety, but all our constructions have been transcendental. It is a natural question therefore whether all these can be carried out in the context of algebraic geometry. Indeed the Riemann-Roch theorem assures us that any line bundle of degree at least g has at least one nonzero section. In other words, any such line bundle is isomorphic to $\mathcal{O}(D)$ where D is an effective divisor. Clearly the set of effective divisors of a given degree d can be identified with the d-fold *symmetric* power $S^d(X)$ of the curve. Hence there is a morphism from $S^d(X)$ into J^d which takes D to $\mathcal{O}(D)$. The Riemann-Roch theorem implies that when $d \geq g$ this map is surjective, for any line bundle L of degree d would then have nonzero H^0. The divisor of zeros of a nonzero section is then an effective divisor D such that $\mathcal{O}(D) \simeq L$. If $d = g$, this is in fact a birational morphism and André Weil used this to give a purely algebraic construction of the Jacobian.

If $d = g - 1$, we get a morphism of S^{g-1} into J^{g-1} and one can check that the image is a divisor θ in J^{g-1} and consequently defines a line bundle on it. Its Chern class is the hermitian form which we explained purely in analytic terms. Thus we might redefine *theta functions* to be sections of powers of the line bundle $\mathcal{O}(\theta)$.

When $g = 1$, this gives an isomorphism of X with J^1. In other words, X is itself an abelian variety of dimension 1. One-dimensional abelian varieties are called *elliptic curves*.

The theory of line bundles, Jacobians and theta functions has a long history and has been developed intensely from the geometric, arithmetic and analytic points of view. We have given above a short account of those aspects which are relevant to the following lectures.

4 Vector bundles

4.1 Locally free sheaves and vector bundles

The notion of locally free sheaves (resp. vector bundles) is entirely similar to that of invertible sheaves (resp. line bundles). We will run through some facts on vector bundles which are easily proved either by reducing them to, or imitating the arguments of, analogous theorems on line bundles.

Definition 4.1. A sheaf of $\mathcal{O}$-Modules which is locally isomorphic to $\oplus^n \mathcal{O}$ is called a *locally free sheaf* on X.

Similarly, let E be a complex manifold (or a variety) with a morphism $\pi : E \to X$ with the structure of vector spaces of rank n on all fibres $\pi^{-1}(x)$, $x \in X$ satisfying the following condition. Every point $x \in X$ has a neighbourhood U such that $\pi^{-1}(U)$ is isomorphic to the product $U \times \mathbf{C}^n$, the isomorphism being linear on all the fibres. Then we say E is a *vector bundle of rank n* over X. The sheaf of sections of E is a locally free sheaf of $\mathcal{O}$-modules and this gives a bijection between the sets of isomorphism classes of vector bundles and those of locally free sheaves. We generally make no distinction between the two.

4.2 Duality and Riemann-Roch theorems

All the linear algebraic operations that one performs on a vector space may be performed on vector bundles as well. For example, if E_1 and E_2 are two vector bundles, then one can form its *direct sum* $E_1 \oplus E_2$, by taking the fibre product of $E_1 \to X$ and $E_2 \to X$ and equipping the fibres with the direct sum structure. We may also construct the tensor product of two bundles and hence also the *tensor power* $\otimes^n E$ of any bundle E. Also, the *symmetric power* and the *exterior power* of a vector bundle are defined similarly. Clearly the rank of $E_1 \oplus E_2$ (resp. $E_1 \otimes E_2$) is $rk(E_1) + rk(E_2)$ (resp. $rkE_1.rkE_2$) and so on. In particular, if n is the rank of E, then the n-th exterior power of E, which we call the *determinant* of E and denote by $det(E)$, is of rank 1, that is to say, a line bundle. As such one can talk of its *Chern class*, or over curves, of its degree.

Definition 4.2. The *degree* of a vector bundle E is the degree of the line bundle $det(E)$.

We can now state the vector bundle analogues of the results on line bundles. Firstly it is true that $H^i(X, E)$ are 0 for all $i \geq 2$. Secondly we have the duality theorem for vector bundles exactly as for line bundles.

Theorem 4.3. *There is a natural duality between* $H^0(X, E)$ *and* $H^1(X, K \otimes E^*)$.

We also have the following more general Riemann-Roch theorem.

Theorem 4.4. $dim\ H^0(X, E) - dim\ H^1(X, E) = deg(E) + rk(E)(1-g)$.

4.3 Extensions

The linear algebraic operations outlined above, give methods of construction of other vector bundles, starting with one. But we have not given thus far examples of vector bundles, other than those obtained by taking direct sums of line bundles. We wish now to discuss more interesting constructions of vector bundles.

Consider short exact sequences of sheaves of $\mathcal{O}$-Modules of the form

$$0 \to E' \to E \to E'' \to 0$$

with E', E'' locally free. Then one can deduce that E is also locally free. So, starting with E' and E'', if we can construct an exact sequence as above, then we obtain a new vector bundle E, which is said to be an *extension* of E'' by E'. The direct sum $E' \oplus E''$ is then a particular case of this. The set of equivalence class of all such extensions can be put in bijective correspondence with the vector space $Ext^1(E'', E') = H^1(X, E''^* \otimes E')$. By taking nonzero elements of this space, we may construct new vector bundles which are not direct sums of E' and E''.

In fact, we will show that starting with line bundles, by successively taking extensions, we may obtain *all* vector bundles. Indeed, let E be *any* vector bundle. One may tensor it with a line bundle L of large degree and ensure (for example using Theorem 4.4 above) that $H^0(E \otimes L) \neq 0$. Take a nonzero section s. Suppose it vanishes at some point $x \in X$. This means that it is actually a section of $\mathcal{M}_x \otimes E \otimes L$, where $\mathcal{M}_x$ is the ideal sheaf of functions vanishing at the point x. But $\mathcal{M}_x$ is an invertible sheaf and is indeed isomorphic to $\mathcal{O}(-x)$. Thus, ultimately s may be regarded as an everywhere nonzero section of $\mathcal{O}(-D) \otimes E \otimes L$ where D is the divisor of zeros of s. Thus we get an exact sequence

$$0 \to \mathcal{O} \to E \otimes M \to F \to 0$$

of locally free sheaves, denoting by M the line bundle $\mathcal{O}(-D) \otimes L$. Tensoring this with M^{-1} we see that E is obtained as an extension of a vector bundle of rank $n-1$ by a line bundle. Iterating it, one concludes that *every vector bundle* is obtained as successive extension of line bundles. One can prove this fact independently and reduce the Riemann-Roch theorem for vector bundles stated above, to the same for line bundles.

Remark 4.5. If we take elements $v, \lambda v \in H^1(E''^* \otimes E') = V$ (with $\lambda \neq 1 \in \mathbf{C}^\times$), the extensions obtained are distinct, but the vector bundles obtained as extensions are isomorphic. Now one can vary λ and get vector bundles on X corresponding to each λ. Note that every nonzero λ corresponds to the 'same' vector bundle (i.e. upto isomorphism) while $\lambda = 0$ corresponds to the direct sum of E' and E'', which is generally speaking not isomorphic to the non-trivial extension. This strange fact (called the 'jump' phenomenon) gives the following negative result. *There cannot be a variety structure on the set of isomorphic classes of all vector bundles of rank ≥ 2 with even the minimal naturality assumptions, such as we explained in our discussion of the Poincaré bundle.* For if there were such a structure, there would be a morphism from $\mathbf{C}v$ into that space which would take the open set of nonzero vectors into a single point and the zero vector to some other point.

4.4 Elementary transformations

Here we start with a vector bundle and 'alter' it at a point to get a new vector bundle of the same rank. For example, one may start with the trivial line bundle $\mathcal{O}$ and alter it at a point $x \in X$ to obtain the line bundle $\mathcal{O}(-x)$. Since the latter consists of functions which vanish at x, the way to go about this construction is to take a surjection $\mathcal{O} \to \mathcal{O}_x$ where the latter is the structure sheaf of the single point x. The kernel is $\mathcal{O}(-x)$. The same procedure can be adopted in general. Take a locally free sheaf E and take any surjection to $\mathcal{O}_x$. It is clear that the kernel is locally free. Since the sheaf $\mathcal{O}_x$ is concentrated at x, in order to give a surjection of E onto $\mathcal{O}_x$, we need take only a nonzero linear form on the fibre E_x of the vector bundle E at x.

4.5 Direct images

Let Y be another compact Riemann surface and $\pi : Y \to X$ a surjective morphism. Take any locally free sheaf on Y and take its direct image on X. It is easy to check that it is a locally free sheaf on X and this is another way of constructing vector bundles on X. For example, if we start with a line bundle L on Y, its direct image is a vector bundle of rank equal to the degree of the map π. Since the Euler characteristic is invariant under direct images by finite maps, we can compute the degree of the direct image. In fact, we have

$$
\begin{aligned}
\chi(E) &= deg(E) + rk(E)(1 - g(Y)) \\
&= deg(\pi_*(E)) + rk(E)deg(\pi)(1 - g(X))
\end{aligned}
$$

This computes the degree of $\pi_*(E)$ in terms of that of $deg(E), deg(\pi)$ and the genera of X and Y.

4.6 Representations of the fundamental group

A transcendental construction of vector bundles on X is as follows. Let ρ be a linear representation of the fundamental group $\pi(X)$ in a vector space V. Now π acts (conventionally on the right) as deck transformations on the universal covering space $\tilde{X}$ of X on the one hand and on V by linear transformations on the other. Let E_ρ be the quotient of the space $\tilde{X} \times V$ by the right action of π on the product by the prescription

$$
(z, v)g = (zg, \rho(g)^{-1}v)
$$

for $g \in \pi, z \in \tilde{X}$ and $v \in V$. There is a natural morphism of $E_\rho \to X$ given by the second projection. This actually makes it a vector bundle. This is called the *vector bundle associated to the representation* π. Obviously, the rank of this vector bundle is the same as that of V. Its degree is 0 as we shall see presently.

If we take characters (i.e. one-dimensional representations) of π, we of course get line bundles by this procedure. Since we have seen that line bundles are classified by $H^1(X, \mathcal{O}^\times)$, we have a natural homomorphism of the group of characters of π into $H^1(X, \mathcal{O}^\times)$. This homomorphism is easy to explain. The natural homomorphism of the constant sheaf $\mathbf{C}^\times$ into the sheaf $\mathcal{O}^\times$ induces a homomorphism of $H^1(X, \mathbf{C}^\times)$ into $H^1(X, \mathcal{O}^\times)$. The former can be identified with the group of characters on π and the latter with the group of isomorphism classes of line bundles. Now consider the following commutative diagram of sheaves (with exact rows).

$$
\begin{array}{ccccccccc}
0 & \to & \mathbf{Z} & \to & \mathbf{C} & \to & \mathbf{C}^\times & \to & 0 \\
 & & \downarrow & & \downarrow & & \downarrow & & \\
0 & \to & \mathbf{Z} & \to & \mathcal{O} & \to & \mathcal{O}^\times & \to & 0
\end{array}
$$

Here the maps $\mathbf{C} \to \mathbf{C}^\times$ and $\mathcal{O} \to \mathcal{O}^\times$ are both given by the exponential map. We are interested in the associated cohomology sequences. Since $H^2(X, \mathbf{Z}) = \mathbf{Z} \to H^2(X, \mathbf{C}) = \mathbf{C}$ is an inclusion, it follows that the homomorphism $H^1(X, \mathbf{C}^\times) \to H^2(X, \mathbf{Z})$ is zero. This implies that the associated line bundle, as an element of $H^1(X, \mathcal{O}^\times)$ gets mapped onto zero by the connecting homomorphism into $H^2(X, \mathbf{Z})$. In other words, the degree of the associated line bundle is always zero. But then since the map $H^1(X, \mathbf{C}) \to H^1(X, \mathcal{O})$ is the Hodge projection, it follows that any line bundle of degree zero, which is the image of an element of $H^1(X, \mathcal{O})$ is associated to a suitable character. Indeed, by using the exact sequence

$$0 \to \mathbf{Z} \to \mathbf{R} \to S^1 \to 0$$

instead and noting that by virtue of Hodge decomposition, $H^1(X, \mathbf{R})$ is mapped isomorphically on $H^1(X, \mathcal{O})$, we can actually conclude that any line bundle of degree zero comes from an element of $H^1(X, S^1)$, namely a *unitary* character of π.

In fact, the above argument gives the following result.

Theorem 4.6. *There is a natural isomorphism between the group of unitary characters of the fundamental group and the group of line bundles of degree zero.*

What is the corresponding result in respect of vector bundles? If ρ is a representation of π then the determinant of the associated bundle E_ρ is clearly the line bundle associated to the character $det(\rho)$. Hence we conclude that $deg(E_\rho)$ is zero. If we restrict ourselves to unitary representations, then we have the following easily proved, but very useful, statement.

Proposition 4.7. *Any homomorphism of E_ρ into $E_{\rho'}$ with ρ and ρ' unitary, is induced by a π-homomorphism of the representation spaces. In particular, E_ρ and $E_{\rho'}$ are isomorphic if and only if the representations ρ and ρ' are equivalent.*

However, it is *not true* that every vector bundle of degree 0 arises from a unitary representation of π. For example, take a nontrivial extension E of $\mathcal{O}$ by $\mathcal{O}$. Such an extension exists since these extensions are classified by $H^1(X, \mathcal{O})$ which is of dimension $g \geq 1$ by assumption. If E were associated to a unitary representation ρ of π, the inclusion of $\mathcal{O}$ in E would arise, by the above result, by a homomorphism of the trivial representation space into the representation of ρ. Since unitary representations are completely reducible, this subrepresentation would split and hence so would the inclusion of $\mathcal{O}$ in E. By assumption this is not the case.

One might of course ask whether there exists a possibly non-unitary representation to which any given bundle of degree 0 is associated. This is also false. In fact, we have a very precise theorem in this regard, due to A. Weil.

Definition 4.8. A vector bundle E is said to be *decomposable* if it is isomorphic to a nontrivial direct sum of subbundles. If not, it is said to be *indecomposable*.

It is obvious that any vector bundle is a direct sum of indecomposable bundles. Such a decomposition is not unique but in any two decompositions the components are isomorphic, upto order. Hence one can talk of *indecomposable components* of a vector bundle.

Theorem 4.9. *A vector bundle is associated to a representation of the fundamental group, if and only if every indecomposable component is of degree zero.*

Thus a direct sum of a line bundle of degree 1 with one of degree -1 has degree zero, but is not associated to any representation of π.

All these show that it is not possible to construct a reasonable structure of a variety on the set of isomorphism classes of *all* vector bundles of a given degree. We may however restrict ourselves to a big subset of vector bundles and then give an affirmative answer to this question. We will now turn to these considerations.

5 Moduli space of vector bundles

In order to construct a variety whose points correspond to isomorphism classes of vector bundles, one would first like to fix some numerical invariants. The rank r of the bundle is one such invariant and the degree d is another.

One might first consider a rough classification and then pass to equivalence classes in order to construct the moduli variety in question. When we do this we encounter the problem of passing to the quotient by a group action on a projective variety.

In the sixties, Mumford studied group actions on projective varieties and this led to the notion of *stability* under group actions. He applied his theory to many constructions in algebraic geometry and in particular to the construction of the *moduli space* of vector bundles.

5.1 Stable and semistable vector bundles

The key to the construction of the moduli of vector bundles is the definition of a stable (resp. semistable) vector bundle.

Definition 5.1. The *slope* $\mu(E)$ of a vector bundle E is the rational number $deg(E)/rk(E)$. A vector bundle on X is *stable* if for every (proper) subbundle F of E, we have the inequality

$$\mu(F) < \mu(E).$$

If the inequality is replaced by $\mu(F) \leq \mu(E)$ then we get the notion of a *semistable* vector bundle.

It is trivial to check that the above condition is equivalent to any one of the following.

1) $\mu(E/F) > \mu(E)$;
2) $\mu(F) < \mu(E/F)$;
3) $\chi(F)/rk(F) < \chi(E)/rk(E)$;
4) $\chi(E/F)/rk(E/F) > \chi(E)/rk(E)$;
5) $\chi(F)/rk(F) < \chi(E/F)/rk(E/F)$.

Remark 5.2 Note that if n and d are coprime, equality is not possible in the definition of semistable bundles. Hence there is no distinction between stable and semistable bundles in this case.

5.2 Elementary properties

Since line bundles do not have any proper subbundles, they are all stable. For any line bundle L, we have $det(E \otimes L) = det(E) \otimes L^{rk(E)}$ and hence $deg\ (E \otimes L) = deg\ E + rk(E).deg\ (L)$. In other words, $\mu(E \otimes L) = \mu(E) + deg(L)$. From this we see that E is stable or semi-stable if and only if $E \otimes L$ is so.

Remark 5.3 Indeed, we will eventually show that the tensor product of two semistable bundles is semistable. But it is not very easy to prove at this stage.

Thanks to 5.2, in order to study stability of bundles and their properties, we can (and often will) assume that the slope is sufficiently large.

Stable vector bundles behave like line bundles in many ways. To start with, a semistable bundle of negative degree cannot admit any nonzero section. In fact, we have remarked (4.3) that any section of

E can be considered to be a section of $E \otimes \mathcal{O}(-D)$ which does not vanish anywhere, where D is an effective divisor. In other words, $\mathcal{O}$ is a subbundle of $E \otimes \mathcal{O}(-D)$. Since the latter is semistable, we have $0 \leq \mu(E) - degD$, but by assumption, $\mu(E) < 0$ and $deg\ (D) \geq 0$. Moreover, using the duality theorem, we can derive a vanishing theorem for H^1 as well.

Proposition 5.4. *If E is a semistable vector bundle of negative degree, then $H^0(E) = 0$. The same conclusion is valid if E is stable, and non-trivial of degree ≤ 0. If E is a semistable bundle of slope greater than $2g - 2$, then $H^1(X, E) = 0$. The same vanishing is true if E is stable of slope at least $2g - 2$ and not isomorphic to K.*

Using this, we can also give a criterion for the stability of a vector bundle in terms of the dimensions of the space of sections of subbundles instead of the Euler characteristics.

Proposition 5.5. *If the slope of E is at least $2g - 2$ then E is stable if $H^1(E) = 0$ and dim $(H^0(X, F))/rk(F) <$ dim $(H^0(X, E))/rk(E)$ for every proper subbundle F.*

Proof: If E is not stable, then take a subbundle F of maximal slope. (Using the fact that every vector bundle is an extension of line bundles one can check the existence of such a bundle). Then F is clearly semistable with $\mu(F) > \mu(E) \geq 2g - 2$. Hence $H^1(X, F) = 0$ and consequently $\chi(F)/rk(F) = \dim H^0(F)/rk(F)$ so that we have

$$
\begin{aligned}
\dim(H^0(X, F))/rk(F) = \chi(X, F)/rk(F) &= \mu(F) + 1 - g \\
&> \mu(E) + 1 - g \\
&= \chi(E)/rk(E) \\
&= \dim(H^0(X, E))/rk(E),
\end{aligned}
$$

a contradiction. $\square$

Remark 5.6 Since given any vector bundle we may tensor it with a line bundle to make H^1 vanish and inflate the slope to $2g - 2$ or more, this is indeed a criterion for stability.

Consider any semistable bundle E of rank r and slope greater than $2g - 1$. Then for any $x \in X$, we have both $H^1(E)$ and $H^1(E \otimes \mathcal{O}(-x))$ are zero. Hence we conclude from Riemann-Roch theorem that the dimension of the space of sections of E that vanish at x (which we identified with $H^0(E \otimes \mathcal{O}(-x)))$ is n less than the dimension of $H^0(E)$. This shows that sections of E generate the fibre of E at all points. In other words, all these bundles occur as quotients of the trivial line bundle (of rank $R = d + r(1 - g)$). Now there is a variety which parametrises all quotients of a fixed bundle (the trivial bundle of rank R in our case), of a fixed rank and degree. Moreover the group $GL(R)$ acts in an obvious way on this variety, and it turns out that points of the 'quotient' correspond to *S-equivalence classes* (see below for definition) of semistable bundles of rank r and degree d.

However the problem of taking quotients is more delicate than simply taking the orbits. At stable points for the action which form an open invariant set, things are ok. In this way, Mumford proved the following theorem.

Theorem 5.7. *There exists a natural structure of a nonsingular variety on the space of isomorphism classes of stable vector bundles of rank r and degree d.*

5.3 Theorem of Narasimhan and Seshadri

From the definition, it follows easily that a stable bundle is indecomposable. For if $E = E_1 \oplus E_2$ then $deg E = deg(E_1) + deg(E_2)$ and also $rk(E) = rk(E_1) + rk(E_2)$. This contradicts the stability of E, since we cannot have both $\mu(E_1) < \mu(E)$, $\mu(E_2) < \mu(E)$ at the same time and the above equalities.

In particular, by Weil's theorem, if E is stable of degree 0, then it is given by a representation of the fundamental group (see 4.9). It turns out that this is not so very significant. There is a deeper fact, due to Narasimhan and Seshadri, (see [NS] below) which is more vital.

In this respect stable bundles behave again like line bundles. In order to state the theorem neatly, we will introduce a related definition.

Definition 5.8. A vector bundle is said to be *polystable* if it is a direct sum of stable bundles all of which have the same slope.

Theorem 5.9. *A vector bundle of degree 0 is polystable if and only if it is associated to a unitary representation of the fundamental group. A vector bundle of degree 0 is stable if and only if it is associated to an irreducible unitary representation of π.*

We have already seen (4.7) that the unitary representation which gives rise to a vector bundle is uniquely determined up to equivalence. So this implies that the set of all polystable bundles can be naturally topologised as follows. Consider the $2g$-fold product $U(n)^{2g}$ of the unitary group $U(n)$. In view of the presentation of π that we have described in (2.1), the space of all unitary matrix representations can be identified with matrices (A_i, B_i), $i = 1, \cdots, g$ with the single relation $\prod A_i B_i A_i^{-1} B_i^{-1} = 1$. In other words, it is the inverse image of Id under 'the product of commutators' map $U(n)^{2g}$ into the special unitary group $SU(n)$. This is therefore a compact space R. The quotient of this space by the action of the (special) unitary group, acting by (diagonal) conjugation gives then a topological model for the set of equivalence classes of n-dimensional unitary representations of π.

This, together with the remarks after Proposition 4.7, means that the category of polystable vector bundles and that of unitary representations of π are equivalent.

1) This shows also that bundles associated to a polystable (resp. semistable) bundle, such as exterior, symmetric, tensor powers, ... are all polystable (resp. semistable).

2) For polystable bundles with other (fixed) slopes there are analogous theorems, but we pass them here for simplicity.

5.4 Moduli space of vector bundles

Definition 5.10. A *family* of vector bundles on X parametrised by a variety (or scheme) T is a vector bundle E on $X \times T$.

The restriction of E to $X \times \{t\}$ with $t \in T$ gives a bundle E_t on X, and we think of the family as $\{E_t\}, t \in T$.

Proposition 5.11. *In any family E of vector bundles, the set consisting of $t \in T$ such that E_t is semi-stable, is a Zariski open subset.*

Proof: We saw that for all i, the bundle $\Lambda^i(E)$ is also semistable. On the other hand, if E has a subbundle F of rank i and higher slope, then $\Lambda^i(F)$ is a line subbundle of $\Lambda^i(E)$ and thus E is semistable if and only for every i, $1 \leq i < n$, there is no line subbundle ξ of $\Lambda^i(E)$ of degree $> \mu(\Lambda^i(E))$. What is the same, we have $\Gamma(\xi^* \otimes \Lambda^i(E)) = 0$, for all line bundles ξ of degree $\mu(\Lambda^i(E))+1$. (In fact, we easily compute $\mu(\Lambda^i(E))$ to be $i\mu(E)$). Consider the family $(p_{13})_*(P) \otimes p_{23}^*(E)$ on $J^{i\mu(E)+1} \times T \times X$, where P is the Poincare bundle on $J^{i\mu(E)+1} \times X$. Let S_i be the support of its direct image on $J^{i\mu(E)+1} \times T$. It is a Zariski closed subset. The set $\{t \in T : E_t$ is semistable$\}$ is the complement of the union of the images of S_i in T, $1 \leq i \leq n-1$ and is hence open. $\quad\square$

This proposition shows that the set of semi-stable vector bundles is quite big in the eventual moduli space.

In the open set consisting of semistable bundles, the subset of stable bundles is again seen to be open in a similar way.

5.5 Connectedness

If E is a vector bundle of rank 2, then by tensoring it with a suitable line bundle we can ensure that it has an everywhere nonvanishing section. In other words E occurs in an exact sequence

$$0 \to \mathcal{O} \to E \to L \to 0$$

Taking determinants we conclude that $L = det(E)$. Thus in the family of vector bundles parametrised by $H^1(L^*)$, there is a Zariski open set which corresponds to stable (resp. semistable) bundles. In particular any two stable bundles with the same degree occur in the same connected family.

A similar argument can also be given for bundles of arbitrary rank. If E_1 and E_2 are any two vector bundles of the same degree d, then both of them occur as extensions of a line bundle of degree d, by $\mathcal{O}^{\oplus n-1}$. As we have seen, if L is given, these form a family parametrised by a vector space, namely $H^1(X, L^* \otimes \mathcal{O}^{\oplus n})$. Since L is any element of $J^d, d = deg(E)$, it follows that all these extensions form a family parametrised by a vector bundle T over J^d. In particular, if they are both stable, then they lie in a Zariski open subset of T. Thus we conclude that E_1 and E_2 occur in a connected family. Therefore, eventually if one constructs a moduli space of stable bundles, it would be connected and indeed irreducible! Besides, if E is *any* bundle, we have seen that it is obtainable as an extension of the above type. Since we may also take a stable bundle and get it as such an extension, it implies that there is a Zariski open set of the affine space, which corresponds to stable bundles, and so in a sense, any bundle can be approximated by stable bundles.

Since polystable bundles form a compact set as we have remarked, although the parameter variety was not a complex variety in that construction, one might even hope to construct a *projective variety* whose points correspond to polystable bundles of a given rank and degree. This is essentially true and was proved by Seshadri. However, there are some delicate points here, which we will presently explain.

Theorem 5.12. *There exists a natural structure of a normal complex projective variety of dimension $n^2(g-1)+1$ on the set of isomorphism classes of polystable vector bundles of rank n and degree d.*

From the fact that we have mentioned normality in the above statement, it may be surmised that it is not smooth. Indeed, it is only smooth if n and d are co-prime, *and when $g = 2, n = 2$ for any degree.* See [NR1].

Although we have said above that it parametrises polystable bundles, there is a better way to think of it. Consider any *semistable* vector bundle. If it is not stable, then it admits a proper subbundle which is also of the same slope. If F is a subbundle of E of least rank and same slope, then it follows that F is stable. By induction then, we obtain a flag of subbundles

$$F_0 = 0 \subset F_1 \subset \cdots \subset F_r = E$$

where all the subbundles F_i have the same slope and F_i/F_{i-1} are stable. A flag with this property is not unique, much as the Jordan-Hölder series of a module of finite length, is not unique. However, again as in Jordan-Hölder theorem, the successive quotients are however isomorphic upto order. In other words, the polystable bundle $GrE = \sum F_i/F_{i-1}$ is uniquely determined, upto isomorphism, by E.

Definition 5.13. Two semistable vector bundles E_1 and E_2 of the same slope are said to be *S-equivalent* if $Gr(E_1)$ and $Gr(E_2)$ are isomorphic.

Remark 5.14 The notion of S-equivalence is relevant only for nonstable polystable bundles. On stable bundles, it reduces to isomorphism.

The openness, valid for stable and semistable bundles, does not hold for polystable bundles. For example, we may take extensions of $\mathcal{O}$ by itself. All the extensions are S-equivalent to the trivial bundle of rank 2 but the trivial extension is the only one which gives a polystable bundle. We gave this earlier as an example to show that a 'good' structure of moduli cannot exist. But with the notion of S-equivalence, all the extensions are S-equivalent to the trivial bundle, so that it is no longer a negative example!

Thus it is more fruitful to think of the moduli space as the set of S-equivalence classes of semistable bundles, rather than as the set of isomorphism classes of polystable bundles.

5.6 Universal property

Let us denote the moduli space we referred to above by $U_X(n, d)$, or simply $U(n, d)$. It has the following universal properties.

Theorem 5.15. *If E is any family of semistable vector bundles on X, parametrised by a variety T, then the (classifying) map φ_E, which maps $t \in T$ on the S-equivalence class of E_t, is a morphism $T \to U(n,d)$.*

Theorem 5.16. *Let M be a variety. Assume given for every family E of semistable bundles, parametrised by T, a morphism $f_E : T \to M$ and that the morphisms f_E are compatible with pull backs in the sense that if E' is a family obtained by pulling back E by the morphism $(g \times Id)$: $T' \times X \to T \times X$ where $g : T' \to T$ is a morphism, then we have $f_{E'} = f_E \circ g$. Then there is a unique morphism $f : U(n,d) \to M$ with the property that $f \circ \varphi_E = f_E$ for any family E.*

The most optimistic expectation would be that there exists a Poincaré bundle on the lines of section (3.1) at least on the open set of stable points. However this turns out to be false in general. The precise theorem that I proved was:

Theorem 5.17. *[R1]. If there exists a Poincaré bundle on any Zariski open set of $U(n,d)$, then n and d are coprime. If they are indeed coprime, there exists a Poincaré bundle on the whole of $U(n,d) \times X$ with the obvious universal property.*

6 Global properties of the moduli spaces $U(n,d)$ and $SU(n,d)$

If $g = 0$, there are no stable bundles of rank ≥ 2. So we will assume hereafter that $g \geq 1$. In order to study the structure of the moduli spaces, one constructs various families of semistable vector bundles and studies their parameter spaces.

6.1 The space $SU(n,d)$

If E is a family of semistable vector bundles parametrised by T, then the line bundle $det(E)$ is a family of line bundles with the same parameter space. The classifying map $\varphi_{det(E)}$ of this family gives a map of T into the Jacobian J^d. These are clearly compatible with pull backs, and so by (3.1) we get a morphism $det : U(n,d) \to J^d$. This is in fact a locally trivial fibration, and so the fibres are all isomorphic. We will denote the fibre by $SU(n,d)$. If it is necessary to specify the determinant to be a given line bundle L, we will denote it by $SU(n,L)$. The group J_n of line bundles ξ of order n in J acts on $SU(n,d)$ through tensor product.

Then one checks easily that $U(n, d)$ can be constructed out of $SU(n, d)$ and J.

Proposition 6.1. *The quotient of $SU(n, d) \times J$ by the group J_n under the action $j(E, \xi) = (j \otimes E, j^{-1} \otimes \xi)$ is canonically isomorphic to $U(n, d)$.*

Proof: In fact, the map $(E, \xi) \mapsto E \otimes \xi$ of $SU(n, d) \times J$ into $U(n, d)$ obviously goes down to the quotient and induces the required isomorphism.   □

This essentially reduces the study of $U(n, d)$ to that of $SU(n, d)$. Moreover, in a sense, the latter is the *non-abelian* analogue of the Jacobian. We will therefore take this up now.

First of all, if $d \equiv d' \mod n$, then $SU(n, d)$ is isomorphic to $SU(n, d')$ under tensor product by any line bundle of degree $\frac{d'-d}{n}$. Thus for every n, there are, upto isomorphism, n varieties to be reckoned with. We have already remarked that if n and d are relatively prime, then these varieties are projective and smooth and admit Poincaré families.

We saw that every vector bundle E (after tensoring by a line bundle) can be written as an extension of $det(E)$ by $\mathcal{O}^{n-1}$. It is easy to see that the required line bundle can be chosen to be independent of E as long as E is semi-stable of a given degree. Now in the family of extensions a (non-empty) open set parametrises semi-stable bundles, so that we have a surjection of an open set of an affine space on $SU(n, d)$. In other words,

Proposition 6.2. $SU(n, d)$ *is unirational.*

By a closer analysis of this method, we may show in some cases, that it is rational [New1]. For example, $SU(n, 1)$ is known to be rational. Recently, King and Schofield have proved that in the relatively prime situation $SU(n, d)$ is rational [KS]. The question of rationality or otherwise of $SU_X(n)$ for example is still open.

One can show that $SU(n, d)$ is simply connected and has second Betti number 1. This implies that there is a line bundle h on it, such that any line bundle is an integral power of h.

When one deforms the Riemann surface, the corresponding $SU(n, d)$ also gets perturbed. One can indeed show that every perturbation of $SU(n, d)$ is indeed obtained this way and that the corresponding infinitesimal deformation of X is unique [NR2]. This is an interesting non-classical fact since a similar result is not true for the Jacobian. The number of parameters for the perturbation of the Jacobian as an Abelian variety (taking into account its polarisation as well) is $g(g+1)/2$, while the number of parameters for the deformation of X itself is $3g - 3$. The proof of the above theorem is heavily dependent on the construction outlined in (5.2).

7 Geometry of $SU_X(2)$ for low genus

The first nontrivial study of the moduli space that one undertakes is naturally that of rank 2 bundles. So it is not surprising that much is known about the moduli spaces in this case.

7.1 Determination of $SU(2,0)$ and $SU(2,1)$ when $g = 2$

M.F. Atiyah [A] had studied vector bundles on elliptic curves, i.e. when $g = 1$, and his results can be summarised in our language as saying that in this case, $SU(n, d)$ reduces to a point for every n and d. Hereafter we will assume that $g \geq 2$.

When $g = 2$, there is a close relationship between $SU(2, \mathcal{O})$ and the geometry of Kummer surfaces. In fact, the first result [NR1] states

Theorem 7.1. *The variety* $S = SU(2, \mathcal{O})$ *is isomorphic to the 3-dimensional complex projective space.*

To every line bundle L of degree zero, associate the polystable bundle $L \oplus L^{-1}$ in S. Denoting by ι the involution $L \mapsto L^{-1}$ of J, we see that this gives an isomorphism of J/ι onto the variety of nonstable points in S. The variety $\mathcal{K} = J/\iota$ is the *Kummer surface*. The imbedding of $\mathcal{K}$ in $S = \mathbf{P}^3$ realises it as a quartic surface with sixteen nodes, namely the points $L \oplus L$ with $L^2 = \mathcal{O}$. Incidentally this is essentially the only case, where the variety $SU(n, d)$ is smooth even at nonstable points, although n and d are not coprime.

Theorem 7.2. *[NR1, New2]. The space* $SU(2, 1)$ *is isomorphic to a complete intersection of two quadrics in* P^5.

7.2 Quadratic complexes and vector bundles

Fixing the determinant to be the line bundle $\mathcal{O}(x)$ for some $x \in X$, one may take one of the quadrics to be the Grassmannian of lines in $\mathbf{P}^3$, since the latter is a quadric in its Plucker imbedding.

Definition 7.3. A proper subvariety of the Grassmannian given by intersection with a hyperplane (resp. quadric) in $\mathbf{P}^5$ is called a *line* (resp. *quadratic*) *complex* of lines in $\mathbf{P}^3$.

We will implicitly assume that the intersection is smooth. Given a quadratic complex, for every point v in $\mathbf{P}^3$ the set of all lines belonging to the quadratic complex and passing through v forms a quadratic cone with v as vertex. The set of points v such that the cone breaks up into a pair of planes, is called the *singular locus* of the quadratic complex. Kummer showed that this singular locus is a quartic surface in $\mathbf{P}^3$ with 16 nodes. Cayley later pointed out that it is actually the quotient of the Jacobian of a curve of genus 2 by the involution $x \mapsto -x$. Klein [K] returned to this question and showed that given a Kummer surface $\mathcal{K}$, there is a one-parameter family (parametrised actually by $\mathbf{P}^1$) of quadratic complexes all of which (except for 6 points) have $\mathcal{K}$ as singular loci. The two sheeted covering of this $\mathbf{P}^1$ ramified at the six exceptional points gives the curve whose Kummer surface is $\mathcal{K}$.

It is amazing that all this beautiful geometry has an interpretation in terms of vector bundle moduli. Start with a compact Riemann surface of genus 2. Let $x \in X$, not a Weierstrass point, and E a vector bundle with $det(E) = \mathcal{O}(x)$. Then by performing elementary transformations on E at x, as outlined in (5.2), we get a family of vector bundles with $det \simeq \mathcal{O}$ parametrised by $P(E_x) = \mathbf{P}^1$. One can show that this imbeds $\mathbf{P}^1$ as a *line* in $SU_X(2) = \mathbf{P}^3$. These lines constitute the quadratic complex. The corresponding map of $SU(2, \mathcal{O}(x))$ into the Grassmannian of lines in $\mathbf{P}^3$ is injective and the image is the intersection of a quadric with the Grassmannian. As we vary x in the curve outside the six Weierstrass points, we get different quadratic complexes. All of these have the Kummer surface of X as their singular loci. One can also show that in this correspondence the points x and its hyperelliptic image ix give rise to the same quadratic complex, so that we have a complete translation of Kummer's and Klein's geometry and can relate it to Cayley's understanding of the relationship betwen Kummer surface and Riemann surfaces of genus 2.

Let us again fix a point x as before. To every element ξ of the Jacobian associate the set of all extensions of $\xi \otimes \mathcal{O}(x)$ by ξ^{-1}. All the bundles corresponding to nontrivial extensions are stable bundles of rank 2 and have determinants isomorphic to $\mathcal{O}(x)$, and two bundles arising as extensions are isomorphic if and only if the corresponding extensions differ by a scalar factor. In other words, we have a family of stable bundles parameterised by $PH^1(X, \xi^{-2}(-x))$. By Riemann Roch theorem we see that H^1 of a line bundle of degree -1 is two dimensional. In other words we have a $\mathbf{P}^1$ inside $SU_X(2, \mathcal{O}(x))$ corresponding to each ξ in the Jacobian. Then one can prove the following theorem.

Theorem 7.4. *The variety of lines in the intersection of two quadrics is isomorphic to the Jacobian.*

This theorem which has a classical flavour, had been proved by Godeaux, again from the point of view of classical projective geometry.

7.3 Kummer three-fold

We saw that the Kummer variety is imbedded naturally (using the linear system of twice θ on the Jacobian) in $\mathbf{P}^{2^g-1}$. Having interpreted classical geometry in terms of the vector bundle moduli, the question arises if we can reverse the procedure and obtain new geometric results from an understanding of the moduli variety.

The geometry of Kummer varieties in the case $g = 3$ from the classical point of view was studied largely by Wirtinger and Coble. In this case, the variety $SU(2, \mathcal{O})$ has been explicitly determined to be a quartic hypersurface in $\mathbf{P}^7$ [NR2]. This easily follows now from the theorem of Brivio and Verra, which gives an imbedding of $SU_C(2)$ into $\mathbf{P}^{2^g-1}$ which extends the imbedding of the Kummer variety in this projective space. The image is a hypersurface, and one easily computes its degree to be 4. This quartic had been earlier written down explicitly by Coble from the point of view of theta functions. He also proved that the quartic contains the Kummer 3-fold in its singular locus and conjectured that it is precisely the singular locus. However since this quartic is identified with $SU_C(2)$, it follows that the singular locus of the quartic is the Kummer variety. In particular, the Kummer variety is defined by the partial derivatives of the quartic form. This shows that the Kummer variety is defined by cubic equations, as was conjectured by Coble and perhaps Wirtinger. Thus vector bundle theory has been useful in proving their conjecture [NR3].

Theorem 7.5. *If X is a non-hyperelliptic curve of genus 3, the associated Kummer three-fold is given by cubic equations.*

8 Study of $SU(2, \mathcal{O})$ for higher genera

Can one extend these results to higher genera?

The pull back of the generator h of the divisor class group of $SU(2, \mathcal{O})$ to the Jacobian by the map $L \mapsto L \oplus L^{-1}$ is 2θ. It turns out that $H^0(SU(2, \mathcal{O}), h)$ is of dimension 2^g. From this one can conclude easily that $H^0(SU(2, \mathcal{O}), h) \to H^0(J, \mathcal{O}(2\theta))$ is an isomorphism. Much of what we said above is the interplay of the geometry of the 2θ-linear system via this isomorphism.

In a direct way, one can associate to every vector bundle E with $det(E) = \mathcal{O}$, a divisor D_E in J^{g-1} whose support is given by $\{L : H^0(E \otimes L) \neq 0\}$. This divisor is the divisor of zeros of a section of $\mathcal{O}(2\theta)$ in J^{g-1}.

8.1 Hyperelliptic curves

When X is hyper-elliptic, one has a natural involution on X, namely the sheet interchange of the map $X \to \mathbf{P}^1$. It is uniquely determined and is referred to as the *hyper-elliptic involution*. It also acts on $SU(2, \mathcal{O})$ via pullbacks. The above morphism factors to the quotient by this action.

Indeed, we have a precise result. The fixed points of the hyper-elliptic involution are called *Weierstrass points* of X. It is easy to see that these are $2g + 2$ in number. Their images in $\mathbf{P}^1$ are also uniquely determined, upto a Möbius transformation. If we take them to be complex numbers, say λ_i, $i = 1, \cdots, 2g + 2$, then we have the following theorem due to Usha Desale and S. Ramanan [DR].

Theorem 8.1. *Consider the two quadrics Q_1 and Q_2 in $\mathbf{P}^{2g+1}$ with homogeneous co-ordinates (X_i) given respectively by*

$$\sum X_i^2 = 0 \quad \text{and} \quad \sum \lambda_i X_i^2 = 0.$$

Then the variety $SU(2, 1)$ is isomorphic to the variety of linear subspaces of dimension $g - 1$ contained in the two quadrics. Linear subspaces of maximal dimension $(= g)$ contained in Q_1 has two connected components. Let us take one of the components and consider the subvariety consisting of the linear subspaces of maximal dimension contained in Q_1 and belonging to the chosen component, which satisfy the condition that the restriction of the quadratic form Q_2 to them has rank ≤ 4. This variety is isomorphic to the variety $SU(2, 0))/\iota$, where ι is the action on $SU(2, 0)$ induced by the hyper-elliptic involution.

8.2 Theorem of Brivio and Verra

We are led to the natural question if this imbeds $SU(2, 0)$ in $PH^0(J^{g-1}, \mathcal{O}(2\theta))$ if X is not hyper-elliptic. The following theorem, due to Brivio and Verra [BV] settled the question.

Theorem 8.2. *The above map is an imbedding if X is not hyper-elliptic.*

Remark 8.3 Actually Brivio and Verra [BV] proved this only at the stable points. Subsequently Laszlo showed that it is an imbedding at all points.

9 Moduli of curves and principally polarised abelian varieties

We saw that one can associate to any Riemann surface X the Jacobian, which is a principally polarised abelian variety. Now one can constuct it – as Mumford did – in the same spirit as one constructs the spaces $SU(n, d)$ for a fixed X.

Definition 9.1. A *family* of smooth projective curves, parametrised by an irreducible variety T is by definition a projective morphism $P \to T$ which is smooth, the fibres being curves.

As in the case of vector bundles, one thinks of a family as the curves C_t parametrised by T, the curve C_t being the fibre over t in P.

Fix an integer g and associate to every T the set of all families of smooth curves of genus g, parametrised by T. This functor can be classified in the coarse sense we talked about in the vector bundle situation. In other words, there is a structure of a variety $\mathcal{M}_g$ called the *Moduli of curves of genus g* on the set of isomorphism classes of all curves of genus g, such that given any family as above, the natural map which associates to t the isomorphism class of C_t is a morphism of T into $\mathcal{M}_g$. Moreover if S is a variety and for every family parametrised by T there is assigned a morphism from $T \to S$ in a functorial fashion, then there is a natural morphism of $\mathcal{M}_g$ into S. This moduli space is normal and quasi-projective. The finer requirement, namely there be a universal family of curves parametrised by $\mathcal{M}_g$ is again false.

We have thus highlighted the similarity of this moduli variety to $SU_C(n, d)$ but the analogy breaks down at a few points. Firstly, while the group of automorphisms of all curves are finite, some curves admit more automorphisms than normal. In fact, 'normal' here means that it is $\mathbf{Z}/2$ for $g = 2$ and $\{1\}$ for $g \geq 3$. The nonexistence of the universal curve is only due to the presence of these curves. Over the remaining curves, which are dense, there does exist such a family. In the vector bundle moduli there does not exist a Poincaré bundle over *any* Zariski open set. All stable vector bundles have the same group of automorphisms.

Much the same kind of theorems are true for principally polarised abelian varieties as well. Let us denote the corresponding moduli variety by $\mathcal{A}_g$. Associating the Jacobian to a curve leads to a morphism of $\mathcal{M}_g$ into $\mathcal{A}_g$. The dimension of $\mathcal{M}_g$ is $3g - 3$, while that of $\mathcal{A}_g$ is $\frac{g(g+1)}{2}$. We now have the following theorem due to Torelli.

Theorem 9.2. *If the Jacobians of two smooth projective curves of genus g are isomorphic as polarised abelian varieties, then the curves are themselves isomorphic.*

Simply stated, the map $\mathcal{M}_g \to \mathcal{A}_g$ is injective.

One might therefore wonder if something similar is true for the moduli of vector bundles. Assume $g \geq 3$. In fact, one might start with one of the moduli spaces $SU_C(n, d)$, preferably one that is smooth, and construct the Jacobian of the curve C in terms of the geometry of the space itself. In fact, one can show that the intermediate Jacobian of the moduli space is actually the Jacobian of the curve. So, thanks to the classical Torelli theorem, one can recover the curve from the moduli space and so the analogue of the Torelli theorem holds. This is in the same spirit of Godeaux' theorem mentioned above.

There is also a classical infinitesimal Torelli theorem, which states that the differential of the map $\mathcal{M}_g \to \mathcal{A}_g$ is injective at the points of the curve C if it is non-hyperelliptic. A similar can also be proved for the moduli space. But a surprising result is the following: The differential is actually an isomorphism. In other words, any infinitesimal deformation of the moduli space arises from that of the curve.

9.1 Generalized Theta divisor

Consider now the moduli variety $SU(n, n(g - 1))$. As we have noted above it is isomorphic to the variety $SU(n, \mathcal{O})$. Then one can show that a divisor in $S = SU(n, n(g-1))$ can be defined whose Chern class is the generator h. On the stable locus, it is given by $\{E \in S : H^0(E) \neq 0\}$. A technical point here is that although S is not a manifold, any subvariety of codimension 1 nevertheless gives rise to a line bundle [DN]. The above divisor is generally referred to as the *generalized theta divisor*.

The next problem is to study the linear system of sections of h^r on the variety S. These sections may regarded as nonabelian analogues of theta functions. In an amazing development about ten years ago, a physicist, Verlinde came up with a conjectural formula for the dimension of these spaces, purely from a physical motivation. It has since been proved to be true [BL, KNR].

9.2 Schottky relation

Let $\pi : C' \to C$ be a two sheeted etale covering. Then we can associate to any divisor class in C' one on C as follows. We can represent it by a divisor $D = \sum m_i x_i$ and associate to it the divisor class of $\pi(D) = \sum m_i \pi(x_i)$. If D is the divisor of a meromorphic function f on C', then define a function g on C by setting $g(x) = \prod_{\pi y = x} f(y)$. Then it is obvious that $div(g) = \pi(div(f))$. Hence the map above gives a homomorphism $J^d(C') \to J^d(C)$. This map is called the *norm* map. Consider the kernel of $J(C') \to J(C)$. It can be shown that it has two connected components. The connected component of the identity element is called the *Prym* variety of the covering.

The norm map can also be identified with the following. Let ξ be any line bundle on C'. Take its direct image $\pi_*(\xi)$ on C. We have remarked that it is a vector bundle of rank 2. We can associate to ξ the determinant of $\pi_*(\xi)$. This is essentially the norm map: 'essentially' because they differ by a translation by a line bundle. In fact, the direct image by π of $\mathcal{O}$ on C' can be identified with $\mathcal{O} \oplus \alpha$ where α is a line bundle satisfying $\alpha^2 \simeq \mathcal{O}$. Conversely we can take any nontrivial line bundle α of order 2 and make the vector bundle $\mathcal{O} \oplus \alpha$ a sheaf of algebras. Note that $\mathcal{O}$ is already a sheaf of algebras, while α is an $\mathcal{O}$-Module. To define multiplication of elements in α, we use the isomorphism $\alpha^2 \simeq \mathcal{O}$. Then *Spec* of this $\mathcal{O}$-Algebra is an étale 2-sheeted covering C'. Every such covering is obtained this way.

Note that the determinant of $\pi_*(\mathcal{O}) = \mathcal{O} \oplus \alpha$ is therefore α. In general we have $det(\pi_*(\xi)) \simeq Nm(\xi) \otimes \alpha$.

The restriction of the natural theta divisor on the Jacobian of the covering to Prym can be shown to be twice a principal polarisation. Thus we have associated to any etale 2-sheeeted covering, or equivalently a nontrivial line bundle of order 2, a polarised abelian variety of dimension $g - 1$.

One can easily check that the direct image on C of any line bundle on C' is semistable and hence the map $\xi \mapsto \pi_*(\xi)$ gives a morphism of Prym into the $SU_C(2, \alpha)$. If we wish to have a morphism into $SU_C(2)$, then we can fix a line bundle η such that $\eta^2 \simeq \alpha$, and consider the map $\xi \mapsto \pi_*(\xi) \otimes \eta^{-1}$. This map of Prym is not an imbedding, for the Galois involution ι on C' leaves the Prym invariant and ξ and $\iota(\xi)$ have the same direct image. It is easy to see that the action of ι on Prym is simply the map $\zeta \mapsto \zeta^{-1}$. So the morphism we gave above goes down to one of the Kummer variety associated to the Prym variety, into $SU_C(2)$. Thus $SU_C(2)$ contains the Kummer variety of C as well as the Kummer variety of the Prym variety of the covering. The moduli variety is of dimension $3g - 3$ and these two Kummer varieties are of dimension g and $g - 1$ respectively.

Nevertheless, these two varieties meet. In fact, we have $\pi_*(\mathcal{O}) = \mathcal{O} \oplus \alpha$. Hence our imbedding of the Kummer of Prym in $SU_C(2)$ takes the point $\mathcal{O}$ to the point $\eta \oplus \eta^{-1}$. In other words, the two Kummer varieties intersect and we have determined the intersecting point in each of these varieties. In particular, after imbedding $SU_C(2)$ in $\mathbf{P}^{2^g-1}$ we may say that the two Kummers meet in their natural imbedding at a point of the projective space. One can more or less trivialise this vector space and state this as an algebraic condition satisfied by the Jacobian of a curve in its natural imbedding. This relation is called the *Schottky equation*.

10 Other applications

10.1 Spectral cover and the moduli of vector bundles

We will start by generalizing the construction given in the last section to what are called spectral covers and derive the Verlinde formula for the first power.

Theorem 10.1. $\dim H^0(U(n, n(g-1)), h) = 1$; $\dim H^0(S, h) = n^g$.

Proof: We have already remarked (4.5) that one way of constructing vector bundles is to take a morphism $Y \to X$ of degree n and take direct images of line bundles. The idea is to construct such a covering (which may be ramified) and get a morphism of an open set R of $J(Y)^{g(Y)-1}$ into $SU(n, d)$. The open set R consists of those line bundles whose direct images are semistable. If we choose a covering in such a way that this map has a dense image, then we would get an injective map of $H^0(U(n, d), h)$ into H^0 of the pull back of h to R. If $J(Y) \setminus R$ is of codimension ≥ 2, then the pullback of h extends to a line bundle on $J(Y)$. We need to adjust things in such a way that $d = n(g(X) - 1)$.

Notice that sections of a line bundle L on Y give rise to sections of its direct image on X and this leads to a bijection of their H^0. In particular, the direct image belongs to the generalized theta divisor Θ if and only if L belongs to the theta divisor. In other words, the inverse image of Θ

is $\theta|R$. Now since $H^0(J^{g(Y)-1}, \mathcal{O}(\theta)) = 1$, this shows that $H^0(U, \Theta) = 1$ also.

The second part of the assertion is similar, but slightly more involved. Firstly one has to take the map $J^{g(Y)-1} \to J^{n(g-1)}$ given by $L \mapsto det(\pi_*(L))$. We may take a fibre, (called $Prym$ of this map) and get a morphism $Prym \to S = SU(n, n(g(X)-1))$. Fortunately $Prym$ is also an abelian variety and the pull back of Θ is only the restriction of θ from the Jacobian of Y to Prym. One can compute the dimension of H^0 of the pull back of Θ and so we get an upper bound for $H^0(S, \Theta)$ as $n^{g(X)}$. Then one has to give an argument to show that it is also a lower bound.

Now we will explain how such a covering may be found. Take a section say s_i each of K^i for $i = 1, \cdots, n-1$. Then one can construct a Riemann surface Y (assuming that s_i are general), and a map of degree n of Y onto X. Indeed take the total space of the line bundle K. Over any point $x \in X$ consider the solutions of the equation $v^n + \sum s_i(x)v^{n-i} = 0$ in the fibre over x. Here $v \in K_x$ and all the terms are in K_x^n. There are of course n distinct solutions at most points $x \in X$, but at some points this equation may have repeated roots. Thus we get a ramified covering of X. By construction, it is easily seen that $\pi_*(\mathcal{O}_Y) = \sum_{i=0}^{i=n-1} K^{-i}$. From this one deduces that the genus of Y is $n^2(g(X)-1) + 1$. It can be shown that this covering has the required properties. See [R2] for details.   □

Remark 10.2 The above procedure gives a covering when we take sections $s_i \in H^0(L^i)$ where L is any line bundle and not necessarily K. Coverings obtained by this procedure are called *spectral covers*.

In fact, when we take direct images of a line bundle from a spectral cover, we not merely obtain a vector bundle E on X, but a homomorphism of E into $E \otimes L$, as well. Note that $\pi^*(L)$ comes with a canonical section. In fact, we have $\pi_*(\pi^*L) = L \otimes \pi_*\mathcal{O} = L \otimes \sum_{0 \le i \le n-1} L^{-i} = L \oplus \mathcal{O} \oplus \sum_{0 \le i \le n-3} L^{-i-1}$ and the second summand has a canonical section. This gives the section of π^*L as claimed.

These induced homomorphisms $E \to E \otimes L$ may be thought of as endomorphisms with a twist. One can also make sense of the characteristic polynomial of this twisted endomorphism. For example, it is clear how to define the *trace*. Considering it as a section of $E^* \otimes E \otimes L$ we may contract the first two factors and get a section of L. Also the determinant is the homomorphism $det(E) \to det(E \otimes L) = det(E) \otimes L^n$, where $n = rk(E)$. In other words, it is a section of L^n. Similarly one can also define all other coefficients of the characteristic polynomial as sections of L^i, $i = 1, \cdots, n$. It turns out these are the same as the sections used to construct the spectral cover. Conversely, given a vector bundle E and a L-twisted endomorphism, we may define a spectral cover, using the coefficients of the characteristic polynomial. In this way one can prove the following theorem rather easily.

Theorem 10.3. *Let s_i be sections of L^i for $i = 1, \cdots, n$. Then there is a natural bijection between line bundles on the spectral cover given by (s_i) and pairs (E, φ) consisting of a vector bundle E of rank n and a L-twisted endomorphism of E with characteristic coefficients s_i.*

11 Higgs pairs

We introduced pairs consisting of a vector bundle E and a L-twisted endomorphism $\alpha : E \to E \otimes L$. We refer to them as *Higgs pairs*, or a *Higgs bundle*. The twisted endomorphism is called *Higgs field* of the pair. We say a Higgs pair is *semistable* (resp. stable) if the usual inequalities hold good for those subbundles F of E which are invariant under α in the sense that $\alpha(F) \subset F \otimes L$. Of course if E is itself stable, the pair is also stable. As in the case of vector bundles, one can define S-equivalence and indeed construct the moduli space of S-equivalence classes of such pairs as well. Then there is a natural map from such pairs into $\Gamma(L^i)$ for $i = 1, \cdots, n$ where $rk(E) = n$. These are obtained as the characteristic coefficients of the Higgs field. We pointed out that for every choice of the coefficients, we get a *spectral covering* and the set of Higgs pairs with these as characteristic coefficients, are bijective with the Jacobian of the covering.

We take now L to be the canonical line bundle K. The tangent space at a stable point E of $U(n, d)$ can be identified with $H^1(X, End(E))$, which is dual to $H^0(X, End(E) \otimes K)$. In other words, the total space of the cotangent bundle is an open subvariety of the variety of Higgs pairs. The fibres of the characteristic map on the cotangent bundle, are then open subsets of abelian varieties. It turns out that this gives rise to a *completely integrable system* on the cotangent bundle.

11.1 Fundamental groups and Higgs pairs

A spectacular application of these ideas is the following generalisation of the theorem of Narasimhan and Seshadri.

Theorem 11.1. *There is a natural bijection between the set of equivalence classes of completely reducible representations of the fundamental group and the set of isomorphism classes of polystable Higgs pairs of degree 0.*

Under this bijection, unitary representations correspond to those pairs which have Higgs field 0 and we recover the theorem of Narasimhan and Seshadri. It should be pointed out however that this bijection is quite involved and unlike the theorem of Narasimhan and Seshadri, where it is at least clear how to associate a holomorphic bundle to a unitary representation, even this part is difficult in this generalization. Given any representation, one can associate to it by the general procedure a vector bundle, but it is *not* the bundle of the Higgs pair associated to the representation by the above correspondence.

Since the definition of Higgs pairs is purely algebraic geometric in nature, the above correspondence throws much light on the nature of the fundamental group. Before we explain this we remark that the correspondence above is not confined only to Riemann surfaces, but can be carried over to arbitary smooth projective varieties M, the only change required being that in the definition of the Higgs field, we should require that $\alpha \wedge \alpha = 0$. This wedging operation uses the Lie algebra structure in $End(E)$ and is a priori an element of $H^0(M, End(E) \otimes \Omega^2)$.

One can define direct sums, duals, tensor products, etc. for Higgs pairs, just as for ordinary vector bundles. Then one checks that the above correspondence respects these operations. Finally we will also fix a point $x \in X$ and consider the assignment to any Higgs pair (E, ω) the fibre of E at x. It is easy to see that this assignment is faithful in the sense that any homomorphism of polystable Higgs pairs is completely determined by its restriction to the fibre. We can summarise this by saying that polystable Higgs pairs form an abelian category which has tensor products, direct sums, etc. Such a category is called a *tensor category*. There is a faithful additive functor from this category to the category of vector spaces, and the functor takes tensor products to tensor products, etc. A tensor category provided with such a functor (called *the fibre functor*) is called a *Tannaka* category.

Tannaka himself considered the following category. Let G be any Lie group. Consider the abelian category of semisimple G-modules. In that category, we have tensor products, direct sums, etc. It also comes equipped with a functor from the category of G-modules to the category of vector spaces, namely that which ignores the G-module structure. In other words, this is also a Tannaka category. Tannaka's observation was that the group can in many cases be recovered from the Tannakian category of G-modules. The idea is very simple. One takes the group of automorphisms of the fibre functor. We will call it the *Tannaka* group associated to the Tannaka category. It is obvious that we have a homomorphism of G into the Tannaka group. If G is a reductive group, then one easily sees that this homomorphism is actually an isomorphism.

But one can start with any group π, consider the category of semisimple π-modules, and associated Tannaka group $Tk(\pi)$. We will call it the Tannakian completion of π. It has also been referred to in the literature as the *pro-reductive algebraic group completion* of π. We will now take the fundamental group π of a smooth projective variety (e.g. a smooth projective curve). We then have a Tannaka equivalence of the category of π-modules with the category of polystable Higgs pairs of degree 0. This means that the Tannakian completion of the fundamental group can be captured purely in terms of the algebraic structure. It is easy to see that the Grothendieck's algebraic fundamental group can be got out of this, but that this group carries considerably more information.

Now, given a polystable Higgs pair and $\lambda \in \mathbf{C}^\times$, we can multiply the Higgs field of the pair by λ and get a new polystable pair. This leads to an 'action of $\mathbf{C}^\times$ on the Tannaka category' of polystable Higgs pairs of degree 0 and hence on the group $Tk(\pi)$. One can think of this as some kind of a *non-abelian Hodge decomposition*. Take the element -1 in $\mathbf{C}^\times$. It acts on a Higgs pair (E, ω) and gives the pair $(E, -\omega)$. Although in general at the representation level, it is impossible to write down the action of $\mathbf{C}^\times$, the action of -1 is simple to describe. It associates to any representation ρ of π the representation $\overline{(\rho^t)^{-1}}$. The existence of an action by $\mathbf{C}^\times$ on a group such that -1 acts as above in its Tannaka category of semisimple representations, is a strong restriction on the group. We may call such a group a *Hodge group*. One can show for example, $SL(3, \mathbf{Z})$ does not admit such an action. As a consequence, this group cannot be the fundamental group of any smooth projective variety!

For a more detailed summary of Higgs pairs, see [R2].

12 The Hitchin connection

12.1 Heisenberg extensions

Let $A = \mathbf{C}^g / \Gamma$ be a complex torus. If h a hermitian positive definite form on A whose imaginary part is integral on Γ, then we saw that A

is an Abelian variety. If L is a line bundle with h as its chern class, consider the subset of A consisting of a such that $T_a^* L \simeq L$. It is clearly a subgroup of A, which we will denote by $K(L)$. If D is a divisor in A whose class is L, then $T_a(D)$ is also in the same class and so we get an action of $K(L)$ on the set of all such divisors, or what is the same, on $PH^0(A, L)$. One may ask whether this projective representation comes from a linear representation of $K(L)$ on $H^0(A, L)$. In fact, it does not, since while $T_a^*(L)$ is isomorphic to L, there is no such *canonical isomorphism*. Indeed, there is no way one can construct such isomorphisms corresponding to every $a \in K(L)$ in a consistent way, except in the trivial case when $K(L) = (0)$.

How do we resolve this problem? We construct a group as follows. Consider the set $Heis(L)$ of all pairs (a, φ) such that $a \in A$ and $\varphi : T_a^* L \to L$ an isomorphism. Then for any $x \in A$, it gives automatically an isomorphism $T_x^*(\varphi) : T_x^*(T_a^*(L)) = T_{a+x}^*(L) \to T_x^*(L)$. In particular, if $(b, \psi) \in Heis(L)$, then $\psi \circ T_b^*(\varphi)$ is an isomorphism of $T_{a+b}^*(L)$ with L. It is easy to check that the law $(b, \psi).(a, \varphi) = (a + b, \psi \circ T_b^*(\varphi))$ gives a group structure on $Heis(L)$. It comes with a homomorphism $(a, \varphi) \mapsto a$ of $Heis(L)$ onto $K(L)$. The kernel is the set of all isomorphisms of L with itself, namely $\mathbf{C}^\times$. In other words, we have an exact sequence of groups

$$1 \to \mathbf{C}^\times \to Heis(L) \to K(L) \to 1.$$

From our construction, it is easy to see that $Heis(L)$ acts on $H^0(L)$ as we wanted. In fact, if $g = (a, \varphi) \in Heis(L)$ and s is a section of L, then gs is defined to be $\varphi(T_a^* s)$. Note that any $\lambda \in \mathbf{C}^\times$ acts as $\lambda.Id$ in this representation and therefore induces a projective representation of $K(L)$, which is the one we started with.

We have written (1) rather than (0) in the exact sequence, because while both the groups at the extremities are abelian, the group in the middle, namely $Heis(L)$, is not so. In fact, $\mathbf{C}^\times$ is the centre of $Heis(L)$. In order to measure the noncommutativity of this group, we consider the map $(g, h) \mapsto ghg^{-1}h^{-1}$ of $Heis(L) \times Heis(L)$ into $Heis(L)$. This goes down to a map e_L of $K(L) \times K(L) \to \mathbf{C}^\times$. This map is bi-multiplicative (i.e. multiplicative in each variable), and is alternating. It is called the *Weil pairing*. The key point is that this form is nondegenerate, which implies that $\mathbf{C}^\times$ is the centre of $Heis(L)$.

In fact, one can actually deduce that the group $Heis(L)$ and the exact sequence above can be made (up to isomorphism) explicit as follows.

The integral alternating form $(Imh)|\Gamma$ takes the following form in terms of a basis of the free abelian group Γ.

$$\begin{pmatrix} 0 & diag(d_1,\ldots,d_g) \\ diag(-d_1,\ldots,-d_g) & 0 \end{pmatrix}.$$

Here d_i are positive integers satisfying $d_1|\cdots d_i|\cdots|d_g$. We call the g-tuple $d = (d_1,\cdots,d_g)$ the *type* of the line bundle L. It is easy to see that the group $K(L)$ is isomorphic to $(\mathbf{Z}/d_1)^2 \times \cdots \times (\mathbf{Z}/d_g)^2$. It is convenient to write this group as $B \times \hat{B}$ where $B = \mathbf{Z}/d_1 \times \cdots \times \mathbf{Z}/d_g$ and $\hat{B}$ its dual group, namely the group of characters of B. Then consider the set $\mathbf{C}^\times \times B \times \hat{B}$ and define multiplication by $(\lambda,b,\chi)(\lambda',b',\chi') = (\lambda\lambda'\chi'(b),bb',\chi\chi')$. It is a routine check to see that this is a group $Heis(d)$ and that the map $(\lambda,b,\chi) \mapsto (b,\chi)$ is a homomorphism of $Heis(d)$ onto $K(d) = B \times \hat{B}$. In other words, we have an exact sequence

$$1 \to \mathbf{C}^\times \to Heis(d) \to K(d) \to 1.$$

Now we have the following theorem of Mumford.

Theorem 12.1. *The group $Heis(L)$ and indeed the associated exact sequence are respectively isomorphic to $Heis(d)$ and the above exact sequence.*

Definition 12.2. A *theta structure* on an abelian variety A provided with an ample line bundle L of type d, is an isomorphism of the Heisenberg exact sequence of L with that of type d described above.

Now any (finite dimensional) representation of $K(d)$, holomorphic on $\mathbf{C}^\times$, is completely reducible. In any irreducible representation, the centre $\mathbf{C}^\times$ is represented by scalars and so we get a character on it. Every character of $\mathbf{C}^\times$ is of the form $\lambda \mapsto \lambda^r$ for some r. Then we have the following easily proved theorem.

Theorem 12.3. *Any two irreducible representations of $Heis(d)$ on which the central character is the identity (i.e. $r = 1$) are equivalent.*

In fact, if we take the vector space $T(d)$ of functions on B with the above notation, then B acts on it by translation. The group $\hat{B}$ can be regarded as functions on B and as such it acts on $T(d)$ by multiplication. Scalars also act by multiplication. It is easy to check that these build an action of $Heis(d)$ on $T(d)$. This is irreducible and gives a model for the unique irreducible representation mentioned above. We will fix this and call it 'the standard model'. In particular the unique irreducible representation that we made mention of above, is of dimension $d_1 \cdots d_g$.

If A is an abelian variety with an ample line bundle, then we have seen that $Heis(L)$ acts on $H^0(L)$, in which the central character is the identity. Hence it is isomorphic to the direct sum of many copies the standard representation. However, the dimension of $\Gamma(L)$ is $d_1 \cdots d_g$ (for example from the Riemann Roch theorem) and so we conclude the following.

Theorem 12.4. *The representation of $Heis(L)$ on $\Gamma(L)$ is isomorphic to the unique irreducible representation with identity as the central character.*

As a corollary we deduce that if we are given a theta structure in addition, then we get an irreducible action of $Heis(d)$ on $H^0(L)$ and so there is an isomorphism of $\Gamma(L)$ with the standard model. By Schur's Lemma, this isomorphism is unique up to a scalar factor. Thus we get, up to a scalar factor *a basis* of $\Gamma(L)$.

This fact can be interpreted as follows.

Consider the moduli $\mathcal{A}_g$ of polarised abelian varieties of dimension g. Then at least over the set of abelian varieties which do not have automorphisms other than ± 1, there is a universal abelian variety with a relatively ample line bundle along the fibres. Its direct image gives a vector bundle (which we will call the *theta bundle*) of rank $d_1 \cdots d_g$. The moduli of polarised abelian varieties together with theta structures is a Galois covering of $\mathcal{A}_g$. Since any two theta structures differ by an automorphism of $Heis(d)$ which is Id on $\mathbf{C}^\times$ it is easy to determine the group D of the covering. It is an extension of the symplectic group of $K(d)$ (automorphisms which leave invariant an alternating bi-multiplicative map into $\mathbf{C}^\times$), by the group $K(d)$ itself. The projective bundle associated to the theta bundle is isomorphic to the bundle associated to the above Galois covering for an action of D on the projective standard model.

In particular, we get a flat connection on the projective theta bundle.

We can do the same thing over the moduli $\mathcal{M}_g$ of curves, by restricting the theta bundle to the space of Jacobians. Note that in this case, the abelian varieties in question are all principally polarised and so the theta bundle is not very interesting from this point of view. However, we can take multiples of the principal theta divisor, and construct theta vector bundles of ranks r^g for every r. The corresponding projective bundles come with canonically defined flat connections.

12.2 The Hitchin connection

Welters gave a direct algebraic approach to the construction of the above connection on the theta bundle. Hitchin adapted his proof to the case of vector bundles. The result may be summarised as follows. To every smooth, projective curve C, associate the vector space of sections of h^r where h is the generator of the Picard group of the moduli space of vector bundles of fixed rank n and degree d. The projective bundle over $\mathcal{M}_g$ has a canonical flat connection. This is called the *Hitchin connection*.

Remark 12.5 The fact that the vector spaces $H^0(SU_C(n,d), h^r)$ depend very little on the particular curve C, is the reason that physicists could identify this space in terms of the topological surface of genus g, without the intermediary of the complex structure.

13 Another application: Syzygies of the canonical curve

We will now give with one more application, which was studied in [PR]. This time we do not bring the moduli space into focus, but only the notion of stability. The question of interest to us here is the nature of the canonical imbedding of the Riemann surface. In other words, one might like to understand the map φ of X into $\mathbf{P}^{g-1}$ given by the linear system of K.

Firstly, φ imbeds X if and only if it is not hyper-elliptic. If it *is* hyper-elliptic, the canonical linear system maps X as a two-sheeted covering over the rational normal curve of degree g in $\mathbf{P}^{g-1}$. A theorem of Max Noether states that in the non-hyperelliptic case, the natural product map

$$Sym^i H^0(X, K) \to H^0(K^i)$$

is surjective, for all $i \geq 0$.

Assuming φ does imbed X, we may ask for the defining equations for X in this imbedding. We have the following theorem due to Petri in this regard.

Theorem 13.1. *The ideal given as the kernel of the map*

$$Sym(H^0(X, K)) \to H^0(X, Sym(K))$$

is generated by elements in Sym^2 except in these cases:
X is trigonal, i.e. it has three-sheeted map onto $\mathbf{P}^3$.
X is imbedded in $\mathbf{P}^2$ as a variety defined by a polynomial of degree 5.
(This can happen only when the genus of X is 6.)

To put these two results in a single framework, one defines a number called the *Clifford index* of X. This is 0 if and only if X is hyper-elliptic, and 1 if and only if one of the the the above conditions is satisfied. The Clifford index is at most equal to the integral part of $\frac{g-1}{2}$. Mark Green formulated a conjecture regarding the algebraic nature of the defining relations, which says that if the Clifford index gets larger, these have also better properties.

These can be related to the following canonically defined vector bundle on X. Consider the pull back of the tangent bundle T of the projective space by φ. The bundle we are interested in is $Q = \varphi^*(T) \otimes K^{-1}$. Then the conjecture of Green can be restated as $dim H^0(X, \Lambda^i Q) = \binom{g}{i}$. It turns out that Q is always polystable and gets more and more stable in a specific sense as the Clifford index grows [PR]. Much progress has been made in this problem, but the conjecture of Green has not been fully proved yet, either by these ideas or any others.

14 Opers and quantization

We will start with some simple preliminary remarks. Let V be a (finite dimensional) vector space. The algebra of polynomial functions on V, namely the symmetric algebra $\sum S^i(V^*)$ over its dual, is a grade algebra. If V itself is graded, i.e. $V = \sum_{j=1}^{j=q} V_j$, then there is a total gradation on this algebra namely that in which the r-th graded piece is given by $\sum_{\sum j i_j = r} S^{i_1}(V_1) \otimes S^{i_2}(V_2) \otimes \ldots \otimes S^{i_q}(V_q)$.

On the other hand, if A is a V-affine space, namely a V-torsor, then we can still talk of the algebra of polynomial functions on A. But this is no longer a graded algebra. It is easily seen that it is actually a filtered algebra. The associated graded algebra can be identified canonically with $Sym(V^*)$. If V is graded as above, then this algebra acquires a total filtration such that the associated graded algebra is $Sym(V^*)$ with the total gradation.

We are going to apply these considerations to $H_r = \sum_{i=1}^{i=r} \Gamma(K^i)$ which is a graded vector space. For every polynomial function of total gradation k, we get on composition with the characteristic coefficients map, a function on $T^*(SU_C(r, d))$ which is homogeneous of degree k along the fibres of the cotangent fibration.

There is a natural affine space called *opers* which is a H_r torsor. Hence the totally filtered algebra of functions on it has as its associated graded algebra, the totally graded algebra of polynomial functions on H_r.

The geometric Langlands theory states that this algebra is actually the algebra of differential operators on $U_C(r, d)$ from the square root of its canonical bundle into itself.

Acknowledgements. I thank the referee for carefully going through the manuscript and pointing out many mistakes.

Bibliography

[A] M. Atiyah. Vector bundles over an elliptic curve, *Proc. London Math. Soc. (3)* **7** (1957), 414–452.

[A2] M. Atiyah. Complex analytic connections in fibre bundles, *Trans. Amer. Math. Soc.* **85** (1957), 181–207.

[BL] A. Beauville and Y. Laszlo. Conformal blocks and theta functions. *Comm. Math. Phys.* **164** (1994) 385–419.

[BNR] A. Beauville, M.S. Narasimhan and S. Ramanan. Spectral curves and the generalised theta divisor, *Jour. fur die reine u. ang. Math.* **398** (1989), 169–179.

[BV] S. Brivio and A. Verra. The theta divisor of $SU_C^s(2, 2d)$ is very ample if C is not hyperelliptic, *Duke Math. J.* **82** (1996), 503–552.

[DR] U.V. Desale and S. Ramanan. Classification of vector bundles of rank 2 on hyperelliptic curves, *Invent. Math.* **38** (1976), 161–185.

[DN] Drezet and M.S. Narasimhan. Groupe de Picard des variétés de modules de fibrés semi-stables sur les courbes algébriques. *Invent. Math.*, **97** (1989), 53–94.

[H1] N.J. Hitchin. The self-duality equations on a Riemann surface, *Proc. London Math. Soc. (3)* **55** (1987), 59–126.

[H2] N.J. Hitchin. Stable bundles and integrable systems, *Duke Math. J.* **54** (1987), 91–114.

[H3] N.J. Hitchin. Flat connection and geometric quantization, *Comm. Math. Phys.* **131** (1990), 347–380.

[KS] A. King and A. Schofield. Rationality of moduli of vector bundles on curves. *Indag. Math. (N.S.)* **10** (1999) 519–535.

[K] F. Klein. Zur theorie der Liniencomplexe der ersten und zweiten Grades. *Math. Ann.* **2** (1870) 198–226.

[KNR] S. Kumar, M.S. Narasimhan and A. Ramanathan. Infinite Grassmannians and moduli spaces of G-bundles. *Math. Ann.* **300** (1994) 41–75.

[Mu1] D. Mumford. *Abelian Varieties*, Oxford University Press, 2nd edition, 1974.

[Mu2] D. Mumford. *Algebraic Geometry I: Complex Projective Varieties*, Springer-Verlag, New York, 1976.

[Mu3] D. Mumford. On the equations defining Abelian varieties, *Invent. Math.*, **1** (1966), 287–384.

[Mu4] D. Mumford. Projective invariants of projective structures, *International Congress of Mathematicians*, Stockholm 1962, (1963), 526–530.

[NR1] M. S. Narasimhan and S. Ramanan. Moduli of vector bundles on a compact Riemann surface, *Ann. Math.* **89** (1969), 14–51.

[NR2] M. S. Narasihman and S. Ramanan. Deformations of the moduli of vector bundles, *Ann. Math.* **101** (1975), 391–417.

[NR3] M. S. Narasimhan and S. Ramanan. 2θ linear system, Proceedings of the Bombay colloquium on Vector Bundles, Tata Institute, 1984.

[NS] M. S. Narasihman and C.S. Seshadri. Stable and unitary bundles vector bundles on a compact Riemann surface, *Ann. Math. (2)* **82** (1965), 540–567.

[New1] P. E. Newstead. Rationality of moduli spaces of stable bundles, *Math. Ann.* **215** (1975), 251–268.

[New2] P. E.Newstead. Stable bundles of rank 2 and odd degree over a curve of genus 2, *Topology*, **7** (1968), 205–215.

[PR] K. Paranjape and S. Ramanan, *On the canonical ring of a curve, Algebraic geomery and commutative algebra, Vol. II*, Kinokuniya, Tokyo (1988), 503–516.

[R1] S. Ramanan, The moduli space of vector bundles over an algebraic curve, *Math. Ann.* **200** (1973), 69–84.

[R2] S. Ramanan, A survey of Higgs pairs, *Volume dedicated to Nigel Hitchin*, to appear.

[S] C. S. Seshadri. Space of unitary vector bundles on a compact Riemann surface, *Ann. Math. (2)* **85** (1967), 303–336.

Part II
Survey Articles

7

Moduli of sheaves from
moduli of Kronecker modules

Luis Álvarez-Cónsul

Instituto de Ciencias Matemáticas CSIC-UAM-UC3M-UCM
Serrano 113 bis, 28006 Madrid, Spain
e-mail: lac@mat.csic.es

Alastair King

Mathematical Sciences, University of Bath
Bath BA2 7AY, UK
e-mail: a.d.king@maths.bath.ac.uk

To Peter Newstead on his 65th birthday

This article is an expository survey of our paper [AK], which provides a new way to think about the construction of moduli spaces of coherent sheaves on projective schemes and the closely related construction of theta functions on such moduli spaces.

More precisely, for any projective scheme X, over an algebraically closed field of arbitrary characteristic, we are interested in the moduli spaces (schemes) $\mathcal{M}_X^{ss}(P)$ of semistable coherent sheaves of $\mathcal{O}_X$-modules, with a fixed Hilbert polynomial P with respect to a very ample invertible sheaf $\mathcal{O}(1)$.

Such moduli spaces were first constructed for vector bundles on smooth projective curves by Mumford [Mu] and Seshadri [S1], and this was the context where the key ideas were first developed, namely the notions of stability, semistability and S-equivalence. Thus Mumford showed that there was a quasi-projective variety parametrising isomorphism classes of stable bundles, while Seshadri showed that this is a dense open set in a projective variety parametrising S-equivalence classes of semistable bundles.

For modern account of moduli spaces of sheaves and their construction in higher dimensions, see [HL]. Recall that every semistable sheaf has a Jordan-Hölder filtration, or *S-filtration*, with stable factors and two semistable sheaves are *S-equivalent* if their associated graded sheaves are isomorphic, i.e. their S-filtrations have isomorphic stable factors (counted with multiplicity). The importance of this notion lies in the fact that, since any semistable sheaf can degenerate to its associated graded sheaf, S-equivalent sheaves must correspond to the same point in a moduli space (see e.g. [HL, Lemma 4.1.2]).

One of the key properties of $\mathcal{M}_X^{ss}(P)$ is then that its (closed) points correspond precisely to S-equivalence classes of semistable sheaves. Indeed, one way to think of the construction of $\mathcal{M}_X^{ss}(P)$ is the specification of a scheme structure on the underlying set of S-equivalence classes with an appropriate universal property with respect to families of semistable sheaves, nowadays called 'corepresenting the moduli functor' (see Section 4 for more details and [Ne, §1.2] or [HL, §4.1] for a full discussion).

Since Mumford and Seshadri's original work, and subsequent generalisations to higher dimensions and arbitrary characteristic by Gieseker [Gi], Maruyama [Ma], Simpson [Si] and Langer [La], the basic method for the construction of $\mathcal{M}_X^{ss}(P)$ has proceeded in two steps. First, 'rigidify' by identifying isomorphism classes of sheaves with orbits in a certain Quot-scheme for a certain action of a reductive group. Second, 'linearise' by finding a projective embedding of the Quot-scheme to obtain a problem in Geometric Invariant Theory (GIT), as developed for precisely such a purposes by Mumford [MF] (see also Newstead's article in this volume). It is the second step where the essential difficulties and variations of approach occur.

Once one has an intrinsic definition of semistability for sheaves, the basic problem is to find a linearisation where semistable sheaves correspond to GIT semistable orbits in (a suitably chosen subscheme of) the Quot-scheme, and furthermore S-equivalence of sheaves corresponds to closure equivalence of orbits.

The widely-accepted intrinsic notion of semistability, that generalises Mumford's for curves, was formulated by Gieseker and refined by Simpson. However, there have been many projective embeddings used to try to capture this notion geometrically. One of the most natural, first used by Simpson in this context, is into the Grassmannian originally used by Grothendieck [Gr] to construct the Quot-scheme.

The fundamental change of view point introduced in [AK] may be encapsulated by saying that, in Simpson's version of the construction, it is possible to linearise before rigidifying. To be more precise, now to 'linearise' means to embed the category of sheaves (subject to some regularity condition) in a simpler and more 'linear' category; in this case, in the category of Kronecker modules for the vector space $H = H^0(\mathcal{O}(n))$, for suitably large n. In Sections 1 and 2, we will explain in detail how this is done.

Such a Kronecker module is a linear map $\alpha\colon V_0 \otimes H \to V_1$, for finite dimensional vector spaces V_i, or equivalently a representation of the quiver

$$\bullet \xrightarrow{\;H\;} \bullet$$

One may also say that $V = V_0 \oplus V_1$ is a module for the path algebra A (see Section 2 for details) and we shall use the language of A-modules and Kronecker modules interchangeably. When we 'rigidify' the problem of classifying Kronecker modules up to isomorphism, by fixing the vector spaces V_i, we obtain the *linear* space $R = \mathrm{Hom}(V_0 \otimes H, V_1)$ with a *linear* action of the reductive group $G = \mathrm{GL}(V_0) \times \mathrm{GL}(V_1)$.

A notable property of the category of Kronecker modules is that there is a unique intrinsic notion of semistability, which corresponds to the unique GIT problem associated to the action of G on R, which thus constructs the moduli space $\mathcal{M}_A^{ss}$ of semistable A-modules (see Section 4 for details). This construction should be considered to be straightforward and transparent, including the correspondence between S-equivalence of A-modules and closure equivalence of orbits in the GIT problem. Indeed it is a simple case of a more general, but equally transparent, theory of moduli of representations of quivers [Ki].

Note that, in the case $\dim V_0 = \dim V_1 = 1$, a (non-zero) Kronecker module is in effect a point in the projective space $\mathbf{P}(H^*)$ in which X itself is embedded by the linear system H. Indeed, the categorical embedding of sheaves in Kronecker modules, when restricted to point sheaves on X gives precisely this embedding.

From the new view point, the basic problem for this 'categorical linearisation' is to show that semistability of sheaves corresponds to semistability of A-modules and that S-filtrations and S-equivalence are preserved. As we will see in Section 3, this helps to clarify the procedure and identify the delicate parts of the problem. We can then, in Section 4, use the construction and properties of $\mathcal{M}_A^{ss}$ to deduce the corresponding construction and properties of $\mathcal{M}_X^{ss}$.

One corollary is that we obtain an 'embedding' of moduli spaces

$$\varphi\colon \mathcal{M}_X^{ss} \to \mathcal{M}_A^{ss}.$$

More precisely, this is a scheme-theoretic embedding in characteristic zero and at stable points in charactistic p. Well-understood subtleties with quotients mean that we only know that it is a set-theoretic embedding at strictly semistable points in charactistic p.

Up to this technical detail, we can then import known definitions and properties of determinantal semi-invariants of quivers, i.e. the natural homogeneous coordinates on $\mathcal{M}_A^{ss}$, to define and study the corresponding coordinates, or 'theta functions', on $\mathcal{M}_X^{ss}$ (see Section 5 for details). In this way, we are able to strenghten, and generalise to arbirary X, results of Faltings [Fa] for smooth curves.

1 Simpson's construction revisited

To explain precisely how our shift in view point occurs, we recall in more detail Simpson's version of the construction of moduli of sheaves.

For the first (rigidification) step, one chooses an integer $n_0 \gg 0$ such that any semistable sheaf E with Hilbert polynomial P is n_0-regular in the sense of Castelnuovo-Mumford (see e.g. [HL, §1.7] for details), which in particular guarantees that the natural evaluation map

$$\epsilon_E \colon H^0(E(n_0)) \otimes \mathcal{O}(-n_0) \to E \tag{1.1}$$

is surjective and $\dim H^0(E(n_0)) = P(n_0)$. Thus, after the choice of an isomorphism $H^0(E(n_0)) \cong V_0$, where V_0 is some fixed $P(n_0)$-dimensional vector space, we may identify E with a point in the Quot-scheme parametrising quotients of $V_0 \otimes \mathcal{O}(-n_0)$ with Hilbert polynomial P. Changing the choice of isomorphism is given by the natural action of the reductive group $\mathrm{SL}(V_0)$ on the Quot-scheme.

For the second (linearisation) step, one chooses $n_1 \gg n_0$ so that applying the functor $H^0(- \otimes \mathcal{O}(n_1))$ to (1.1) yields a surjective map

$$\alpha_E \colon H^0(E(n_0)) \otimes H \to H^0(E(n_1)) \tag{1.2}$$

where $H = H^0(\mathcal{O}(n_1 - n_0))$ and $\dim H^0(E(n_1)) = P(n_1)$. More precisely, this construction is applied after choosing the isomorphism

$$H^0(E(n_0)) \cong V_0$$

and thus α_E determines a point in the Grassmannian of $P(n_1)$-dimensional quotients of $V_0 \otimes H$. Note that the kernel $\beta_E \colon U \hookrightarrow V_0 \otimes H$ of α_E determines E, and indeed the corresponding point in the Quot-scheme, as the cokernel of the corresponding map

$$U \otimes \mathcal{O}(-n_1) \xrightarrow{\beta_E} V_0 \otimes \mathcal{O}(-n_0) \to E \to 0. \tag{1.3}$$

This is how to see that the map from the Quot-scheme to the Grassmannian is an embedding.

Now we may observe that, by not choosing the rigidifying isomorphism $H^0(E(n_0)) \cong V_0$, one may interpret Simpson's method as a single functorial procedure whereby the sheaf E determines the Kronecker module α_E, from which the sheaf E may in turn be recovered.

From this point of view, the importance of regularity is also clear. Applying Serre's construction to a Veronese embedding of X, we know that a sheaf E is determined by the graded module

$$V_\bullet(E) = \bigoplus_{k \geq 0} H^0\left(E(n_0 + nk)\right)$$

for the algebra $\mathrm{Sym}^\bullet(H)$, for any n_0 and $n = n_1 - n_0 > 0$. The regularity of E and of $\mathcal{O}_X$ determine how large n_0 and n must be to guarantee that the generators and relations of $V_\bullet$ are all in degree $k = 0$, so that $V_\bullet$ (and hence E) is determined by the Kronecker module $V_0 \otimes H \to V_1$.

2 The functorial point of view

We may describe the above procedure more formally as follows. Let $T = \mathcal{O}(n_0) \oplus \mathcal{O}(n_1)$ and let $A \subset \mathrm{End}_X(T)$ be the algebra spanned by the two projections e_0, e_1 and $H = \mathrm{Hom}_X(\mathcal{O}(n_0), \mathcal{O}(n_1))$. Indeed, in most cases, when $H^0(\mathcal{O})$ consists just of scalars, we actually have $A = \mathrm{End}_X(T)$. An A-module V is equivalent to a Kronecker module $\alpha \colon V_0 \otimes H \to V_1$, where $V_i = e_i V$ and conversely $V = V_0 \oplus V_1$. The *dimension vector* of V is $v = (\dim V_0, \dim V_1)$.

Now, the assignment of α_E to E is achieved by the functor

$$\Phi \colon \mathrm{Coh}(X) \to \mathrm{Mod}(A) \colon E \mapsto H^0(T \otimes E) = \mathrm{Hom}_X(T^\vee, E). \qquad (2.1)$$

Moreover, the recovery of E from α_E is achieved by the adjoint functor

$$\Phi^\vee \colon \mathrm{Mod}(A) \to \mathrm{Coh}(X) \colon V \mapsto T^\vee \otimes_A V.$$

Explicitly, $T^\vee \otimes_A V$ may be described in terms of the Kronecker module $\alpha \colon V_0 \otimes H \to V_1$ as the pushout of the natural diagram

$$
\begin{array}{ccc}
V_0 \otimes H \otimes \mathcal{O}(-n_1) & \xrightarrow{1 \otimes \mu} & V_0 \otimes \mathcal{O}(-n_0) \\
{\scriptstyle \alpha \otimes 1} \downarrow & & \\
V_1 \otimes \mathcal{O}(-n_1) & &
\end{array}
$$

When α is surjective, this is equivalent to the procedure as in (1.3). The fact that this procedure works, when it does, may then be formulated as follows [AK, Thm 3.4].

Theorem 2.1. *Suppose that $\mathcal{O}_X$ is n-regular and that $T = \mathcal{O}(n_0) \oplus \mathcal{O}(n_1)$ for $n_1 - n_0 \geq n$. Then the functor Φ of (2.1) is fully faithful on the full subcategory of n_0-regular sheaves. In other words, if E is n_0-regular, then the natural map $\varepsilon_E : \Phi^\vee \Phi(E) \to E$ is an isomorphism.*

Note that the natural 'counit' map ε_E in the theorem is the evaluation map

$$\varepsilon_E : T^\vee \otimes_A \operatorname{Hom}_X(T^\vee, E) \to E.$$

We may paraphrase Theorem 2.1 by saying that Φ gives a functorial embedding of n_0-regular sheaves into A-modules. In fact, regularity is also crucial to extending this embedding to families of sheaves.

Let $\mathcal{M}_X^{reg}(n_0; P)$ be the moduli functor of n_0-regular sheaves on X with Hilbert polynomial P, that is, the (contravariant) functor that assigns to any scheme S the set of isomorphism classes of flat families over S of such sheaves. Similarly, let $\mathcal{M}_A(v)$ be the moduli functor of A-modules with dimension vector v. Then regularity also implies that Φ preserves flat families [AK, Prop. 4.1] and so induces an embedding of moduli functors, i.e. a natural transformation

$$[\Phi] : \mathcal{M}_X^{reg}(n_0; P) \to \mathcal{M}_A(P(n_0), P(n_1)) \qquad (2.2)$$

such that $[\Phi]_S$ is injective for all S. Note in particular that this means that we only need to look among A-modules of dimension vector

$$(P(n_0), P(n_1))$$

for the images of all n_0-regular sheaves of Hilbert polynomial P.

The general machinery of adjunction provides an explicit condition that determines when an A-module V is in the image of this functorial embedding. The adjunction also has a natural 'unit' map

$$\eta_V : V \to \Phi\Phi^\vee(V) = \operatorname{Hom}_X(T^\vee, T^\vee \otimes_A V) \qquad (2.3)$$

and $V \cong \Phi(E)$, for some sheaf E for which ε_E is an isomorphism, if and only if η_V is an isomorphism, in which case $E = \Phi^\vee(V)$ is the appropriate sheaf and whether E is n_0-regular may be considered a property of V.

Now, the set of (isomorphism classes of) Kronecker modules with dimension vector $v = (v_0, v_1)$ is in natural bijection with the set of orbits in the *representation space*

$$R = R_A(v) = \operatorname{Hom}(V_0 \otimes H, V_1) \qquad (2.4)$$

for the action of the symmetry group $G = \operatorname{GL}(V_0) \times \operatorname{GL}(V_1)$ by conjugation, where V_i is some fixed vector space of dimension v_i. Note that

R carries a tautological G-equivariant family $\mathbb{V}$ of A-modules, which is 'equivariantly locally universal' in the sense that the induced natural transformation from the quotient functor $\underline{R}/\underline{G} \to \mathcal{M}_A$ is a local isomorphism, i.e. an isomorphism after sheafification [AK, Prop. 4.4]. Here $\underline{Z}$ denotes the functor of points of a scheme Z, i.e. $\mathrm{Hom}(-, Z)$.

Now, using $\Phi^\vee$ and its associated flattening stratification (see [AK, Prop. 4.2] for details), we can determine a locally closed G-invariant subscheme $Q \subset R$ over which $\mathbb{E} = \Phi^\vee(\mathbb{V})$ is a flat family of n_0-regular sheaves with Hilbert polynomial P. This family is also equivariantly locally universal in same sense as above [AK, Thm 4.5]. The functorial embedding (2.2) of moduli functors can naturally be enhanced to an embedding of moduli stacks, which is thus modelled on the embedding of quotient stacks $[Q/G] \to [R/G]$.

It is the space Q with G-action that plays the role in our story of the Quot-scheme with $\mathrm{SL}(V_0)$-action, or, more strictly speaking, of the open set of n_0-regular sheaves in the Quot-scheme. Then, the embedding $Q \subset R$ plays the role of the embedding of the Quot scheme in the Grassmannian.

3 Semistability

We now turn to the essential goal of the 'categorical linearisation' of sheaves by Kronecker modules, namely of demonstrating the relationship between the semistability of a sheaf E and the semistability of the corresponding A-module $\Phi(E)$. It is this which will enable us to use the moduli spaces $\mathcal{M}_A^{ss}$ to construct the moduli spaces $\mathcal{M}_X^{ss}$.

Recall the usual (Gieseker-Simpson) definition of semistability for sheaves. Note that this notion depends just on the Hilbert polynomial P_E of a sheaf E and of its subsheaves. The 'multiplicity' r_E of E is the leading coefficient of P_E and the dimension of the support of E is the degree of P_E. Then E is 'pure' if it has no proper subsheaves with lower dimensional support.

Definition 3.1. *A sheaf E is* semistable *if E is pure and, for all $E' \subset E$,*

$$\frac{P_{E'}(n)}{r_{E'}} \leq \frac{P_E(n)}{r_E} \quad \text{for } n \gg 0.$$

For our purposes, this definition has a crucial reformulation, which gives a cleaner dependence on the Hilbert polynomial and for which purity is an automatic consequence (see also [Ru] for another formulation).

Lemma 3.2. *A sheaf E is semistable if and only if, for all $E' \subset E$,*

$$\frac{P_{E'}(n_0)}{P_{E'}(n_1)} \leq \frac{P_E(n_0)}{P_E(n_1)} \quad \textit{for } n_1 \gg n_0 \gg 0.$$

This formulation is manifestly related to the (essentially unique) notion of semistablity for Kronecker modules.

Definition 3.3. *An A-module V is semistable if, for all $V' \subset V$,*

$$\frac{\dim V'_0}{\dim V'_1} \leq \frac{\dim V_0}{\dim V_1}.$$

Thus E is semistable if and only if for all $E' \subset E$, $\Phi(E')$ does not destabilise $\Phi(E)$ for $n_1 \gg n_0 \gg 0$. Note that this is still some way from saying the E is semistable if and only if $\Phi(E)$ is semistable, but this would seem to be the result that one could hope for. In fact, we do not prove this ideal result, and it is quite possible that it is not true, demonstrating the more subtle role of purity in this problem. What we do show is the following [AK, Thm 5.10a],

Theorem 3.4. *Given P, for $n_1 \gg n_0 \gg 0$, suppose that E is n_0-regular and pure with Hilbert polynomial P. Then E is semistable if and only if $\Phi(E)$ is semistable.*

Note that n_0 is, in particular, chosen large enough that all semistable sheaves with Hilbert polynomial P are n_0-regular, and, of course, all semistable sheaves are pure.

For the proof, we need to show that n_0, n_1 can be chosen so that Φ and $\Phi^\vee$ provide a one-one correspondence between the critical (i.e. most destabilising) subsheaves of E and the critical submodules of $\Phi(E)$ and furthermore that the numbers n_0, n_1 in Lemma 3.2 can be chosen uniformly in P, i.e. independently of E and E'. It is this that makes the proof very delicate and in particular seems to require purity as an explicit condition.

This result shows that the Kronecker module $\Phi(E)$ is semistable whenever the sheaf E is semistable and thus the embedding of moduli functors (2.2) restricts to an embedding

$$[\Phi] \colon \mathscr{M}_X^{ss}(P) \to \mathscr{M}_A^{ss}(P(n_0), P(n_1)), \tag{3.1}$$

where $\mathscr{M}_X^{ss}(P)$ and $\mathscr{M}_A^{ss}(v)$ are respectively the moduli functors of semistable sheaves with Hilbert polynomial P and semistable modules with dimension vector v.

The other key result [AK, Thm 5.10b-d, Cor 5.11] is that Φ and $\Phi^\vee$ provide mutually inverse identifications between the S-filtrations of a semistable sheaf E (and the associated graded sheaf of stable factors) and the S-filtrations of $\Phi(E)$ (and the associated graded module).

Thus one may see that there is a well-defined and injective map from S-equivalence classes of semistable sheaves with Hilbert polynomial P to S-equivalence classes of semistable A-modules with dimension vector $v = (P(n_0), P(n_1))$. In other words, we have at least a set-theoretic embedding of moduli spaces $\mathcal{M}_X^{ss}(P) \to \mathcal{M}_A^{ss}(v)$.

4 Moduli spaces

We now consider more carefully how it is that $\mathcal{M}_X^{ss}$ and $\mathcal{M}_A^{ss}$ are defined as moduli spaces, that is, as schemes that 'corepresent' the corresponding moduli functors. This means that there is a natural transformation $\mathscr{M}^{ss} \to \underline{\mathcal{M}^{ss}}$ through which any other natural transformation $\mathscr{M}^{ss} \to \underline{Z}$ uniquely factorises. Note that such a universal property uniquely characterises $\mathcal{M}^{ss}$ as a scheme. In particular, we will show how this property for $\mathcal{M}_X^{ss}$ follows from the same property for $\mathcal{M}_A^{ss}$, thereby justifying the claim that $\mathcal{M}_X^{ss}$ is 'constructed' using the functor Φ.

Recall that the set of isomorphism classes of A-modules of dimension vector v is in natural one-one correspondence with the set of G-orbits in the representation space $R = R_A(v)$, as defined in (2.4). Consider the character χ on G given by

$$\chi(g_0, g_1) = (\det g_0)^{-k_0} (\det g_1)^{k_1}, \qquad (4.1)$$

where $k_1/k_0 = v_0/v_1$ with k_0, k_1 coprime. This determines the graded ring

$$\mathrm{SI}_\chi^\bullet = \bigoplus_{d \geq 0} \mathrm{SI}_\chi^d$$

of associated semi-invariants, i.e. a polynomial f on R is in SI_χ^d if and only if $f(gx) = \chi(g)^d f(x)$ for all $g \in G$ and $x \in R$. Then the following hold [Ki]:

(i) a point $x \in R$ is χ-semistable in the sense of GIT, i.e. $f(x) \neq 0$ for some $f \in \mathrm{SI}_\chi^d$ with $d > 0$, if and only if the corresponding A-module is semistable, in the sense of Definition 3.3,

(ii) two points x, y in the open subset R^{ss} of semistable points are closure equivalent, i.e. $\overline{Gx}$ and $\overline{Gy}$ intersect in R^{ss}, if and only if the corresponding A-modules are S-equivalent.

The general machinery of GIT implies that the projective variety

$$M_A = \text{Proj } \text{SI}_\chi^\bullet$$

is a 'good' quotient of R^{ss} by G, meaning in particular that $R^{ss} \to M_A$ is a categorical quotient, whose fibres are closure equivalence classes.

Thus M_A is a scheme whose points correspond to S-equivalence classes of semistable A-modules and M_A corepresents the quotient functor

$$\underline{R^{ss}/G}$$

(this is the definition of categorical quotient). But it follows from the observations in Section 2 that this quotient functor is locally isomorphic to the moduli functor $\mathcal{M}_A^{ss}(v)$ of semistable A-modules. In other words, M_A is the moduli space $\mathcal{M}_A^{ss}(v)$.

Now, also recall that there is a locally closed G-invariant subscheme $Q \subset R$, which parametrises n_0-regular sheaves E with Hilbert polynomial P. Supposing that n_0 is large enough that all semistable E with Hilbert polynomial P are n_0-regular, it follows from Theorem 3.4 that the open subset $Q^{[ss]} \subset Q$, parametrising semistable sheaves, is a locally closed subscheme of R^{ss}.

Since the moduli functor of semistable sheaves is locally isomorphic to the quotient functor $\underline{Q^{[ss]}/G}$, the problem of 'construction' of the moduli space $\mathcal{M}_X^{ss}(P)$ amounts to firstly showing that $Q^{[ss]}$ has a good quotient by G and secondly showing that the closure equivalence classes in $Q^{[ss]}$ are in one-one correspondence with the S-equivalence classes of sheaves.

For the first, the fact that $Q^{[ss]}$ has a good quotient follows from the fact that R^{ss} does, *provided* we also know that, for any G-orbit O in $Q^{[ss]}$, if O' is the closed orbit in the closure of O in R^{ss}, then $O' \subset Q^{[ss]}$ (see [AK, Lemma 6.2]). This follows because we know that, if O corresponds to $\Phi(E)$ for a semistable sheaf E, with associated graded sheaf E', then O' corresponds to the associated graded module of $\Phi(E)$, which is equal to $\Phi(E')$, and thus it is indeed in $Q^{[ss]}$.

The second follows for almost the same reason. Two semistable sheaves E' and E'' are S-equivalent if and only if $\Phi(E')$ and $\Phi(E'')$ are S-equivalent, i.e. the corresponding orbits have the same closed orbit in their closure within R^{ss} or equally within $Q^{[ss]}$ and thus correspond to the same point in the good quotient of $Q^{[ss]}$.

Thus we have constructed the moduli space $\mathcal{M}_X^{ss}(P)$ and it remains to show that it is a projective variety. Let $Z \subset R$ be the closure of $Q^{[ss]}$. Then the projectivity would follow immediately if we could show *a priori* that the inclusion $Q^{[ss]} \subset Z^{ss}$ is an equality, where $Z^{ss} = Z \cap R^{ss}$. However, knowing only this inclusion we can only deduce that $\mathcal{M}_X^{ss}(P)$ is quasi-projective, being a dense open subset of the GIT quotient of Z.

On the other hand, we can show that $\mathcal{M}_X^{ss}(P)$ is proper using Langton's method [AK, Prop. 6.5] (cf. [Ma, §5]), a well-known application of the valuative criterion for properness. Hence we can deduce that $\mathcal{M}_X^{ss}(P)$ is projective and thus *a posteriori* that $Q^{[ss]} = Z^{ss}$ and therefore that $Q^{[ss]}$ is closed in R^{ss}.

In conclusion, we have therefore reproved the existence of a projective moduli space of semistable sheaves [AK, Thm 6.4, Prop. 6.6].

Theorem 4.1. *There is a projective scheme $\mathcal{M}_X^{ss}(P)$ which is the moduli space of semistable sheaves on X with Hilbert polynomial P, i.e. it corepresents the moduli functor $\mathscr{M}_X^{ss}(P)$. The closed points of $\mathcal{M}_X^{ss}$ are in one-one correspondence with the S-equivalence classes of semistable sheaves.*

We have also obtained an explicit map $\varphi \colon \mathcal{M}_X^{ss} \to \mathcal{M}_A^{ss}$ induced by the inclusion $Q^{[ss]} \subset R^{ss}$ (see [AK, Prop. 6.3] for details), which fits into the commuting diagram of natural transformations,

$$
\begin{array}{ccc}
\mathscr{M}_X^{ss} & \xrightarrow{\ [\Phi]\ } & \mathscr{M}_A^{ss} \\
\psi_X \downarrow & & \downarrow \psi_A \\
\mathcal{M}_X^{ss} & \xrightarrow{\ \varphi\ } & \mathcal{M}_A^{ss}
\end{array}
\tag{4.2}
$$

where ψ_X and ψ_A are the corepresenting transformations. Note that the corepresenting property of ψ_X means that such a map φ must exist and be uniquely determined by $[\Phi]$.

Because we know that a semistable sheaf E is stable if and only if the semistable Kronecker module $\Phi(E)$ is stable [AK, Theorem 5.10b], we see that there is an open subscheme $\mathcal{M}_X^s(P) \subset \mathcal{M}_X^{ss}(P)$ corepresenting the moduli functor $\mathscr{M}_X^s(P) \subset \mathscr{M}_X^{ss}(P)$ of stable sheaves. Indeed, we have $\mathcal{M}_X^s = \varphi^{-1}\mathcal{M}_A^s$, where $\mathcal{M}_A^s \subset \mathcal{M}_A^{ss}$ is the corresponding (open) moduli space of stable A-modules.

Remark 4.2. As already observed at the end of the previous section, the map φ is set-theoretically injective, but, since we now know that it is induced by the closed embedding $Q^{[ss]} \subset R^{ss}$ we also know that the image of φ is a closed subset of $\mathcal{M}_A^{ss}$. Indeed, in characteristic zero, this further shows that φ is a scheme-theoretic embedding. However, the characteristic zero assumption is crucial to this deduction, and we cannot obtain the same conclusion in characteristic $p > 0$, although we can still prove (see [AK, Prop. 6.7] for details) that the restriction to $\mathcal{M}_X^s$ is a scheme-theoretic embedding.

One point we would like to emphasize is that the spaces $\mathcal{M}_A^{ss}(v)$ are a family of well-behaved (essentially linear) projective varieties, which naturally generalise projective spaces as potential targets for embedding moduli spaces. Because they are constucted as GIT quotients of linear spaces by classical reductive group actions, the spaces $\mathcal{M}_A^{ss}(v)$ have well-controlled singularities (only at strictly semistable points) and well-understood homogeneous coordinates.

They are also a good test case for developing a theory of 'non-commuta-tive moduli spaces' which should in particular carry appropriately universal families of A-modules. Our functorial construction should then adapt naturally to construct non-commutative moduli of sheaves.

5 Theta functions

We finish by explaining how the natural homogeneous coordinates on $\mathcal{M}_A^{ss}$ are obtained from Schofield's general theory of determinantal semi-invariants of quivers and how the adjunction between Φ and $\Phi^\vee$ enables us to restrict them to $\mathcal{M}_X^{ss}$ to obtain natural coordinates that we will call 'theta functions'.

Let $P_i = Ae_i$, for $i = 0, 1$, be the two indecomposable projective A-modules and note that, for any A-module V of dimension vector v, we have $\mathrm{Hom}_A(P_i, V) = e_i V = V_i$. Suppose we are given any map

$$\gamma \colon P_1^{k_1 d} \to P_0^{k_0 d}$$

where $k_1/k_0 = v_0/v_1$, with k_0, k_1 coprime. Then the linear map

$$\mathrm{Hom}_A(\gamma, V) \colon V_0^{k_0 d} \to V_1^{k_1 d}$$

is between vector spaces of the same dimension and hence there is naturally defined an element

$$\theta_\gamma(V) = \det \mathrm{Hom}_A(\gamma, V)$$

in the line $\lambda(V)^d$, where

$$\lambda(V) = (\det V_0)^{-k_0} \otimes (\det V_1)^{k_1}.$$

If V is a family of A-modules over a scheme S, then naturality means that $\lambda(V)$ is a line bundle over S and $\theta_\gamma(V)$ is a global section of $\lambda(V)^d$.

In particular, if we consider V_0 and V_1 to be fixed, giving rise to the G-equivariant family $\mathbb{V}$ on the representation space R, and if we trivialise $\lambda(\mathbb{V})$, then naturality also means that $\theta_\gamma(\mathbb{V})$ is a 'determinantal' semi-invariant in SI_χ^d (cf. (4.1) and after).

Furthermore, one may show (see [AK, Prop. 7.5] for details) that $\lambda(\mathbb{V})$ satisfies Kempf's descent criterion over R^{ss} and so descends to a line bundle $\lambda(v)$ on $\mathcal{M}_A^{ss}(v)$. Since $\theta_\gamma(\mathbb{V})$ is an invariant section, it descends to a global section $\theta_\gamma(v)$ of $\lambda(v)^d$. Because $\mathcal{M}_A^{ss}(v)$ is constructed by GIT as Proj $\mathrm{SI}_\chi^\bullet$, we see that $\lambda(v)$ is ample and the global sections of $\lambda(v)^d$ may be naturally identified with SI_χ^d.

The main result about semi-invariants of quivers [DW, SV] is that the functions θ_γ span SI_χ^d. Thus choosing d large enough that $\lambda(v)^d$ is very ample, we see that it is possible to find finitely many

$$\gamma_0, \ldots, \gamma_N : P_1^{k_1 d} \to P_0^{k_0 d}$$

such that the map

$$\Theta_\gamma : \mathcal{M}_A^{ss}(v) \to \mathbb{P}^N : [V] \mapsto (\theta_{\gamma_0}(V) : \cdots : \theta_{\gamma_N}(V)) \qquad (5.1)$$

is a scheme-theoretic closed embedding.

We will now see how the 'embedding' $\varphi : \mathcal{M}_X^{ss}(P) \to \mathcal{M}_A^{ss}(v)$ enables us to deduce similar results for $\mathcal{M}_X^{ss}(P)$. More precisely, suppose that $V = \Phi(E)$ for some sheaf (or family of sheaves) E which is n_0-regular with Hilbert polynomial P. Then the adjunction

$$\mathrm{Hom}_A(\gamma, \Phi(E)) = \mathrm{Hom}_X(\Phi^\vee(\gamma), E),$$

enables us to write $\theta_\gamma(V)$ entirely in terms of E. Indeed, if

$$\delta = \Phi^\vee(\gamma) : \mathcal{O}(-n_1)^{k_1 d} \to \mathcal{O}(-n_0)^{k_0 d},$$

with k_0 and k_1 coprime and such that $k_1/k_0 = P(n_0)/P(n_1)$, then

$$\mathrm{Hom}_X(\delta, E) : H^0(E(n_0))^{k_0 d} \to H^0(E(n_1))^{k_1 d}$$

is a linear map between vector spaces of the same dimension and we can define

$$\theta_\delta(E) = \det \mathrm{Hom}_X(\delta, E),$$

as a natural element of the line $\lambda(E)^d$, where

$$\lambda(E) = (\det H^0(E(n_0)))^{-k_0} \otimes (\det H^0(E(n_1)))^{k_1}.$$

If $\mathbb{E} = \Phi^\vee(\mathbb{V})$ is the tautological family of semistable sheaves over $Q^{[ss]}$, then $\lambda(\mathbb{E})$ is the restriction of $\lambda(\mathbb{V})$ and thus it descends to an ample line bundle $\lambda(P) = \varphi^* \lambda(v)$ on $\mathcal{M}_X^{ss}(P)$. Furthermore, the invariant sections $\theta_\delta(\mathbb{E})$ descend to global sections $\theta_\delta(P) = \varphi^* \theta_\gamma(v)$ of $\lambda(P)^d$, which may be called 'determinantal' theta functions. Note that, even when X is a smooth curve, this determinant line bundle $\lambda(P)$ is already an uncontrollably large power (depending on n_0, n_1) of the fundamental determinant line bundle on $\mathcal{M}_X^{ss}(P)$.

A first consequence of the spanning property of determinantal semi-invariants of Kronecker modules, is that a module V is semistable if and only if there is a γ such that $\theta_\gamma(V) \neq 0$, that is, $\mathrm{Hom}(\gamma, V)$ is an isomorphism. Thus, using Theorem 3.4, we deduce the following determinantal characterisation of semistable sheaves [AK, Thm 7.2].

Theorem 5.1. *Given P, for $n_1 \gg n_0 \gg 0$, let E be n_0-regular and pure with Hilbert polynomial P. Then E is semistable if and only there is a map $\delta \colon \mathcal{O}(-n_1)^{m_1} \to \mathcal{O}(-n_0)^{m_0}$ with $\mathrm{Hom}_X(\delta, E)$ invertible, i.e. $\theta_\delta(E) \neq 0$.*

Note that the invertibility of $\mathrm{Hom}_X(\delta, E)$ and the regularity of E automatically imply that $m_1/m_0 = P(n_0)/P(n_1)$ and so we do not need to impose that condition explicitly here.

If we now combine Theorem 3.4 with the full force of the spanning property, i.e. the projective embedding (5.1), we obtain the following embedding theorem [AK, Thm 7.10], modulo the technical detail of Remark 4.2.

Theorem 5.2. *There exist m_0 and m_1, satisfying $m_1/m_0 = P(n_0)/P(n_1)$, and finitely many*

$$\delta_0, \ldots, \delta_N \colon \mathcal{O}(-n_1)^{m_1} \to \mathcal{O}(-n_0)^{m_0},$$

such that the map

$$\Theta_\delta \colon \mathcal{M}_X^{ss}(P) \to \mathbb{P}^N : [E] \mapsto (\theta_{\delta_0}(E) : \cdots : \theta_{\delta_N}(E))$$

is a set-theoretic closed embedding. This embedding is scheme-theoretic in characteristic zero, while in characteristic $p > 0$ it is scheme-theoretic on the stable locus.

These two theorems may be compared to two results of a similar flavour proved by Faltings [Fa] (see also [S2]) for vector bundles on a smooth projective curve C. Firstly, Faltings showed that a vector bundle E is semistable if and only if there exists a non-zero bundle F such that

$$\mathrm{Hom}_C(F, E) = 0 = \mathrm{Ext}_C^1(F, E). \tag{5.2}$$

This condition easily implies that E is semistable [S2, Lemma 8.3], so the main point is to show that, for any semistable E, such an F exists.

Note that the condition (5.2) is equivalent to the condition that $\theta_F(E) \neq 0$, for a suitably defined theta function θ_F (see [Fa, §1], [S2, §2] or [AK, §7.4] for details). Such theta functions have a broader 'domain of definition' than our theta functions θ_δ, in the sense that $\theta_F(E)$ is a well-defined section of a line bundle on the base of *any* family of sheaves E with $\chi(F, E) = 0$, whereas the definition of $\theta_\delta(E)$ requires that the family also be n_0-regular. On the other hand, note that the construction of θ_F requires crucially that X, or at least the support of F, is a smooth curve, whereas θ_δ requires no such restriction.

Secondly, Faltings showed that it is possible to find finitely many bundles $F_0, \ldots, F_N$ which can be used to give a morphism

$$\Theta_F : \mathcal{M}_C^{ss}(P) \to \mathbb{P}^N : [E] \mapsto (\theta_{F_0}(E) : \cdots : \theta_{F_N}(E))$$

which is the normalisation of its image. Esteves [Es] improved this to show that one could arrange that Θ_F is injective on points and, in characteristic zero, a scheme-theoretic embedding on the stable locus.

To see more closely the relation to Theorems 5.1 and 5.2, note that, for any X, because $\mathcal{O}(n_1 - n_0)$ is very ample and $m_1 > m_0$, the generic map $\delta : \mathcal{O}(-n_1)^{m_1} \to \mathcal{O}(-n_0)^{m_0}$ is injective as a map of sheaves and thus provides an acyclic resolution of its cokernel F, with respect to the functor $\mathrm{Hom}_X(-, E)$ for any n_0-regular sheaf E.

Note further that the condition that $\mathrm{Hom}_X(\delta, E)$ is invertible is an open condition on δ, for a fixed n_0-regular E. Hence, if this condition holds for some δ, then it also holds for some injective δ, in which case it is equivalent to the condition, for $F = coker\,\delta$, that $\mathrm{Ext}_X^i(F, E) = 0$ for $i = 0, 1$ and hence indeed for all $i \geq 0$, because the others vanish automatically by the nature of F and the regularity of E.

Thus, over a smooth curve, Theorem 5.1 reproves the first result of Faltings. Furthermore, we may also suppose that all δ_i in Theorem 5.2 are injective and then, over any family of semistable (and hence n_0-regular) sheaves on a smooth curve, $\theta_{\delta_i} = \theta_{F_i}$, for $F_i = coker\,\delta_i$. Thus, Theorem 5.2 further strengthens Faltings second result.

In fact, we can also use our methods to show that theta functions θ_δ span sufficiently high powers of the line bundle $\lambda(P)$ on $\mathcal{M}_X^{ss}$. Hence, in the curve case, theta functions θ_F span sufficiently high powers of the fundamental line bundle, although this result is now superceded by the proof of the strange duality conjecture by Marian and Oprea [MO], which shows that theta functions θ_F span all powers of the fundamental line bundle.

Thus, by restricting our attention to regular sheaves, we have been able to define theta functions θ_δ, which satisfy the natural generalisations of Faltings results in higher dimensions, where there are theoretical obstructions to defining $\theta_F(E)$ for general sheaves E and F. Note that the line bundles $\mathcal{O}(-n_i)$ for $n_i \geq n$ are effectively projective objects with respect to n-regular sheaves and thus it is in keeping with Schofield's philosophy that maps between them should provide the natural source of 'homogeneous functions of moduli of sheaves'.

Acknowledgements. Supported by the Ministerio de Educación y Ciencia under Grant MTM-67623. LAC is also supported by the "Programa Ramón y Cajal".

Bibliography

[AK] L. Álvarez-Cónsul and A. King: *A functorial construction of moduli of sheaves*, Invent. Math. **168** (2007), 613–666.

[DW] H. Derksen and J. Weyman: *Semi-invariants of quivers and saturation of Littlewood-Richardson coefficients*, J. Amer. Math. Soc. **13** (2000), 467–479.

[Es] E. Esteves: *Separation properties of theta functions*, Duke Math. J. **98** (1999), 565–593.

[Fa] G. Faltings: *Stable G-bundles and projective connections*, J. Alg. Geom. **2** (1993), 507–568.

[Gi] D. Gieseker: *On the moduli of vector bundles on an algebraic surface*, Ann. of Math. **106** (1977), 45–60.

[Gr] A. Grothendieck: *Techniques de construction et théorèmes d'existence en géométrie algébrique IV: Les schémas de Hilbert*, Séminaire Bourbaki Exposé **221** (1961) Paris.

[HL] D. Huybrechts and M. Lehn: *The Geometry of Moduli Spaces of Sheaves* (Braunschweig: Friedr. Vieweg & Sohn, 1997).

[Ki] A. D. King: *Moduli of representations of finite dimensional algebras*, Quart. J. Math. Oxford **45** (1994), 515–530.

[La] A. Langer: *Moduli spaces of semistable sheaves in mixed characteristic*, Duke Math. J. **124** (2004), 571–586.

[MO] A. Marian and D. Oprea: *The rank-level duality for non-abelian theta functions*, Invent. Math. **168** (2007), 225–247.

[Ma] M. Maruyama: *Moduli of stable sheaves. II*, J. Math. Kyoto Univ. **18** (1978), 557–614.

[Mu] D. Mumford: *Projective invariants of projective structures and applications*, In Proc. Internat. Congr. Mathematicians (Stockholm, 1962) (Djursholm: Inst. Mittag-Leffler, 1963), pp. 526–530.

[MF] D. Mumford and J. Fogarty: *Geometric Invariant Theory*, 2nd edn (Berlin: Springer-Verlag, 1982).

[Ne] P. E. Newstead: *Introduction to Moduli Problems and Orbit Spaces*, Tata Inst. Fund. Res., Mumbai (New Delhi: Narosa Publishing House, 1978).

[Ru] A. Rudakov: *Stability for an abelian category*, J. Algebra **197** (1997), 231–245.

[SV] A. Schofield and M. Van den Bergh: *Semi-invariants of quivers for arbitrary dimension vectors*, Indag. Math. **12** (2001), 125–138.

[S1] C. S. Seshadri: *Space of unitary vector bundles on a compact Riemann surface*, Ann. of Math. **85** (1967), 303–336.

[S2] C. S. Seshadri: *Vector bundles on curves*, In Linear algebraic groups and their representations (Los Angeles, CA, 1992). Contemp. Math. **153** (Providence, RI: Amer. Math. Soc., 1993), pp. 163–200.

[Si] C. Simpson: *Moduli of representations of the fundamental group of a smooth projective variety, I*, Inst. Hautes Études Sci. Publ. Math. **79** (1994), 47–129.

8

Coherent Systems: a brief survey

S. B. Bradlow

Department of Mathematics
University of Illinois
Urbana IL 61801 USA
e-mail: bradlow@uiuc.edu

With an Appendix by H. Lange

Mathematisches Institut
Bismarckstr. 1 1/2
D-91054 Erlangen, Germany
e-mail: lange@mi.uni-erlangen.de

Dedicated to Peter Newstead on the occasion of his 65th birthday.

1 Introduction

A coherent system is a pair (E, V) where E is a holomorphic bundle and V is a linear subspace of its space of holomorphic sections. If E is a semistable bundle, then the existence of such objects is equivalent to the non-emptiness of a higher rank Brill-Noether locus. This connection to higher rank Brill-Noether theory provides one of the motivations for studying coherent systems. It certainly motivated Peter Newstead's guiding role in the development of the subject, and thus makes a volume in honor of his 65th birthday a fitting place for a survey.

Interest in coherent systems extends beyond Brill-Noether theory, mainly because there is a stability notion for a pair (E, V), distinct from the stability of the bundle E. The natural definition of such stability depends on a real parameter (denoted by α) and leads to a finite family of moduli spaces of α-stable coherent systems. These moduli spaces present a rich display of topological and geometric phenomena, most of which have yet to be fully explored.

This survey will be limited in scope because of space constraints and because of a survey in preparation by Peter Newstead based on his lectures at a Clay Institute workshop held in October 2006 ([N]). In both [N] and in this survey we consider only bundles over smooth algebraic curves over $\mathbf{C}$, though the definitions make sense in more general contexts. Indeed when LePotier first introduced coherent systems in [LeP] his main interest was in algebraic surfaces, where the moduli spaces of coherent systems could be used as a tool to study Donaldson invariants.†

Our choice of topics is designed to complement those in [N], where the emphasis is on basic features of the moduli spaces, especially the conditions for non-emptiness. While Newstead restricted attention to curves of genus $g \geq 2$ and considered coherent systems only from the point of view of algebraic geometry, we will emphasize

- structure results, both for individual coherent systems and for their moduli spaces,

- some non-algebraic aspects, specifically gauge theoretic and symplectic features

We also include, in an Appendix by Herbert Lange, a discussion of coherent systems on curves of genus zero or one.

After a brief review (in Section 2) of basic definitions and properties, we show in Section 3 how the structures of stable coherent systems depends on the parameter which appears in the definition of stability. In Section 4 we consider the difference between moduli spaces defined using different notions of stability, i.e. using different values of the parameter. We examine both the structure of loci where the spaces differ (the so-called flip loci), and also examine the structure of the coherent systems in these loci. In some cases, where the differences between the moduli spaces are small and well understood, the available topological information extends much further than non-emptiness. In Section 5 we discuss homotopy groups and Poincaré polynomials.

Section 6 is a brief description of some non-algebraic aspects of coherent systems, i.e. we look at things from the perspective of gauge theory and symplectic geometry. This leads naturally to consideration of some related objects called holomorphic k-pairs.

The material in Sections 3–5 comes mostly from the collaborative results reported in [BGMN, BGMMN1, BGMMN2]; because of space constraints, the reader seeking proofs and other technical details will frequently be referred to these sources. I take this opportunity to thank my collaborators, especially Peter Newstead, for all they have taught me.

---

† see also [He] for more on coherent systems over higher dimensional varieties, and [Ba] for coherent systems on curves over finite fields, and [Bh] for coherent systems over nodal curves.

2 Definitions, basic facts and general features

Let X be a smooth projective curve of genus g and let E denote a holomorphic vector bundle on X. We denote the rank and degree of E by the integers n and d respectively. Recall that the *slope* $\mu(E)$ is defined by $\mu(E) := \frac{d}{n}$.

Definition 2.1. A *coherent system* on X of *type* (n, d, k) is a pair (E, V), where E is a vector bundle on X of rank n and degree d and $V \subseteq H^0(E)$ is a linear subspace of dimension k. A *homomorphism* $f : (E, V) \longrightarrow (F, W)$ *between coherent systems* is a bundle homomorphism $f : E \longrightarrow F$ together with the induced map on the space of sections. The space of all such homomorphisms is denoted as usual by $\mathrm{Hom}((E, V), (F, W))$. A *coherent subsystem* of (E, V) is a coherent system (E', V') such that E' is a subbundle of E and $V' \subseteq V \cap H^0(E')$. A *quotient coherent system* of (E, V) is a coherent system (E'', V'') together with a homomorphism $(E, V) \rightarrow (E'', V'')$ such that both $E \rightarrow E''$ and $V \rightarrow V''$ are surjective.

Definition 2.2. *Let (E, V) be a coherent system of type (n, d, k). For any $\alpha \in \mathbf{R}$, the α-slope $\mu_\alpha(E, V)$ is defined by*

$$\mu_\alpha(E, V) := \frac{d}{n} + \alpha \frac{k}{n}.$$

A coherent system (E, V) is α-stable (α-semistable) if, for every proper coherent subsystem (E', V'),

$$\mu_\alpha(E', V') < (\leq) \mu_\alpha(E, V).$$

Remark: Strictly speaking (see [KN]), in order to get an abelian category, one must replace the coherent systems of Definition 2.1 with triples $(E, \mathbf{V}, \Phi)$ where E is a coherent sheaf on X, $\mathbf{V}$ is a vector space of dimension k and $\Phi : \mathbf{V} \otimes \mathcal{O}_X \longrightarrow E$ is a sheaf map. If E is locally free and V is the image in $H^0(E)$ of the map on global sections induced by Φ, then (E, V) is a coherent system as defined in Definition 2.1. The definition of α-(semi)stability extends to this category. It follows from results in [KN] (especially Lemma 3.3) that the α-semistable coherent sytems of type (n, d, k) with fixed α-slope may be regarded as a subcategory of the category of triples $(E, \mathbf{V}, \Phi)$. In the sequel we will not emphasize this distinction.

Proposition 2.3. *([KN, Corollary 2.5.1]) The α-semistable coherent systems of any fixed α-slope form a noetherian and artinian abelian category in which the simple objects are precisely the α-stable systems. In particular the following statements hold.*

(i) (Jordan-Hölder Theorem) *For any α-semistable coherent system (E, V), there exists a filtration by α-semistable coherent systems (E_j, V_j),*

$$0 = (E_0, V_0) \subset (E_1, V_1) \subset \ldots \subset (E_m, V_m) = (E, V),$$

with $(E_j, V_j)/(E_{j-1}, V_{j-1})$ an α-stable coherent system and

$$\mu_\alpha((E_j, V_j)/(E_{j-1}, V_{j-1})) = \mu_\alpha(E, V) \quad \text{for } 1 \leq j \leq m.$$

(ii) *If (E, V) is an α-stable coherent system, then $\text{End}(E, V) \cong \mathbf{C}$.*

Any filtration as in (i) is called a *Jordan-Hölder* filtration of (E, V). It is not necessarily unique, but the associated graded object is uniquely determined by (E, V).

In order for a coherent system of type (n, d, k) to be α-semistable but not α-stable, it must have a subsystem of type (n', d', k') such that $\mu_\alpha(E', V') = \mu_\alpha(E, V)$. If $\frac{k'}{n'} = \frac{k}{n}$ then the equality $\mu_\alpha(E', V') = \mu_\alpha(E, V)$ holds *independently of the value for α*, while if $\frac{k'}{n'} \neq \frac{k}{n}$ then there is a unique value for α at which the equality holds, namely

$$\alpha = \frac{n'd - nd'}{nk' - n'k}. \tag{2.1}$$

Definition 2.4. *A* **critical value** *for coherent systems of type (n, d, k) is a value of $\alpha > 0$ for which there exists a coherent system (E, V) of type (n, d, k) with a coherent subsystem (E', V') of (E, V) of type (n', d', k') such that $\frac{k'}{n'} \neq \frac{k}{n}$ but $\mu_\alpha(E', V') = \mu_\alpha(E, V)$. We also regard $\alpha = 0$ as a critical value.*

Remark: We say that α is *generic* if it is not a critical value. If $\text{GCD}(n, d, k) = 1$ and α is generic, then α-semistability is equivalent to α-stability.

Proposition 2.5 (Lemma 4.2 [BGMN]). *A necessary condition for the existence of an α-stable coherent system of type (n, d, k) is that α lies in the range*

$$0 < \alpha < \begin{cases} \frac{d}{n-k} & \text{if } k < n \\ \infty & \text{if } k \geq n \end{cases} \tag{2.2}$$

For fixed type (n, d, k), and for generic α in the allowed range, there exists a moduli space of α-stable coherent systems of type (n, d, k). If (n, d, k) are not coprime then there is also a moduli space of α-semistable coherent systems.

Definition 2.6. *We denote the moduli space of α-stable (respectively semistable) coherent systems of type (n, d, k) by $G(\alpha; n, d, k)$ (respectively $\widetilde{G}(\alpha; n, d, k)$).*

First constructed in general† by LePotier (in [LeP]) and King and Newstead (see [KN]) the moduli spaces $G(\alpha; n, d, k)$ are projective varieties constructed by the methods of Geometric Invariant Theory, i.e. they are GIT quotients. Apart from this, many basic properties of these moduli spaces remain open questions.

We refer the reader to the survey by Peter Newstead ([N]) for a discussion of recent progress on the non-emptiness question, i.e. on the problem of identifying the necessary and sufficient conditions on (n, d, k) and α in order for $G(\alpha; n, d, k)$ to be non-empty‡, and move on to address the smoothness and dimension of the spaces.

Coherent systems form an abelian category and the functors $\mathrm{Hom}((E, V), .)$ are left exact. Hence their derived functors, denoted by $Ext^i((E, V), .)$, are well defined. Standard deformation theory (see for instance [He, LeP] and also [Th, RV]) shows that the Zariski tangent space at a point defined by coherent system (E, V) is isomorphic to $\mathrm{Ext}^1((E, V), (E, V))$, and that the moduli space is smooth in a neighborhood of the point if $\mathrm{Ext}^2((E, V), (E, V)) = 0$. We regard

$$\mathrm{Ext}^q((E, V), (E, V))$$

as special cases of the more general $\mathrm{Ext}^q((E_1, V_1), (E_2, V_2))$ where the two coherent systems are of type (n_1, d_1, k_1) and (n_2, d_2, k_2) respectively. With

$$
\begin{aligned}
\mathbf{H}_{21}^0 &:= \mathrm{Hom}((E_2, V_2), (E_1, V_1)) \,, & (2.3) \\
\mathbf{H}_{21}^2 &:= \mathrm{Ext}^2((E_2, V_2), (E_1, V_1)) \,, \text{ and} & (2.4) \\
C_{21} &:= k_2 \chi(E_1) - \chi(E_2^* \otimes E_1) - k_1 k_2 \\
&= n_1 n_2 (g - 1) - d_1 n_2 + d_2 n_1 + k_2 d_1 - k_2 n_1 (g - 1) - k_1 k_2 \,,
\end{aligned}
$$
$$(2.5)$$

we get

Proposition 2.7 (Propositions 3.1 and 3.2 in [BGMN]). *Let (E_1, V_1) and (E_2, V_2) be two coherent systems on X. The space of equivalence*

---

† Earlier constructions were given for small α in [RV] and also for large values of d in [Be]

‡ See also [BP, BO, Te, BBN] for recent non-emptiness results in the case $k > n$

classes of extensions

$$0 \longrightarrow (E_1, V_1) \longrightarrow (E, V) \longrightarrow (E_2, V_2) \longrightarrow 0$$

is isomorphic to $\mathrm{Ext}^1((E_2, V_2), (E_1, V_1))$ *and has dimension*

$$\dim \mathrm{Ext}^1((E_2, V_2), (E_1, V_1)) = C_{21} + \dim \mathbf{H}_{21}^0 + \dim \mathbf{H}_{21}^2, \qquad (2.6)$$

Moreover,

$$\begin{aligned} \mathbf{H}_{21}^2 &= \ \mathrm{Ker}(H^0(E_1^* \otimes K) \otimes V_2 \to H^0(E_1^* \otimes E_2 \otimes K))^* \\ &= \ H^0(E_1^* \otimes N_2 \otimes K)^*, \qquad (2.7) \end{aligned}$$

where N_2 is the kernel of the evaluation map $V_2 \otimes \mathcal{O} \to E_2$.

Specializing to the case $(E_1, V_1) = (E_2, V_2) = (E, V)$, we see that smoothness at a point (E, V) in the moduli spaces $G(\alpha; n, d, k)$ is governed by the **Petri map**

$$V \otimes H^0(E^* \otimes K) \longrightarrow H^0(\mathrm{End}E \otimes K), \qquad (2.8)$$

where the map is multiplication of sections.

Proposition 2.8 (Proposition 3.10 in [BGMN]). *Let (E, V) be an α-stable coherent system of type (n, d, k). Then the moduli space $G(\alpha; n, d, k)$ is smooth at the point corresponding to (E, V) and has dimension*

$$\beta(n, d, k) = n^2(g - 1) + 1 - k(k - d + n(g - 1)) \qquad (2.9)$$

if and only if the Petri map is injective.

The number $\beta(n, d, k)$ is often called the Brill-Noether number. In view of Proposition 2.8 we also refer to it as the *expected dimension* of the moduli spaces $G(\alpha; n, d, k)$. By standard deformation theory, the dimension of every irreducible component of $G(\alpha; n, d, k)$ is at least $\beta(n, d, k)$.

As defined above, the parameter α is real-valued, making it look like there is a continuous 1-parameter family of moduli spaces $G(\alpha; n, d, k)$. In fact, since all other parameters in Definition 2.2 are integer-valued, the dependence on α is discrete in the following sense:

Proposition 2.9. *For any (n, d, k), there are finitely many critical values*

$$0 = \alpha_0 < \alpha_1 < \ldots < \alpha_L < \begin{cases} \frac{d}{n-k} & \text{if } k < n \\ \infty & \text{if } k \geq n. \end{cases}$$

The α-range is thus divided by the critical values into a finite set of intervals such that

(i) *Within the interval (α_i, α_{i+1}) the property of α-stability is independent of α, that is if $\alpha, \alpha' \in (\alpha_i, \alpha_{i+1})$, then*
$$G(\alpha; n, d, k) = G(\alpha'; n, d, k).$$

(ii) *if $k \geq n$, the moduli spaces coincide for any two different values of α in the range (α_L, ∞).*

Proof The first statement is obvious if $k < n$. The rest, for $k \geq n$, is Proposition 4.6 in [BGMMN1]. $\square$

Remark: In the case $k \geq n$ the critical value α_L can be characterized as the critical value with the property that any coherent system is either α-stable for all $\alpha > \alpha_L$ or it fails to be α-stable for any α in this range. Moreover, in the second case it either fails to be α-semistable whenever $\alpha > \alpha_L$, or it is α-semistable for all $\alpha > \alpha_L$.

By Proposition 2.9 we can define

Definition 2.10.

$$G_i(n, d, k) = G(\alpha; n, d, k) \text{ for any } \alpha \in (\alpha_i, \alpha_{i+1}) \qquad (2.10)$$

where α_i and α_{i+1} are consecutive critical values for α. Whenever no confusion will result we denote $G_i(n, d, k)$ by G_i. In particular, we define

$$G_L = G(\alpha; n, d, k) \text{ for any } \alpha \in \begin{cases} (\alpha_L, \frac{d}{n-k}) & \text{if } k < n \\ (\alpha_L, \infty) & \text{if } k \geq n \end{cases} \qquad (2.11)$$

2.1 Transformations at critical values of α

The difference between G_{i-1} and G_i is due to the difference between the α-stability conditions for $\alpha < \alpha_i$ and $\alpha > \alpha_i$.

Definition 2.11. *We define $G_i^+ \subseteq G_i = G_i(n, d, k)$ to be the set of (equivalence classes of) all (E, V) in G_i which are not α-stable if $\alpha < \alpha_i$. Similarly, we define $G_i^- \subseteq G_{i-1}$ by the condition that (E, V) is not α-stable if $\alpha > \alpha_i$.*

Thus the set theoretic difference between G_{i-1} and G_i is described by

$$G_{i-1} - G_i^- + G_i^+ = G_i \quad \text{or} \quad G_{i-1} - G_i^- = G_i - G_i^+ . \qquad (2.12)$$

The points in $G_i^+ \subset G_i$ represent coherent systems which are not α_i-stable; they therefore form a closed subscheme of G_i. Similarly G_i^- is a closed subscheme of G_{i-1}. We describe the transformation from

$G_{i-1}(n,d,k)$ to $G_i(n,d,k)$ (or vice versa) as a *flip*, although it is not necessarily a flip in any technical sense. We refer to the subschemes $G_i^{\pm}$ as the *flip loci*. Away from these loci, i.e. at $G_i - G_i^+$ and $G_{i-1} - G_i^-$, the moduli spaces are in bijective correspondence as sets and have isomorphic scheme structures.

2.2 Relation to Brill-Noether theory

Proposition 2.12 (see [BDGW, KN]). *Fix (n,d,k) and let α_1 be the smallest critical value in the range (2.2). Let α lie in the range*

$$0 < \alpha < \alpha_1 . \tag{2.13}$$

If (E,V) is an α-stable coherent system of type (n,d,k) then E is a semistable bundle. Conversely, if E is stable bundle of degree d and rank n, and V is any rank k linear subspace of $H^0(E)$, then (E,V) is α-stable for any α in the range (2.13).

Proposition 2.12 establishes the relation between coherent systems of type (n,d,k) and the *Brill-Noether loci* in $M(n,d)$, the moduli space of rank n degree d stable bundles. These loci are defined by

$$B(n,d,k) := \{E \in M(n,d) \mid \dim H^0(E) \geq k\}.$$

Similarly, if n and d are not coprime, then the Brill-Noether loci of semistable bundles are defined by

$$\widetilde{B}(n,d,k) := \{[E] \in \widetilde{M}(n,d) \mid \dim H^0(\mathrm{Gr}(E)) \geq k\},$$

where $\widetilde{M}(n,d)$ is the moduli space of S-equivalence classes of semistable bundles, $[E]$ is the S-equivalence class of E and $\mathrm{Gr}(E)$ is the polystable bundle defined by a Jordan-Hölder filtration of E. By proposition 2.12 the projection $(E,V) \mapsto E$ defines a map

$$G_0(n,d,k) \longrightarrow \widetilde{B}(n,d,k) . \tag{2.14}$$

Moreover the map is one-to-one over $B(n,d,k) - B(n,d,k+1)$ and $B(n,d,k)$ lies in the image. If $\mathrm{GCD}(n,d,k) \neq 1$, then this map extends to a (not necessarily surjective) map from $\widetilde{G}_0(n,d,k)$ to $\widetilde{B}(n,d,k)$. We see that the non-emptiness and irreducibility of $G_0(n,d,k)$ implies the same for the Brill-Noether locus $\widetilde{B}(n,d,k)$. In cases where $G_0(n,d,k)$ is easier to study than $B(n,d,k)$ (see for instance [BGMN, BGMMN1]), this makes coherent systems a useful tool in Brill-Noether Theory.

3 Structure results for α-stable coherent systems

We describe two kinds of structure results. The first depend on α but for given α they apply to all α-stable objects; the second are α-independent but apply only to the generic element of each moduli space. The latter type of result describes the 'intersection' of all the moduli spaces, while the former sheds light on their differences.

3.1 Boundary critical values and related structure results

In this section we describe how the structure of α-stable coherent system changes qualitatively at certain critical values for α which we call *boundary critical values*. The results divide naturally into cases according to whether $k < n$ or $k \geq n$. The main distinction is between cases in which the evaluation map $V \otimes \mathcal{O} \longrightarrow E$ can be injective and those in which it can be surjective.

We begin by using the evaluation map $V \otimes \mathcal{O} \longrightarrow E$ to repackage the defining data for a coherent system (E, V) in the form of an exact sequence of sheaves

$$0 \longrightarrow N \longrightarrow V \otimes \mathcal{O} \longrightarrow E \longrightarrow F \oplus T \longrightarrow 0 \tag{3.1}$$

where N and F are vector bundles, T is a torsion sheaf and $H^0(N) = 0$. In all cases we find that as α increases, the sequence (3.1) describing α-stable coherent system simplifies until in a suitable it is sense as simple as possible.

3.1.1 The case $k < n$

For α-stable coherent systems of type (n, d, k) with $k < n$, the most basic dependence on α is that there are no such coherent systems if α lies outside the range $(0, \frac{d}{n-k})$. Within this range there are three critical values of special significance, with the structure of α-stable coherent systems becoming successively simpler as α increases from 0 and crosses these boundary critical values. The smallest boundary critical value, denoted α_I, is the **injectivity bound**.

Proposition 3.1. *(see Lemma 2.4 (i) in [BGMMN1]) Fix (n, d, k) with $k < n$. There is a critical value α_I with $0 \leq \alpha_I < \frac{d}{n-k}$ such that if a coherent system (E, V) is α-stable for $\alpha \geq \alpha_I$ then the map $\phi : V \otimes \mathcal{O} \to E$ is injective, i.e. $N = 0$ in the the sequence (3.1), which thus has the form*

$$0 \longrightarrow V \otimes \mathcal{O} \longrightarrow E \longrightarrow F \oplus T \longrightarrow 0 . \tag{3.2}$$

Moreover

$$\alpha_I \leq \begin{cases} max\{\frac{(k-1)(d-2n)}{k(n-k+1)} \ , \ 0\} & \text{if } 1 \leq k \leq g+1 \\ max\{\frac{(k-1)(d-n)-ng}{k(n-k+1)} \ , \ 0\} & \text{if } k > g+1 \end{cases} \tag{3.3}$$

The following result reveals the importance of the injectivity bound.

Proposition 3.2. *Let (E, V) be an α-stable coherent system such that $k \leq n$. If $V \otimes \mathcal{O} \to E$ is injective then the moduli space $G(\alpha; n, d, k)$ is smooth of dimension $\beta(n, d, k)$ at the point corresponding to (E, V).*

Proof. If $V \otimes \mathcal{O} \to E$ is injective then the Petri map (2.8) is injective. The result thus follows from Proposition 2.8.

□

The next boundary critical value, denoted by α_T and called the **torsion bound**, can be specified precisely:

Proposition 3.3 (Lemma 2.4 (ii) in [BGMMN1]). *Fix (n, d, k) with $k < n$. Let α_T be given by*

$$\alpha_T = max\{\frac{d-n}{n-k}, 0\} \ . \tag{3.4}$$

If a coherent system (E, V) is α-stable for $\alpha \geq \alpha_T$ then $T = 0$ in the sequence (3.2), which thus has the form

$$0 \longrightarrow V \otimes \mathcal{O} \longrightarrow E \longrightarrow F \longrightarrow 0 \ , \tag{3.5}$$

Finally, as α crosses the last critical value less than $\frac{d}{n-k}$, we get:

Proposition 3.4 (Proposition 3.1 (iii) in [BGMMN1]). *Let α_L be the greatest critical value less than $\frac{d}{n-k}$ and suppose that $\alpha_L < \alpha < \frac{d}{n-k}$. Then any α-stable coherent system (E, V) has the form given by the sequence (3.5), with the extra features that*

- *F is semistable, and*
- *if $\vec{e} = (e_1, \ldots, e_k) \in H^1(F^* \otimes \mathcal{O}^{\oplus k}) = H^1(F^*)^{\oplus k}$ denotes the class of the extension, then $e_1, \ldots, e_k$ are linearly independent as vectors in $H^1(F^*)$.*

Conversely, any extension of this type in which F is stable gives rise to a coherent system which is α-stable for any α in the range $(\alpha_L \ , \ \frac{d}{n-k})$.

Remark: Extensions of the form (3.5) and satisfying the two properties listed in Proposition 3.4 are called BGN extensions. The name comes from the role they play in [BGN]. Proposition 3.4 is the basis for an explanation using coherent systems of the results in [BGN] (see [BG2]).

In all cases we get

$$0 \leq \alpha_I \leq \alpha_T \leq \alpha_L < \frac{d}{n-k} \, . \qquad (3.6)$$

The true significance of α_T and α_L becomes apparent (see section 4.2) when one considers the moduli spaces $G_i(n,d,k)$: for $\alpha_i > \alpha_T$ all $G_i(n,d,k)$ are birationally equivalent to each other, and $G_L(n,d,k)$ can be explicitly described (at least birationally) as a Grassman bundle over a moduli space of semistable bundles.

3.1.2 The case $k \geq n$

If $k \geq n$ then the range for α is unbounded above. Moreover, if $dim(V)$ is bigger than $rank(E)$ then the evaluation map $V \otimes \mathcal{O} \longrightarrow E$ can never be injective. In this case there are two special critical values for α. One is α_L, which marks an upper bound for changes in the moduli spaces. The other, described below, is a true boundary critical value in the sense that it marks a value in the α-range at which the structure of α-stable coherent systems changes qualitatively.

Proposition 3.5 (Proposition 4.4 in [BGMN]). *Fix (n,d,k) with $k \geq n$. There is a critical value α_{gg} such that if a coherent system (E,V) is α-stable for $\alpha \geq \alpha_{gg}$ then the map $\phi : V \otimes \mathcal{O} \rightarrow E$ is generically surjective, i.e. the sequence (3.1) has the form*

$$0 \longrightarrow N \longrightarrow \mathcal{O}^{\oplus k} \xrightarrow{\phi} E \longrightarrow T \longrightarrow 0$$

where T is a torsion sheaf (possibly 0) and $\mathrm{rk}\, N = k - n$.
The value of α_{gg} and α_L satisfy the following constraints

$$\alpha_{gg} \leq \frac{d(n-1)}{k} \qquad (3.7)$$

$$\alpha_{gg} \leq \alpha_L \leq d(n-1) \, . \qquad (3.8)$$

Definition 3.6. *If $\phi : V \otimes \mathcal{O} \rightarrow E$ is generically surjective we say that the coherent system is **generically generated**; if the map is actually surjective we say that the coherent system is **generated** (or altogether generated).*

3.2 Generic α-stable objects with small slope

The results of Section 3.1 show that the structure of α-stable coherent systems simplifies as α increases. The simplification results from stripping away inessential parts of the kernel and cokernel of the evaluation map $V \otimes \mathcal{O} \longrightarrow E$. When α is below the boundary critical values at which this happens, we can analyze the α-stable coherent systems in which the simplification fails. By showing that they occur in families which depend on fewer parameters than the dimensions of the moduli spaces $G(\alpha; n, d, k)$ we can obtain structure results for the generic members of the moduli spaces.

Theorem 3.7. [Theorem 4.4 in [BGMMN2]]

Suppose that $0 < d \leq 2n$ and $\alpha > 0$. If $G(\alpha; n, d, k)$ is non-empty then it is irreducible. Moreover

(a) if $k < n$, the generic element of $G(\alpha; n, d, k)$ has the form

$$0 \to V \otimes \mathcal{O} \to E \to F \to 0,$$

where F is a vector bundle with $h^0(F^*) = 0$;

(b) if $k = n$, the generic element of $G(\alpha; n, d, k)$ has the form

$$0 \to V \otimes \mathcal{O} \to E \to T \to 0,$$

where T is a torsion sheaf;

(c) if $k > n$, the generic element of $G(\alpha; n, d, k)$ has the form

$$0 \to N \to V \otimes \mathcal{O} \to E \to 0,$$

i.e. (E, V) is generated;

Remark: Combining Theorem 3.7 (c) with Proposition 3.5 we thus get:

(c') If $k > n$ and $d \leq 2n$, then for sufficiently large α all α-semistable coherent systems are generically generated and the generic such α-semistable coherent system is altogether generated.

4   Structure results for destabilizing patterns and flip loci

In this section we look at a structure result which allows us to describe the flip loci $G_i^{\pm}$ (see Definition 2.11) and also to estimate their codimensions in the moduli spaces G_i. In cases where the codimensions are all positive, so that the moduli spaces are all birationally equivalent, we can transfer useful information across critical values from G_L to other moduli spaces in the family $\{G_0, G_1, \ldots, G_L\}$.

4.1 Destabilizing patterns

The basic result is the following (see Lemma 6.5 in [BGMN] and Proposition 2.6 in [BGMMN2]for details)

Lemma 4.1. *Let α_i be a critical value of α with $1 \le i \le L$. Denote values of α in the intervals below and above α_i by α_i^- and α_i^+ respectively. Let (E, V) be a coherent system of type (n, d, k) which is $\alpha_i^\pm$-stable but $\alpha_i^\mp$-unstable. Then (E, V) appears as the middle term in an extension*

$$0 \to (E_1, V_1) \to (E, V) \to (E_2, V_2) \to 0 \qquad (4.1)$$

in which

(i) *(E_1, V_1) and $(E_2, V_2)$are both α_i-semistable, with*

$$\mu_{\alpha_i}(E_1, V_1) = \mu_{\alpha_i}(E_2, V_2) = \mu_{\alpha_i}(E,) ,$$

(ii) *(E_1, V_1) and (E_2, V_2) are both $\alpha_i^\pm$-stable, with $\mu_{\alpha_i^\pm}(E_1, V_1) < \mu_{\alpha_i^\pm}(E_2, V_2)$,*

(iii) *In case (E, V) is α_i^+-stable, $\frac{k_1}{n_1}$ is a maximum among all proper subsystems $(E_1, V_1) \subset (E, V)$ which satisfy (ii), while in case (E, V) is α_i^--stable, $\frac{k_1}{n_1}$ is a minimum among all proper subsystems $(E_1, V_1) \subset (E, V)$ which satisfy (ii)*

(iv) *n_1 is a minimum among all subsystems which satisfy (iii).*

Conversely, if the extension is nontrivial, and

(i) *(E_1, V_1) and (E_2, V_2) are both α_i-stable,*
(ii) *$\mu_{\alpha_i}(E_1, V_1) = \mu_{\alpha_i}(E_2, V_2)$, and*
(iii) *$\mu_{\alpha_i^-}(E_1, V_1) < \mu_{\alpha_i^-}(E_2, V_2)$,*

then (E, V) is α_i^--stable and α_i^+-unstable.

Notice that, unlike in the Jordan-Hölder filtrations for semistable objects, the descriptions of the α_i-semistable objects require only 1-step filtrations. This simplification results from a careful exploitation of the stability parameter.

Definition 4.2. *We refer to the extensions of the form (4.1) with the properties of Lemma 4.1 as the **destabilising patterns** of the coherent systems.*

For fixed α_i and (n, d, k) we can organize these destabilizing patterns into families indexed by n_1 and k_1, or equivalently n_1 and $\lambda_1 = \frac{k_1}{n_1}$:

Definition 4.3. *Let* $W_i^{\pm}(\lambda_1, n_1; n, d, k)$ *(abbreviated to* $W_i^{\pm}(\lambda_1, n_1)$ *whenever possible) denote the set of all destabilizing patterns of the form (4.1) with* $\mathrm{rk}(E_1) = n_1$, $\dim(V_1) = \lambda_1 n_1$ *and such that the coherent system* (E, V) *is* $\alpha_i^{\pm}$*-stable. Define*

$$W_i^{\pm}(n, d, k) = \bigsqcup_{\lambda_1, n_1 < k_1, n_1 < n} W_i^{\pm}(\lambda_1, n_1; n, d, k),$$

If $k < n$ and α is sufficiently large then we can be more explicit in our description of the coherent systems in the families $W_i^{\pm}(n, d, k)$.

Lemma 4.4 (Lemma 4.1 (Diagram Lemma) in [BGMMN1]). *Suppose that* $k < n$ *and that* α_i *is a critical value in the range* $\alpha_T \leq \alpha_i \leq \frac{d}{n-k}$. *Let* (E, V) *be a torsion-free coherent system in* $W_i^{\pm}(\lambda_1, n_1; n, d, k)$. *Then there exists a diagram*

$$
\begin{array}{ccccccccc}
 & & 0 & & 0 & & 0 & & \\
 & & \downarrow & & \downarrow & & \downarrow & & \\
0 & \to & V_1 \otimes \mathcal{O} & \to & E_1 & \to & F_1 & \to & 0 \\
 & & \downarrow & & \downarrow & & \downarrow & & \\
0 & \to & V \otimes \mathcal{O} & \to & E & \to & F & \to & 0 \qquad (4.2)\\
 & & \downarrow & & \downarrow & & \downarrow & & \\
0 & \to & V_2 \otimes \mathcal{O} & \to & E_2 & \to & F_2 & \to & 0 \\
 & & \downarrow & & \downarrow & & \downarrow & & \\
 & & 0 & & 0 & & 0 & &
\end{array}
$$

where

(a) *the quotients* F_1, F, *and* F_2 *are all locally free with positive rank,*
(b) $h^0(F_1^*) = h^0(F^*) = h^0(F_2^*) = 0$,
(c) *the extension classes of* E_1, E, E_2 *are given respectively by* k_1, k, k_2 *linearly independent vectors in* $H^1(F_1^*)$, $H^1(F^*)$, $H^1(F_2^*)$.

As an application of this result consider coherent systems of type (n, d, k) with $k = n - 1$. One consequence of Lemma 4.4 is that either there are no critical values between α_T and $\frac{d}{n-k}$ or $k \leq n - 2$. In particular, if $k = n - 1$ then there are no critical values in the range

$$max\{d - n, 0\} < \alpha < d . \qquad (4.3)$$

Hence for all α in this range we can identify the moduli space $G(\alpha; n, d, n - 1)$ with the moduli space $G_L(n, d, n - 1)$. Moreover, we can explicitly describe $G_L(n, d, n - 1)$ in terms of a Poincaré bundle $\mathcal{P}$ on $X \times J^d$, where J^d denotes the Jacobian of degree d line bundles. Let

p_J and p_X be the projections from $X \times J^d$ onto its factors. If $d > 0$ then $H^0(L^*) = 0$ for all $L \in J^d$ and $R^1 p_{J_*} \mathcal{P}^*$ is locally free of rank $d + g - 1$. The fiber over $L \in J^d$ describes extensions

$$0 \to \mathcal{O} \to E \to L \to 0 \ .$$

The structure result Proposition 3.4 thus leads directly to the following:

Proposition 4.5 ([BGMMN1] Cor. 5.5). *$G_L(n, d, n - 1)$ is non empty if and only if $d \geq \max\{n - g, 0\}$. If $d \geq \max\{n - g, 1\}$ then G_L – and hence $G(\alpha; n, d, n - 1)$ for all α in the range (4.3) – can be identified with the Grassmann bundle $Gr(n - 1, R^1 p_{J_*} \mathcal{P}^*)$.*

Using relative Serre duality, we can identify $Gr(n - 1, R^1 p_{J_*} \mathcal{P}^*)$ with the Grassmann bundle $Gr(d + g - n, p_{J_*}(\mathcal{P} \otimes p_X^* K_X))$ where K_X is the canonical bundle, and this in turn can be identified with the variety of linear systems of degree $d + 2g - 2$ and dimension $d + g - n - 1$, i.e. with the variety denoted in classical Brill-Noether theory by $G_{d+2g-2}^{d+g-n-1}$.

4.2 Counting codimensions: good, better, and best flips

For all allowable values of (λ_1, n_1), we denote the image of $W_i^\pm(\lambda_1, n_1)$ in the flip loci $G_i^\mp$ by $G_i^\pm(\lambda_1, n_1)$. Since the flip loci $G_i^\mp$ are covered by the unions of these images, the dimensions of the families $W_i^\pm(\lambda_1, n_1)$ give an upper bound on the dimensions of the flip loci, and hence a lower bound on their codimensions in the appropriate moduli spaces. If the codimensions of the flip loci are all positive, then the moduli spaces $G_{i-1}(n, d, k)$ and $G_i(n, d, k)$ are birationally equivalent.

Lemma 4.1 implies that for fixed (n, d, k) and α_i, each set $W_i^\pm(\lambda_1, n_1)$ is contained in the union of family of extensions, in which the sub and quotient objects are $\alpha_i^\pm$-stable coherent systems. The dimensions of such families are determined by the dimensions of the moduli spaces $G(\alpha_i^\pm; n_1, d_1, k_1)$ and $G(\alpha_i^\pm; n_2, d_2, k_2)$, and by the dimension of the space which parameterizes extensions. Using Proposition 2.7 for the latter, we get

Proposition 4.6. *Fix (n, d, k) and also α_i. Let $\mathbf{H}_{21}^0$, $\mathbf{H}_{21}^2$ and C_{21} be as in (2.3)–(2.5). Then each set $W_i^\pm(\lambda_1, n_1)$ is contained in a family whose dimension is bounded above by*

$$w_i^\pm(\lambda_1, n_1) := \dim G(\alpha_i^\pm; n_1, d_1, k_1) + \dim G(\alpha_i^\pm; n_2, d_2, k_2) \qquad (4.4)$$

$$+ \max\{C_{21} + \dim \mathbf{H}_{21}^0 + \dim \mathbf{H}_{21}^2 - 1\} \ .$$

where $n_1 + n_2 = n$, $d_1 + d_2 = d$, $k_1 + k_2 = k$ and the maximum is taken over all (E_1, V_1), (E_2, V_2) which satisfy the relevant part of Definition 4.3.

The set W_i^+ is thus contained in a family whose dimension is bounded above by the maximum of $w_i^+(\lambda_1, n_1)$ for all $\lambda_1 < \frac{k}{n}$ and $n_1 < n$. Similarly, the set W_i^- is contained in a family whose dimension is bounded above by the maximum of $w_i^-(\lambda_1, n_1)$ for all $\lambda_1 > \frac{k}{n}$ and $n_1 < n$.

In general it is difficult to determine these dimensions, but useful estimates can sometimes be computed (see Section 3 of [BGMN]). For instance, it follows from (2.7) that if $k_2 > 0$ and $h^0(E_1^* \otimes K) \neq 0$ then the dimension of $\mathbf{H}_{21}^2$ is bounded above by $(k_2 - 1)(h^0(E_1^* \otimes K) - 1)$. Also, $\mathbf{H}_{21}^0 = 0$ whenever the map $V_2 \otimes \mathcal{O} \to E_2$ is injective. In the simplest cases, i.e. when the moduli spaces have their expected dimensions and $\mathbf{H}_{21}^0 = 0$ for all destabilizing patterns, we can thus compute

$$\mathrm{Codim}(G_i^{\pm}) \geq \min\{n_1 n_2(g-1) - d_2 n_1 + d_1 n_2 + k_1 d_2 - k_1 n_2(g-1) - k_1 k_2\},$$
$$(4.5)$$

where the minimum is taken over all possible destabilizing patterns.

If all these codimensions are positive, we say that the transformation from $G_i(n, d, k)$ to $G_{i-1}(n, d, k)$ is a *good flip*. Of course, flips are not always good flips, but under special conditions this can be true, as the next results illustrate.

Theorem 4.7. *Suppose that $k \leq n$ and that α_i is a critical point in the range*

$$\alpha_I < \alpha_i < \frac{d}{n-k} \,. \tag{4.6}$$

Then the transformation from G_i to G_{i-1} is a good flip, i.e. the codimension of G_i^+ in G_i and of G_i^- in G_{i-1} are both positive.

This is proved in [[BGMMN1]] but the proof is indirect, i.e. it is not by direct analysis of the flip loci. Instead, we show that the moduli spaces $G_i(n, d, k)$ and $G_{i-1}(n, d, k)$ are smooth, irreducible and of dimension $\beta(n, d, k)$, and that the flip loci are contained in Zariski closed subsets. In fact they lie in the complement of the Zariski-open subsets $U \bigcap G_i(n, d, k)$ and $U \bigcap G_{i-1}(n, d, k)$, where

$$U = \{(E, V) \in G_L(n, d, k) \mid E \text{ is stable}\} \,.$$

If we impose the slightly stronger restriction

$$max\{\frac{d-n}{n-k}, 0\} = \alpha_T < \alpha_i < \frac{d}{n-k} \,, \tag{4.7}$$

then the Diagram Lemma (Lemma 4.4) implies that the flips are better than good:

Proposition 4.8 (Proposition 6.4 [BGMMN1]). *Let α_i be a critical value in the range (4.7). Then*

$$\mathrm{Codim}(G_i^-) \geq g \tag{4.8}$$

$$\mathrm{Codim}(G_i^+) \geq g+1 \tag{4.9}$$

In some situations we can explicitly describe the flip loci. This will be true for a given critical value α_i if the following assumptions hold.

Assumptions 4.9. *Let n_1, d_1, k_1 be any triple of positive integers corresponding to the critical value α_i, i.e. satisfying*

$$\frac{n_1}{d_1} + \alpha_i \frac{k_1}{d_1} = \frac{n}{d} + \alpha_i \frac{k}{d} . \tag{4.10}$$

and let $n_2 = n - n_1$, $d_2 = d - d_1$, $k_2 = k - k_1$. Then

(a) $\mathrm{GCD}(n_1, d_1, k_1) = \mathrm{GCD}(n_2, d_2, k_2) = 1$;
(b) α_i is not a critical value for (n_1, d_1, k_1), (n_2, d_2, k_2) ;
(c) $G_1 := G(\alpha_i; n_1, d_1, k_1)$ and $G_2 := G(\alpha_i; n_2, d_2, k_2)$ are smooth of the expected dimensions ;
(d) $\mathrm{Ext}^2((E_1, V_1), (E_2, V_2)) = \mathrm{Ext}^2((E_2, V_2), (E_1, V_1)) = 0$.

Remarks:

- Assumptions (a) and (b) ensure that α_i-stability coincides with α_i-semistability for coherent systems of type (n_1, d_1, k_1) or of type (n_2, d_2, k_2).
- Assumptions(c) and (d) always hold if α_i is in the injective range for both (n_1, d_1, k_1) and (n_2, d_2, k_2).
- In Section 5.2 we describe one example situation in which all assumptions are satisfied.

Under the above conditions (see Appendix A in [BGMMN1]) each flip locus $G_i^+(\lambda_1, n_1) \subset G_i(n, d, k)$ can be identified as a projective bundle $p : \mathbf{P}\mathcal{W}_i^+ \longrightarrow G_1 \times G_2$, with fibers

$$\mathbf{P}\mathcal{W}_i^+|_{((E_1, V_1), (E_2, V_2))} = W_i^+(\lambda_1, n_1) . \tag{4.11}$$

Similarly $G_i^-(\lambda_2, n_2) \subset G_{i-1}(n, d, k)$ can be identified as a projective bundle $\mathbf{P}\mathcal{W}_i^-$ over $G_1 \times G_2$, with fibers $\mathbf{P}W_i^-$ where $W_i^- = W_i^-(\lambda_2, n_2)$.

Furthermore, $G_i^+(\lambda_1, n_1)$ is a smoothly embedded subvariety in $G_i(n, d, k)$ with normal bundle

$$N_{G_{i-1}(n,d,k)} G_i^+(\lambda_1, n_1) = p^* W_i^- \otimes \mathcal{O}_{\mathbf{P}W^+}(-1) . \qquad (4.12)$$

If we blow up $G_i(n, d, k)$ along $G_i^+(\lambda_1, n_1)$ we thus get a variety with exceptional divisor

$$\mathbf{P}\mathcal{W}_i^+ \times_{G_1 \times G_2} \mathbf{P}\mathcal{W}_i^- .$$

Similarly, if we blow up $G_{i-1}(n, d, k)$ along $G_i^-(\lambda_2, n_2)$ we get a variety with *the same exceptional divisor*. This works for each allowable choice of (n_1, d_1) and all the blow-ups can be done simultaneously. Since the blow-ups of $G_{i-1}(n, d, k)$ and $G_i(n, d, k)$ agree outside of the exceptional divisors, and the exceptional divisors are the same, the two blow-ups can be identified. We thus obtain a smooth variety which is both the blow up of $G_i(n, d, k)$ along the flip loci $G_i^+(\lambda_1, n_1)$, and the blow up of $G_i^-(\lambda_2, n_2)$ along the flip loci $G_i^-(\lambda_2, n_2)$, and in which the exceptional divisors of the two blow ups coincide. This allows us to describe the transformation from $G_{i-1}(n, d, k)$ to $G_i(n, d, k)$ as a true **flip**, i.e. a blow-up of $G_{i-1}(n, d, k)$ along the disjoint union of the loci $G_i^-(\lambda_2, n_2)$ followed by a blow-down of the exceptional divisor onto the disjoint union of the loci $G_i^+(\lambda_1, n_1)$ in $G_i(n, d, k)$.

Remark In the case $k = 1$, this is precisely the picture described by Thaddeus in [Th].

5 Information beyond non-emptiness

If the codimensions of the flip loci are greater than one, or if assumptions (4.9) are valid, then the moduli spaces on either side of a critical value share topological and geometric features that go beyond non-emptiness. In this section we describe some situations where this is so.

5.1 Homotopy groups

Suppose that $0 < k < n$ and that *(4.7)* holds. Then the structural information from the Diagram Lemma allows us to compute the Picard groups and also certain homotopy groups of the coherent systems moduli spaces. Proposition (4.8), which follows from the Diagram Lemma, leads directly to

Theorem 5.1 (Proposition 7.1[BGMMN1]). *Suppose that $0 < k < n$ and that (4.7) holds. Then $G(\alpha; n, d, k)$ and $G_L(n, d, k)$, if non-empty,* *are smooth and irreducible and isomorphic outside subvarieties of codimension at least g. Hence*

(i) $Pic(G(\alpha; n, d, k)) \cong Pic(G_L(n, d, k))$;
(ii) $\pi_i(G(\alpha; n, d, k)) \cong \pi_i(G_L(n, d, k))$ *for* $i \leq 2g - 2$.

Using our structure results for the coherent systems in $G_L(n, d, k)$ we can build a description of the moduli space. The description is complete if $n - k$ is coprime to d, but only birationally complete otherwise. In all cases it is good enough so that, combined with Theorem 5.1, we get

Theorem 5.2 ([BGMMN1] Theorem C). *Suppose that* $0 < k < n$, *that (4.7) holds, and that* $d \geq \max\{1, kg + n(1 - g)\}$. *Except possibly when* $g = 2$, $k = n - 2$ *and* d *is even:*

(a) $Pic(G(\alpha; n, d, k)) \cong \begin{cases} Pic(M(n - k, d)) \times \mathbf{Z} & \text{if } k < n + \frac{(d-n)}{g} \\ Pic(M(n - k, d)) & \text{if } k = n + \frac{(d-n)}{g} \end{cases}$,

(b) $Pic^0(G(\alpha; n, d, k))$ *is isomorphic to the Jacobian* $J(C)$,
(c) $\pi_1(G(\alpha; n, d, k)) \cong \pi_1(M(n - k, d)) \cong H_1(C, \mathbf{Z})$;
(d) *If* $d + n(g - 1) - kg > 0$ *and* $k \neq n - 1$, *then* $\pi_2(G(\alpha; n, d, k))$ *fits in an exact sequence*

$$0 \longrightarrow \mathbf{Z} \longrightarrow \pi_2(G(\alpha; n, d, k)) \longrightarrow \mathbf{Z} \times \mathbf{Z}_p \longrightarrow 0 ,$$

where $p = \mathrm{GCD}(n - k, d)$.
(e) *If* $d + n(g - 1) - kg = 0$ *then* $\pi_2(G(\alpha; n, d, k)) \cong \mathbf{Z} \times \mathbf{Z}_p$.
(d) *If* $k = n - 1$ *then*

$$\pi_i(G(\alpha; n, d, n - 1)) \cong \pi_i(\mathrm{Gr}(n - 1, d + g - 1)) \text{ for } i \geq 0, i \neq 1.$$

5.2 Poincaré polynomial

If $k = n - 2$ and α_i is a critical value above the torsion bound, i.e. lying in the range (4.7), then the Diagram Lemma shows that in any destabilizing pattern the parameters (n_1, d_1, k_1) and (n_2, d_2, k_2) must satisfy

$$k_1 = n_1 - 1 \quad and \quad k_2 = n_2 - 1 . \tag{5.1}$$

It follows that Assumptions (4.9) are valid (see Section 8 in [BGMMN1] for more details). We can thus regard $G_i(n, d, n-2)$ as the result of a sequence of true flips (in the sense of Section 4.2) applied to $G_L(n, d, n - 2)$, one for each critical α-value between α_L and α_i . This allows us to pass information from $G_L(n, d, n - 2)$ to $G_i(n, d, n - 2)$. In particular, using the formula for the Poincaré polynomial of a blow-up, it allows us to relate the Poincaré polynomial of $G_i(n, d, n-2)$ to that of $G_L(n, d, n-2)$.

Denote the Poincaré polynomial of $G_i(n, d, k)$ by $P_i(n, d, k)(t)$. In particular if $\alpha_i = \alpha_T$ or α_L denote the Poincaré polynomials by $P_T(n, d, k)(t)$ and $P_L(n, d, k)(t)$ respectively.

Proposition 5.3 (Th 8.5 in [BGMMN1]). *Let $k = n - 2$ and let α_i be any critical value in the interval $(\alpha_T, \frac{d}{2})$. Then*

$$P_i(n, d, n-2)(t) - P_L(n, d, n-2)(t) =$$

$$\sum \frac{t^{2C_{21}} - t^{2C_{12}}}{1 - t^2} P_L(n_1, d_1, n_1 - 1)(t) P_L(n_2, d_2, n_2 - 1)(t), \qquad (5.2)$$

where the summation is over all (n_1, d_1) such that

$$2n_1 < n \qquad\qquad\qquad\qquad\qquad\qquad\qquad (5.3)$$

$$max\{(d - n)(n - 2n_1), 0\} < 2(nd_1 - n_1 d) < d(n - 2n_1) \qquad (5.4)$$

$$\frac{nd_1 - n_1 d}{n - 2n_1} = \alpha_j \ for \ some \ critical \ value \ \alpha_j > \alpha_i \ , \qquad (5.5)$$

and where

$$C_{21} = n_1(g - 1) + dn_1 - d_1(n_1 + 1) - (n_1 - 1)(n - n_1 - 1) \qquad (5.6)$$

$$C_{12} = (n - n_1)(g - 1) - d + d_1(n - n_1 + 1) - (n_1 - 1)(n - n_1 - 1). \qquad (5.7)$$

In particular

$$P_T(n, d, n-2)(t) - P_L(n, d, n-2)(t) =$$

$$\sum \frac{t^{2C_{21}} - t^{2C_{12}}}{1 - t^2} P_L(n_1, d_1, n_1 - 1)(t) P_L(n_2, d_2, n_2 - 1)(t),$$

the summation being over all solutions of the above three equations for which $\alpha_j > \alpha_T$.

We can use this to compute a fully explicit formula for $P_i(n, d, n-2)(t)$ if we have closed forms for the $P_L(n, d, k)(t)$ which appear in the formula. Such formulae exist if $n - k$ and d are coprime since then $G_L(n, d, k)$ is a Grassmann fibration over $M(n - k, d)$ (the moduli space of stable bundles of rank $n-k$ and degree d). Using known results for the Poincaré polynomial of $M(2, d)$ (with d odd) and of Grassmannians, we get:

Proposition 5.4. *Let* $0 < k < n$ *and* $d > 0$. *Suppose that* $k \leq n + \frac{1}{g}(d - n)$ *and that* $n - k$ *and* d *are coprime. Then the Poincaré polynomial of* G_L *is given by*

$$P_L(n, d, k)) = P(M(n - k, d)) \cdot P(\mathrm{Gr}(k, N)),$$

where $N = d + (n - k)(g - 1)$. *In particular, if* $k = n - 2$ *and* d *is odd then*

$$P_L(n, d, n - 2)) = \frac{(1 + t)^{2g}((1 + t^3)^{2g} - t^{2g}(1 + t)^{2g})}{(1 - t^2)(1 - t^4)}$$

$$\times \frac{(1 - t^{2(d+2g-n+1)})(1 - t^{2(d+2g-n+2)}) \cdots (1 - t^{2(d+2g-2)})}{(1 - t^2)(1 - t^4) \cdots (1 - t^{2n-4})}.$$

For example, when $n = 3$ or $n = 4$, the number of terms in the summation (5.2) is very small (in fact it is zero or one!) and we can easily compute the results (see Corollaries 8.7 and 8.8 in [BGMMN1]).

6 Gauge theoretic aspects

The α-stability property for coherent systems has an analytic description as an existence condition for solutions to a set of gauge theoretic equations. Correspondences of this sort, i.e. between stability and existence of solutions to gauge theoretic equations are known as Hitchin-Kobayashi correspondences. The original such correspondence relates slope stability for a holomorphic bundle over a Kahler manifold to the existence of solutions to the Hermitian-Einstein equations. This leads to a complex analytic description of the moduli space of stable objects, and also reveals the moduli space to be a symplectic reduced space. In this section we describe briefly the analogous constructions for coherent systems. We refer the interested reader to [BDGW] for a more detailed overview.

6.1 Analytic description of Coherent Systems

We begin by distinguishing between a holomorphic bundle E and its underlying smooth complex bundle $\mathbf{E}$. This is analogous to the distinction between a complex manifold and its underlying real manifold. Indeed, just as the complex structure on the manifold is specified analytically by an integrable $\bar{\partial}$-operator, so is the holomorphic structure on E defined by a choice of a $\bar{\partial}$-operator on $\mathbf{E}$. Denoted by $\bar{\partial}_E$, such an operator is a linear map

$$\bar{\partial}_E : \Omega^0(\mathbf{E}) \longrightarrow \Omega^{0,1}(\mathbf{E}) \tag{6.1}$$

satisfying a $\bar{\partial}$-Leibniz formula

$$\bar{\partial}_E(fs) = \bar{\partial}(f)s + f\bar{\partial}_E(s) \tag{6.2}$$

for every smooth function f on X and section $s \in \Omega^0(E)$. Here $\Omega^{p,q}(\mathbf{E})$ denotes the space of smooth complex forms of holomorphic type (p, q) with values in $\mathbf{E}$.

In general, in order to define a holomorphic structure, the operator must satisfy the integrability condition $\bar{\partial}_E^2 = 0$ but this is automatically satisfied if the base manifold X is a Riemann surface. From this point of view

- a holomorphic bundle E on a Riemann surface (or smooth algebraic curve) is a smooth bundle, say $\mathbf{E}$, together with a holomorphic structure defined by a given operator $\bar{\partial}_E$, i.e. E is a pair $(\mathbf{E}, \bar{\partial}_E)$,
- a smooth section $\phi \in \Omega^0(\mathbf{E})$ is a holomorphic section of E, i.e. $\phi \in H^0(E)$, if and only if $\bar{\partial}_E(\phi) = 0$,
- a rank k linear subspace of $H^0(E)$ can be specified by means of a k-frame, i.e. k linearly independent holomorphic sections.

Thus a coherent system (E, V) on a fixed smooth bundle $\mathbf{E}$ is described by a $(k + 1)$ tuple of data $(\bar{\partial}_E, \phi_1, \ldots, \phi_k)$ in which the ϕ_i are linearly independent elements in $\Omega^0(E)$ satisfying $\bar{\partial}_E(\phi_i) = 0$: the pair $(\mathbf{E}, \bar{\partial}_E)$ defines the holomorphic bundle E, and V is the linear span of the holomorphic sections. Setting

$$\chi_k = \Omega^{0,1}(X, E) \times \Omega^0(X, E) \times \cdots \times \Omega^0(X, E),$$

we define the *coherent systems configuration space* to be the subspace

$$\mathcal{H}_{CS} = \{(\bar{\partial}_E, \phi_1, \ldots, \phi_k) \in \chi_k \mid \bar{\partial}_E(\phi_i) = 0 \text{ for } i = 1, \ldots, k$$
$$\text{and } \phi_1, \ldots, \phi_k \text{ are linearly independent}\}.$$

Two groups, namely the complex gauge group and $GL(k)$, act on $\mathcal{H}_{CS}$. The complex gauge group of $\mathbf{E}$, denoted by $\mathfrak{G}_C$, is the group of complex bundle automorphisms covering the identity map on the base. The action of $\mathfrak{G}_C$ on $\mathcal{H}_{CS}$ is given by

$$g(\bar{\partial}_E, \phi_1, \ldots, \phi_k) = (g \circ \bar{\partial}_E \circ g^{-1}, g\phi_1, \ldots, g\phi_k).$$

In addition, $GL(k)$ acts on the k-tuples of sections, i.e. it changes the k-frame for the rank k subspace of $\Omega^0(X,E) \times \cdots \times \Omega^0(X,E)$ spanned by the sections.

Proposition 6.1. *(see [BDGW])*
(1) A coherent system (E,V) corresponds to an orbit of $GL(k)$ in $\mathcal{H}_k$, or equivalently to a point in the orbit space $\mathcal{H}_k/GL(k)$.
(2) Isomorphism classes of coherent systems correspond to $\mathfrak{G}_C \times GL(k)$-orbits in $\mathcal{H}_k$.

6.2 Equations

Given a Hermitian metric (H) and a holomorphic structure $(\bar{\partial}_E)$ on a smooth bundle **E** there is a unique bundle connection, called the Chern connection, which is both compatible with the complex structure and unitary with respect to the metric. The curvature, denoted by $F_{\bar{\partial}_E,H}$, is a complex differential 2-form of holomorphic type $(1,1)$, with values in the endomorphism bundle of **E**, i.e. $F_{\bar{\partial}_E,H} \in \Omega^{1,1}(\mathrm{End}E)$. If we fix the holomorphic structure, $F_{\bar{\partial}_E,H}$ is a function only of the metric and in that case we denote it by F_H.

We assume that X has a fixed Kahler metric, which, for convenience, is normalized so that X has unit volume. We can then formulate the following set of equations for a coherent system (E,V). The equations, known as the *orthonormal vortex equations* may be regarded as equations for a metric on the underlying smooth bundle **E** and a k-frame for V. If $\{\phi_1, \ldots, \phi_k\}$ is a set of linearly independent holomorphic sections spanning V, the orthonormal vortex equations can be written as

$$i\Lambda F_H + \Sigma_{i=1}^k \phi_i \otimes \phi_i^* = \tau\mathbf{I} \tag{6.3}$$

$$<\phi_i, \phi_j> = \alpha\mathbf{I}_k$$

Here α and τ are real parameters, **I** is the identity section of $\mathrm{End}E$ while $\mathbf{I}_k$ is the unit $k \times k$ matrix, the term ϕ^* indicates the adjoint with respect to the metric H, the left hand side of the second equation is the $k \times k$ matrix whose ij-entry is the L^2 inner product of ϕ_i and ϕ_j in $\Omega^0(X,E)$, and ΛF_H denotes the contraction of the curvature with the Kähler form on X. Since the curvature is an $\mathrm{End}E$-valued 2-form, ΛF_H is a 0-form with values in $\mathrm{End}E$, i.e. it is a section of the bundle $\mathrm{End}E$. Taking the trace of the first equation and integrating over X, we see that τ and α must be related by

$$d + \alpha k = n\tau . \tag{6.4}$$

Definition 6.2. *If we fix the value of α (and hence of τ) we refer to the resulting set of equations as the α-**orthonormal vortex equations**.*

6.3 The Hitchin-Kobayashi correspondence

The relevance of the orthonormal equations comes from its relation to stability.

Theorem 6.3. *Let $\mathbf{E}$ be a smooth complex vector bundle of rank n and degree d. Fix α and τ such that (6.4) holds. Suppose that with this choice of α and τ a coherent system (E, V) of type (n, d, k) on $\mathbf{E}$ admits a solution to the α-orthonormal vortex equation. That is, suppose there is a frame $\{\phi_1, \ldots, \phi_k\}$ for V and a smooth metric on $\mathbf{E}$, say H, such that equations (6.3) are satisfied. Then (E, V) is a direct sum of α-stable coherent systems. If n is coprime to either d or k, and if τ is generic, then (E, V) is α-stable.*

Conversely, if (E, V) is an α-stable coherent system of type (n, d, k) on $\mathbf{E}$, then it admits a solution to the α-orthonormal vortex equations, i.e. there is a smooth metric on $\mathbf{E}$, say H, and frame $\{\phi_1, \ldots, \phi_k\}$ for V such that the equations hold. Moreover, the solution, i.e. the set $(\{\phi_1, \ldots \phi_k\}, H)$, is unique up to the action of $U(k)$ on the basis $\{\phi_1, \ldots \phi_k\}$ and up to the rescaling

$$(\{\phi_1, \ldots \phi_k\}, H) \mapsto (\{\frac{\phi_1}{\lambda}, \ldots, \frac{\phi_k}{\lambda}\}, |\lambda|^2 H)$$

by any $\lambda \in \mathbf{C}^*$

The first part of this Theorem is proved in [BDGW] (Proposition 2.9), the second part is Theorem 3.13 in [BG1].

If we denote by $\mathcal{H}_{CS}(\alpha)$ the subset of points in $\mathcal{H}_{CS}$ which support a solution to the α-orthonormal vortex equations, then the above theorem says that we can identify the moduli space of α-semistable coherent systems as

$$G(\alpha; n, d, k) = \mathcal{H}_{CS}(\alpha)/\mathfrak{G}_C \times GL(k). \tag{6.5}$$

6.4 Symplectic interpretation

Fixing a Hermitian metric on $\mathbf{E} \longrightarrow X$ induces hermitian inner products on the infinite dimensional linear space $\Omega^{0,1}(X, E)$ and $\Omega^0(X, E)$. This gives rise to a Kähler structures – and hence a symplectic structure – on each of these spaces, and hence on the product χ_k. Moreover, the fixed bundle metric, say H_0 allows one to define the *unitary gauge group of* $\mathbf{E}$ i.e. the group of sections of $\mathrm{End}(E)$ which define unitary automorphisms of E. We will denote this unitary gauge group by $\mathfrak{G}$. This unitary gauge group acts by gauge transformations on χ_k and the action preserves the symplectic structure. In addition, the $GL(k)$ action on χ restricts to a symplectic action of the unitary group $U(k)$. This leads to the definition of a symplectic moment map for the action of the product group $\mathfrak{G} \times U(k)$. For more details, and the proofs of the following propositions, see Section 2.2. in [BDGW].

Proposition 6.4. *The moment map for the action of the product group $\mathfrak{G} \times U(k)$ on χ_k, with respect to the above symplectic structure,*

$$\Psi_{CS} : \chi_k \longrightarrow \mathfrak{g} \oplus \mathfrak{u}(k) \ ,$$

is given by

$$\Psi_{CS}(\overline{\partial}_E, \phi_1, \ldots, \phi_k) = (\Lambda F_{\overline{\partial}_E, H_0} - i\Sigma_{i=1}^k \phi_i \otimes \phi_i^*, -i < \phi_i, \phi_j >)$$

where $F_{\overline{\partial}_E, H_0}$ is the curvature of the Chern connection determined by $\overline{\partial}_E$ and the fixed metric H_0.

Everything – the symplectic structure, the group actions, and the moment map – restricts to the subspace $\mathcal{H}_{CS} \subset \chi_k$. The pre-image $\Psi^{-1}(-i\alpha I, -i\tau I_k) \subset \mathcal{H}_{CS}$ thus consists of all coherent systems which satisfy the orthonormal vortex equations with $H = H_0$. By virtue of the Hitchin-Kobayashi correspondence we thus get identifications

Proposition 6.5.

$$\Psi^{-1}(-i\alpha I, -i\tau I_k)/\mathfrak{G} \times U(k) = \mathcal{H}_{CS}(\alpha)/\mathfrak{G}_C \times GL(k) = G(\alpha; n, d, k) \ .$$
$$(6.6)$$

6.5 Holomorphic k-pairs

If we drop the $U(k)$-action on χ_k, the moment map for the remaining $\mathfrak{G}$-action is given by

$$\Psi_k(\overline{\partial}_E, \phi_1, \ldots, \phi_k) = \Lambda F_H - i\Sigma_{i=1}^k \phi_i \otimes \phi_i^* \ . \tag{6.7}$$

We can restrict this to the subspace

$$\mathcal{H}_k = \{(\overline{\partial}_E, \phi_1, \ldots, \phi_k) \in \chi_k \ | \overline{\partial}_E(\phi_i) = 0 \text{ for } i = 1, \ldots, k\} \ . \tag{6.8}$$

Notice that $\mathcal{H}_k$ differs from $\mathcal{H}_{CS}$ in that we do not require that the sections be linearly independent. The points in $\mathcal{H}_k$ define a holomorphic bundle together with k (not necessarily linearly independent) holomorphic sections. These objects, known as *holomorphic k-pairs*, turn out to be interesting in their own right (see especially [BDW]) and also in relation to coherent systems and to Brill-Noether theory.

Just as for coherent systems, there is a notion of stability for k-pairs and a corresponding set of gauge theoretic equations. We refer the interested reader to [BDGW, BDW, DW] for the details. The stability notion depends on a parameter and results in a finite family of moduli spaces. The equations, dubbed the k-τ-vortex equations (see [BDW]) and given by the first of the equations in (6.3), define a metric on the bundle. The parameter τ on the right-hand side corresponds to the parameter in the definition of stability. Several key features of k-pairs are analogous to those for coherent systems:

- A necessary condition for the existence of a τ-semistable k-pair of type (n, d) is that τ lies in a range bounded below for all k, and bounded above if $k < n$,
- The range for τ is partitioned into intervals bounded by critical values at which strict semistability is possible.
- Within each of the resulting intervals τ-semistability is equivalent to τ-stability and is independent of τ.
- If $k \geq n$ there is a largest finite critical value. In fact if $k \geq n$ and $\tau > d$, then all τ-semistable k-pairs are τ-stable.
- There are moduli spaces, one for each subinterval in the τ-range.

If we denote by $\mathcal{H}_k(\tau)$ the subset of points in $\mathcal{H}_k$ which support a solutions to the k-τ-vortex equations, and let $G^{(k)}(\tau; n, d)$ be the moduli space of τ-stable k-pairs, then we get

$$G^{(k)}(\tau; n, d) = \mathcal{H}_k(\tau)/\mathfrak{G}_C = \Psi_k^{-1}(-i\tau I_k)/\mathfrak{G} , \qquad (6.9)$$

where Ψ_k is the moment map described in (6.7). A comparison of (6.6) and (6.9) shows that the coherent system moduli space may be viewed as a symplectic reduction from the moduli space of k-pairs.

The analytic description of the moduli spaces (of either coherent systems or k-pairs) opens the way for tools from geometric analysis to be applied. For example (see [DW]) one can define real-valued functionals (called Yang-Mills-Higgs energies) on χ_k which satisfy the requisite properties for Morse theory and for which the k-τ-vortex equations define the absolute minima. The gradient of the Yang-Mills-Higgs energy defines a vector field whose flow lines retract $\mathcal{H}_k$ onto critical submanifolds. In particular, all points with sufficiently small energy lie on gradient-flow lines which terminate in the minimal submanifold, i.e. in $\mathcal{H}_k(\tau)$. This idea is used in [DW] to prove the non-emptiness of $G^{(k)}(\tau; n, d)$ under the conditions $0 < d \leq n$, $k \leq (\frac{g-1}{g})n + \frac{d}{n}$ and for sufficiently small τ. This led to an alternative analytic proof of the landmark results in [BGN] on the non-emptiness of Brill-Noether loci. The full scope of these analytic techniques has yet to be explored.

Appendix: Coherent systems on elliptic curves and the projective line

by H. Lange

Many of the results for $g \geq 2$ also hold for $g = 0, 1$, though these special cases require different proofs. Surprisingly, these cases turn out to be more interesting than one might guess. In this Appendix we include results about non-emptiness and irreducibility of the moduli spaces, since they are not contained in the survey [N]. Although these results are complete in the case of elliptic curves, this is far from true in the case of the projective line. The reason for this is, essentially, that on an elliptic curve there are semistable vector bundles for every rank and degree, whereas on the projective line there are stable coherent systems of any rank but there are no stable vector bundles of rank ≥ 2.

6.6 Elliptic curves

Let C be an elliptic curve and consider a triple (n, d, k). We assume $k \geq 1$, the case $k = 0$ being well understood by [At]. In this case the Brill-Noether number (see (2.9) in Proposition 2.8) is independent of n and equal to $\beta(d, k) := k(d - k) + 1$.

Theorem 6.6. [LN] *Suppose $d \geq 1, k \geq 1$ and $\alpha > 0$. Then there are α-stable coherent systems of type (n, d, k) if and only if*

$$(n - k)\alpha < d \quad \text{and either} \quad k < d \quad \text{or} \quad k = d \text{ and } \gcd(n, d) = 1.$$

In these cases the moduli space $G(\alpha; n, d, k)$ is smooth and irreducible of dimension $\beta(d, k)$.

Note that this is a complete result, since there are no stable coherent systems with $\alpha \leq 0$ and the only stable coherent system with $d \leq 0$, $k \geq 1$ is $(\mathcal{O}_C, H^0(\mathcal{O}_C))$.

The proof relies heavily on Atiyah's classification [At] of vector bundles on C. For the bounds of α it uses results of [BG2] and [BGMN]. Injectivity of the Petri map implies that the moduli spaces are smooth and of the right dimension. Finally, non-emptiness and irreducibility are consequences of the following structure result (cf. Theorem 3.7 in the case $g \geq 2$), since the families constructed in it are of the right dimension and all other families of extensions, which might provide subvarieties of $G(\alpha; n, d, k)$, are of smaller dimension.

Proposition 6.7. *Suppose (E, V) is a general element of a component of $G(\alpha; n, d, k)$. Then*
(a) *If $0 < k < n$, there is an exact sequence*

$$0 \to \mathcal{O}^k \to E \to G \to 0,$$

where $V = H^0(\mathcal{O}^k) \subset H^0(E)$ and G is a polystable vector bundle with pairwise non-isomorphic indecomposable direct summands.
(b) *If $k = n \geq 2$, there is an exact sequence*

$$0 \to \mathcal{O}^n \to E \to T \to 0,$$

with $V = H^0(\mathcal{O}^n) \subset H^0(E)$ and T a torsion sheaf of length d.
(c) *If $k > n$, there is an exact sequence*

$$0 \to H \to \mathcal{O}^k \to E \to 0$$

such that $H^0(\mathcal{O}^k) \to H^0(E)$ is an isomorphism onto V and H is polystable with pairwise non-isomorphic indecomposable direct summands.

There are also some results concerning the variations of the moduli spaces $G(\alpha; n, k, d)$ with α, for which we refer to [LN].

Hernández Ruipérez and Tejero Prieto define for any positive integer a a Fourier-Mukai transform Φ_a for coherent systems on an elliptic curve and use this to show that certain moduli spaces are isomorphic. To be more precise, as in (Definition 2.10) let $G_0(n, d, k)$ and $G_L(n, d, k)$ denote the moduli space of smallest and largest value of α. Their main result is the following theorem.

Theorem 6.8. *[HT, Theorems 4.5 and 4.15] Let a be any positive integer. The Fourier-Mukai transform Φ_a induces an isomorphism of moduli spaces*

$$G_0(n, d, k) \to G_0(n + ad, d, k) \quad and \quad G_L(n, d, k) \to G_L(n + ad, d, k).$$

For G_L it is assumed that $k < n$.

A consequence of this is that the birationality type of $G(\alpha; n, d, k)$ depends only on the class $[n] \in \mathbf{Z}/d\mathbf{Z}$.

6.7 The projective line

According to a Theorem of Grothendieck (proved in a different language already by Dedekind and Weber in [DeWe, Section 22]) every vector bundle E on $\mathbf{P}^1$ can be written uniquely as

$$E \simeq \oplus_{i=1}^n \mathcal{O}(a_i) \quad \text{with} \quad a_1 \geq a_2 \geq \cdots \geq a_n. \tag{6.10}$$

In particular there are no stable vector bundles of rank ≥ 2 on $\mathbf{P}^1$.

6.7.1 Generic splitting type

Stable coherent systems on curves of positive genus can be constructed starting with suitable extensions of stable or at least semistable vector bundles. Here one starts with vector bundles which are "as semistable as possible".

Definition 6.9. *A vector bundle* $E \simeq \oplus_{i=1}^n \mathcal{O}(a_i)$ *as in* (6.10) *is* **of generic splitting type** *if* $a_1 \leq a_n + 1$ *or equivalently if* $h^1(\mathrm{End}(E)) = 0$.

Define integers a and t by

$$d = an - t \quad \text{with} \quad 0 \leq t < n. \tag{6.11}$$

Then E is of generic splitting type if and only if

$$E \simeq \mathcal{O}(a)^{n-t} \oplus \mathcal{O}(a-1)^t.$$

Theorem 6.10. [LN1, Theorem 3.2] *Suppose* $G(\alpha; n, d, k) \neq \emptyset$ *and* $k > 0$. *Then* $G(\alpha; n, d, k)$ *is smooth and irreducible of expected dimension*

$$\beta(n, d, k) = n^2 + 1 - k(k - d - n) .$$

Indication of Proof. The Petri map (see (2.8)) is injective, which implies smoothness. Let $(E, V) \in G(\alpha; n, d, k)$. The set $U = \{V \in \mathrm{Gr}(k, H^0(E)) \mid (E, V) \text{ is } \alpha\text{-stable}\}$ is open and dense in $\mathrm{Gr}(k, H^0(E))$. Working out the dimension of the image of the canonical map $U \to G(\alpha; n, d, k)$ gives irreducibility.

In order to study non-emptiness and bounds for α, we need the following statement [LN1, Proposition 3.5].

Proposition 6.11. *Suppose $G(\alpha; n, d, k) \neq \emptyset$ and $0 < k < n$. Then for a general $(E, V) \in G(\alpha; n, d, k)$, E is of generic splitting type and there is an exact sequence*

$$0 \to \mathcal{O}^k \to E \to G \to 0, \qquad (6.12)$$

where $V = H^0(\mathcal{O}^k) \subset H^0(E)$ and G is a vector bundle of generic splitting type.

Let a and t be as in (6.11). Define integers ℓ and m by

$$ka - t = \ell(n - k) + m \qquad \text{with} \qquad 0 \leq m < n - k. \qquad (6.13)$$

Then G is of generic splitting type if and only if

$$G \simeq \mathcal{O}(a + \ell + 1)^m \oplus \mathcal{O}(a + \ell)^{n-k-m}.$$

6.7.2 Bounds for α

For $g = 0$ and $0 < k < n$ the bounds for α are stronger than the bounds of Proposition 2.5 because of the way the integers t and m come in:

Proposition 6.12. *Suppose $0 < k < n$ and let t and m be as in equations (6.11) and (6.13). A necessary condition for the existence of an α-stable coherent system of type (n, d, k) is*

$$\frac{t}{k} < \alpha < \frac{d}{n - k} - \frac{mn}{k(n - k)} \qquad (6.14)$$

For the proof see [LN1, Propositions 4.1 and 4.2].

6.7.3 Existence of α-stable coherent systems

For $g = 0$ and $0 < k < n$ there are complete results concerning existence of α-stable coherent systems only for $k = 1, 2, 3$ and $\alpha = 0^+$, i.e. α close to but larger than 0. There are also some scattered results, which we omit here, for $k \geq n$.

Theorem 6.13. *[LN2, Theorem 4.5] Suppose $0 < k < n$. There exists a 0^+-stable coherent system of type (n, d, k) if and only if*

$$d = na \quad \text{and} \quad ka \geq n - k + \frac{k^2 - 1}{n}. \qquad (6.15)$$

Note that according to Proposition 6.12 a 0^+-stable coherent system can exist only if d is a multiple of n. The inequality in (6.15) is equivalent to the Brill-Noether inequality $\beta(n, na, k) \geq 0$. The existence for small positive values of α is reduced to a problem in projective geometry which can be solved completely.

Theorem 6.14. *[LN1, Theorems 5.1 and 5.4] Let t and m be again as in* (6.11) *and* (6.13). (a) *Suppose* $n \geq 2$. *There is an α-stable coherent system of type* $(n, d, 1)$ *if and only if*

$$t < \alpha < \frac{d}{n-1} - \frac{mn}{n-1}.$$

(b) *Suppose* $n \geq 3$. *There is an α-stable coherent system of type* $(n, d, 2)$ *if and only if*

$$d \geq \frac{1}{2}n(n-2) + \frac{3}{2}, \quad (n, d) \neq (4, 6)$$

and

$$\frac{t}{2} < \alpha < \frac{d}{n-2} - \frac{mn}{2(n-2)}.$$

The inequality for d is just the Brill-Noether inequality. The inequalities for α are special cases of (6.14). Although there exist no α-stable coherent system of type $(4, 6, 2)$, there exist α-semistable coherent systems of this type for $1 \leq \alpha \leq 3$.

Already for $k = 3$ the situation is more complicated.

Theorem 6.15. *[LN2, Theorem 8.4] Suppose* $n \geq 4$ *and let* t, ℓ *and* m *be as in* (6.11) *and* (6.13). *There is an α-stable coherent system of type* $(n, d, 3)$ *if and only if*

$$\ell \geq 1, \quad d \geq \frac{1}{3}n(n-3) + \frac{8}{3}, \quad (n, d) \neq (6, 9)$$

and

$$\frac{t}{3} < \alpha < \frac{d}{n-3} - \frac{mn}{3(n-3)},$$

except for the following pairs (n, d) *where the range of α is as stated*

$$\text{for } (4, 7): \frac{3}{5} < \alpha < 7; \quad \text{for } (5, 9): \frac{3}{4} < \alpha < \frac{11}{3};$$

$$\text{for } (6, 11): 1 < \alpha < \frac{7}{3}; \quad \text{for } (7, 13): \frac{3}{2} < \alpha < \frac{8}{3}.$$

The strategy to prove Theorems 6.14 and 6.15 is as follows (the original proof of Theorem 6.14 given in [LN1] was different): As in the proof of Theorem 6.13, for small $\alpha > \frac{t}{k}$ the existence of an α-stable coherent systems is reduced to a problem of projective geometry, which can be solved. Then the flip loci (see section 2.1) are analyzed. By estimating the numbers C_{21} (see (2.5)) and C_{12} (i.e. the number obtained by switching the indices in (2.5)) these loci are shown to be of positive codimension.

In the exceptional cases of Theorem 6.15, the upper bound of α coincides with the upper bound of (6.14) and $t = 1$. So the lower bound is strictly bigger that the lower bound of (6.14). This is due to the fact that $C_{21} = 0$ in these cases, which means that the corresponding flip locus coincides with the whole moduli space for the corresponding critical value.

For higher k further complications arise. In [LN2, Section 10] an example of a critical value is given with $C_{12} = 0$. This means that the upper bound for α is smaller than expected by (6.14). However, for a fixed k, these exceptions can occur at most finitely many times. To be more precise,

Theorem 6.16. *[LN2, Corollary 5.9] Let k be a positive integer. Then, for all but finitely many pairs (n,d) with $n > k$, one of the following possibilities hold:*

- *there is no α-stable coherent system of type (n,d,k) for any α;*
- *there is an α-stable coherent system for any α satisfying (6.14).*

6.7.4 Flips for $k = 1$

The paper [LN3] gives an explicit inductive description of all moduli spaces $G(\alpha; n, d, 1)$ as disjoint unions of locally closed subvarieties determined by the first moduli space and the flip loci. Here we collect the main results.

Fix a triple $(n, d, 1)$ and let the integers a, t, ℓ, m be defined as in (6.11) and (6.13). The critical values are the numbers

$$\alpha_e = en + t \qquad \text{for} \qquad 1 \le e \le \ell - 1.$$

For any such integer e we write for brevity

$$G_e = G_e(n, d, 1) := G(\alpha; n, d, 1)$$

for any α in the range $en + t < \alpha < (e+1)n + t$. The corresponding flip loci are denoted by G_e^+ and G_e^-.

Now denote $a_0 := a, t_0 := t, \ell_0 := \ell, m_0 := m$ and $e_0 := e$ and define inductively, for $1 \le r \le n - 2$,

- s_r and t_r by $s_r(n - r) + t_r = t_{r-1} + e_{r-1}$ with $0 \le t_r \le n - r - 1$,
- $a_r := a_{r-1} - s_r$,

- l_r and m_r by $a_r = l_r(n-r-1) + m_r + t_r$ with $0 \le m_r \le n-r-2$ and

- $e_r := e_{r-1} + s_r$.

Writing

$$G_e^r := G_{e_r}(n-r, d-ra-re, 1) \quad \text{and} \quad G_e^{r\pm} := G_{e_r}^{\pm}(n-r, d-ra-re, 1)$$

for $1 \le r \le n-1$, we have the following inductive description of the moduli space $G_0 = G_0(n, d, 1)$ (see [LN3, Propositions 4.1 and 4.2]):

Proposition 6.17. (a) *For $t = 0$ the moduli space G_0 is isomorphic to* $\mathrm{Gr}(n, a+1)$;
(b) *For $t > 0$ the variety G_0 is the disjoint union of the following locally closed subvarieties:*

$$G_0 = \bigsqcup_{r=1}^{n-1} V_e^{r+},$$

where V_0^{r+} is a $\mathrm{Gr}(r, a-t+r-n+1)$-bundle over $G_0^r \setminus G_0^{r+}$.

There is a similar description of the last moduli space $G_{\ell-1}$. The following Theorem gives an inductive description of the intermediate moduli spaces G_e (see [LN3, Theorem 3.5]).

Theorem 6.18. *For $1 \le e \le l-1$, the moduli space G_e can be obtained from G_0 by a series of flips at the critical values $\alpha_1, \ldots, \alpha_e$.*

The flip at $\alpha_{e'}$ consists of the removal of disjoint locally closed subvarieties $V_{e'}^{r-}$ and the insertion of disjoint locally closed subvarieties $V_{e'}^{r+}$ for $1 \le r \le n-1$, where $V_{e'}^{r+}$ is a $\mathrm{Gr}(r, a-t-(n-1)(e'+1)+r)$-bundle over $G_{e'}^r \setminus G_{e'}^{r+}$ and where $V_{e'}^{r-}$ is a $\mathrm{Gr}(r, r-n+e'n+t)$-bundle over $G_{e'_r-1} \setminus G_{e'_r}^-$.

The elements of $V_{e'}^{r+}$ are given by extensions of the form

$$0 \to (\mathcal{O}(a+e')^r, 0) \to (E, V) \to (E_r, V) \to 0,$$

where $V \subset H^0(E)$ is a 1-dimensional vector space, and (E_r, V) is an element of $G_{e'}^r \setminus G_{e'}^{r+}$. This explains the bundle structure. There is an analogous description for $V_{e'}^{r-}$.

In the case $n = 2$ one obtains the genus 0-version of a result of Thaddeus [Th], which says that the moduli space G_e can be constructed out of G_0 by a sequence of blowing-ups of flip loci and blowing-downs to flip loci. Already for $n = 3$ the flips are not any more of this type.

6.7.5 Hodge polynomials

The Hodge polynomial of a smooth projective variety X over the field of complex numbers is defined by

$$\epsilon(X)(u,v) := \sum_{p,q} h^{p,q}(X) u^p v^q,$$

where $h^{p,q}(X)$ are the usual Hodge numbers. In particular, $P(X)(u) = \epsilon(X)(u,u)$ is the usual Poincaré polynomial. For any quasiprojective variety X over the field of complex numbers Deligne defined a mixed Hodge structure on the cohomology groups $H_c^k(X, \mathbf{C})$ with compact support with associated Hodge polynomial $\epsilon(X)(u,v)$. These polynomials satisfy the following two properties
(1) If X is a finite disjoint union $X = \sqcup_i X_i$ of locally closed subvarieties X_i, then $\epsilon(X) = \sum_i \epsilon(X_i)$.
(2) If $Y \to X$ is an algebraic fibre bundle with fibre F which is locally trivial in the Zariski topology, then $\epsilon(Y) = \epsilon(X) \cdot \epsilon(F)$.

According to Proposition 6.17, Theorem 6.18 and [LN3, Proposition 3.4] the moduli spaces $G_e(n,d,1)$, $G_e^+(n,d,1)$ and $G_e^-(n,d,1)$ with $0 \leq e \leq \ell - 1$ are disjoint unions of Grassmannian bundles over Zariski open sets of Grassmannians. Since the Hodge polynomials of Grassmannians are well-known, this implies that these Hodge polynomials can be explicitly computed. In fact, for coherent systems of rank 2 there is a closed formula (see [LN3, Proposition 6.8]).

Proposition 6.19. *Suppose $n = 2$ and $0 \leq e \leq \ell - 1$. Then*

$$\epsilon(G_e)(u,v) = \frac{(1 - (uv)^{a-t-e})(1 - (uv)^{a-t-e+1})(1 - (uv)^{2e+t+1})}{(1 - uv)^2(1 - (uv)^2)}.$$

In the case of rank 3 the formulas are already more complicated and we refer to [LN3, Corollary 7.2 and Proposition 7.3].

Bibliography

[At] M. F. Atiyah, *Vector bundles over an elliptic curve*, Proc. London Math. Soc. (3) **7** (1957) 414–452.

[Ba] E. Ballico, *Coherent systems on smooth elliptic curves over a finite field* Int. J. Math. Anal. **2** (2006) 31–42

[Be] A. Bertram, *Stable pairs and stable parabolic pairs*, J. Alg. Geom. **3** (1994) 703–724.

[BDW] A. Bertram, G. Daskalopoulos and R. Wentworth, *Gromov invariants for holomorphic maps from Riemann surfaces to Grassmannians*, J. Amer. Math. Soc. **9** (1996) 529–571.

[BBN] U. Bhosle, L. Brambila-Paz, P. Newstead, *On coherent systems of type* $(n, d, n + 1)$ *on Petri curves* Manuscripta Mathematica **126** (2008), 409–441.

[BDGW] S.B. Bradlow, G. Daskalopoulos, O. García-Prada and R. Wentworth, *Stable augmented bundles over Riemann surfaces* Vector Bundles in Algebraic Geometry, Durham 1993, ed. N.J. Hitchin, P.E. Newstead and W.M. Oxbury, LMS Lecture Notes Series **208**, 15–67, Cambridge University Press, 1995.

[BG1] S.B. Bradlow and O. García-Prada, *A Hitchin-Kobayashi correspondence for coherent systems on Riemann surfaces*, J. London Math. Soc. **(2) 60** (1999) 155–170.

[BG2] S.B. Bradlow and O. García-Prada, *An application of coherent systems to a Brill-Noether problem*, J. Reine Angew. Math., (2002), 123–143.

[BGG] S.B. Bradlow, O. García-Prada and P. Gothen, *Representations of the fundamental group of a surface in $PU(p, q)$ and holomorphic triples*, C. R. Acad. Sci. Paris, **333**, Série I (2001), 347–352.

[BGMN] S. B. Bradlow, O. García-Prada, V. Muñoz and P. E. Newstead, *Coherent Systems and Brill-Noether theory*, Internat. J. Math. **14** (2003), 683–733.

[BGMMN1] S. B. Bradlow, O. García-Prada, V. Mercat, V. Muñoz and P. E. Newstead, *On the geometry of moduli spaces of coherent systems on algebraic curves*, Internat. J. Math. **18** (2007), 411–453.

[BGMMN2] S. B. Bradlow, O. García-Prada, V. Mercat, V. Muñoz and P. E. Newstead, *Moduli spaces of coherent systems of small slope on algebraic curves*, Comm. Alg. (to appear).

[Bh] U. Bhosle, *Coherent systems on a nodal curve* Moduli spaces and vector bundles, ed. L. Brambila-Paz, S.Bradlow, O. Garcia-Prada, and S. Ramanan, London Math. Soc. Lecture Notes Series 359, Cambridge University Press.

[BP] L. Brambila-Paz, *Non-emptiness of moduli spaces of coherent systems* Internat. J. Math. **18**, (2008), 777–799.

[BGN] L. Brambila-Paz, I. Grzegorczyk and P. Newstead, *Geography of Brill-Noether loci for small slope*, J. Alg. Geom. **6** (1997) 645–669.

[BMNO] L. Brambila-Paz, V. Mercat, P. Newstead and F. Ongay, *Nonemptiness of Brill-Noether loci*, Internat. J. Math. **11** (2000) 737–760.

[BO] L. Brambila-Paz and Angela Ortega, *Brill-Noether bundles and coherent systems on special curves* Moduli spaces and vector bundles, ed. L. Brambila-Paz, S.Bradlow, O. Garcia-Prada, and S. Ramanan, London Math. Soc. Lecture Notes Series 359, Cambridge University Press.

[DW] G. Daskalopoulos and R. Wentworth, *On the Brill-Noether problem for vector bundles* Forum Math. **11** (1999) 63–77

[DeWe] R. Dedekind and H. Weber, *Theorie der algebraischen Functionen einer Veränderlichen*, Crelle J. **92** (1882) 181–290.

[He] M. He, *Espaces de modules de systèmes cohérents*, Internat. J. Math. **9** (1998) 545–598.

[HT] D. Hernández Ruipérez and C. Tejero Prieto, *Fourier-Mukai transforms for coherent systems on elliptic curves*, J. London Math. Soc. **77** (2008), 15–32.

[KN] A. King and P. Newstead, *Moduli of Brill-Noether pairs on algebraic curves*, Internat. J. Math. **6** (1995) 733–748.

[LN] H. Lange and P. Newstead, *Coherent sytems on elliptic curves*, Int. J. Math **16** (2005) 787–805

[LN1] H. Lange and P. E. Newstead, *Coherent systems of genus 0*, Internat. J. Math. **15** (2004), 409–424.

[LN2] H. Lange and P. E. Newstead, *Coherent systems of genus 0 II: Existence results for k ≥ 3*, Internat. J. Math. **18** (2007), 363–393.

[LN3] Lange and P. E. Newstead, *Coherent systems of genus 0 III: Computation of flips for k = 3*, Internat. J. Math. **19** (2008), 1103–1119.

[LeP] J. Le Potier, *Faisceaux semi-stables et systèmes cohérents*, Vector Bundles in Algebraic Geometry, Durham 1993, ed. N.J. Hitchin, P.E. Newstead and W.M. Oxbury, LMS Lecture Notes Series **208**, 179–239, Cambridge University Press, 1995.

[N] P. Newstead, Existence of α-stable coherent systems on algebraic curves, in preparation

[RV] N. Raghavendra and P.A. Vishwanath, *Moduli of pairs and generalized theta divisors*, Tôhoku Math. J. **46** (1994) 321–340.

[Te] M. Teixidor i Bigas, *Existence of Coherent Systems* Int. J. Math. (to appear)

[Th] M. Thaddeus, *Stable pairs, linear systems and the Verlinde formula*, Invent. Math. **117** (1994) 317–353.

9

Higgs bundles and surface group representations

Oscar García-Prada

Instituto de Ciencias Matemáticas
CSIC-UAM-UC3M-UCM, Serrano 121, 28006 Madrid, Spain
e-mail: oscar.garcia-prada@uam.es

Dedicated to Peter Newstead in celebration of his 65th birthday.

1 Introduction

Let G be a real reductive Lie group, let $H \subset G$ be a maximal compact subgroup and let $\mathfrak{g} = \mathfrak{h} \oplus \mathfrak{m}$ be a Cartan decomposition. Let X be a compact Riemann surface and let K be its canonical line bundle. A G-Higgs bundle over X is a pair (E, φ) consisting of a principal holomorphic $H^{\mathbf{C}}$-bundle E over X and a holomorphic section of $E(\mathfrak{m}^{\mathbf{C}}) \otimes K$, where $E(\mathfrak{m}^{\mathbf{C}})$ is the $\mathfrak{m}^{\mathbf{C}}$-bundle associated to E via the adjoint representation. There are appropriate notions of stability and polystability and a corresponding moduli space of polystable G-Higgs bundles $\mathcal{M}(G)$, which is a complex algebraic variety. These objects were introduced by Hitchin for $G = \mathrm{SL}(2, \mathbf{C})$ in [Hi1] in his study of the self-duality equations on Riemann surfaces. He showed that the moduli space of $\mathrm{SL}(2, \mathbf{C})$-Higgs bundles has a very rich geometric structure, namely, it is a hyperKähler manifold and an algebraically completely integrable system.

In these notes we review the theory of G-Higgs bundles on X and their relation to representations of the fundamental group of X in G. This is the content of non-abelian Hodge theory, developed by Hitchin, Donaldson, Simpson, Corlette and others. We consider the moduli space of representations of $\pi_1(X)$ in G which is defined as the set

$$\mathcal{R}(G) = \mathrm{Hom}^+(\pi_1(X), G)/G$$

of reductive homomorphisms from $\pi_1(X)$ to G modulo conjugation. The reductiveness condition ensures that this orbit space is Hausdorff, and in fact $\mathcal{R}(G)$ is a real analytic variety. The key fact is that if G is a connected semisimple Lie group, the moduli spaces $\mathcal{M}(G)$ and $\mathcal{R}(G)$

are homeomorphic. A similar correspondence exists when G is reductive replacing the fundamental group by a certain central extension of the fundamental group. In these notes we focus on the semisimple case.

After explaining the main ingredientes of the non-abelian Hodge theory correspondence, we describe some aspects of the theory when G is complex, and also when G is a split real form of a classical complex simple Lie group, or a simple classical group of Hermitian type, that is when G/H is a Hermitian symmetric space.

We have not attempted to give a comprehensive account, and we do not give complete proofs of most results. Instead we refer to the main sources on the subject.

2 G-Higgs bundles

2.1 Definition of G-Higgs bundle

Let G be a real reductive Lie group, let $H \subset G$ be a maximal compact subgroup and let $\mathfrak{g} = \mathfrak{h} \oplus \mathfrak{m}$ be a Cartan decomposition, so that the Lie algebra structure on $\mathfrak{g}$ satisfies

$$[\mathfrak{h}, \mathfrak{h}] \subset \mathfrak{h}, \qquad [\mathfrak{h}, \mathfrak{m}] \subset \mathfrak{m}, \qquad [\mathfrak{m}, \mathfrak{m}] \subset \mathfrak{h}.$$

The group H acts linearly on $\mathfrak{m}$ through the adjoint representation, and this action extends to a linear holomorphic action of $H^{\mathbf{C}}$ on $\mathfrak{m}^{\mathbf{C}} = \mathfrak{m} \otimes \mathbf{C}$ — the isotropy representation.

Let X be a compact Riemann surface and let K be its canonical line bundle. A G-**Higgs bundle** over X is a pair (E, φ) consisting of a principal holomorphic $H^{\mathbf{C}}$-bundle E over X and a holomorphic section of $E(\mathfrak{m}^{\mathbf{C}}) \otimes K$, where $E(\mathfrak{m}^{\mathbf{C}})$ is the $\mathfrak{m}^{\mathbf{C}}$-bundle associated to E via the isotropy representation.

When G is compact $\mathfrak{m} = 0$ and hence a G-Higgs bundle is simply a holomorphic principal $G^{\mathbf{C}}$-bundle. When G is complex, if $U \subset G$ is a maximal compact subgroup, the Cartan decomposition of $\mathfrak{g}$ is $\mathfrak{g} = \mathfrak{u} + i\mathfrak{u}$, where $\mathfrak{u}$ is the Lie algebra of U. Then a G-Higgs bundle (E, φ) consists of a a holomorphic G-bundle E and $\varphi \in H^0(X, E(\mathfrak{g}) \otimes K)$, where $E(\mathfrak{g})$ is the $\mathfrak{g}$-bundle associated to E via the adjoint representation. These are the objects introduced originally by Hitchin [Hi1] when $G = \mathrm{SL}(2, \mathbf{C})$ (see Section 5).

2.2 Stability of G-Higgs bundles

The notion of stability, semistability and polystability of G-Higgs bundles depends on an element of $i\mathfrak{h} \cap \mathfrak{z}$, where $\mathfrak{z}$ is the centre of $\mathfrak{h}^{\mathbf{C}}$. However, in order to relate G-Higgs bundles to representations of the fundamental group of X in G, one requires this element to lie also in the centre of $\mathfrak{g}$. Since we will be mostly concerned with G-Higgs bundles for G semisimple, we will take this element to be 0.

Before defining stability, we recall first some basic facts about parabolic subgroups of a complex reductive Lie group $H^{\mathbb{C}}$, where H is a compact Lie group and $\mathfrak{h}$ is its Lie algebra. Let $\mathfrak{h}_s^{\mathbb{C}}$ be the semisimple part of $\mathfrak{h}^{\mathbb{C}}$, that is, $\mathfrak{h}_s^{\mathbb{C}} = [\mathfrak{h}^{\mathbb{C}}, \mathfrak{h}^{\mathbb{C}}]$, $\mathfrak{c}$ the Cartan subalgebra of $\mathfrak{h}_s^{\mathbb{C}}$ and $\mathfrak{z}$ the centre of $\mathfrak{h}^{\mathbb{C}}$. For $\alpha \in \mathfrak{c}^*$, let $\mathfrak{h}_\alpha^{\mathbb{C}}$ be the corresponding root space of $\mathfrak{h}_s^{\mathbb{C}}$. Let R be the set of all roots and Δ a fundamental system of roots.

For $A \subseteq \Delta$, define

$$R_A = \left\{ \alpha \in R \ : \ \alpha = \sum_{\beta \in \Delta} m_\beta \beta \text{ with } m_\beta \geq 0 \text{ for every } \beta \in A \right\}.$$

One has that, for each $A \subseteq \Delta$,

$$\mathfrak{p}_A = \mathfrak{z} \oplus \mathfrak{c} \oplus \bigoplus_{\alpha \in R_A} \mathfrak{h}_\alpha^{\mathbb{C}}$$

is a parabolic subalgebra of $\mathfrak{h}^{\mathbb{C}}$ and all parabolic subalgebras can be obtained in this way. Denote by P_A the corresponding parabolic subgroup.

Similarly, we define, for $A \subseteq \Delta$,

$$R_A^0 = \left\{ \alpha \in R \ : \ \alpha = \sum_{\beta \in \Delta} m_\beta \beta \text{ with } m_\beta = 0 \text{ for every } \beta \in A \right\}.$$

The vector space

$$\mathfrak{l}_A = \mathfrak{z} \oplus \mathfrak{c} \oplus \bigoplus_{\alpha \in R_A^0} \mathfrak{h}_\alpha^{\mathbb{C}}$$

is a Lie subalgebra of $\mathfrak{p}_A$. In fact, one can compute that it is a Levi subalgebra of $\mathfrak{p}_A$, that is, a maximal reductive subalgebra of $\mathfrak{p}_A$. Let L_A be the only connected subgroup of P_A with Lie algebra $\mathfrak{l}_A$. Then, L_A is a Levi subgroup of P_A (i.e. a maximal reductive subgroup of P_A).

An antidominant character of $\mathfrak{p}_A$ is an element of $\mathfrak{z}^* \oplus \mathfrak{c}^*$ of the form

$$\chi = z + \sum_{\alpha \in A} n_\alpha \lambda_\alpha$$

where $z \in \mathfrak{z}^*$, $n_\alpha \in \mathbf{R}$, $n_\alpha \leq 0$ and, for each $\alpha \in A$,

$$\lambda_\alpha = \frac{2\alpha}{\langle \alpha, \alpha \rangle}$$

(here $\langle , \rangle$ denotes the Killing form). When $n_\alpha < 0$ for all $\alpha \in A$, we say that χ is a strictly antidominant character.

Via the Killing form, we have an isomorphism $\mathfrak{z}^* \oplus \mathfrak{c}^* \cong \mathfrak{z} \oplus \mathfrak{c}$. For each antidominant character χ of $\mathfrak{p}_A$, call $s_\chi \in \mathfrak{z} \oplus \mathfrak{c}$ the corresponding element via this isomorphism. One has that $s_\chi \in i\mathfrak{h}$.

For $s \in i\mathfrak{h}$, define the sets

$$\mathfrak{p}_s = \{x \in \mathfrak{h}^{\mathbf{C}} \ : \ Ad(e^{ts})x \text{ is bounden as } t \to \infty\}$$
$$P_s = \{g \in H^{\mathbf{C}} \ : \ e^{ts}ge^{-ts} \text{ is bounden as } t \to \infty\}$$
$$\mathfrak{l}_s = \{x \in \mathfrak{h}^{\mathbf{C}} \ : \ [x,s] = 0\}$$
$$L_s = \{g \in H^{\mathbf{C}} \ : \ Ad(g)(s) = s\}.$$

One has the following.

Proposition 2.1. *For $s \in i\mathfrak{h}$, $\mathfrak{p}_s$ is a parabolic subalgebra of $\mathfrak{h}^{\mathbf{C}}$, P_s is a parabolic subgroup of $H^{\mathbf{C}}$ and the Lie algebra of P_s is $\mathfrak{p}_s$, $\mathfrak{l}_s$ is a Levi subalgebra of $\mathfrak{p}_s$ and L_s is a Levi subgroup of P_s with Lie algebra $\mathfrak{l}_s$. If χ is an antidominant character of $\mathfrak{p}_A$, then $\mathfrak{p}_A \subseteq \mathfrak{p}_{s_\chi}$ and $L_A \subseteq L_{s_\chi}$ and, if χ is strictly antidominant, $\mathfrak{p}_A = \mathfrak{p}_s$ and $\mathfrak{l}_A = \mathfrak{l}_{s_\chi}$.*

Let E be a principal $H^{\mathbf{C}}$-bundle and $A \subseteq \Delta$. Let χ be an antidominant character of P_A. Let σ be a holomorphic section of $E(G/P_A)$ — the G/P_A-bundle associated to E via the natural action of G on G/P_A. That is, σ is a reduction of the structure group of E to P_A. Denote by E_σ the corresponding P_A-bundle. We define the **degree** of E with respect to σ and χ by

$$\deg E(\sigma, \chi) = \deg \chi_* E_\sigma.$$

Recall that G is a real reductive Lie group, $H \subset G$ is a maximal compact subgroup and $\mathfrak{g} = \mathfrak{h} \oplus \mathfrak{m}$ is a Cartan decomposition. Let $\iota : H^{\mathbf{C}} \to \mathrm{GL}(\mathfrak{m}^{\mathbf{C}})$ be the isotropy representation. We define

$$\mathfrak{m}_\chi^- = \{v \in \mathfrak{m}^{\mathbf{C}} \ : \ \iota(e^{ts_\chi})v \text{ is bounded as } t \to \infty\}$$
$$\mathfrak{m}_\chi^0 = \{v \in \mathfrak{m}^{\mathbf{C}} \ : \ \iota(e^{ts_\chi})v = v \text{ for every } t\}.$$

One proves that $\mathfrak{m}_\chi^-$ is invariant under the action of P_{s_χ} and $\mathfrak{m}_\chi^0$ is invariant under the action of L_{s_χ} (this follows from Proposition 2.1). If G is complex, $\mathfrak{m}^{\mathbf{C}} = \mathfrak{g}$ and ι is the adjoint representation, then $\mathfrak{m}_\chi^- = \mathfrak{p}_{s_\chi}$ and $\mathfrak{m}_\chi^0 = \mathfrak{l}_{s_\chi}$.

A G-Higgs bundle (E, φ) is called **semistable** if for any parabolic subgroup P_A of $H^{\mathbf{C}}$, any antidominant character χ of P_A and any reduction of the structure group of E to P_A, σ, such that $\varphi \in H^0(X, E_\sigma(\mathfrak{m}_\chi^-) \otimes K)$, we have

$$\deg E(\sigma, \chi) \geq 0.$$

The Higgs bundle (E, φ) is called **stable** if it is semistable and for any P_A, χ and σ as above such that $\varphi \in H^0(X, E_\sigma(\mathfrak{m}_\chi^-) \otimes K)$ and $A \neq 0$,

$$\deg E(\sigma, \chi) > 0.$$

The Higgs bundle (E, φ) is called **polystable** if it is semistable and for each P_A, σ and χ as in the definition of semistable G-Higgs bundle such that $\deg E(\sigma, \chi) = 0$, there exists a holomorphic reduction of the structure group of E_σ to the Levi subgroup L_A of P_A, $\sigma_L \in \Gamma(E_\sigma(P_A/L_A))$, where by E_σ we mean the principal P_A-bundle obtained by reducing the structure group of E to the parabolic subgroup P_A. Moreover, in this case, we require $\varphi \in H^0(X, E(\mathfrak{m}_\chi^0) \otimes K)$.

We shall assume that G is connected. Then the topological classification of $H^{\mathbf{C}}$-bundles E on X is given by an invariant $c(E) \in \pi_1(H^{\mathbf{C}}) = \pi_1(H) = \pi_1(G)$. For a fixed $d \in \pi_1(G)$, the **moduli space of polystable G-Higgs bundles** $\mathcal{M}_d(G)$ is the set of isomorphism classes of polystable G-Higgs bundles (E, φ) such that $c(E) = d$. When G is compact, the moduli space $\mathcal{M}_d(G)$ coincides with $M_d(G^{\mathbf{C}})$, the moduli space of polystable $G^{\mathbf{C}}$-bundles with topological invariant d. Sometimes we will denote by $\mathcal{M}(G)$ the moduli space of G-Higgs bundles including all possible values of $d \in \pi_1(G)$.

The moduli space $\mathcal{M}_d(G)$ has the structure of a complex analytic variety. This can be seen by the standard slice method (see, e.g., Kobayashi [K]). Geometric Invariant Theory constructions are available in the literature for G compact algebraic (Ramanathan [R1, R2]) and for G complex reductive algebraic (Simpson [Si3, Si4]). The case of a real form of a complex reductive algebraic Lie group follows from the general constructions of Schmitt [Sch1, Sch2]. We thus have the following.

Theorem 2.2. *The moduli space $\mathcal{M}_d(G)$ is a complex analytic variety, which is algebraic when G is algebraic.*

2.3 Deformation theory of G-Higgs bundles

In this section, we recall some standard facts about the deformation theory of G-Higgs bundles. A convenient reference for this material is Biswas–Ramanan [BR].

Definition 2.3. Let (E, φ) be a G-Higgs bundle. The *deformation complex* of (E, φ) is the following complex of sheaves:

$$C^\bullet(E, \varphi): E(\mathfrak{h}^{\mathbf{C}}) \xrightarrow{ad(\varphi)} E(\mathfrak{m}^{\mathbf{C}}) \otimes K. \tag{2.1}$$

Note that this definition makes sense because $[\mathfrak{m}^{\mathbf{C}}, \mathfrak{h}^{\mathbf{C}}] \subseteq \mathfrak{m}^{\mathbf{C}}$.

The following result generalizes the fact that the infinitesimal deformation space of a holomorphic vector bundle V is isomorphic to $H^1(\mathrm{End}V)$.

Proposition 2.4. *The space of infinitesimal deformations of a G-Higgs bundle (E, φ) is naturally isomorphic to the hypercohomology group*

$$\mathbf{H}^1(C^\bullet(E, \varphi)).$$

In particular, if (E, φ) represents a non-singular point of the moduli space $\mathcal{M}_d(G)$ then the tangent space at this point is canonically isomorphic to $\mathbf{H}^1(C^\bullet(E, \varphi))$.

For any G-Higgs bundle there is a natural long exact sequence

$$
\begin{aligned}
0 \to \mathbf{H}^0(C^\bullet(E, \varphi)) &\to H^0(E(\mathfrak{h}^{\mathbf{C}})) \xrightarrow{ad(\varphi)} H^0(E(\mathfrak{m}^{\mathbf{C}}) \otimes K) \\
&\to \mathbf{H}^1(C^\bullet(E, \varphi)) \to H^1(E(\mathfrak{h}^{\mathbf{C}})) \xrightarrow{ad(\varphi)} H^1(E(\mathfrak{m}^{\mathbf{C}}) \otimes K) \\
&\to \mathbf{H}^2(C^\bullet(E, \varphi)) \to 0.
\end{aligned}
\tag{2.2}
$$

A G-Higgs bundle (E, φ) is said to be **simple** if $\mathrm{Aut}\,(E, \varphi) = \ker \iota \cap Z(H^{\mathbf{C}})$, where $Z(H^{\mathbf{C}})$ is the centre of $H^{\mathbf{C}}$. The following result on smoothness of the moduli space can be proved, for example, from the standard slice method construction referred to above.

Proposition 2.5. *Let (E, φ) be a stable G-Higgs bundle. If (E, φ) is simple and*

$$\mathbf{H}^2(C_G^\bullet(E, \varphi)) = 0,$$

then (E, φ) is a smooth point in the moduli space.

Suppose now that we are in the situation of Proposition 2.5. Then a local universal family exists and hence the dimension of the component of the moduli space containing (E, φ) equals the dimension of the infinitesimal deformation space $\mathbf{H}^1(C_G^\bullet(E, \varphi))$. We shall refer to this dimension as the **expected dimension** of the moduli space. The expected dimension of the moduli space is $\dim \mathbf{H}^1(C_G^\bullet(E, \varphi)) = -\chi(C_G^\bullet(E, \varphi))$. The long exact sequence (2.2) gives us

$$\chi(C_G^\bullet(E,\varphi)) - \chi(E(\mathfrak{h}^{\mathbf{C}})) + \chi(E(\mathfrak{m}^{\mathbf{C}}) \otimes K) = 0.$$

From the Riemann–Roch formula we therefore obtain

$$\dim \mathbf{H}^1(C_G^\bullet(E,\varphi)) = \deg(E(\mathfrak{m}^{\mathbf{C}})) + (g-1)\operatorname{rk}(E(\mathfrak{m}^{\mathbf{C}})) - \big(\deg(E(\mathfrak{h}^{\mathbf{C}})) \\ + (1-g)\operatorname{rk}(E(\mathfrak{h}^{\mathbf{C}})).$$

But $\deg(E(\mathfrak{m}^{\mathbf{C}})) = \deg(E(\mathfrak{h}^{\mathbf{C}})) = 0$, and we thus have that the expected dimension of the moduli space of G-Higgs bundles when G is semisimple is $(g-1)\dim G^{\mathbf{C}}$.

Remark 2.6. Note that the actual dimension of the moduli space (if non-empty) can be smaller than the expected dimension. This happens for example when $G = \mathrm{SU}(p,q)$ with $p \neq q$ and maximal Toledo invariant (this follows from the study of $\mathrm{U}(p,q)$-Higgs bundles in [BGG1]) — in this case there are in fact no stable $\mathrm{SU}(p,q)$-Higgs bundles (see Section 7.2).

2.4 G-Higgs bundles and Hitchin equations

Let G be a connected semisimple real Lie group. Let (E,φ) be a G-Higgs bundle over a compact Riemann surface X. By a slight abuse of notation, we shall denote the C^∞-objects underlying E and φ by the same symbols. In particular, the Higgs field can be viewed as a $(1,0)$-form: $\varphi \in \Omega^{1,0}(E(\mathfrak{m}^{\mathbf{C}}))$. Let $\tau \colon \Omega^1(E(\mathfrak{g}^{\mathbf{C}})) \to \Omega^1(E(\mathfrak{g}^{\mathbf{C}}))$ be the compact conjugation of $\mathfrak{g}^{\mathbf{C}}$ combined with complex conjugation on complex 1-forms. Given a reduction h of structure group to H in the smooth $H^{\mathbf{C}}$-bundle E, we denote by F_h the curvature of the unique connection compatible with h and the holomorphic structure on E.

Theorem 2.7. *A reduction h of structure group of E from $H^{\mathbf{C}}$ to H satisfies the Hitchin equation*

$$F_h - [\varphi, \tau(\varphi)] = 0$$

if and only if (E,φ) is polystable.

Theorem 2.7 was proved by Hitchin [Hi1] for $G = \mathrm{SL}(2,\mathbf{C})$ and Simpson [Si1, Si2] for an arbitrary semisimple complex Lie group G. The proof for an arbitrary reductive real Lie group G when (E,φ) is stable is given in [BGM], and the general polystable case follows as a particular case of a more general Hitchin–Kobayashi correspondence proved in [GGM].

From the point of view of moduli spaces it is convenient to fix a C^∞ principal H-bundle $\mathbf{E}_H$ with fixed topological class $d \in \pi_1(H)$ and study the moduli space of solutions to **Hitchin's equations** for a pair (A, φ) consisting of an H-connection A and $\varphi \in \Omega^{1,0}(X, \mathbf{E}_H(\mathfrak{m}^{\mathbf{C}}))$:

$$F_A - [\varphi, \tau(\varphi)] = 0$$
$$\bar{\partial}_A \varphi = 0. \tag{2.3}$$

Here d_A is the covariant derivative associated to A and $\bar{\partial}_A$ is the $(0,1)$ part of d_A, which defines a holomorphic structure on $\mathbf{E}_H$. The gauge group $\mathscr{H}$ of $\mathbf{E}_H$ acts on the space of solutions and the moduli space of solutions is

$$\mathcal{M}_d^{\mathrm{gauge}}(G) := \{(A, \varphi) \text{ satisfying } (2.3)\}/\mathscr{H}.$$

Now, Theorem 2.7 can be reformulated as follows.

Theorem 2.8. *There is a homeomorphism*

$$\mathcal{M}_d(G) \cong \mathcal{M}_d^{\mathrm{gauge}}(G)$$

To explain this correspondence we interpret the moduli space of G-Higgs bundles in terms of pairs $(\bar{\partial}_E, \varphi)$ consisting of a $\bar{\partial}$-operator (holomorphic structure) on the $H^{\mathbf{C}}$-bundle $\mathbf{E}_{H^{\mathbf{C}}}$ obtained from $\mathbf{E}_H$ by the extension of structure group $H \subset H^{\mathbf{C}}$, and $\varphi \in \Omega^{1,0}(X, \mathbf{E}_{H^{\mathbf{C}}}(\mathfrak{m}^{\mathbf{C}}))$ satisfying $\bar{\partial}_E \varphi = 0$. Such pairs are in correspondence with G-Higgs bundles (E, φ), where E is the holomorphic $H^{\mathbf{C}}$-bundle defined by the operator $\bar{\partial}_E$ on $\mathbf{E}_{H^{\mathbf{C}}}$ and $\bar{\partial}_E \varphi = 0$ is equivalent to $\varphi \in H^0(X, E(\mathfrak{m}^{\mathbf{C}}) \otimes K)$. The moduli space of polystable G-Higgs bundles $\mathcal{M}_d(G)$ can now be identified with the orbit space

$$\{(\bar{\partial}_E, \varphi) \quad : \quad \bar{\partial}_E \varphi = 0 \text{ which are polystable}\}/\mathscr{H}^{\mathbf{C}},$$

where $\mathscr{H}^{\mathbf{C}}$ is the gauge group of $\mathbf{E}_{H^{\mathbf{C}}}$, which is in fact the complexification of $\mathscr{H}$. Since there is a one-to-one correspondence between H-connections on $\mathbf{E}_H$ and $\bar{\partial}$-operators on $\mathbf{E}_{H^{\mathbf{C}}}$, the correspondence given in Theorem 2.8 can be interpreted by saying that in the $\mathscr{H}^{\mathbf{C}}$-orbit of a polystable G-Higgs bundle $(\bar{\partial}_{E_0}, \varphi_0)$ we can find another Higgs bundle $(\bar{\partial}_E, \varphi)$ whose corresponding pair (d_A, φ) satisfies $F_A - [\varphi, \tau(\varphi)] = 0$, and this is unique up to H-gauge transformations.

3 Surface group representations and Higgs bundles

3.1 Surface group representations

Let X be a closed oriented surface of genus g and let

$$\pi_1(X) = \langle a_1, b_1, \ldots, a_g, b_g \ : \ \prod_{i=1}^{g} [a_i, b_i] = 1 \rangle$$

be its fundamental group. Let G be a connected reductive real Lie group. By a **representation** of $\pi_1(X)$ in G we understand a homomorphism $\rho \colon \pi_1(X) \to G$. The set of all such homomorphisms, $\mathrm{Hom}(\pi_1(X), G)$, can be naturally identified with the subset of G^{2g} consisting of $2g$-tuples $(A_1, B_1 \ldots, A_g, B_g)$ satisfying the algebraic equation $\prod_{i=1}^{g} [A_i, B_i] = 1$. This shows that $\mathrm{Hom}(\pi_1(X), G)$ is a real analytic variety, which is algebraic if G is algebraic.

The group G acts on $\mathrm{Hom}(\pi_1(X), G)$ by conjugation:

$$(g \cdot \rho)(\gamma) = g\rho(\gamma)g^{-1}$$

for $g \in G$, $\rho \in \mathrm{Hom}(\pi_1(X), G)$ and $\gamma \in \pi_1(X)$. If we restrict the action to the subspace $\mathrm{Hom}^+(\pi_1(X), g)$ consisting of *reductive representations*, the orbit space is Hausdorff. By a **reductive representation** we mean one that composed with the adjoint representation in the Lie algebra of G decomposes as a sum of irreducible representations. If G is algebraic this is equivalent to the Zariski closure of the image of $\pi_1(X)$ in G to be a reductive group. (When G is compact every representation is reductive.) Define the *moduli space of representations* of $\pi_1(X)$ in G to be the orbit space

$$\mathcal{R}(G) = \mathrm{Hom}^+(\pi_1(X), G)/G.$$

One has the following (see e.g. Goldman [Go3])

Theorem 3.1. *The moduli space $\mathcal{R}(G)$ has the structure of a real analytic variety, which is algebraic if G is algebraic and is a complex variety if G is complex.*

Given a representation $\rho \colon \pi_1(X) \to G$, there is an associated flat G-bundle on X, defined as $E_\rho = \tilde{X} \times_\rho G$, where $\tilde{X} \to X$ is the universal cover and $\pi_1(X)$ acts on G via ρ. This gives in fact an identification between the set of equivalence classes of representations $\mathrm{Hom}(\pi_1(X), G)/G$ and the set of equivalence classes of flat G-bundles, which in turn is parametrized by the cohomology set $H^1(X, G)$. We can then assign a topological invariant to a representation ρ given by the characteristic class $c(\rho) := c(E_\rho) \in \pi_1(G)$ corresponding to E_ρ. To define this, let $\widetilde{G}$ be the universal covering group of G. We have an exact sequence

$$1 \longrightarrow \pi_1(G) \longrightarrow \widetilde{G} \longrightarrow G \longrightarrow 1$$

which gives rise to the (pointed sets) cohomology sequence

$$H^1(X, \widetilde{G}) \longrightarrow H^1(X, G) \xrightarrow{c} H^2(X, \pi_1(G)). \tag{3.1}$$

Since $\pi_1(G)$ is abelian, we have

$$H^2(X, \pi_1(G)) \cong \pi_1(G),$$

and $c(E_\rho)$ is defined as the image of E under the last map in (3.1). Thus the class $c(E_\rho)$ measures the obstruction to lifting E_ρ to a flat $\widetilde{G}$-bundle, and hence to lifting ρ to a representation of $\pi_1(X)$ in $\widetilde{G}$. For a fixed $d \in \pi_1(G)$, the *moduli space of reductive representations* $\mathcal{R}_d(G)$ with topological invariant d is defined as the subvariety

$$\mathcal{R}_d(G) := \{\rho \in \mathcal{R}(G) \ : \ c(\rho) = d\}. \tag{3.2}$$

3.2 Representations and G-Higgs bundles

We assume now that G is connected and semisimple. With the notation of the previous sections, we have the following.

Theorem 3.2. *Let G be a connected semisimple real Lie group. There is a homeomorphism $\mathcal{R}_d(G) \cong \mathcal{M}_d(G)$.*

Remark 3.3. On the open subvarieties defined by the smooth points of $\mathcal{R}_d$ and $\mathcal{M}_d$, this correspondence is in fact an isomorphism of real analytic varieties.

The proof of Theorem 3.2 is the combination of two existence theorems for gauge-theoretic equations. To explain this, let $\mathbf{E}_G$ be, as above, a C^∞ principal G-bundle over X with fixed topological class $d \in \pi_1(G) = \pi_1(H)$. Every G-connection D on $\mathbf{E}_G$ decomposes uniquely as

$$D = d_A + \psi,$$

where d_A is an H-connection on $\mathbf{E}_H$ and $\psi \in \Omega^1(X, \mathbf{E}_H(\mathfrak{m}))$. Let F_A be the curvature of d_A. We consider the following set of equations for the pair (d_A, ψ):

$$\begin{aligned}
F_A + \tfrac{1}{2}[\psi, \psi] &= 0 \\
d_A \psi &= 0 \\
d_A^* \psi &= 0.
\end{aligned} \tag{3.3}$$

These equations are invariant under the action of $\mathcal{H}$, the gauge group of $\mathbf{E}_H$. A theorem of Corlette [C], and Donaldson [D] for $G = \mathrm{SL}(2, \mathbf{C})$, says the following.

Theorem 3.4. *There is a homeomorphism between*

$$\{Reductive\ G\text{-}connections\ D\ :\ F_D = 0\}/\mathcal{G}$$

and

$$\{(d_A, \psi)\ satisfying\ (3.3)\}/\mathcal{H}.$$

The first two equations in (3.3) are equivalent to the flatness of $D = d_A + \psi$, and Theorem 3.4 simply says that in the $\mathcal{G}$-orbit of a reductive flat G-connection D_0 we can find a flat G-connection $D = g(D_0)$ such that if we write $D = d_A + \psi$, the additional condition $d_A^* \psi = 0$ is satisfied. This can be interpreted more geometrically in terms of the reduction $h = g(h_0)$ of $\mathbf{E}_G$ to an H-bundle obtained by the action of $g \in \mathcal{G}$ on h_0. The equation $d_A^* \psi = 0$ is equivalent to the harmonicity of the $\pi_1(X)$-equivariant map $\widetilde{X} \to G/H$ corresponding to the new reduction of structure group h.

To complete the argument, leading to Theorem 3.2, we just need Theorem 2.7 and the following simple result.

Proposition 3.5. *The correspondence* $(d_A, \varphi) \mapsto (d_A, \psi := \varphi - \tau(\varphi))$ *defines a homeomorphism*

$$\{(d_A, \varphi)\ satisfying\ (2.3)\}/\mathcal{H} \cong \{(d_A, \psi)\ satisfying\ (3.3)\}/\mathcal{H}.$$

4 Moment maps and moduli spaces

4.1 Symplectic and Kähler quotients

In this section we review some standard facts about the moment map for the symplectic action of a Lie group G on a symplectic manifold, and the special situation in which the manifold has a Kähler structure which is preserved by the action of the group.

A symplectic manifold is by definition a differentiable manifold M together with a nondegenerate closed 2-form ω. A Kähler manifold with its Kähler form is an example of a symplectic manifold. A transformation f of M is called *symplectic* if it leaves invariant the 2-form , i.e., $f^*\omega = \omega$.

Suppose now that a Lie group G acts symplectically on (M, ω). If v is a vector field generated by the action, then the Lie derivative $L_v \omega$

vanishes. Now for ω, as for any differential form,

$$L_v\omega = i(v)d\omega + d(i(v)\omega);$$

hence $d(i(v)\omega) = 0$, and so, if $H^1(M,\mathbf{R}) = 0$, there exists a function $\mu_v : M \to \mathbf{R}$ such that

$$d\mu_v = i(v)\omega.$$

The function μ_v is said to be a *Hamiltonian function* for the vector field v. As v ranges over the set of vector fields generated by the elements of the Lie algebra $\mathfrak{g}$ of G, these functions can be chosen to fit together to give a map to the dual of the Lie algebra,

$$\mu : M \longrightarrow \mathfrak{g}^*,$$

defined by

$$\langle \mu(x), a \rangle = \mu_{\widetilde{a}}(x),$$

where $\widetilde{a}$ is the vector field generated by $a \in \mathfrak{g}$, $x \in M$ and $\langle \cdot, \cdot \rangle$ is the natural pairing between $\mathfrak{g}$ and its dual. There is a natural action of G on both sides and a constant ambiguity in the choice of μ_v. If this can be adjusted so that μ is G-equivariant, i.e.

$$\mu(g(x)) = (Adg)^*(\mu(x)) \quad \text{for} \ g \in G \ x \in M,$$

then μ is called a *moment map* for the action of G on M. The remaining ambiguity in the choice of μ is the addition of a constant abelian character in $\mathfrak{g}^*$. If μ is a moment map then

$$d\mu_{\widetilde{a}}(x)(v) = \omega(\widetilde{a}(x), v) \quad \text{for} \ a \in \mathfrak{g} \ , \ v \in T_x Mx \ , \ x \in M.$$

An important feature of the moment map is that it gives a way of constructing new symplectic manifolds. More precisely, suppose that G acts freely and discontinuously on $\mu^{-1}(0)$ (recall that $\mu^{-1}(0)$ is G-invariant), then

$$\mu^{-1}(0)/G$$

is a symplectic manifold of dimension $\dim M - 2\dim G$. This is the Marsden–Weinstein *symplectic quotient* of a symplectic manifold acted on by a group [MW]. There is a more general construction by taking μ^{-1} of a coadjoint orbit. In particular if λ is a central element in $\mathfrak{g}^*$ we can consider the symplectic quotient

$$\mu^{-1}(\lambda)/G.$$

Suppose now that M has a Kähler structure. It is convenient to describe a Kähler structure on the manifold M as a triple (g, J, ω) consisting of a Riemannian metric g, an integrable almost complex structure (a complex structure) J and a symplectic form ω on M which satisfies

$$\omega(u, v) = g(Ju, v), \quad \text{for } x \in M \text{ and } u, v \in T_x M.$$

Any two of these structures determines the third one.

Let G now be a Lie group acting on (M, g, J, ω) preserving the Kähler structure. Then if $\mu : M \longrightarrow \mathfrak{g}^*$ is a moment map, and G acts freely and discontinuously on $\mu^{-1}(\lambda)$, for a central element $\lambda \in \mathfrak{g}^*$, the quotient $\mu^{-1}(\lambda)/G$ is also a Kähler manifold. This process is called *Kähler reduction* [MFK].

A very basic example is the following. Let $M = \mathbf{C}^n$ be equipped with its natural Kähler structure and let $U(1)$ act on M by multiplication. The action of $U(1)$ preserves the symplectic structure and has a moment map $\mu : \mathbf{C}^n \to \mathbf{R}$ given by $z \mapsto \sum |z_i|^2$. We can consider $\mu^{-1}(1) = S^{2n-1}$. We then have the symplectic quotient

$$\mu^{-1}(1)/U(1) = S^{2n-1}/U(1) \cong \mathbf{P}_{n-1}(\mathbf{C}).$$

Since the action of $U(1)$ preserves also the complex structure, this construction exhibits $\mathbf{P}_{n-1}(\mathbf{C})$ as a Kähler quotient whose induced Kähler structure is in fact the standard one. Note that $\mu^{-1}(1)/U(1)$ is hence isomorphic to the "good" quotient $\mathbf{C}^n - \{0\}/\mathbf{C}^*$. This relation turns out to be true in a more general context as we will see below.

When M is a projective algebraic manifold there is a very important relation between the symplectic quotient and the algebraic quotient defined by Mumford's Geometric Invariant Theory (GIT) [MFK]. Let $i : M \subset \mathbf{P}_{n-1}(\mathbf{C})$ be a projective algebraic manifold acted on by a reductive algebraic group. We can assume this group to be the complexification $G^{\mathbf{C}}$ of a compact subgroup $G \subset U(n)$. Then, following [MFK], we say that $x \in M$ is *semistable* if there is a non-constant invariant polynomial f with $f(x) \neq 0$. This is equivalent to saying that if $\tilde{x} \in \mathbf{C}^n$ is any representative of x, then the closure of the $G^{\mathbf{C}}$-orbit of $\tilde{x}$ does not contain the origin. Let $M^{ss} \subset M$ the set of all semistable points. There is a subset $M^s \subset M^{ss}$ of *stable* points which satisfy the stronger condition that the $G^{\mathbf{C}}$-orbit of $\tilde{x}$ is closed in $\mathbf{C}^n$. The *algebraic quotient* is by definition the orbit space $M^{ss}/G^{\mathbf{C}}$. That this is the right quotient in this setup is confirmed by the fact that if $A(M)$ is the graded coordinate ring of M, then the invariant subring $A(M)^{G^{\mathbf{C}}}$ is finitely generated and has $M^{ss}/G^{\mathbf{C}}$ as its corresponding projective variety. The quotient $M^s/G^{\mathbf{C}}$ gives a dense open set of $M^{ss}/G^{\mathbf{C}}$.

To relate to symplectic quotients, consider the action of $U(n)$ on $\mathbf{P}_{n-1}(\mathbf{C})$ induced by the standard action on $\mathbf{C}^n$. This action is symplectic and has a moment map $\mu : \mathbf{P}_{n-1}(\mathbf{C}) \to \mathfrak{u}(n)^*$ given by

$$\mu(x) = \frac{1}{2\pi} \frac{xx^*}{\|x\|^2},$$

where we are using the Killing form of $\mathfrak{u}(n)$ to identify $\mathfrak{u}(n)$ with $\mathfrak{u}(n)^*$. Let M and G be as above. Then $p \circ \mu \circ i$, where $p : \mathfrak{u}(n)^* \to \mathfrak{g}^*$ is the projection induced by the inclusion $\mathfrak{g} \subset \mathfrak{u}(n)$, is a moment map for the action of G on M. We can then consider the symplectic quotient $\mu^{-1}(0)/G$. The relation between this quotient and the algebraic quotient is given by the following result due to Mumford, Kempf–Ness, Guillemin and Sternberg and others (see [MFK]).

Theorem 4.1. *There is a homeomorphism*

$$\mu^{-1}(0)/G \cong M^{ss}/G^{\mathbf{C}}.$$

4.2 Moduli spaces of Higgs bundles as Kähler quotients

The symplectic and Kähler quotient constructions explained above can also be extended to the context of infinite dimensional manifolds (see [MW]). We show now how this can be used to endow our moduli spaces with symplectic and Kähler structures.

Coming back to the setup of Section 2.4, let $\mathbf{E}_{H^{\mathbf{C}}}$ be a smooth principal $H^{\mathbf{C}}$-bundle over a compact Riemann surface X, and let $\mathbf{E}_H$ be a reduction of $\mathbf{E}_{H^{\mathbf{C}}}$ to an H-bundle. The set $\mathscr{A}$ of H-connections on $\mathbf{E}_H$ is an affine space modelled on $\Omega^1(X, \mathbf{E}(\mathfrak{h}))$, which is equipped with a symplectic structure defined by

$$\omega_{\mathscr{A}}(\psi, \eta) = \int_X tr(\psi \wedge \eta), \quad \text{for } A \in \mathscr{A} \text{ and } \psi, \eta \in T_A\mathscr{A} = \Omega^1(X, \mathbf{E}_H(\mathfrak{h})).$$

This is obviously closed since it is independent of $A \in \mathscr{A}$.

Now, the set $\mathscr{C}$ of holomorphic structures on $\mathbf{E}_{H^{\mathbf{C}}}$ is an affine space modelled on $\Omega^{0,1}(X, \mathbf{E}_{H^{\mathbf{C}}}(\mathfrak{h}^{\mathbf{C}}))$, and it has a complex structure $J_{\mathscr{C}}$, induced by the complex structure of the Riemann surface, which is defined by

$$J_{\mathscr{C}}(\alpha) = i\alpha, \quad \text{for } \bar{\partial}_E \in \mathscr{C} \text{ and } \alpha \in T_{\bar{\partial}_E}\mathscr{C} = \Omega^{0,1}(X, \mathbf{E}_{H^{\mathbf{C}}}(\mathfrak{h}^{\mathbf{C}})).$$

The complex structure $J_{\mathscr{C}}$ defines a complex structure $J_{\mathscr{A}}$ on $\mathscr{A}$ via the identification $\mathscr{A} \cong \mathscr{C}$ given by the map

$$
\begin{aligned}
\mathscr{A} &\longrightarrow \mathscr{C} \\
d_A &\longmapsto d''_A
\end{aligned} \tag{4.1}
$$

where d''_A is the projection of d_A into the $(0,1)$-part. This map is an isomorphism (see e.g. [W, DK]).

The symplectic structure $\omega_{\mathscr{A}}$ and the complex structure $J_{\mathscr{A}}$ define a Kähler structure on $\mathscr{A}$, which is preserved by the action of the gauge group $\mathscr{H}$ of $\mathbf{E}_H$. We have that $\operatorname{Lie}\mathscr{H} = \Omega^0(X, \mathbf{E}_H(\mathfrak{h}))$ and hence its dual $(\operatorname{Lie}\mathscr{H})^*$ can be identified with $\Omega^2(X, \mathbf{E}_H(\mathfrak{h}))$. One has the following [AB].

Proposition 4.2. *There is a moment map for the action of $\mathscr{H}$ on $\mathscr{A}$ given by*

$$
\begin{aligned}
\mathscr{A} &\longrightarrow \Omega^2(X, \mathbf{E}(\mathfrak{h})) \\
A &\longmapsto F_A.
\end{aligned}
$$

To prove this, let $a \in \operatorname{Lie}\mathscr{H} = \Omega^0(X, \mathbf{E}_H(\mathfrak{h}))$, and let $\tilde{a}$ be the vector field generated by a. We have to show that the function $\mu_{\tilde{a}} : \mathscr{A} \to \mathbf{R}$ given by

$$
\mu_{\tilde{a}}(A) = \int_X tr(a \wedge F_A)
$$

is Hamiltonian. But this follows simply from the following:

$$
\begin{aligned}
d\mu_{\tilde{a}}(A)(v) &= \int_X tr(a \wedge d_A v) \\
&= -\int_X tr(d_A a \wedge v) \\
&= \omega_{\mathscr{A}}(v, \tilde{a}),
\end{aligned}
$$

where we have used that $\tilde{a} = d_A a$.

Now, let us denote $\Omega = \Omega^{1,0}(X, \mathbf{E}_{H^{\mathbb{C}}}(\mathfrak{h}^{\mathbf{C}}))$. The linear space Ω has a natural complex structure J_Ω defined by multiplication by i, and a symplectic structure given by

$$
\omega_\Omega(\psi, \eta) = i \int_X tr(\psi \wedge \eta^*), \quad \text{for } \Phi \in \Omega \text{ and } \psi, \eta \in T_\Phi \Omega = \Omega.
$$

We can now consider $\mathscr{X} = \mathscr{A} \times \Omega$ with the symplectic structure $\omega_{\mathscr{X}} = \omega_{\mathscr{A}} + \omega_{\mathscr{E}}$ and complex structure $J_{\mathscr{X}} = J_{\mathscr{A}} + J_\Omega$. The action of $\mathscr{G}$ on $\mathscr{X}$ preserves $\omega_{\mathscr{X}}$ and $J_{\mathscr{X}}$ and there is a moment map given by (see e.g.[Hi1, BGM])

$$
\begin{aligned}
\mu_{\mathscr{X}} : \mathscr{X} &\longrightarrow \Omega^2(X, \mathbf{E}_H(\mathfrak{h})) \\
(A, \varphi) &\mapsto F_A - [\varphi, \tau(\varphi)].
\end{aligned} \tag{4.2}
$$

We now consider the subvariety

$$\mathcal{N} = \{(d_A, \varphi) \in \mathcal{X} \ : \ d''_A \varphi = 0\}. \tag{4.3}$$

The subvariety $\mathcal{N}^* \subset \mathcal{N}$ of smooth points inherits a Kähler structure from $\mathcal{X}$ and, since it is $\mathcal{H}$-invariant, the moment map is the restriction $\mu = \mu_{\mathcal{X}}|^*_{\mathcal{N}} : \mathcal{N}^* \to \Omega^2(X, \mathbf{E}_H(\mathfrak{h}))$. Now, the moduli space of smooth solutions to Hitchin equations (2.3) is the Kähler quotient $\mu^{-1}(0)/\mathcal{H}$.

5 G-Higgs bundles for complex groups

5.1 Stability

One has the following (see e.g. [BGM]).

Proposition 5.1. *Let G be a complex reductive Lie group. A G-Higgs bundle (E, φ) is stable if, and only if, for every maximal parabolic subgroup P, every reduction $\sigma : E_P \to E$ of E to P and every antidominant character χ of P such that $\varphi \in H^0(X, E_P(\mathfrak{p}) \otimes K)$, we have that $\deg E(\sigma, \chi) > 0$.*

If $G = \mathrm{GL}(n, \mathbf{C})$ (with its underlying real structure), we recover the original notion of Higgs bundle introduced by Hitchin [Hi1], consisting of a holomorphic vector bundle $V := E(\mathbf{C}^n)$ — associated to a principal $\mathrm{GL}(n, \mathbf{C})$-bundle E via the standard representation — and a homomorphism

$$\varphi : V \longrightarrow V \otimes K.$$

Applying Proposition 5.1 we recover the notion of stability introduced by Hitchin [Hi1], i.e., (V, φ) is said to be *stable* if

$$\frac{\deg V'}{\mathrm{rank}\, V'} < \frac{\deg V}{\mathrm{rank}\, V} \tag{5.1}$$

for every proper subbundle $V' \subset V$ such that $\varphi(V') \subset V' \otimes K$. The Higgs bundle (V, φ) is *polystable* if $(V, \varphi) = \oplus_i (V_i, \varphi_i)$ where (V_i, φ_i) is a stable Higgs bundles and $\deg V_i/\mathrm{rank}\, V_i = \deg V/\mathrm{rank}\, V$.

If we take $G = \mathrm{SL}(n, \mathbf{C})$, a G-Higgs bundle can be described as a pair (V, φ) consisting of a vector bundle V with trivial determinant and a traceless Higgs field $\varphi : V \longrightarrow V \otimes K$. Stability simplifies now to the condition $\deg V' < 0$ for every proper subbundle $V' \subset V$ such that $\varphi(V') \subset V' \otimes K$.

Similarly, if G is $\mathrm{SO}(n, \mathbf{C})$ or $\mathrm{Sp}(2n, \mathbf{C})$, a G-Higgs bundle is a pair consisting of a vector bundle V with trivial determinant of rank n and

$2n$, respectively, equipped with a quadratic form and symplectic form, Q and ω, respectively, and a Higgs field $\varphi \in H^0(X, \mathfrak{g}(V) \otimes K)$, where $\mathfrak{g}(V)$ is the Lie algebra bundle whose fibre at $x \in X$ is $\mathfrak{so}(V_x)$ and $\mathfrak{sp}(V_x)$, respectively. Now, the stability condition is that $\deg V' < 0$ for every proper totally isotropic subbundle $V' \subset V$ such that $\varphi(V') \subset V' \otimes K$.

Using Morse-theoretic techniques (see [Hi1, BGG3]) one can reduce the connectedness properties of the moduli space of G-Higgs bundles to those of the moduli space of G-bundles and then, using [R2], to prove the following.

Theorem 5.2. *Let* $d \in \pi_1(G)$. *Then* $\mathcal{M}_d(G)$ *is non-empty and connected.*

5.2 HyperKähler quotients and moduli spaces

We will see now that when G is complex the moduli space has a hyperKähler structure. To see this, recall first that a hyperKähler manifold is a differentiable manifold M equipped with a Riemannian metric g and complex structures J_i, $i = 1, 2, 3$, satisfying the quaternion relations $J_i^2 = -I$, $J_3 = J_1 J_2$, etc., such that if we define $\omega_i(\cdot, \cdot) = g(J_i \cdot, \cdot)$, then (g, J_i, ω_i) is a Kähler structure on M. As for Kähler manifolds, there is a natural quotient construction for hyperKähler manifolds [HKLR].

Let G be a Lie group acting on M preserving the Kähler structure (g, J_i, ω_i) and having moment maps $\mu_i : M \to \mathfrak{g}^*$ for $i = 1, 2, 3$. We can combine these moment maps in a map

$$\boldsymbol{\mu} : M \longrightarrow \mathfrak{g}^* \otimes \mathbf{R}^3$$

defined by $\boldsymbol{\mu} = (\mu_1, \mu_2, \mu_3)$. Let $\lambda_i \in \mathfrak{g}^*$ for $i = 1, 2, 3$ be central elements and consider the G-invariant submanifold $\boldsymbol{\mu}^{-1}(\boldsymbol{\lambda})$ where $\boldsymbol{\lambda} = (\lambda_1, \lambda_2, \lambda_3)$. Then, if G acts on $\boldsymbol{\mu}^{-1}(\boldsymbol{\lambda})$ freely and properly, the quotient

$$\boldsymbol{\mu}^{-1}(\boldsymbol{\lambda})/G$$

is a hyperKähler manifold.

One way to understand the non-abelian Hodge theory correspondence explainned in Sections 2 and 3 when G is complex is through the analysis of the hyperKähler structures of the moduli spaces involved. We explain how these can be obtained as hyperKähler quotients. For this, let us go back to the setup of Section 4.2. In our situation now, $G = H^{\mathbf{C}}$. Let $\mathbf{E}_G$ be a smooth G-bundle over a compact Riemann surface X, and let $\mathbf{E}_H$ be a fixed reduction of $\mathbf{E}_G$ to an H-bundle. As we have seen in

Section 4.2, the space $\mathscr{X} = \mathscr{A} \times \Omega$ has a Kähler structure defined by $J_{\mathscr{X}}$ and $\omega_{\mathscr{X}}$. Let us rename $J_1 = J_{\mathscr{X}}$. Via the identification $\mathscr{A} \cong \mathscr{C}$, we have for $\alpha \in \Omega^{0,1}(X, \mathbf{E}_G(\mathfrak{g}))$ and $\psi \in \Omega^{1,0}(X, \mathbf{E}_G(\mathfrak{g}))$ the following three complex structures on $\mathscr{X}$:

$$
\begin{aligned}
J_1(\alpha, \psi) &= (i\alpha, i\psi) \\
J_2(\alpha, \psi) &= (-i\tau(\psi), i\tau(\alpha)) \\
J_3(\alpha, \psi) &= (\tau(\psi), -\tau(\alpha)),
\end{aligned}
$$

where τ is the conjugation on $\mathfrak{g}$ defining its compact form $\mathfrak{h}$ (determined fibrewise by the reduction to $\mathbf{E}_H$), combined with complex conjugation on complex 1-forms.

Clearly, J_i, $i = 1, 2, 3$ satisfy the quaternion relations, and define a hyperKähler structure on $\mathscr{X}$, with symplectic structures ω_i, $i = 1, 2, 3$, where $\omega_1 = \omega_{\mathscr{X}}$. The action of the gauge group $\mathscr{H}$ on $\mathscr{X}$ preserves the hyperKähler structure and there are moment maps given by

$$
\mu_1(A, \varphi) = F_A - [\varphi, \tau(\varphi)], \quad \mu_2(A, \varphi) = \mathrm{Re}(\bar{\partial}_E \varphi), \quad \mu_3(A, \varphi) = \mathrm{Im}(\bar{\partial}_E \varphi).
$$

We have that $\boldsymbol{\mu}^{-1}(0)/\mathscr{H}$ is the moduli space of solutions to Hitchin equations (2.3). In particular, if we consider the irreducible solutions (equivalently, smooth) $\boldsymbol{\mu}_*^{-1}(0)$ we have that

$$
\boldsymbol{\mu}_*^{-1}(0)/\mathscr{H}
$$

is a hyperKähler manifold which, by Theorem 2.8, is homeomorphic to the subvariety of smooth points in moduli space $\mathcal{M}_d(G)$ of stable G-Higgs bundles with fixed topological class $d \in \pi_1(G)$.

Let us now see how the moduli of harmonic flat connections on $\mathbf{E}_H$ can be realized as a hyperKähler quotient. Let $\mathscr{D}$ be the set of G-connections on $\mathbf{E}_G$. This is an affine space modelled on $\Omega^1(X, \mathbf{E}_G)(\mathfrak{g})) = \Omega^0(X, T^*X \otimes_{\mathbf{R}} \mathbf{E}_G(\mathfrak{g}))$. The space $\mathscr{D}$ has a complex structure $I_1 = 1 \otimes i$, which comes from the complex structure of the bundle. Using the complex structure of X we have also the complex structure $I_2 = i \otimes \tau$. We can finally consider the complex structure $I_3 = I_1 I_2$.

The reduction to H of the G-bundle $\mathbf{E}_G$ together with a Riemannian metric in the conformal class of X defines a flat Riemannian metric $g_{\mathscr{D}}$ on $\mathscr{D}$ which is Kähler for the above three complex structures. Hence $(\mathscr{D}, g_{\mathscr{D}}, I_1, I_2, I_3)$ is also a hyperKähler manifold. As in the previous case, the action of the gauge group $\mathscr{H}$ on $\mathscr{D}$ preserves the hyperKähler structure and there are moment maps

$$
\mu_1(D) = \nabla^* \Psi, \quad \mu_2(D) = \mathrm{im}(F_D), \quad \mu_3(D) = \mathrm{Re}(F_D),
$$

where $D = \nabla + \Psi$ is the decomposition of D defined by

$$\mathbf{E}_G(\mathfrak{g}) = \mathbf{E}_H(\mathfrak{h}) \oplus \mathbf{E}_H(i\mathfrak{h}).$$

Hence the moduli space of solutions to the harmonicity equations (3.3) is the hyperKähler quotient defined by

$$\boldsymbol{\mu}^{-1}(0)/\mathscr{H},$$

where $\boldsymbol{\mu} = (\mu_1, \mu_2, \mu_3)$. The homeomorphism between the moduli spaces of solutions to the Hitchin and the harmonicity equations is induced from the affine map

$$
\begin{array}{ccc}
\mathscr{A} \times \Omega & \longrightarrow & \mathscr{D} \\
(d_A, \varphi) & \longmapsto & d_A + \varphi - \tau(\varphi).
\end{array}
$$

One can see easily, for example, that this map sends $\mathscr{A} \times \Omega$ with complex structure J_2 to $\mathscr{D}$ with complex structure I_1 (see [Hi1]).

Now, Theorems 2.8 and 3.4 for G complex can be regarded as existence theorems, establishing the non-emptiness of the hyperKähler quotient, obtained by focusing on different complex structures. For Theorem 2.8 one gives a special status to the complex structure J_1. Combining the symplectic forms determined by J_2 and J_3 one has the J_1-holomorphic symplectic form $\omega_c = \omega_2 + i\omega_3$ on $\mathscr{A} \times \Omega$. The gauge group $\mathscr{G} = \mathscr{H}^{\mathbf{C}}$ acts on $\mathscr{A} \times \Omega$ preserving ω_c. The symplectic quotient construction can also be extended to the holomorphic situation (see e.g. [K]) to obtain the holomorphic symplectic quotient $\{(\bar{\partial}_E, \varphi) \ : \ \bar{\partial}_E \varphi = 0\}/\mathscr{G}$. What Theorem 2.8 says is that for a class $[(\bar{\partial}_E, \varphi)]$ in this quotient to have a representative (unique up to H-gauge) satisfying $\mu_1 = 0$ it is necessary and sufficient that the pair $(\bar{\partial}_E, \varphi)$ be polystable. This identifies the hyperKähler quotient to the set of equivalence classes of polystable G-Higgs bundles on $\mathbf{E}_G$. If one now takes J_2 on $\mathscr{A} \times \Omega$ or equivalently $\mathscr{D}$ with I_1 and argues in a similar way, one gets Theorem 3.4 identifying the hyperKähler quotient to the set of equivalence classes of reductive flat connections on $\mathbf{E}_G$.

5.3 The Hitchin system

Let p be an G-invariant homogeneous polynomial of degree d of the Lie algebra $\mathfrak{g}$. Then, if (E, φ) is a G-Higgs bundle, the evaluation of p on φ gives an element of $H^0(X, K^d)$, as explained in [Hi2], so p induces a map

$$p : \mathcal{M}(G) \longrightarrow H^0(X, K^d).$$

Let $p_1, \ldots, p_r$ ($r = \operatorname{rk} \mathfrak{g}$) be a basis of the ring of G-invariant homogeneous polynomials of $\mathfrak{g}$. This basis defines a map

$$\mathcal{M}(G) \to \bigoplus_{i=1}^{r} H^0(X, K^{d_i})$$

where, for $i = 1, 2, \ldots, r$, d_i is the degree of p_i. This map is called the **Hitchin map**, and the vector space

$$B = \bigoplus_{i=1}^{r} H^0(X, K^{d_i})$$

is called the **base** of the Hitchin map. It turns out that if E is a stable G-bundle, one has (see [Hi2]) that

$$\dim B = \dim H^0(X, E(\mathfrak{g}) \otimes K)$$

and hence the dimension of B coincides with the dimension of the moduli space of G-bundles, which is half the dimension of $\mathcal{M}(G)$. In [Hi2] Hitchin shows that $\mathcal{M}(G)$ is a completely algebraically integrable system and that the generic fibre is either a Jacobian or a Prym variety of a curve covering X, which is called the **spectral curve**.

Note that if E is a stable G-bundle, then (E, φ) is a stable G-Higgs bundle for any $\varphi \in H^0(X, E(\mathfrak{g}) \otimes K)$. This means that $\mathcal{M}(G)$ naturally contains (as an open set) the cotangent bundle of the smooth locus of the moduli space of G-bundles. This is clear since the tangent space at smooth point E in the moduli space of G-bundles can be identified with $H^1(X, E(\mathfrak{g}) \otimes K)$, which, by Serre duality, is isomorphic to the dual of $H^0(X, E(\mathfrak{g}) \otimes K)$.

6 G-Higgs bundles for split real forms

6.1 Some general facts

In [Hi3], Hitchin showed the following.

Theorem 6.1. *Let X be a compact oriented surface of genus $g > 1$. Let $G^{\mathbb{C}}$ be a complex semisimple Lie group and let G be the split real form of $G^{\mathbb{C}}$. Then the moduli space $\mathcal{R}(G)$ of representations of the fundamental group of X in G has a connected component homeomorphic to a Euclidean space of dimension $(2g - 2) \dim G$.*

When $G = \mathrm{SL}(2, \mathbf{R})$, this component can be identified with Teichmüller space, and this is why Hitchin calls these Teichmüller components. We will refer to them as **Hitchin components**. A geometric interpretation of these components for $G = \mathrm{SL}(3, \mathbf{R})$ has been given by Choi–Goldman in [CG] and recently for more general split groups by Labourie [L] and Fock–Goncharov [FG].

In [Hi3], Hitchin constructs explictly these components as sections of the Hitchin map. In the following sections we will illustrate his construction for the split forms of the classical simple Lie groups. In [Hi3] he gives the general construction and the details for $G = \mathrm{SL}(n, \mathbf{R})$. For the details for the other split real forms see [A].

6.2 $\mathrm{SL}(n, \mathbf{R})$-*Higgs bundles*

The Cartan decomposition of the Lie algebra is given by

$$\mathfrak{sl}(n, \mathbf{R}) = \mathfrak{so}(n) + \mathfrak{m},$$

where $\mathfrak{m} = \{\text{symmetric real matrices of trace } 0\}$. A $\mathrm{SL}(n, \mathbf{R})$-**Higgs bundle** is thus a pair (E, φ), where E is a principal holomorphic $\mathrm{SO}(n, \mathbf{C})$-bundle over X and the Higgs field is a holomorphic section

$$\varphi \in H^0(E(\mathfrak{m}^{\mathbf{C}}) \otimes K),$$

where K is the canonical line bundle over X.

Using the standard representations of $\mathrm{SO}(n, \mathbf{C})$ in $\mathbf{C}^n$ one can associate to E a holomorphic vector bundle V of rank n with $\det V = \mathcal{O}$ together with a nondegenerate quadratic form $Q \in H^0(S^2 V^*)$.

A $\mathrm{SL}(n, \mathbf{R})$-Higgs bundle is then in correspondence with a triple

$$(V, Q, \varphi),$$

where the Higgs field is a symmetric and traceless endomorphism $\varphi : V \to V \otimes K$.

The simplest case is to consider the complex Lie group $\mathrm{SL}(2, \mathbf{C})$ and its split real form $\mathrm{SL}(2, \mathbf{R})$. The Lie algebra $\mathfrak{sl}(2, \mathbf{C})$ has rank 1 and the algebra of invariant polynomials on it is generated by p_2 of degree 2 obtained from the characteristic polynomial

$$\det(x1 - A) = x^2 + p_2(A)$$

of a trace-free matrix. We are going to define a section of the Hitchin map

$$\begin{aligned} p : \mathcal{M} &\to H^0(K^2) \\ (E, \varphi) &\mapsto p_2(\varphi) \end{aligned}$$

where here $\mathcal{M}$ denotes the moduli space of polystable $SL(2, \mathbf{C})$-Higgs bundles. This section will give an isomorphism between the vector space $H^0(K^2)$ and a connected component of the moduli space $\mathcal{M}(SL(2, \mathbf{R})) \subset \mathcal{M}$ of polystable $SL(2, \mathbf{R})$-Higgs bundles. To construct the section, we consider the elements

$$\langle x = \begin{pmatrix} 1 & 0 \\ 0 & -1 \end{pmatrix}, e = \begin{pmatrix} 0 & 1 \\ 0 & 0 \end{pmatrix}, \tilde{e} = \begin{pmatrix} 0 & 0 \\ 1 & 0 \end{pmatrix} \rangle \cong \mathfrak{sl}(2, \mathbf{C})$$

that satisfy

$$[x, e] = 2e, \ [x, \tilde{e}] = -2\tilde{e} \text{ and } [e, \tilde{e}] = x,$$

where x is an element of the Cartan subalgebra (a semisimple element) and e, $\tilde{e}$ are nilpotent. The pair $(K^{1/2} \oplus K^{-1/2}, \varphi = \tilde{e} - \alpha e = \begin{pmatrix} 0 & -\alpha \\ 1 & 0 \end{pmatrix})$, where $\alpha \in H^0(K^2)$, is a $SL(2, \mathbf{R})$-Higgs bundle. In the vector bundle $K^{1/2} \oplus K^{-1/2}$ we have the orthogonal structure $Q = \begin{pmatrix} 0 & 1 \\ 1 & 0 \end{pmatrix}$ and the Higgs field is symmetric with respect to this orthogonal form. The section is finally defined by

$$s(\alpha) = \left(K^{1/2} \oplus K^{-1/2}, \varphi = \begin{pmatrix} 0 & -\alpha \\ 1 & 0 \end{pmatrix} \right).$$

That is, the pairs

$$\left\{ (K^{1/2} \oplus K^{-1/2}, \varphi = \begin{pmatrix} 0 & -\alpha \\ 1 & 0 \end{pmatrix}) \right\}_{\alpha \in H^0(K^2)}$$

form a connected component of $\mathcal{M}(SL(2, \mathbf{R}))$ which is the **Hitchin component**. This component has dimension $6g - 6$ and there are 2^{2g} connected components isomorphic to this one, the number of possible choices of the square $K^{1/2}$.

Now consider the general case $SL(n, \mathbf{R})$ which is the split real form of $SL(n, \mathbf{C})$. The Lie algebra $\mathfrak{sl}(n, \mathbf{C})$ has rank $n - 1$ and a basis for the invariant polynomials on $\mathfrak{sl}(n, \mathbf{C})$ is provided by the coefficients of the characteristic polynomial of a trace-free matrix,

$$\det(x - A) = x^n + p_1(A)x^{n-2} + \ldots + p_{n-1}(A),$$

where $\deg(p_i) = i + 1$. We can consider the Hitchin map

$$p : \mathcal{M}(SL(n, \mathbf{C})) \to \bigoplus_{i=1}^{n-1} H^0(K^{i+1})$$

defined by

$$p(E, \varphi) = (p_1(\varphi), \ldots, p_{n-1}(\varphi)),$$

where $\mathcal{M}(\mathrm{SL}(n, \mathbf{C}))$ is the moduli space of polystable $\mathrm{SL}(n, \mathbf{C})$-Higgs bundles.

We are going to define a section of this map that will give an isomorphism between the vector space $\bigoplus_{i=1}^{n-1} H^0(K^{i+1})$ and a connected component of the moduli space $\mathcal{M}(\mathrm{SL}(n, \mathbf{R})) \subset \mathcal{M}(\mathrm{SL}(n, \mathbf{C}))$ of polystable $\mathrm{SL}(n, \mathbf{R})$-Higgs bundles.

A nilpotent element $e \in \mathfrak{sl}(n, \mathbf{C})$ is called regular if its centralizer is $(n-1)$-dimensional. In $\mathfrak{sl}(n, \mathbf{C})$ a regular nilpotent element is conjugate to an element

$$e = \sum_{\alpha \in \Delta} X_\alpha,$$

where $\Delta = \{\alpha_i = e_i - e_{i+1}, 1 \leq i \leq n-1\}$ and $X_{\alpha_i} = E_{i,i+1}$ is a root vector for α_i, that is,

$$e = \begin{pmatrix} 0 & 1 & 0 & \cdots & \cdots & 0 \\ \vdots & 0 & 1 & 0 & \cdots & 0 \\ \vdots & \vdots & & \ddots & & \vdots \\ \vdots & \vdots & & & \ddots & 0 \\ \vdots & \vdots & & & & 1 \\ 0 & 0 & \cdots & \cdots & \cdots & 0 \end{pmatrix}.$$

Any nilpotent element can be embedded in a 3-dimensional simple subalgebra $\langle x, e, \widetilde{e} \rangle \cong \mathfrak{sl}(2, \mathbf{C})$, where x is semisimple, e and $\widetilde{e}$ are nilpotent, and they satisfy

$$[x, e] = 2e; \qquad [x, \widetilde{e}] = -2\widetilde{e}; \qquad [e, \widetilde{e}] = x.$$

The adjoint action

$$\langle x, e, \widetilde{e} \rangle \cong \mathfrak{sl}(2, \mathbf{C}) \to \mathrm{End}(\mathfrak{sl}(n, \mathbf{C}))$$

of this subalgebra breaks up the Lie algebra $\mathfrak{sl}(n, \mathbf{C})$ as a direct sum of irreducible representations

$$\mathfrak{sl}(n, \mathbf{C}) = \bigoplus_{i=1}^{n-1} V_i,$$

with $\dim(V_i) = 2i + 1$. That is, each V_i is the irreducible representation

$S^{2i}\mathbf{C}^2$, where $\mathbf{C}^2$ is the standard representation of $\mathfrak{sl}(2,\mathbf{C})$, and the eigenvalues of adx on V_i are $-2i, -2i+2, \ldots, 2i-2, 2i$.

The highest weight vector of V_i, defined as a vector $e_i \in V_i$ that is an eigenvector for the action of x and is in the kernel of $ad(e)$, has eigenvalue $2i$ for adx. We take $V_1 = \langle x, e, \tilde{e} \rangle$ and $e = e_1$.

Given $(\alpha_1, \ldots, \alpha_{n-1}) \in \bigoplus_{i=1}^{n-1} H^0(K^{i+1})$, we define the Higgs field in the Hitchin component by

$$\varphi = \tilde{e}_1 + \alpha_1 e_1 + \ldots + \alpha_{n-1} e_{n-1},$$

and the vector bundle is given by

$$S^{n-1}(K^{-1/2} \oplus K^{1/2}) = K^{-(n-1)/2} \oplus K^{-(n-3)/2} \oplus \cdots \oplus K^{(n-3)/2} \oplus K^{(n-1)/2}.$$

The field φ is given in the following order of the basis, $K^{(n-1)/2} \oplus K^{(n-3)/2} \oplus \cdots \oplus K^{-(n-3)/2} \oplus K^{-(n-1)/2}$.

6.3 $SO_0(p,q)$-*Higgs bundles*

We denote by $SO_0(p,q)$ the connected component of the identity in $SO(p,q)$. We begin considering $\mathfrak{so}(p,q)$ in $\mathfrak{so}(p+q,\mathbf{C}) \subset \mathfrak{sl}(p+q,\mathbf{C})$. If $I_{p,q} = \begin{pmatrix} -I_p & \\ & I_q \end{pmatrix}$, we have

$$\begin{aligned} \mathfrak{so}(p,q) &= \{X \in \mathfrak{so}(p+q,\mathbf{C}) \mid I_{p,q}\bar{X}I_{p,q} = X\} \\ &= \{X \in \mathfrak{sl}(p+q,\mathbf{C}) \mid X + X^t = 0, I_{p,q}\bar{X}I_{p,q} = X\} \\ &= \left\{ \begin{pmatrix} X_1 & iX_2 \\ -iX_2^t & X_3 \end{pmatrix} \right\}, \end{aligned}$$

where X_1 and X_3 are real skew-symmetric matrices of rank p and q, respectively, and X_2 is a real $(p \times q)$-matrix.

The Cartan decomposition of the complex Lie algebra is

$$\mathfrak{so}(p+q,\mathbf{C}) = (\mathfrak{so}(p,\mathbf{C}) \times \mathfrak{so}(q,\mathbf{C})) \oplus \mathfrak{m}^{\mathbf{C}},$$

where

$$\mathfrak{m}^{\mathbf{C}} = \left\{ \begin{pmatrix} 0 & X_2 \\ -X_2^t & 0 \end{pmatrix} \mid X_2 \text{ complex}(p \times q) - \text{matrix} \right\}.$$

The complexified isotropy representation is the following

$$\iota : SO(p,\mathbf{C}) \times SO(q,\mathbf{C}) \to GL(\mathfrak{m}^{\mathbf{C}}),$$

where

$$\iota \begin{pmatrix} a & 0 \\ 0 & b \end{pmatrix} \begin{pmatrix} 0 & X_2 \\ -X_2^t & 0 \end{pmatrix}$$

equals

$$\begin{pmatrix} a & 0 \\ 0 & b \end{pmatrix} \begin{pmatrix} 0 & X_2 \\ -X_2^t & 0 \end{pmatrix} \begin{pmatrix} a^{-1} & 0 \\ 0 & b^{-1} \end{pmatrix},$$

which is equal to

$$\begin{pmatrix} 0 & aX_2b^{-1} \\ -bX_2^t a^{-1} & 0 \end{pmatrix} \in \mathfrak{m}^{\mathbf{C}}.$$

An $SO_0(p,q)$-**Higgs bundle** is a pair (E,φ), where E is a principal holomorphic $SO(p,\mathbf{C}) \times SO(q,\mathbf{C})$-bundle over X and the Higgs field is a holomorphic section $\varphi \in H^0(E(\mathfrak{m}^{\mathbf{C}}) \otimes K)$, where K is the canonical line bundle over X.

The principal $SO(p,\mathbf{C}) \times SO(q,\mathbf{C})$-bundle E is in fact the fibred product of two principal bundles,

$$E = E_{SO(p,\mathbf{C})} \times E_{SO(q,\mathbf{C})}.$$

Using the standard representations of $SO(p,\mathbf{C})$ and $SO(q,\mathbf{C})$ in $\mathbf{C}^p$ and $\mathbf{C}^q$ we can associate to $E_{SO(p,\mathbf{C})}$ and $E_{SO(q,\mathbf{C})}$ two holomorphic vector bundles V and W of rank p and q respectively with $\det V = \det W = \mathcal{O}$, together with two associated nondegenerate quadratic forms $Q_V \in H^0(S^2V^*)$ and $Q_W \in H^0(S^2W^*)$.

Since the forms Q_V and Q_W are nondegenerate, they induce two isomorphisms $q_V : V \cong V^*$ and $q_W : W \cong W^*$. We have

$$E(\mathfrak{m}^{\mathbf{C}}) = \{(\eta,\nu) \in \mathrm{Hom}(W,V) \oplus \mathrm{Hom}(V,W) \mid \nu = -\eta^\top\},$$

where $\eta^\top = q_W^{-1} \circ \eta^t \circ q_V$,

$$\begin{array}{ccc} V & \xrightarrow{\ \eta^\top\ } & W \\ {\scriptstyle q_V}\downarrow & & \downarrow{\scriptstyle q_W} \\ V^* & \xrightarrow{\ \eta^t\ } & W^* \end{array} \quad ,$$

i.e., $E(\mathfrak{m}^{\mathbf{C}}) \cong \mathrm{Hom}(W,V)$. Then, a $SO_0(p,q)$-Higgs bundle (E,φ) is in correspondence with a tuple (V, Q_V, W, Q_W, η).

The following are the simplified semistability and stability conditions for this kind of Higgs bundles (see [A]).

Proposition 6.1. *A* $SO_0(p,q)$-*Higgs bundle* (V, Q_V, W, Q_W, η) *is said to be **semistable** if and only if for any pair of isotropic subbundles* $V' \subset V$, $W' \subset W$ *such that* $\eta(W') \subseteq V' \otimes K$, *the inequality* $\deg V' + \deg W' \leq 0$ *holds. It is **stable** if and only if it is semistable and for any pair of isotropic subbundles* $V' \subset V$, $W' \subset W$, *at least one of them proper, such that* $\eta(W') \subseteq V' \otimes K$, *we have* $\deg V' + \deg W' < 0$.

6.4 Hitchin component for $SO_0(n,n)$

The group $SO_0(n,n)$ is the (connected) split real form of $SO(2n, \mathbf{C})$. A basis for the invariant polynomials on $\mathfrak{so}(2n, \mathbf{C})$ is provided by a modification of the coefficients of the characteristic polynomial of a skew-symmetric matrix, which is of the form

$$\det(x - A) = x^{2n} + p_1(A)x^{2n-2} + \ldots + p_n(A),$$

where $\deg p_i = 2i$. The polynomial $p_n = \det A$ of degree $2n$ is the square of the Pfaffian polynomial p'_n, which has degree n. Then, a basis is given by $p_1, \ldots, p_{n-1}, p'_n$, (the rank of $\mathfrak{so}(2n, \mathbf{C})$ is n), and the corresponding Hitchin map is

$$p : \mathcal{M}(SO(2n, \mathbf{C})) \to \bigoplus_{i=1}^{n-1} H^0(K^{2i}) \oplus H^0(K^n)$$

defined by

$$p(E, \varphi) = (p_1(\varphi), \ldots, p_{n-1}(\varphi), p'_n(\varphi)).$$

We fix the nilpotent regular element

$$e = \sum_{\alpha \in \Delta} X_\alpha \in \mathfrak{so}(2n, \mathbf{C}),$$

where

$$\Delta = \{\alpha_i = e_i - e_{i+1}(1 \leq i \leq n-1), \alpha_n = e_{n-1} + e_n\}$$

and the corresponding root vectors are

$$X_{e_i - e_j} = E_{ij} - E_{n+j, n+i} \text{ and } X_{e_i + e_j} = E_{i, n+j} - E_{j, n+i}.$$

The element x, that we can consider in the Cartan subalgebra, is of the form $x = \sum_{i=1}^{n} h_i(E_{i,i} - E_{n+i, n+i})$. Imposing $[x, e] = 2e$ we obtain

$$x = \sum_{i=1}^{n} 2(n-i)(E_{i,i} - E_{n+i, n+i}).$$

Finally, the conditions $[x, \tilde{e}] = -2\tilde{e}$ and $[e, \tilde{e}] = x$ determine $\tilde{e}$.

The adjoint action

$$\langle x, e, \widetilde{e} \rangle \cong \mathfrak{sl}(2, \mathbf{C}) \to \mathrm{End}(\mathfrak{so}(2n, \mathbf{C}))$$

gives the decomposition

$$\mathfrak{so}(2n, \mathbf{C}) = \bigoplus_{i=1}^{n-1} V_i \oplus V_n,$$

with $\dim V_i = 4i - 1$, for $1 \le i \le n - 1$ and $\dim V_n = 2n - 1$. That is, for $1 \le i \le n - 1$, $V_i = S^{4i-2}\mathbf{C}^2$ with eigenvalues $4i - 2, 4i - 4 \ldots, -4i + 4, -4i + 2$ for the action of adx and $V_n = S^{2n-2}\mathbf{C}^2$ with eigenvalues $2n - 2, 2n - 4, \ldots, -2n + 4, -2n + 2$.

The highest weight vectors in this case are $e_1, \ldots, e_{n-1}, e_n$, where e_i has eigenvalue $4i - 2$ for $1 \le i \le n - 1$, and e_n has eigenvalue $2n - 2$. We take $V_1 = \langle x, e, \widetilde{e} \rangle$ and $e = e_1$.

Given $(\alpha_1, \ldots, \alpha_{n-1}, \alpha_n) \in \bigoplus_{i=1}^{n-1} H^0(K^{2i}) \oplus H^0(K^n)$, the Higgs field in the Hitchin component is the sum $\varphi = \widetilde{e} + \alpha_1 e + \ldots + \alpha_{n-1} e_{n-1} + \alpha_n e_n$.

To find the bundle we consider the representation

$$\mathfrak{sl}(2, \mathbf{C}) \to \mathfrak{so}(2n, \mathbf{C}) = \Lambda^2(S^{2n-2}\mathbf{C}^2 + 1).$$

The vector bundle is $\mathbf{E} = (S^{2n-2} + 1)(K^{1/2} \oplus K^{-1/2}) = K^{n-1} \oplus K^{n-2} \oplus \ldots \oplus K \oplus \mathcal{O} \oplus K^{-1} \oplus \ldots \oplus K^{-n+2} \oplus K^{-n+1} \oplus \mathcal{O}_1 = V \oplus W$, where the subindex 1 is only to distinguish this trivial bundle from the other. The vector bundle $\mathbf{E}$ has the following orthogonal structure

$$Q = \left(\begin{array}{ccc|c} 0 & 0 & 1 & 0 \\ 0 & \ldots & 0 & 0 \\ 1 & 0 & 0 & 0 \\ \hline 0 & 0 & 0 & 1 \end{array} \right).$$

The field φ is given in the following order of the basis,

$$K^{n-1} \oplus K^{n-2} \oplus \ldots \oplus K \oplus \mathcal{O} \oplus K^{-n+1} \oplus K^{-n+2} \oplus \ldots \oplus K^{-1} \oplus \mathcal{O}_1$$

$$\downarrow$$

$$K^{n-1} \oplus K^{n-2} \oplus \ldots \oplus K \oplus \mathcal{O}_1 \oplus K^{-n+1} \oplus K^{-n+2} \oplus \ldots \oplus K^{-1} \oplus \mathcal{O}$$

(or the other way around, changing $\mathcal{O} \leftrightarrow \mathcal{O}_1$) because we are considering the algebra $\mathfrak{so}(2n, \mathbf{C})$ defined by the orthogonal structure $Q = \left(\begin{array}{cc} 0 & I_n \\ I_n & 0 \end{array} \right)$.

6.5 *Hitchin component for* $\mathrm{SO}_0(n, n+1)$

Consider now $\mathrm{SO}_0(n, n+1)$ which is the split real form of $\mathrm{SO}(2n+1, \mathbf{C})$. A basis for the invariant polynomials on $\mathfrak{so}(2n+1, \mathbf{C})$ is provided by the coefficients of the characteristic polynomial

$$\det(x - A) = x(x^{2n} + p_1(A)x^{2n-2} + \ldots + p_n(A)),$$

where $\deg p_i = 2i$. That is, $p_1, \ldots, p_{n-1}, p_n$, (the rank of $\mathfrak{so}(2n+1, \mathbf{C})$ is n), and the corresponding Hitchin map is

$$p : \mathcal{M}(\mathrm{SO}(2n+1, \mathbf{C})) \rightarrow \bigoplus_{i=1}^{n} H^0(K^{2i})$$

defined by

$$p(E, \varphi) = (p_1(\varphi), \ldots, p_n(\varphi)).$$

The nilpotent regular element is now

$$e = \sum_{\alpha \in \Delta} X_\alpha \in \mathfrak{so}(2n+1, \mathbf{C}),$$

where

$$\Delta = \{\alpha_i = e_i - e_{i+1}(1 \leq i \leq n-1), \alpha_n = e_n\}$$

and the corresponding root vectors are

$$X_{e_i - e_j} = E_{ij} - E_{n+j, n+i}, \qquad X_{e_i + e_j} = E_{i, n+j} - E_{j, n+i},$$
$$X_{e_i} = E_{i, 2n+1} - E_{2n+1, n+i}, \qquad X_{-e_i} = E_{n+i, 2n+1} - E_{2n+1, i}.$$

The element x, that we can consider in the Cartan subalgebra, is of the form $x = \sum_{i=1}^{n} h_i(E_{i,i} - E_{n+i, n+i})$. Imposing $[x, e] = 2e$ we obtain

$$x = \sum_{i=1}^{n} 2(n+1-i)(E_{i,i} - E_{n+i, n+i}).$$

Finally, the conditions $[x, \widetilde{e}] = -2\widetilde{e}$ and $[e, \widetilde{e}] = x$ determine $\widetilde{e}$.

The adjoint action

$$\langle x, e, \widetilde{e} \rangle \cong \mathfrak{sl}(2, \mathbf{C}) \rightarrow \mathrm{End}(\mathfrak{so}(2n+1, \mathbf{C}))$$

gives the decomposition

$$\mathfrak{so}(2n+1, \mathbf{C}) = \bigoplus_{i=1}^{n} V_i,$$

with $\dim V_i = 4i - 1$, for $1 \leq i \leq n$. That is, $V_i = S^{4i-2}\mathbf{C}^2$, $1 \leq i \leq n$, with eigenvalues $4i - 2, 4i - 4, \ldots, -4i + 4, -4i + 2$ for the action of adx.

The highest weight vectors in this case are $e_1, \ldots, e_{n-1}, e_n$, where e_i has eigenvalue $4i - 2$ for $1 \leq i \leq n$. We take $V_1 = \langle x, e, \tilde{e} \rangle$ and $e = e_1$.

Given $(\alpha_1, \ldots, \alpha_n) \in \bigoplus_{i=1}^{n} H^0(K^{2i})$, the Higgs field in the Hitchin component is the sum $\varphi = \tilde{e} + \alpha_1 e + \ldots + \alpha_n e_n$.

To find the bundle we consider the representation

$$\mathfrak{sl}(2, \mathbf{C}) \to \mathfrak{so}(2n+1, \mathbf{C}) = \Lambda^2(S^{2n}\mathbf{C}^2).$$

The vector bundle is $\mathbf{E} = V \oplus W = S^{2n}(K^{1/2} \oplus K^{-1/2}) = K^n \oplus \ldots \oplus K \oplus \mathcal{O} \oplus K^{-1} \oplus \ldots \oplus K^{-n}$.

The field φ is given in the following order of the basis, $K^n \oplus K^{n-1} \oplus \ldots \oplus K \oplus K^{-n} \oplus K^{-n+1} \oplus \ldots \oplus K^{-1} \oplus \mathcal{O}$.

6.6 Hitchin component for $\mathrm{Sp}(2n, \mathbf{R})$

The group $\mathrm{Sp}(2n, \mathbf{R})$ is the split real form of $\mathrm{Sp}(2n, \mathbf{C})$. In this case $H = \mathrm{U}(n)$, and hence $H^{\mathbf{C}} = \mathrm{GL}(n, \mathbf{C})$ and $\mathfrak{m}^{\mathbf{C}} = S^2(\mathbf{C}^n) + S^2(\mathbf{C}^{n*})$. We have that a $\mathrm{Sp}(2n, \mathbf{R})$-Higgs bundle over X is thus a triple (V, β, γ) consisting of a rank n holomorphic vector bundle V and holomorphic sections $\beta \in H^0(X, S^2V \otimes K)$ and $\gamma \in H^0(X, S^2V^* \otimes K)$. A simplified stability condition for such objects is given by Theorem 7.9.

A basis for the invariant polynomials on $\mathfrak{sp}(2n, \mathbf{C})$, which has rank n is provided by the coefficients of the characteristic polynomial

$$\det(x - A) = x^{2n} + p_1(A)x^{2n-2} + \ldots + p_n(A),$$

where $\deg p_i = 2i$. The corresponding Hitchin map is

$$p : \mathcal{M}(\mathrm{Sp}(2n, \mathbf{C})) \to \bigoplus_{i=1}^{n} H^0(K^{2i})$$

defined by

$$p(E, \varphi) = (p_1(\varphi), \ldots, p_n(\varphi)).$$

We fix the nilpotent regular element

$$e = \sum_{\alpha \in \Delta} X_\alpha \in \mathfrak{sp}(2n, \mathbf{C}),$$

where

$$\Delta = \{\alpha_i = e_i - e_{i+1}(1 \leq i \leq n-1), \alpha_n = 2e_n\}$$

and the corresponding root vectors are

$$X_{e_i - e_j} = E_{ij} - E_{n+j,n+i} \text{ and } X_{e_i} = E_{i,n+i}.$$

The element x, that we can consider in the Cartan subalgebra, is of the form $x = \sum_{i=1}^{n} h_i(E_{i,i} - E_{n+i,n+i})$. Imposing $[x, e] = 2e$ we obtain x and the conditions $[x, \tilde{e}] = -2\tilde{e}$ and $[e, \tilde{e}] = x$ determine $\tilde{e}$.

The adjoint action

$$\langle x, e, \tilde{e} \rangle \cong \mathfrak{sl}(2, \mathbf{C}) \to \text{End}(\mathfrak{sp}(2n, \mathbf{C}))$$

gives the decomposition

$$\mathfrak{sp}(2n, \mathbf{C}) = \bigoplus_{i=1}^{n} V_i,$$

with $\dim V_i = 4i - 1$, for $1 \leq i \leq n$. That is, for $1 \leq i \leq n$, $V_i = S^{4i-2}\mathbf{C}^2$ with eigenvalues $4i - 2, 4i - 4, \ldots, -4i + 4, -4i + 2$ for the action of adx.

The highest weight vectors in this case are $e_1, \ldots, e_n$, where e_i has eigenvalue $4i - 2$. We take $V_1 = \langle x, e, \tilde{e} \rangle$ and $e = e_1$.

Given $(\alpha_1, \ldots, \alpha_n) \in \bigoplus_{i=1}^{n} H^0(K^{2i})$, the Higgs field in the Hitchin component is the sum $\varphi = \tilde{e} + \alpha_1 e + \ldots + \alpha_n e_n$.

To find the bundle we consider the representation

$$\mathfrak{sl}(2, \mathbf{C}) \to \mathfrak{sp}(2n, \mathbf{C}) = S^2(S^{2n-1}\mathbf{C}^2).$$

The vector bundle is $\mathbf{E} = (S^{2n-1})(K^{1/2} \oplus K^{-1/2}) = K^{-(2n-1)/2} \oplus K^{-(2n-3)/2} \oplus \ldots \oplus K^{(2n-3)/2} \oplus K^{(2n-1)/2} = V \oplus V^*$.

7 G-Higgs bundles for classical groups of Hermitian type

7.1 G-Higgs bundles for groups of Hermitian type

We study now G-Higgs bundles for a semisimple real Lie group G where G/H is a Hermitian symmetric space. This means that G/H admits a complex structure compatible with the Riemannian structure of G/H, making G/H a Kähler manifold. If G/H is irreducible, the centre of $\mathfrak{h}$ is one-dimensional and the almost complex structure on G/H is defined by a generating element in $J \in Z(\mathfrak{h})$ (acting through the isotropy representation on $\mathfrak{m}^{\mathbf{C}}$). This complex structure defines a decomposition

$$\mathfrak{m}^{\mathbf{C}} = \mathfrak{m}_+ + \mathfrak{m}_-,$$

where $\mathfrak{m}_+$ and $\mathfrak{m}_-$ are the $(1,0)$ and the $(0,1)$ part of $\mathfrak{m}^{\mathbf{C}}$ respectively. Table 9.1 shows the main ingredients for the irreducible classical Hermitian symmetric spaces:

G	H	$H^{\mathbf{C}}$	$\mathfrak{m}^{\mathbf{C}} = \mathfrak{m}_+ + \mathfrak{m}_-$
$\mathrm{SU}(p,q)$	$S(\mathrm{U}(p) \times \mathrm{U}(q))$	$S(\mathrm{GL}(p,\mathbf{C}) \times \mathrm{GL}(q,\mathbf{C}))$	$\mathrm{Hom}(\mathbf{C}^q, \mathbf{C}^p) + \mathrm{Hom}(\mathbf{C}^p, \mathbf{C}^q)$
$\mathrm{Sp}(2n,\mathbf{R})$	$\mathrm{U}(n)$	$\mathrm{GL}(n,\mathbf{C})$	$S^2(\mathbf{C}^n) + S^2(\mathbf{C}^{n*})$
$\mathrm{SO}^*(2n)$	$\mathrm{U}(n)$	$\mathrm{GL}(n,\mathbf{C})$	$\Lambda^2(\mathbf{C}^n) + \Lambda^2(\mathbf{C}^{n*})$
$\mathrm{SO}_0(2,n)$	$\mathrm{SO}(2) \times \mathrm{SO}(n)$	$\mathrm{SO}(2,\mathbf{C}) \times \mathrm{SO}(n,\mathbf{C})$	$\mathrm{Hom}(\mathbf{C}^n, \mathbf{C}) + \mathrm{Hom}(\mathbf{C}, \mathbf{C}^n)$

Table 9.1. *Irreducible classical Hermitian symmetric spaces G/H*

Let now (E, φ) be a G-Higgs bundle over a compact Riemann surface X. The decomposition $\mathfrak{m}^{\mathbf{C}} = \mathfrak{m}_+ + \mathfrak{m}_-$ gives a vector bundle decomposition $E(\mathfrak{m}^{\mathbf{C}}) = E(\mathfrak{m}_+) \oplus E(\mathfrak{m}_-)$ and hence

$$\varphi = (\beta, \gamma) \in H^0(X, E(\mathfrak{m}_+) \otimes K) \oplus H^0(X, E(\mathfrak{m}_-) \otimes K).$$

When G is a classical group it is sometimes convenient to replace the $H^{\mathbf{C}}$-bundle E by a vector bundle associated to the standard representation of $H^{\mathbf{C}}$. For example, for $G = \mathrm{SU}(p,q)$,

$$H^{\mathbf{C}} = S(\mathrm{GL}(p,\mathbf{C}) \times \mathrm{GL}(q,\mathbf{C}))$$
$$:= \{(A,B) \in \mathrm{GL}(p,\mathbf{C}) \times \mathrm{GL}(q,\mathbf{C}) \ : \ \det B = (\det A)^{-1}\},$$

and the $H^{\mathbf{C}}$-bundle E is replaced by two holomorphic vector bundles V and W or rank p and q, respectively such that $\det W = (\det V)^{-1}$.

Via the natural inclusion $G \subset G^{\mathbf{C}} \subset \mathrm{SL}(N, \mathbf{C})$ for G in Table 9.1, to a G-Higgs bundle (E, φ) we can naturally associate an $\mathrm{SL}(N, \mathbf{C})$-Higgs bundle $(\mathcal{E}, \Phi)$, that is $\mathcal{E}$ is a holomorphic vector bundle of rank N with trivial determinant and $\Phi : \mathcal{E} \to \mathcal{E} \otimes K$ is traceless . This is a very useful correspondence that we will use below.

If G/H is an irreducible Hermitian symmetric space, then the torsion-free part of $\pi_1(H)$ is isomorphic to $\mathbf{Z}$ and hence the topological invariant of either a representation of $\pi_1(X)$ in G, or of a G-Higgs bundle, is measured by an integer $d \in \mathbf{Z}$, known as the **Toledo invariant**. In the classical cases described in Table 9.1 this coincides with the degree of a certain vector bundle. In fact, besides $G = \mathrm{SO}_0(2,n)$ with $n \geq 3$, for which $\pi_1(H) \cong \mathbf{Z} \oplus \mathbf{Z}_2$, for all the other groups in Table 9.1 $\pi_1(H) \cong \mathbf{Z}$. From the polystability of a G-Higgs bundle one can show that

$$|d| \leq \operatorname{rank}(G/H)(g-1), \tag{7.1}$$

where $\operatorname{rank}(G/H)$ is the rank of the symmetric space and g is the genus of X. This is a **Milnor–Wood type inequality** which for representations of $\pi_1(X)$ in G has been proved by Domic and Toledo [DT] and Clerk and Ørsted [CO], generalizing the classical inequality of Milnor–Wood for $G = \mathrm{SU}(1,1)$ [M]. We will denote the bound by $d_{\max} := \operatorname{rank}(G/H)(g-1)$.

There is a duality isomorphism $\mathcal{M}_d(G) \cong \mathcal{M}_{-d}(G)$ for every G in Table 9.1, and there is hence no loss of generality in considering only the case with positive Toledo invariant, which we will do from now on.

Particularly relevant is the case when the Toledo invariant d is maximal, that is $|d| = d_{\max}$. We will denote by

$$\mathcal{M}_{\max}(G) := \mathcal{M}_{d_{\max}}(G)$$

the moduli space of G-Higgs bundles with maximal Toledo invariant, and similarly $\mathcal{R}_{\max}(G) := \mathcal{R}_{d_{\max}}(G)$ for the corresponding moduli space of representations.

From many points of view maximal representations are the most interesting ones. Indeed, they have been the object of intense study in recent years, using methods from diverse branches of geometry, and it has become clear that they enjoy very special properties. In particular, at least in many cases, maximal representations have a close relationship to geometric structures on certain bundles related to the surface. The prototype of this philosophy is Goldman's theorem [Go1, Go4] that the maximal representations in $\mathrm{SL}(2,\mathbf{R}) \cong \mathrm{SU}(1,1)$ are exactly the Fuchsian ones. For example, Burger, Iozzi and Wienhard have proved [BIW1, BILW, BIW2] that maximal representations are discrete embeddings, and in fact quasi-isometric embeddings of the surfac group into the Lie group.

In the following sections we briefly review $G = \mathrm{SU}(p,q)$ and $G = \mathrm{Sp}(2n,\mathbf{R})$. For $G = \mathrm{SO}^*(2n)$ and $G = \mathrm{SO}_0(2,n)$ see [BGG2].

7.2 SU(p,q)-*Higgs bundles*

The details can be seen in [BGG1, BGG2].

Let $G = \mathrm{SU}(p,q)$. As we can see from Table 9.1, an $\mathrm{SU}(p,q)$-Higgs bundle over X is defined by a 4-tuple (V, W, β, γ) consisting of holomorphic vector bundles V and W of rank p and q, respectively, such that $\det W = (\det V)^{-1}$, and homomorphisms

$$\beta : W \longrightarrow V \otimes K \quad \text{and} \quad \gamma : V \longrightarrow W \otimes K.$$

Let

$$\mathcal{M}(p,q,d) := \mathcal{M}_d(\mathrm{SU}(p,q))$$

be the moduli space of polystable $\mathrm{SU}(p,q)$-Higgs bundles such that $\deg V = -\deg W = d$.

If (V, W, β, γ) let $(\mathcal{E}, \varPhi)$ with

$$\mathcal{E} = V \oplus W \quad \text{and} \quad \varPhi = \begin{pmatrix} 0 & \beta \\ \gamma & 0 \end{pmatrix}$$

be the associated $\mathrm{SL}(n, \mathbf{C})$-Higgs bundle with $n = p + q$. One has (see [BGG1]) that (V, W, β, γ) is stable if an only if $(\mathcal{E}, \varPhi)$ is stable. Now applying the semistability numerical criterion to special Higgs subbundles defined by the kernel and image of $\varPhi$ (see [BGG1]) we obtain

$$d \le \operatorname{rank}(\gamma)(g-1) \tag{7.2}$$

$$-d \le \operatorname{rank}(\beta)(g-1), \tag{7.3}$$

which gives the Milnor–Wood type inequality

$$|d| \le \min\{p, q\}(g-1). \tag{7.4}$$

Remark 7.1. There is a duality map

$$(V, W, \beta, \gamma) \mapsto (V^*, W^*, \gamma^t, \beta^t),$$

giving an isomorphism between $\mathcal{M}_d$ and $\mathcal{M}_{-d}$. There is hence no loss of generality in considering only the case with positive Toledo invariant, which we will do from now on.

The moduli space of representations of $\pi_1(X)$ in $\mathrm{SL}(2, \mathbf{R}) \cong \mathrm{SU}(1,1)$ was studied by Goldman [Go1, Go2]. In this case d is the Euler class of the flat $\mathrm{SL}(2, \mathbf{R})$-bundle and (7.4) for $p = q = 1$ is the Milnor inequality [M]. In [Go1] Goldman showed that for d satisfying $|d| = g - 1$ there are 2^{2g} isomorphic connected components that can be identified with Teichmüller space, and in [Go2] he showed that for d such that $|d| < g-1$ there is only one component. In [Hi1] this was also proved by Hitchin, who also gave a very explicit description of each component, which we explain now.

From the description of $\mathcal{M}(1,1,d)$ as the set of equivalence classes (L,β,γ) where L is a line bundle of degree d and

$$\beta \in H^0(X, L^2 K), \qquad \gamma \in H^0(X, L^{-2} K),$$

we deduce that $\mathcal{M}(1,1,d)$ for $0 < |d| \leq g - 1$ is the total space of a holomorphic complex vector bundle of rank $g + 2|d| - 1$ over a 2^{2g}-fold covering of the $2g - 2 - 2|d|$-symmetric power

$$\mathrm{Sym}^{2g-2-2|d|} X$$

of X. To see this, assume that $d > 0$ (the case $d < 0$ is similar), then the pair (L,γ) defines an element in $\mathrm{Sym}^{2g-2-2d}(X)$, given by the zeros of γ. But if L_0 is a line bundle of degree 0 of order two, that is, $L_0^2 = \mathcal{O}$, then the element (LL_0, γ) defines also the same divisor in $\mathrm{Sym}^{2g-2-2d}(X)$. Hence the set of pairs (L,γ) gives a point in the 2^{2g}-fold covering of $\mathrm{Sym}^{2g-2-2d}(X)$. The section β now gives the fibre of the vector bundle. Note that by Riemann-Roch, $H^1(X, L^2 K) = 0$. If $\deg L = g - 1$, the line bundle $L^{-2} \otimes K$ is of zero degree and hence has a section (unique up to multiplication by a scalar) if and only if $L^{-2} \otimes K \cong \mathcal{O}$, i.e. if L is a square root of K. For each of the 2^{2g} choices of square root $L = K^{1/2}$, one has a connected component which is parametrized by $\beta \in H^0(X, K^2)$. We thus conclude the following ([Hi1]).

Theorem 7.2. *The moduli space* $\mathcal{M}(1,1,d)$ *is a smooth complex algebraic variety of dimension* $3g - 3$, *which is connected if* $0 < |d| < g - 1$, *and has* 2^{2g} *connected components if* $|d| = g - 1$, *each isomorphic to* $\mathbf{C}^{3g-3}$.

Remark 7.3. The moduli space $\mathcal{M}(1,1,0)$ (for $d = 0$) is also connected but has singularities due to jumping phenomena.

We now study the moduli spaces $\mathcal{M}(p,q,d)$ when $p \neq 1$ and $q \neq 1$. We will consider first the case $p = q = m$. An element of $\mathcal{M}(m,m,d)$ is defined by a 4-tuple (V,W,β,γ) consisting of two holomorphic vector bundles V and W of rank m such that $\det W = (\det V)^{-1}$, and homomorphisms

$$\beta : W \longrightarrow V \otimes K \quad \text{and} \quad \gamma : V \longrightarrow W \otimes K,$$

with $d = \deg V = -\deg W$.

Assume now that the Toledo invariant d is maximal and positive, that is, $d = m(g-1)$. From (7.2) we deduce that γ must be an isomorphism. Let $\theta : W \to W \otimes K^2$ be defined as $\theta = (\gamma \otimes I_K) \circ \beta$, where $I_K : K \to K$ is the identity map.

The condition $\det W = (\det V)^{-1}$, together with the isomorphism γ imply that $(\det W)^2 \cong K^{-m}$. Now, if we choose a square root of the canonical bundle, $L_0 = K^{1/2}$, and define $\widetilde{W} = W \otimes L_0$, we have that $(\det \widetilde{W})^2 = \mathcal{O}$ and hence the structure group of $\widetilde{W}$ is the kernel of the group homomorphism $\mathrm{GL}(m, \mathbf{C}) \to \mathbf{C}^*$ given by $A \mapsto (\det A)^2$. This kernel is isomorphic to the semidirect product $\mathrm{SL}(m, \mathbf{C}) \rtimes \mathbf{Z}_2$, where $\mathbf{Z}_2 = \{\pm I\}$ and has then two connected components. The choice of a 2-torsion element in the Jacobian of X for $\det \widetilde{W}$ defines an invariant that takes 2^{2g} values.

Let $\widetilde{\theta} : \widetilde{W} \to \widetilde{W} \otimes K^2$ be defined as $\widetilde{\theta} = \theta \otimes I_{L_0}$. The map

$$(V, W, \beta, \gamma) \mapsto (\widetilde{W}, \widetilde{\theta}) \tag{7.5}$$

defines an isomorphism between $\mathcal{M}(m, m, m(g-1))$ and the moduli space of K^2-twisted pairs $(\widetilde{W}, \widetilde{\theta})$. From this and a Morse-theoretic analysis of the moduli space one deduces the following ([BGG2]).

Theorem 7.4. *If* $|d| = d_{\max} = m(g-1)$*, the moduli space* $\mathcal{M}(m, m, d)$ *has* 2^{2g} *non-empty connected components of dimension* $(4m^2 - 1)(g-1)$*.*

Let us consider now the maximal Toledo invariant case $d = d_{\max}$ when $p \neq q$. Without loss of generality we may assume that $p < q$. The maximal value of the Toledo invariant is then $d_{\max} = p(g-1)$. As shown in [BGG1, BGG2], it turns out that there are no stable $\mathrm{SU}(p, q)$-Higgs bundles with Maximal Toledo invariant. In fact, every polystable $\mathrm{SU}(p, q)$-Higgs bundle (V, W, β, γ) is strictly semistable and decomposes as a direct sum

$$(V, W, \beta, \gamma) \cong (V, W', \beta, \gamma) \oplus (0, W'', 0, 0), \tag{7.6}$$

of a maximal polystable $\mathrm{U}(p, p)$-Higgs bundle and a polystable rank $q - p$ vector bundle with degree zero, where $\gamma \colon V \xrightarrow{\cong} W' \otimes K$, with $W' = \operatorname{im} \gamma \otimes K^{-1}$ and $W'' = W/W'$. Since

$$\det(V) \otimes \det(W') \otimes \det(W'') \cong \mathcal{O},$$

this means that the $\mathrm{SU}(p, q)$-Higgs bundle reduces to an $\mathrm{S}(\mathrm{U}(p, p) \times \mathrm{U}(q - p))$-Higgs bundle

We have the exact sequence

$$1 \to \mathrm{SU}(p, p) \to \mathrm{S}(\mathrm{U}(p, p) \times \mathrm{U}(q - p)) \to \mathrm{S}(\mathrm{U}(1) \times \mathrm{U}(q - p)) \to 1$$
$$(A, B) \mapsto (\det(A), B),$$

from which we conclude the following (see [BGG1, BGG2]).

Theorem 7.5. *Let $p < q$. Then the moduli space $\mathcal{M}(p, q, d_{\max})$ is connected and fibres over the moduli space of polystable vector bundles of rank $q - p$ and zero degree, with fibre isomorphic to $\mathcal{M}(p, p, d_{\max})$. In particular, its dimension is smaller than expected.*

Remark 7.6. The rigidity phenomenon of Theorem 7.5 for the corresponding maximal Toledo invariant representations is obtained for $p = 1$ by Toledo [T] and $p = 2$ by Hernández [He], and for general groups of Hermitian type by Burger–Iozzi–Wienhard [BIW2]. That the dimension is smaller than expected was also noted by Goldman in [Go4].

For the non-maximal Toledo invariant case one has the following (see [BGG1]).

Theorem 7.7. *If $d < d_{\max}$ the moduli space $\mathcal{M}(p, q, d)$ is a non-empty, connected, complex variety of dimension $((p + q)^2 - 1)(g - 1)$.*

7.3 $\mathrm{Sp}(2n, \mathbf{R})$-*Higgs bundles*

The details can be seen in [GGM].

As we have seen in Section 6.6, a $\mathrm{Sp}(2n, \mathbf{R})$-Higgs bundle over X is thus a triple (V, β, γ) consisting of a rank n holomorphic vector bundle V and holomorphic sections $\beta \in H^0(X, S^2V \otimes K)$ and $\gamma \in H^0(X, S^2V^* \otimes K)$, where K is the canonical line bundle of X.

Proposition 7.8. *A $\mathrm{Sp}(2n, \mathbf{R})$-Higgs bundle (V, φ) is stable if, for any filtration*

$$0 \subset V_1 \subset V_2 \subset V$$

such that

$$\beta \in H^0(K \otimes (S^2V_2 + V_1 \otimes_S V)), \quad \gamma \in H^0(K \otimes (S^2V_1^\perp + V_2^\perp \otimes_S V^*)), \tag{7.7}$$

the following holds: if at least one of the subbundles V_1 and V_2 is proper, then the inequality

$$\deg(V) - \deg(V_1) - \deg(V_2) > 0 \tag{7.8}$$

holds and, in any other case,

$$\deg(V) - \deg(V_1) - \deg(V_2) \geq 0. \tag{7.9}$$

The condition for (V, φ) to be semistable is obtained by omitting the strict inequality (7.8).

Since $\mathrm{Sp}(2n, \mathbf{C}) \subset \mathrm{SL}(2n, \mathbf{C})$, every $\mathrm{Sp}(2n, \mathbf{C})$-Higgs bundle $((\mathcal{E}, \Omega), \Phi)$ gives rise to an $\mathrm{SL}(2n, \mathbf{C})$-Higgs bundle $(\mathcal{E}, \Phi)$. If $((\mathcal{E}, \Omega), \Phi)$ is obtained from an $\mathrm{Sp}(2n, \mathbf{R})$-Higgs bundle (V, φ) we denote the associated $\mathrm{SL}(2n, \mathbf{C})$-Higgs bundle by

$$(\mathcal{E}, \Phi) = \left(V \oplus V^*, \begin{pmatrix} 0 & \beta \\ \gamma & 0 \end{pmatrix} \right).$$

Theorem 7.9. *Let $(V, \varphi = (\beta, \gamma))$ be a $\mathrm{Sp}(2n, \mathbf{R})$-Higgs bundle and let $(\mathcal{E}, \Phi)$ be the corresponding $\mathrm{SL}(2n, \mathbf{C})$-Higgs bundle. Then*

  (i) *if $(\mathcal{E}, \Phi)$ is stable then (V, φ) is stable;*

  (ii) *if (V, φ) is stable and simple then $(\mathcal{E}, \Phi)$ is stable unless there is an isomorphism $f : V \xrightarrow{\simeq} V^*$ such that $\beta f = f^{-1}\gamma$, in which case $(\mathcal{E}, \Phi)$ is polystable;*

 (iii) *$(\mathcal{E}, \Phi)$ is semistable if and only if (V, φ) is semistable.*

 (iv) *$(\mathcal{E}, \Phi)$ is polystable if and only if (V, φ) is polystable;*

In particular, if $\deg V \neq 0$ then $(\mathcal{E}, \Phi)$ is stable if and only if (V, φ) is stable.

The topological invariant attached to a $\mathrm{Sp}(2n, \mathbf{R})$-Higgs bundle (V, φ) is an element in the fundamental group of $\mathrm{U}(n)$. Since $\pi_1(U(n)) \cong \mathbf{Z}$, this is an integer, which coincides with the degree of V. From the inclusion $\mathrm{Sp}(2n, \mathbf{R}) \subset \mathrm{SU}(n, n)$ and 7.2, we have the following.

Proposition 7.10. *Let (V, β, γ) be a semistable $\mathrm{Sp}(2n, \mathbf{R})$-Higgs bundle and let $d = \deg V$. Then*

$$
\begin{aligned}
d &\leq \mathrm{rank}(\gamma)(g - 1) \\
-d &\leq \mathrm{rank}(\beta)(g - 1).
\end{aligned}
\tag{7.10}
$$

As a consequence of Proposition 7.10 we have the following.

Proposition 7.11. *Let (V, β, γ) be a semistable $\mathrm{Sp}(2n, \mathbf{R})$-Higgs bundle and let $d = \deg(V)$. Then*

$$|d| \leq n(g - 1).$$

Furthermore,

(1) *$d = n(g-1)$ holds if and only if $\gamma \colon V \to V^* \otimes K$ is an isomorphism;*

(2) *$d = -n(g-1)$ holds if and only if $\beta \colon V^* \to V \otimes K$ is an isomorphism.*

We will denote by

$$\mathcal{M}(n,d) := \mathcal{M}_d(\mathrm{Sp}(2n,\mathbf{R}))$$

the moduli space of $\mathrm{Sp}(2n,\mathbf{R})$-Higgs bundles (V,β,γ) with $\deg V = d$. One has the following immediate duality result.

Proposition 7.12. *The map*

$$(V,\beta,\gamma) \mapsto (V^*,\gamma^t,\beta^t)$$

gives an isomorphism $\mathcal{M}(n,d) \cong \mathcal{M}(n,-d)$.

As a corollary of Proposition 7.11, we obtain the following.

Proposition 7.13. *The moduli space* $\mathcal{M}(n,d)$ *is empty unless*

$$|d| \leq n(g-1).$$

We shall now describe the $\mathrm{Sp}(2n,\mathbf{R})$ moduli space for the extreme value $|d| = n(g-1)$. In fact, for the rest of this section we shall assume that $d = n(g-1)$. This involves no loss of generality, since, by Proposition 7.12, $(V,\varphi) \mapsto (V^*,\varphi^t)$ gives an isomorphism between the $\mathrm{Sp}(2n,\mathbf{R})$ moduli spaces for d and $-d$. The main result is Theorem 7.17, which we refer to as the *Cayley correspondence*.

When γ is an isomorphism, the stability condition for $\mathrm{Sp}(2n,\mathbf{R})$-Higgs bundles, given by Proposition 7.8 simplifies further (see [GGM]). Here is a key observation:

Proposition 7.14. *Let* (V,γ,β) *be a* $\mathrm{Sp}(2n,\mathbf{R})$-*Higgs bundle and assume that* $\gamma\colon V \to V^* \otimes K$ *is an isomorphism. If* $0 \subseteq V_1 \subseteq V_2 \subseteq V$ *is a filtration such that* $\gamma \in H^0(K \otimes (S^2 V_1^\perp + V_2^\perp \otimes_S V^*))$, *then* $V_2 = V_1^{\perp_\gamma}$.

Proposition 7.15. *Let* (V,β,γ) *be a* $\mathrm{Sp}(2n,\mathbf{R})$-*Higgs bundle and assume that* $\gamma\colon V \to V^* \otimes K$ *is an isomorphism. Let* $\widetilde{\beta} = (\beta \otimes 1) \circ \gamma\colon V \to V \otimes K^2$. *Then* (V,β,γ) *is stable if and only if for any* $V_1 \subset V$ *such that* $V_1 \subseteq V_1^{\perp_\gamma}$ *(i.e.,* V_1 *is isotropic with respect to* γ*) and* $\widetilde{\beta}(V_1) \subseteq V_1 \otimes K^2$, *the condition*

$$\mu(V_1) < g-1$$

is satisfied.

Let (V,β,γ) be a $\mathrm{Sp}(2n,\mathbf{R})$-Higgs bundle with $d = n(g-1)$ such that $\gamma \in H^0(K \otimes S^2 V^*)$ is an isomorphism. Let $L_0 = K^{1/2}$ be a fixed square root of K, and define $W = V^* \otimes L_0$. Then $Q := \gamma \otimes I_{L_0^{-1}} : W^* \to W$ is a symmetric isomorphism defining an orthogonal structure on W, in other words, (W, Q) is an $O(n, \mathbf{C})$-holomorphic bundle. The K^2-twisted endomorphism $\psi : W \to W \otimes K^2$ defined by $\psi = (\gamma \otimes I_{K \otimes L_0}) \circ \beta \otimes I_{L_0}$ is Q-symmetric and hence (W, Q, ψ) defines a K^2-twisted $\mathrm{GL}(n, \mathbf{R})$-Higgs pair, from which we can recover the original $\mathrm{Sp}(2n, \mathbf{R})$-Higgs bundle.

Theorem 7.16. *Let (V, β, γ) be a $\mathrm{Sp}(2n, \mathbf{R})$-Higgs bundle with $d = n(g - 1)$ such that γ is an isomorphism. Let (W, Q, ψ) be the corresponding K^2-twisted $\mathrm{GL}(n, \mathbf{R})$-Higgs pair. Then (V, β, γ) is semistable (resp. stable, polystable) if and only if (W, Q, ψ) is semistable (resp. stable, polystable).*

This follows from the simplified stability conditions (see [GGM]) and Proposition 7.15, using the translation $W_1 = V_1^* \otimes L_0$. Similarly for semistability and polystability.

Theorem 7.17. *Let $\mathcal{M}_{\max}$ be the moduli space of polystable $\mathrm{Sp}(2n, \mathbf{R})$-Higgs bundles with $d = n(g - 1)$ and let $\mathcal{M}'$ be the moduli space of polystable K^2-twisted $\mathrm{GL}(n, \mathbf{R})$-Higgs pairs. The map $(V, \beta, \gamma) \mapsto (W, Q, \psi)$ defines an isomorphism of complex algebraic varieties*

$$\mathcal{M}_{\max} \cong \mathcal{M}'.$$

Proof Let (V, β, γ) be a semistable $\mathrm{Sp}(2n, \mathbf{R})$-Higgs bundle with $d = n(g - 1)$. By Proposition 7.11, γ is an isomorphism and hence the map $(V, \beta, \gamma) \mapsto (W, Q, \psi)$ is well defined. The result follows now from Theorem 7.16 and the existence of local universal families. $\qquad\square$

One has the following results on the number of connected components.

Theorem 7.18 (Gothen [Got]). *The moduli space $\mathcal{M}(2, 0)$ is non-empty and connected, and $\mathcal{M}(2, \pm(2g - 2))$ has $3.2^{2g} + 2g - 4$ non-empty connected components.*

Theorem 7.19 ([GM]). *Let d be any integer satisfying $0 < |d| < 2g - 2$. The moduli space $\mathcal{M}(2, d)$ is connected and the subspace of irreducible representations is non-empty.*

Theorem 7.20 ([GGM]). *Let X be a compact Riemann surface of genus g. Let $\mathcal{M}(n, d)$ be the moduli space of polystable $\mathrm{Sp}(2n, \mathbf{R})$-Higgs bundles of degree d. Let $n \geq 3$. Then*

(i) *$\mathcal{M}(n, 0)$ is non-empty and connected;*
(ii) *$\mathcal{M}(n, \pm n(g - 1))$ has 3.2^{2g} non-empty connected components.*

8 Involutions in $\mathcal{M}(G)$ and real forms

8.1 Involution $(E, \varphi) \mapsto (E, -\varphi)$ in $\mathcal{M}(\mathrm{SL}(n, \mathbf{C}))$

Consider the involution $\iota : (E, \varphi) \mapsto (E, -\varphi)$ in the moduli space $\mathcal{M} := \mathcal{M}(\mathrm{SL}(n, \mathbf{C}))$ of $\mathrm{SL}(n, \mathbf{C})$-Higgs bundles. In this section we study the fixed points of ι. These are of two types.

Type 1: $(E, \varphi) \in \mathcal{M}$ such that $\varphi = 0$. This subvariety is isomorphic to the moduli space of polystable vector bundles of rank n and trivial determinant.

Type 2: $(E, \varphi) \in \mathcal{M}$ such that $E = V \oplus W$, where V and W are holomorphic vector bundles of rank $p \neq 0$ and $q \neq 0$, respectively, where $p + q = n$ and $\det W = (\det V)^{-1}$, and

$$\varphi = \begin{pmatrix} 0 & \beta \\ \gamma & 0 \end{pmatrix}, \tag{8.1}$$

where $\beta : W \to V \otimes K$ and $\gamma : V \to W \otimes K$. It is clear that for such Higgs bundles $(E, \varphi) \cong (E, -\varphi)$ since the gauge transformation $\begin{pmatrix} iI_p & 0 \\ 0 & -iI_q \end{pmatrix}$ brings (E, φ) to $(E, -\varphi)$. To see that these are the only fixed points for ι, suppose for simplicity that (E, φ) is stable (the polystable case can be derived from this since (E, φ) is a direct sum of stable Higgs bundles).

If $(E, \varphi) \cong (E, -\varphi)$ then there exists an automorphism $f : E \to E$ such that the diagram

$$\begin{array}{ccc} E & \xrightarrow{\varphi} & E \otimes K \\ {\scriptstyle f}\downarrow & & \downarrow{\scriptstyle f \otimes I_K} \\ E & \xrightarrow{-\varphi} & E \otimes K \end{array} \tag{8.2}$$

is commutative. Now, f^2 is an automorphism of (E, φ) which, since the Higgs bundle is stable it is simple and hence we must have $f^2 = \lambda I$ for λ in the centre of $\mathrm{SL}(n, \mathbf{C})$, i.e. an nth root of unity. At every point $x \in X$, the automorphism f_x of E_x has eigenvalues $\pm\sqrt{\lambda}$, and hence E decomposes as the sum of the eigenbundles V and W corresponding to $+\sqrt{\lambda}$ and $-\sqrt{\lambda}$, respectively. From the commutativity of (8.2) we deduce that φ must be of the form (8.1). So, we see how the assumption that $\varphi \neq 0$ implies the existence of the gauge transformation f and that φ now appears as an eigenvector for f.

We thus conclude that the fixed points of ι coincide with the moduli space of $\mathrm{SL}(n, \mathbf{C})$-bundles — or equivalently $\mathrm{SU}(n)$-Higgs bundles — (type 1) and the image of the moduli spaces $\mathcal{M}(p, q, d)$ of

$SU(p, q)$-Higgs bundles under the natural inclusion of these in the moduli space of $SL(p + q, \mathbf{C})$-Higgs bundles (type 2).

8.2 Involution $(E, \varphi) \mapsto (E^*, \varphi^t)$ in $\mathcal{M}(SL(n, \mathbf{C}))$

Now, $SU(n)$ and $SU(p, q)$ with $p + q = n$ are real forms of $SL(n, \mathbf{C})$, but for $n > 2$ there are other real forms of $SL(n, \mathbf{C})$ that do not emerge from the involution ι. Namely, there is the split real form $SL(n, \mathbf{R})$, and when n is even, say $n = 2m$, there is the real form $SU^*(2m)$ (see [H, S]). These two real forms do not arise from ι because the corresponding conjugations σ do not arise from an involution $\sigma \circ \tau$ which is inner. Indeed, the involution

$$\mu(A) = -A^t \tag{8.3}$$

is a nontrivial outer automorphism such that the conjugation $\mu \circ \tau$ defines the real form $\mathfrak{sl}(n, \mathbf{R})$. Concretely (see [H, S]), the conjugation for $\mathfrak{sl}(n, \mathbf{R})$ is given by $\sigma(A) = \overline{A}$, and that of $\mathfrak{su}^*(2m)$ is $\sigma_*(A) = J_m \overline{A} J_m^{-1}$, where

$$J_m = \begin{pmatrix} 0 & I_m \\ -I_m & 0 \end{pmatrix}.$$

In the Higgs bundle moduli space, these two conjugations induce the holomorphic involution

$$\iota_* : \mathcal{M}(SL(n, \mathbf{C})) \to \mathcal{M}(SL(n, \mathbf{C}))$$
$$(E, \varphi) \mapsto (E^*, \varphi^t).$$

If (E, φ) is a fixed point of ι_*, then there exists an isomorphism $f : E \to E^*$ such that the diagram

$$\begin{array}{ccc} E & \xrightarrow{\varphi} & E \otimes K \\ \downarrow{f} & & \downarrow{f \otimes I_K} \\ E^* & \xrightarrow{\varphi^t} & E^* \otimes K \end{array} \tag{8.4}$$

commutes. It is then clear that $(f^{-1})^t \circ f$ is an automorphism of (E, φ) which by stability (the polystable case reduces to the stable one — details are given in [GR]) must satisfy

$$(f^{-1})^t \circ f = \lambda I \tag{8.5}$$

for an nth root of unity λ. From (8.5) we have that $f = \lambda f^t$. On the other hand, transposing (8.5), we have that $f^t \circ f^{-1} = \lambda I$ and hence $f = \lambda^{-1} f^t$. From this, we deduce that $\lambda^2 = 1$, thus having $\lambda = \pm 1$.

If $\lambda = 1$, f defines on E a nondegenerate quadratic form. In this case, the commutativity of (8.4) indicates that the Higgs bundle (E, φ) corresponds to a representation of $\pi_1(X)$ in $\mathrm{SL}(n, \mathbf{R})$ (see Section 6.2). In [Hi3] Hitchin counted the number of connected components of the moduli space of representations of $\pi_1(X)$ in $\mathrm{SL}(n, \mathbf{R})$ and showed (as mentioned above) that there is one component which is homeomorphic to $\mathbf{C}^{(n^2-1)(g-1)}$, that he calls Teichmüller component, since it contains the space of discrete embeddings of $\pi_1(X)$ in $\mathrm{SL}(2, \mathbf{R})$, which by the uniformization theorem, identifies with the Teichmüller space of marked Riemann surfaces homeomorphic to X.

If $\lambda = -1$, f equips E with a symplectic structure. This can only happen of course if $n = 2m$. Again, the commutativity of (8.4) says that the Higgs bundle (E, φ) corresponds to a representation of $\pi_1(X)$ in $\mathrm{SU}^*(2m)$. The number of connected components in this case is unknown, although one may conjecture that it is one since $\mathrm{SU}^*(2m)$ is simply connected.

Remark 8.1. Note that for $n = 2$, the conjugations τ and σ are conjugate to each other and hence $(E, \varphi) \cong (E^*, -\varphi^t)$ and thus $\iota = \iota_*$ (of course the group of outer automorphisms of $\mathrm{SL}(2, \mathbf{C})$ is trivial).

We conclude then that the fixed points subvariety of ι_* coincides with the image of $\mathcal{M}(\mathrm{SL}(n, \mathbf{R}))$ in $\mathcal{M}(\mathrm{SL}(n, \mathbf{C}))$, and if $n = 2m$ includes also the image of $\mathcal{M}(\mathrm{SU}^*(2m))$.

8.3 Involutions in $\mathcal{M}(G)$

In [GR] we extend the analysis above to arbitrary semisimple Lie groups. Let G be a complex connected semisimple Lie group. The group $Int(G)$ of inner automorphisms of G is a normal subgroup of the group $\mathrm{Aut}\,(G)$ of automorphisms of G. The quotient defines the group $Out(G)$ of outer automorphisms of G, giving an exact sequence of groups

$$1 \longrightarrow Int(G) \longrightarrow \mathrm{Aut}\,(G) \longrightarrow Out(G) \longrightarrow 1. \qquad (8.6)$$

The group $\mathrm{Aut}\,(G)$ acts on the moduli space of G-Higgs bundles, with $Int(G)$ acting trivially, and hence defining an action of $Out(G)$. Let $Out_2(G) \subset Out(G)$ be the set of elements of $Out(G)$ of order 2. Let

$a \in Out_2(G)$. We consider the involution

$$\iota_a : \mathcal{M}(G) \to \mathcal{M}(G)$$
$$(E, \varphi) \mapsto (\theta(E), -\theta(\varphi)), \tag{8.7}$$

where $\theta \in \text{Aut}\,(G)$ is a lift of a and $\theta(E)$ is the G-bundle associated to E via the automorphism $\theta : G \to G$. In [GR] we identify the fixed points of ι_a for any $a \in Out_2(G)$ when G is a connected and simply connected semisimple complex Lie group.

Theorem 8.2. *Let G be a connected and simply connected semisimple complex Lie group. Let X be a compact Riemann surface and let $\mathcal{M}(G)$ be the moduli space of polystable G-Higgs bundles. Let $a \in Out_2(G)$. The fixed point set $\mathcal{M}(G)^a$ of the involution ι_a is given by*

$$\mathcal{M}(G)^a = \bigsqcup_{[\theta] \in \pi^{-1}(a)} \text{im}(\mathcal{M}(G_\theta) \to \mathcal{M}(G)),$$

where $\pi : \text{Aut}_2(G)/\!\sim_i \to Out_2(G)$, and $\theta \sim_i \theta'$ if $\theta' = f\theta f^{-1}$ with $f = Int(g) \in Int(G)$. Here G_θ is the real form of G defined by the Cartan involution θ.

To explain this result from the point of view of $\mathcal{R}(G)$, we have to recall (see [H]) the relation between Cartan involutions and conjugations of $\mathfrak{g}$. We can choose the conjugation $\tau : \mathfrak{g} \to \mathfrak{g}$ with respect to the compact form of $\mathfrak{g}$ in such a way that it commutes with the conjugation $\sigma : \mathfrak{g} \to \mathfrak{g}$ of any other real form. The complexification of the Cartan involution of the real form $(\mathfrak{g}, \sigma)$ is then $\theta = \tau\sigma$.

Two conjugations σ and σ' of $\mathfrak{g}$ are inner equivalent if there is an element $f \in Int(\mathfrak{g})$ such that $\sigma' = f\sigma f^{-1}$. The map $\sigma \mapsto \theta = \tau\sigma$ induces a bijection between the set $\mathscr{C}$ of equivalence classes of conjugations under inner equivalence and the set $Out_2(\mathfrak{g})$ of elements of order 2 in $Out(\mathfrak{g})$.

Clearly, two elements σ and σ' defining the same class $c \in \mathscr{C}$ induce the same antiholomorphic involution of $\mathcal{R}(G)$ that we will denote by $\iota_c : \mathcal{R}(G) \to \mathcal{R}(G)$. Let $a \in Out_2(\mathfrak{g})$ be element corresponding to $c \in \mathscr{C}$. We have the following.

Proposition 8.3. *There is a commutative diagram*

$$\begin{array}{ccc} \mathcal{M}(G) & \xrightarrow{\ \iota_a\ } & \mathcal{M}(G) \\ \cong \Big\downarrow & & \Big\downarrow \cong \\ \mathcal{R}(G) & \xrightarrow{\ \iota_c\ } & \mathcal{R}(G) \end{array} \tag{8.8}$$

where the vertical arrows correspond to the homeomorphism between $\mathcal{M}(G)$ *and* $\mathcal{R}(G)$.

As a corollary of Proposition 8.3, Theorem 8.2 and Theorem 3.2 we have the following.

Theorem 8.4. *Let G be a connected and simply connected semisimple complex Lie group. Let X be a compact Riemann surface and let $\mathcal{M}(G)$ be the moduli space of polystable G-Higgs bundles. Let $c \in \mathscr{C}$. The fixed point set $\mathcal{R}(G)^c$ of the involution ι_c is given by*

$$\mathcal{R}(G)^c = \bigsqcup_{[\sigma] \in \pi^{-1}(c)} \mathrm{im}(\mathcal{R}(G_\sigma) \to \mathcal{R}(G)),$$

where $\pi : \{Conjugations\} \to \mathscr{C}$.

Acknowledgements. Author partially supported by Ministerio de Educación y Ciencia (Spain) through Project MTM2007-67623.

Bibliography

[A] M. Aparicio, SO(p,q)-Higgs bundles, PhD Thesis, in preparation.

[AB] M. F. Atiyah and R. Bott, *The Yang-Mills equations over Riemann surfaces*, Philos. Trans. Roy. Soc. London Ser. A **308** (1982), 523–615.

[BR] I. Biswas and S. Ramanan, *An infinitesimal study of the moduli of Hitchin pairs*, J. London Math. Soc. (2) **49** (1994), 219–231.

[BGG1] S.B. Bradlow, O. García-Prada and P.B. Gothen, *Surface group representations and* U(p,q)-*Higgs bundles*, J. Differential Geom. **64** (2003), 111–170.

[BGG2] _____, *Maximal surface group representations in isometry groups of classical hermitian symmetric spaces*, Geometriae Dedicata **122** (2006), 185–213.

[BGG3] _____, *Homotopy groups of moduli spaces of representations*, Topology **47** (2008), 203–224.

[BGM] S.B. Bradlow, O. García-Prada and I. Mundet i Riera *Relative Hitchin-Kobayashi correspondences for principal pairs*, Quart. J. Math. **54** (2003), 171–208.

[BIW1] M. Burger, A. Iozzi, and A. Wienhard, *Surface group representations with maximal Toledo invariant*, C. R. Math. Acad. Sci. Paris **336** (2003), no. 5, 387–390.

[BIW2] _____, *Surface group representations with maximal Toledo invariant*, Ann. Math., to appear.

[BILW] M. Burger, A. Iozzi, F. Labourie, and A. Wienhard, *Maximal representations of surface groups: symplectic Anosov structures*, Pure Appl. Math. Q. **1** (2005), no. 3, 543–590.

[CG] S. Choi and W. M. Goldman, *Convex real projective structures on closed surfaces are closed*, Proc. Amer. Math. Soc. **118** (1993), 657–661.

[CO] J.L. Clerk and B. Ørsted, *The Gromov norm of the Kähler class and the Maslov index*, Asian J. Math. **7** (2003), no. 2, 269–295.

[C] K. Corlette, *Flat G-bundles with canonical metrics*, J. Differential Geom. **28** (1988), 361–382.

[DT] A. Domic and D. Toledo, *The Gromov norm of the Kaehler class of symmetric domains*, Math. Ann. **276** (1987), 425–432.

[D] S. K. Donaldson, *Twisted harmonic maps and the self-duality equations*, Proc. London Math. Soc. (3) **55** (1987), 127–131.

[DK] S.K. Donaldson and P.B. Kronheimer, *The Geometry of Four Manifolds*, Oxford Mathematical Monographs, Oxford University Press, 1990.

[FG] V. V. Fock and A. B. Goncharov, *Moduli spaces of local systems and higher Teichmuller theory*, Publ. Math. Inst. Hautes Études Sci. **103** (2006), 1–211.

[G] O. García-Prada, *Moduli spaces and geometric structures*, Appendix in: R.O. Wells, *Differential Analysis on Complex manifolds*, GTM 65, Springer, 3rd edition, 2007.

[GGM] O. García-Prada, P.B. Gothen and I. Mundet i Riera *Representations of surface groups in the real symplectic group*, preprint 2008.

[GM] O. García-Prada and I. Mundet i Riera, *Representations of the fundamental group of a closed oriented surface in* Sp(4, **R**), Topology **43** (2004), 831–855.

[GR] O. García-Prada and S. Ramanan, *Involutions of the moduli space of Higgs bundles*, in preparation.

[Go1] W. M. Goldman, *Discontinuous groups and the Euler class*, Ph.D. thesis, University of California, Berkeley, 1980.

[Go2] _____, *Topological components of spaces of representations*, Invent. Math. **93** (1988), 557–607.

[Go3] _____, *The symplectic nature of fundamental groups of surfaces*, Adv. Math. **54** (1984), No. 2, 200–225.

[Go4] _____, *Representations of fundamental groups of surfaces*, Springer LNM 1167, 1985, pp. 95–117.

[Got] P. B. Gothen, *Components of spaces of representations and stable triples*, Topology **40** (2001), 823–850.

[H] S. Helgason, *Differential geometry, Lie groups, and symmetric spaces*, Mathematics, vol. 80, Academic Press, San Diego, 1998.

[He] L. Hernández, *Maximal representations of surface groups in bounded symmetric domains*, Trans. Amer. Math. Soc. **324** (1991), 405–420.

[Hi1] N. J. Hitchin, *The self-duality equations on a Riemann surface*, Proc. London Math. Soc. (3) **55** (1987), 59–126.

[Hi2] N. J. Hitchin, *Stable bundles and integrable systems*, Duke Math. J. **54** (1987), 91–114.

[Hi3] _____, *Lie groups and Teichmüller space*, Topology **31** (1992), 449–473.

[HKLR] N.J. Hitchin, A. Karlhede, U. Lindström, and M. Roçek, *Hyperkähler metrics and supersymmetry*, *Comm. Math. Phys.* **108** (1987), 535–589.

[K] S. Kobayashi, *Differential Geometry of Complex Vector Bundles*, Princeton University Press, 1987.

[L] _____, *Anosov flows, surface groups and curves in projective space*, Invent. Math. **165** (2006), no. 1, 51–114.

[MW] J. Marsden and A.D. Weinstein, Reduction of symplectic manifolds with symmetry, *Reports on Math. Physics* **5** (1974), 121–130.

[M] J. Milnor, *On the existence of a connection with curvature zero* Comment. Math. Helv. **32** (1958) 215–223.

[MFK] D. Mumford, J. Fogarty and F. Kirwan, *Geometric Invariant Theory*, 3rd edition, Springer, 1994.

[NS] M. S. Narasimhan and C. S. Seshadri, *Stable and unitary vector bundles on a compact Riemann surface*, Ann. Math. **82** (1965), 540–567.

[R1] A. Ramanathan, *Stable principal bundles on a compact Riemann surface*, Math. Ann. **213** (1975), 129–152.

[R2] _____, *Moduli for principal bundles over algebraic curves: I and II*, Proc. Indian Acad. Sci. Math. Sci. **106** (1996), 301–328 and 421–449.

[S] H. Samelson, *Notes on Lie Algebras*, Universitext (1990) Springer-Verlag, New York Inc.

[Sch1] A. H. W. Schmitt, *Moduli for decorated tuples for sheaves and representation spaces for quivers*, Proc. Indian Acad. Sci. Math. Sci. **115** (2005), 15–49.

[Sch2] _____, *Geometric invariant theory and decorated principal bundles*, Zürich Lectures in Advanced Mathematics, European Mathematical Society, 2008.

[Si1] C. T. Simpson, *Constructing variations of Hodge structure using Yang-Mills theory and applications to uniformization*, J. Amer. Math. Soc. **1** (1988), 867–918.

[Si2] _____, *Higgs bundles and local systems*, Inst. Hautes Études Sci. Publ. Math. **75** (1992), 5–95.

[Si3] _____, *Moduli of representations of the fundamental group of a smooth projective variety I*, Publ. Math., Inst. Hautes Étud. Sci. **79** (1994), 47–129.

[Si4] _____, *Moduli of representations of the fundamental group of a smooth projective variety II*, Publ. Math., Inst. Hautes Étud. Sci. **80** (1995), 5–79.

[T] D. Toledo, *Representations of surface groups in complex hyperbolic space*, J. Differential Geom. **29** (1989), 125–133.

[W] R.O. Wells Jr. *Differential Analysis on Complex Manifolds*, GTM 65, Springer, 3rd edition (2007).

10

Quotients by non-reductive algebraic group actions

Frances Kirwan

Mathematical Institute,
24-29 St Giles, Oxford OX1 3LB, UK
kirwan@maths.ox.ac.uk

This paper is dedicated to Peter Newstead, from whose Tata Institute Lecture Notes [Ne] I learnt about GIT and moduli spaces some decades ago, with much appreciation for all his help and support over the years since then.

1 Introduction

Geometric invariant theory (GIT) was developed in the 1960s by Mumford in order to construct quotients of reductive group actions on algebraic varieties and hence to construct and study a number of moduli spaces, including, for example, moduli spaces of bundles over a nonsingular projective curve [MFK, Ne, Ne2]. Moduli spaces often arise naturally as quotients of varieties by algebraic group actions, but the groups involved are not always reductive. For example, in the case of moduli spaces of hypersurfaces (or, more generally, complete intersections) in toric varieties (or, more generally, spherical varieties), the group actions which arise naturally are actions of the automorphism groups of the varieties [Co, CK]. These automorphism groups are not in general reductive, and when they are not reductive we cannot use classical GIT to construct (projective completions of) such moduli spaces as quotients for these actions.

In [DK1] (following earlier work including [Fa, Fa2, GP, GP2, W] and references therein) a study was made of ways in which GIT might be generalised to non-reductive group actions; some more recent developments and applications can be found in [AD, AD2]. Since every affine algebraic group H has a unipotent radical $U \trianglelefteq H$ such that H/U is reductive, [DK1] concentrates on unipotent actions. It is shown that when a unipotent group U acts linearly (with respect to an ample line bundle $\mathcal{L}$) on a complex projective variety X, then X has invariant open subsets $X^s \subseteq X^{ss}$, consisting of the 'stable' and 'semistable' points for the action, such that X^s has a geometric quotient X^s/U and X^{ss} has a canonical 'enveloping quotient' $X^{ss} \to X/\!/U$ which restricts to $X^s \to X^s/U$ where X^s/U is an open subset of $X/\!/U$. (When it is necessary to distinguish between stability and semistability for different group actions on X we shall denote X^s and X^{ss} by $X^{s,U}$ and $X^{ss,U}$.) However, in contrast to the reductive case, the natural morphism $X^{ss} \to X/\!/U$ is not necessarily surjective; indeed its image is not necessarily a subvariety of $X/\!/U$, so we do not in general obtain a categorical quotient of X^{ss}. Moreover $X/\!/U$ is in general only quasi-projective, not projective, though when the ring of invariants $\hat{\mathcal{O}}_L(X)^U = \bigoplus_{k\geq0} H^0(X, L^{\otimes k})^U$ is finitely generated as a **C**-algebra then $X/\!/U$ is the projective variety Proj $(\hat{\mathcal{O}}_L(X)^U)$.

In order to construct a projective completion of the enveloping quotient $X/\!/U$ when the ring of invariants $\hat{\mathcal{O}}_L(X)^U$ is not finitely generated, and to understand its geometry, it is convenient to transfer the problem of constructing a quotient for the U-action to the construction of a quotient for an action of a reductive group G which contains U as a subgroup, by finding a 'reductive envelope'. This is a projective completion

$$\overline{G \times_U X}$$

of the quasi-projective variety $G \times_U X$ (that is, the quotient of $G \times X$ by the free action of U acting diagonally on the left on X and by right multiplication on G), with a linear G-action on $\overline{G \times_U X}$ which restricts to the induced G-action on $G \times_U X$, such that 'sufficiently many' U-invariants on X extend to G-invariants on $\overline{G \times_U X}$. If (as is always possible) we choose the linearisation on $\overline{G \times_U X}$ to be ample (or more generally to be 'fine'), then the classical GIT quotient

$$\overline{G \times_U X}/\!/G$$

is a (not necessarily canonical) projective completion $\overline{X/\!/U}$ of $X/\!/U$, and hence also of its open subset X^s/U if $X^s \neq \emptyset$, and we have

$$X^{\bar{s}} \subseteq X^s \subseteq X^{ss} \subseteq X^{\bar{s}s}$$

where $X^{\bar{s}}$ (respectively $X^{\bar{s}s}$) denotes the open subset of X consisting of points of X which are stable (respectively semistable) for the G-action on $\overline{G \times_U X}$ under the inclusion

$$X \hookrightarrow G \times_U X \hookrightarrow \overline{G \times_U X}.$$

In principle, at least, we can apply the methods of classical GIT to study the geometry of this projective completion $\overline{X/\!/U}$ in terms of the geometry of the G-action on $\overline{G \times_U X}$; for example, techniques from symplectic geometry can be used to study the topology of GIT quotients of complex projective varieties by complex reductive group actions [AB, JK, JKKW, Ki, Ki2, Ki3, Ki4].

Given a linear H-action on X where H is an affine algebraic group with unipotent radical U, if an ample reductive envelope $\overline{G \times_U X}$ is chosen in a sufficiently canonical way that the GIT quotient $\overline{G \times_U X}/\!/G = \overline{X/\!/U}$ inherits an induced linear action of the reductive group $R = H/U$, then

$$(\overline{X/\!/U})/\!/R$$

is a projective completion of the geometric quotient

$$X^{s,H}/H = (X^{s,H}/U)/R$$

where $X^{s,H}$ is the inverse image in $X^{s,U}$ of the open set of points in $X^{s,U}/U \subseteq X/\!/U \subseteq \overline{X/\!/U}$ which are stable for the action of $R = H/U$. Moreover we can study the geometry of $(\overline{X/\!/U})/\!/R$ in terms of that of $\overline{X/\!/U}$ using classical GIT and symplectic geometry.

The aim of this paper is to discuss, for suitable actions on projective varieties X of a non-reductive affine algebraic group H with unipotent radical U, how to choose a reductive group $G \geq U$ and reductive envelopes $\overline{G \times_U X}$. In particular we will study the family of examples given by moduli spaces of hypersurfaces in the weighted projective plane $\mathbf{P}(1,1,2)$, obtained as quotients by linear actions of the automorphism group H of $\mathbf{P}(1,1,2)$ (cf. [Fa3]). This automorphism group is a semidirect product $H = R \ltimes U$ where $R \cong GL(2; \mathbf{C})$ is reductive and $U \cong (\mathbf{C}^+)^3$ is the unipotent radical of H, acting on $\mathbf{P}(1,1,2)$ as

$$[x:y:z] \mapsto [x:y:z + \lambda x^2 + \mu xy + \nu y^2]$$

for $\lambda, \mu, \nu \in \mathbf{C}$.

For simplicity we will work over $\mathbf{C}$ throughout. The layout of the paper is as follows. §2 gives a very brief review of classical GIT for reductive group actions, while §3 describes the results of [DK1] on non-reductive actions and the construction of reductive envelopes. §4 discusses the choice of reductive envelopes for actions of unipotent groups of the form

$(\mathbf{C}^+)^r$. Finally §5 considers the case of the automorphism group of $\mathbf{P}(1,1,2)$ acting on spaces of hypersurfaces.

This paper is based heavily on joint work with Brent Doran [DK1, DK2]. I would like to thank him for many helpful discussions, and also Keith Hannabuss for pointing me to some very useful references.

2 Mumford's geometric invariant theory

In the preface to the first edition of [MFK], Mumford states that his goal is "to construct moduli schemes for various types of algebraic objects" and that this problem "appears to be, in essence, a special and highly non-trivial case" of the problem of constructing orbit spaces for algebraic group actions. More precisely, when a family $\mathcal{X}$ of objects with parameter space S has the local universal property (that is, any other family is locally equivalent to the pullback of $\mathcal{X}$ along a morphism to S) for a given moduli problem, and a group acts on S such that objects parametrised by points in S are equivalent if and only if the points lie in the same orbit, then the construction of a coarse moduli space is equivalent to the construction of an orbit space for the action (cf. [Ne, Proposition 2.13]). Here, as in [Ne], by an orbit space we mean a G-invariant morphism $\varphi : S \to M$ such that every other G-invariant morphism $\psi : S \to M$ factors uniquely through φ and $\varphi^{-1}(m)$ is a single G-orbit for each $m \in M$.

Of course such orbit spaces do not in general exist, in particular because of the jump phenomenon: there may be orbits contained in the closures of other orbits, which means that the set of all orbits cannot be endowed naturally with the structure of a variety. This is the situation with which Mumford's geometric invariant theory [MFK] attempts to deal, when the group acting is reductive, telling us (in suitable circumstances) both how to throw out certain (unstable) orbits in order to be able to construct an orbit space, and how to construct a projective completion of this orbit space. Mumford's GIT is reviewed briefly next; for more details see [Ne2] in these volume, or [Do, MFK, Ne, PV].

Example 2.1. Let $G = SL(2; \mathbf{C})$ act on $(\mathbf{P}^1)^4$ in the standard way. Then

$$\{(x_1, x_2, x_3, x_4) \in (\mathbf{P}^1)^4 : x_1 = x_2 = x_3 = x_4\}$$

is a single orbit which is contained in the closure of every other orbit. On the other hand, the open subset of $(\mathbf{P}^1)^4$ where x_1, x_2, x_3, x_4 are distinct has an orbit space which, using the cross-ratio, can be identified with $\mathbf{P}^1 - \{0, 1, \infty\}$.

2.1 Classical geometric invariant theory

Let X be a complex projective variety and let G be a complex reductive group acting on X. To apply geometric invariant theory we require a linearisation of the action; that is, a line bundle L on X and a lift of the action of G to L. Usually L is assumed to be ample, and then we lose little generality in supposing that for some projective embedding $X \subseteq \mathbf{P}^n$ the action of G on X extends to an action on $\mathbf{P}^n$ given by a representation

$$\rho : G \to GL(n+1),$$

and taking for L the hyperplane line bundle on $\mathbf{P}^n$. We have an induced action of G on the homogeneous coordinate ring

$$\hat{\mathcal{O}}_L(X) = \bigoplus_{k \geq 0} H^0(X, L^{\otimes k})$$

of X. The subring $\hat{\mathcal{O}}_L(X)^G$ consisting of the elements of $\hat{\mathcal{O}}_L(X)$ left invariant by G is a finitely generated graded complex algebra because G is reductive [MFK], and so we can define the GIT quotient $X/\!/G$ to be the variety Proj $(\hat{\mathcal{O}}_L(X)^G)$. The inclusion of $\hat{\mathcal{O}}_L(X)^G$ in $\hat{\mathcal{O}}_L(X)$ defines a rational map q from X to $X/\!/G$, but because there may be points of $X \subseteq \mathbf{P}^n$ where every G-invariant polynomial vanishes this map will not in general be well-defined everywhere on X. The set X^{ss} of *semistable* points in X is the set of those $x \in X$ for which there exists some $f \in \hat{\mathcal{O}}_L(X)^G$ not vanishing at x. Then the rational map q restricts to a surjective G-invariant morphism from the open subset X^{ss} of X to the quotient variety $X/\!/G$. However $q : X^{ss} \to X/\!/G$ is still not in general an orbit space: when x and y are semistable points of X we have $q(x) = q(y)$ if and only if the closures $\overline{O_G(x)}$ and $\overline{O_G(y)}$ of the G-orbits of x and y meet in X^{ss}.

A *stable* point of X ('properly stable' in the terminology of [MFK]) is a point x of X^{ss} with a neighbourhood in X^{ss} such that every G-orbit meeting this neighbourhood is closed in X^{ss}, and is of maximal dimension equal to the dimension of G. If U is any G-invariant open subset of the set X^s of stable points of X, then $q(U)$ is an open subset of $X/\!/G$ and the restriction $q|_U : U \to q(U)$ of q to U is an orbit space for the action of G on U, so we will write U/G for $q(U)$. In particular there is an orbit space X^s/G for the action of G on X^s, and $X/\!/G$ is a projective completion of this orbit space.

$$
\begin{array}{ccccc}
X^s & \underset{\text{open}}{\subseteq} & X^{ss} & \underset{\text{open}}{\subseteq} & X \\
\downarrow & & \downarrow & & \\
X^s/G & \underset{\text{open}}{\subseteq} & X/\!/G = X^{ss}/\sim &
\end{array}
\tag{2.1}
$$

X^s, X^{ss}, and $X/\!/G$ are unaltered if the line bundle L is replaced by $L^{\otimes k}$ for any $k > 0$ with the induced action of G, so it is convenient to allow fractional linearisations.

Recall that a categorical quotient of a variety X under an action of G is a G-invariant morphism $\varphi : X \to Y$ from X to a variety Y such that any other G-invariant morphism $\widetilde{\varphi} : X \to \widetilde{Y}$ factors as $\widetilde{\varphi} = \chi \circ \varphi$ for a unique morphism $\chi : Y \to \widetilde{Y}$ [Ne, Chapter 2, §4]. An orbit space for the action is a categorical quotient $\varphi : X \to Y$ such that each fibre $\varphi^{-1}(y)$ is a single G-orbit, and a geometric quotient is an orbit space $\varphi : X \to Y$ which is an affine morphism such that

(i) if U is an open affine subset of Y then

$$
\varphi^* : \mathcal{O}(U) \to \mathcal{O}(\varphi^{-1}(U))
$$

induces an isomorphism of $\mathcal{O}(U)$ onto $\mathcal{O}(\varphi^{-1}(U))^G$, and

(ii) if W_1 and W_2 are disjoint closed G-invariant subvarieties of X then their images $\varphi(W_1)$ and $\varphi(W_2)$ in Y are disjoint closed subvarieties of Y.

If U is any G-invariant open subset of the set $X^s = X^s(L)$ of stable points of X, then $q(U)$ is an open subset of $X/\!/G$ and the restriction $q|_U : U \to q(U)$ of q to U is a geometric quotient for the action of G on U. In particular $X^s/G = q(X^s)$ is a geometric quotient for the action of G on X^s, while $q : X^{ss} \to X/\!/G$ is a categorical quotient of X^{ss} under the action of G.

The subsets X^{ss} and X^s of X are characterised by the following properties (see Chapter 2 of [MFK] or [Ne]).

Proposition 2.2. *(Hilbert-Mumford criteria) (i) A point $x \in X$ is semistable (respectively stable) for the action of G on X if and only if for every $g \in G$ the point gx is semistable (respectively stable) for the action of a fixed maximal torus of G.*

(ii) A point $x \in X$ with homogeneous coordinates $[x_0 : \ldots : x_n]$ in some coordinate system on $\mathbf{P}^n$ is semistable (respectively stable) for the action of a maximal torus of G acting diagonally on $\mathbf{P}^n$ with weights $\alpha_0, \ldots, \alpha_n$ if and only if the convex hull

$$\mathrm{Conv}\{\alpha_i : x_i \neq 0\}$$

contains 0 (respectively contains 0 in its interior).

In [MFK] the definitions of X^s and X^{ss} are extended as follows to allow L to be not ample and X not projective. However it is not necessarily the case that $U^{ss} = U \cap X^{ss}$ or that $U^s = U \cap X^s$ when U is a G-invariant open subset of X.

Definition 2.3 Let X be a quasi-projective complex variety with an action of a complex reductive group G and linearisation L on X. Then $y \in X$ is *semistable* for this linear action if there exists some $m \geq 0$ and $f \in H^0(X, L^{\otimes m})^G$ not vanishing at y such that the open subset

$$X_f := \{x \in X \mid f(x) \neq 0\}$$

is affine, and y is *stable* if also the action of G on X_f is closed with all stabilisers finite. $\triangle$

When X is projective and L is ample and $f \in H^0(X, L^{\otimes m})^G \smallsetminus \{0\}$ for some $m \geq 0$, then X_f is affine when f is nonconstant, so this is equivalent to the previous definition.

Remark 2.4. The reason for introducing the requirement that X_f must be affine in Definition 2.3 above is to ensure that X^{ss} has a categorical quotient $X^{ss} \to X/\!/G$, which restricts to a geometric quotient $X^s \to X^s/G$ (see [MFK] Theorem 1.10); the quotient $X/\!/G$ is quasi-projective, but need not be projective even when X is projective, if L is not ample. However when X is projective we can define 'naively stable' and 'naively semistable' points by omitting the condition that X_f should be affine. More precisely, let $I = \bigcup_{m>0} H^0(X, L^{\otimes m})^G$ and for $f \in I$ let X_f be the G-invariant open subset of X where f does not vanish. Then a point $x \in X$ is *naively semistable* if there exists some $f \in I$ which does not vanish at x, so that the set of naively semistable points is

$$X^{nss} = \bigcup_{f \in I} X_f,$$

whereas $X^{ss} = \bigcup_{f \in I^{ss}} X_f$ where

$$I^{ss} = \{f \in I \mid X_f \text{ is affine }\}.$$

The set of *naively stable* points of X is

$$X^{ns} = \bigcup_{f \in I^{ns}} X_f$$

where

$I^{ns} = \{f \in I \mid \text{the action of } G \text{ on } X_f \text{ is closed with all stabilisers finite}\}$

while $X^s = \bigcup_{f \in I^s} X_f$ where

$I^s = \{f \in I^{ss} \mid \text{the action of } G \text{ on } X_f \text{ is closed with all stabilisers finite}\}$.

If

$$\hat{\mathcal{O}}_L(X) = \bigoplus_{k \geq 0} H^0(X, L^{\otimes k})$$

is finitely generated as a complex algebra, then so is its ring of invariants $\hat{\mathcal{O}}_L(X)^G$ because G is reductive, and then

$$\overline{X /\!/ G} = \operatorname{Proj}\left(\hat{\mathcal{O}}_L(X)^G\right)$$

is a projective completion of $X /\!/ G$ and the categorical quotient $X^{ss} \to X /\!/ G$ is the restriction of a natural G-invariant morphism

$$X^{nss} \to \overline{X /\!/ G},$$

which by analogy with Definition 3.3 below we will call an *enveloping quotient* of X^{nss}.

Throughout the remainder of §2 we will assume that X is projective and L is ample.

2.2 Partial desingularisations of quotients

When X is nonsingular then X^s/G has only orbifold singularities, but if $X^{ss} \neq X^s$ the GIT quotient $X /\!/ G$ is likely to have more serious singularities. However if $X^s \neq \emptyset$ there is a canonical procedure (see [Ki2]) for constructing a partial resolution of singularities $\widetilde{X} /\!/ G$ of the quotient $X /\!/ G$. This involves blowing X up along a sequence of nonsingular G-invariant subvarieties, all contained in the complement of the set X^s of stable points of X, to obtain eventually a nonsingular projective variety $\widetilde{X}$ with a linear G-action, lifting the action on X, for which every semistable point of $\widetilde{X}$ is stable. The blow-down map $\pi : \widetilde{X} \to X$ induces a birational morphism $\pi_G : \widetilde{X} /\!/ G \to X /\!/ G$ which is an isomorphism over the dense open subset X^s/G of $X /\!/ G$, and if X is nonsingular the quotient $\widetilde{X} /\!/ G$ has only orbifold singularities.

This construction works as follows [Ki2]. Let V be any nonsingular G-invariant closed subvariety of X and let $\pi : \hat{X} \to X$ be the blowup of X along V. The linear action of G on the ample line bundle L over X lifts to a linear action on the line bundle over $\hat{X}$ which is the pullback of $L^{\otimes k}$ tensored with $\mathcal{O}(-E)$, where E is the exceptional divisor and k is a fixed positive integer. When k is large the line bundle $\pi^* L^{\otimes k} \otimes \mathcal{O}(-E)$ is ample on $\hat{X}$, and this linear action satisfies the following properties:

(i) if y is semistable in $\hat{X}$ then $\pi(y)$ is semistable in X;

(ii) if $\pi(y)$ is stable in X then y is stable in $\hat{X}$;

(iii) if k is large enough then the sets $\hat{X}^s$ and $\hat{X}^{ss}$ of stable and semistable points of $\hat{X}$ with respect to this linearisation are independent of k (cf. [Rei]).

Now X has semistable points which are not stable if and only if there exists a nontrivial connected reductive subgroup of G which fixes some semistable point. If so, let $r > 0$ be the maximal dimension of the reductive subgroups of G fixing semistable points of X, and let $\mathcal{R}(r)$ be a set of representatives of conjugacy classes in G of all connected reductive subgroups R of dimension r such that

$$Z_R^{ss} = \{x \in X^{ss} : R \text{ fixes } x\}$$

is nonempty. Then

$$\bigcup_{R \in \mathcal{R}(r)} G Z_R^{ss}$$

is a disjoint union of nonsingular closed subvarieties of X^{ss}, and

$$G Z_R^{ss} \cong G \times_{N^R} Z_R^{ss}$$

where N^R is the normaliser of R in G.

By Hironaka's theorem we can resolve the singularities of the closure of $\bigcup_{R \in \mathcal{R}(r)} G Z_R^{ss}$ in X by performing a sequence of blow-ups along nonsingular G-invariant closed subvarieties of $X - X^{ss}$. We then blow up along the proper transform of the closure of $\bigcup_{R \in \mathcal{R}(r)} G Z_R^{ss}$ to get a nonsingular projective variety $\hat{X}_1$. The linear action of G on X lifts to an action on this blow-up $\hat{X}_1$ which can be linearised using suitable ample line bundles as above, and it is shown in [Ki2] that the set $\hat{X}_1^{ss}$ of semistable points of $\hat{X}_1$ with respect to any of these suitable linearisations of the lifted action is the complement in the inverse image of X^{ss}

of the proper transform of the subset

$$\varphi^{-1}\left(\varphi\left(\bigcup_{R\in\mathcal{R}(r)} GZ_R^{ss}\right)\right)$$

of X^{ss}, where $\varphi : X^{ss} \to X//G$ is the canonical map. Moreover no point of $\hat{X}_1^{ss}$ is fixed by a reductive subgroup of G of dimension at least r, and a point in $\hat{X}_1^{ss}$ is fixed by a reductive subgroup R of G of dimension less than r if and only if it belongs to the proper transform of the subvariety Z_R^{ss} of X^{ss}.

The same procedure can now be applied to $\hat{X}_1$ to obtain $\hat{X}_2$ such that no reductive subgroup of G of dimension at least $r - 1$ fixes a point of $\hat{X}_2^{ss}$. After repeating enough times we obtain $\widetilde{X}$ satisfying $\widetilde{X}^{ss} = \widetilde{X}^s$. If we are only interested in $\widetilde{X}^{ss}$ and the partial resolution $\widetilde{X}//G$ of $X//G$, rather than in $\widetilde{X}$ itself, then there is no need in this procedure to resolve the singularities of the closure of $\bigcup_{R\in\mathcal{R}(r)} GZ_R^{ss}$ in X. Instead we can simply blow X^{ss} up along $\bigcup_{R\in\mathcal{R}(r)} GZ_R^{ss}$ (or equivalently along each GZ_R^{ss} in turn) and let $\hat{X}_1^{ss}$ be the set of semistable points in the result, and then repeat the process.

Thus the geometric quotient X^s/G has *two* natural compactifications $X//G$ and $\widetilde{X//G} = \widetilde{X}//G$, which fit into a diagram

$$\begin{array}{ccccccc}
X^s/G & \hookrightarrow & \widetilde{X//G} = \widetilde{X}^{ss}/G & \leftarrow & \widetilde{X}^{ss} = \widetilde{X}^s & \hookrightarrow & \widetilde{X} \\
\| & & \downarrow & & \downarrow & & \downarrow \\
X^s/G & \hookrightarrow & X//G & \leftarrow & X^{ss} & \hookrightarrow & X.
\end{array}$$

2.3 Variation of GIT

The GIT quotient $X//G$ depends not just on the action of G on X but also on the choice of linearisation $\mathcal{L}$ of the action; that is, the choice of the line bundle L and the lift of the action to L. This should of course be reflected in the notation; to avoid ambiguity we will sometimes add appropriate decorations, as in $X//_{\mathcal{L}}G$ and also $X^{s,\mathcal{L}}$ and $X^{ss,\mathcal{L}}$. There is thus a natural question: how does the GIT quotient $X//_{\mathcal{L}}G$ vary as the linearisation $\mathcal{L}$ varies? This has been studied by Brion and Procesi [BP] and Goresky and MacPherson [GM] (in the abelian case) and by Thaddeus [Th], Dolgachev and Hu [DH] and Ressayre [Res] for general reductive groups G.

A very simple case is when the line bundle L is fixed, but the lift $\tau : G \times L \to L$ of the action of G on X to an action on L varies. The only possible such variation is to replace τ with

$$(g, \ell) \mapsto \chi(g)\tau(g, \ell)$$

where $\chi : G \to \mathbf{C}^*$ is a character of G. The Hilbert-Mumford criteria (Proposition 2.2 above) can be used to see how X^s and X^{ss} are affected by such a variation.

Example 2.5. Consider the linear action of $\mathbf{C}^*$ on $X = \mathbf{P}^{r+s}$, with respect to the hyperplane line bundle L on X, where the linearisation $\mathcal{L}_+$ is given by the representation

$$t \mapsto \operatorname{diag}(t^3, \ldots, t^3, t, t^{-1}, \ldots, t^{-1})$$

of $\mathbf{C}^*$ in $GL(r + s + 1; \mathbf{C})$ in which t^3 occurs with multiplicity r and t^{-1} occurs s times. For this linearisation we have

$$X^{ss, \mathcal{L}_+} = X^{s, \mathcal{L}_+} = \{[x_0 : \ldots : x_{r+s}] \in \mathbf{P}^{r+s} :$$

$$x_0, \ldots, x_r \text{ are not all } 0 \quad \text{and} \quad x_{r+1}, \ldots, x_{r+s} \text{ are not all } 0\}$$

and $X /\!/_{\mathcal{L}_+} \mathbf{C}^*$ is isomorphic to the product of a weighted projective space $\mathbf{P}(3, \ldots, 3, 1)$ of dimension r and the projective space $\mathbf{P}^{s-1}$. The same action of $\mathbf{C}^*$ on X has other linearisations with respect to the hyperplane line bundle; let $\mathcal{L}_0$ and $\mathcal{L}_-$ denote the linearisations given by multiplying the representation above by the characters $\chi(t) = t^{-1}$ and $\chi(t) = t^{-2}$. Then $\mathcal{L}_-$ is given by the representation

$$t \mapsto \operatorname{diag}(t, \ldots, t, t^{-1}, t^{-3}, \ldots, t^{-3})$$

of $\mathbf{C}^*$ in $GL(r + s + 1; \mathbf{C})$ and for this linearisation

$$X^{ss, \mathcal{L}_-} = X^{s, \mathcal{L}_-} = \{[x_0 : \ldots : x_{r+s}] \in \mathbf{P}^{r+s} :$$

$$x_0, \ldots, x_{r-1} \text{ are not all } 0 \quad \text{and} \quad x_r, \ldots, x_{r+s} \text{ are not all } 0\}$$

while $X /\!/_{\mathcal{L}_-} \mathbf{C}^*$ is isomorphic to the product of the projective space $\mathbf{P}^{r-1}$ and a weighted projective space $\mathbf{P}(1, 3, \ldots, 3)$ of dimension s. Finally for the linearisation $\mathcal{L}_0$ given by the representation

$$t \mapsto \operatorname{diag}(t^2, \ldots, t^2, 1, t^{-2}, \ldots, t^{-2})$$

of $\mathbf{C}^*$ in $GL(r + s + 1; \mathbf{C})$ we have

$$X^{ss, \mathcal{L}_0} \neq X^{s, \mathcal{L}_0}$$

(semistability does not imply stability) and the quotient $X /\!/_{\mathcal{L}_0} \mathbf{C}^*$ has more serious singularities than the orbifolds $X/\!/_{\mathcal{L}_{\pm}} \mathbf{C}^*$. It can be identified with the result of collapsing $[0 : \ldots : 0 : 1] \times \mathbf{P}^{s-1}$ to a point in $\mathbf{P}(3, \ldots, 3, 1) \times \mathbf{P}^{s-1}$, and also with the result of collapsing $\mathbf{P}^{r-1} \times [1 : 0 : \ldots : 0]$ to a point in $\mathbf{P}^{r-1} \times \mathbf{P}(1, 3, \ldots, 3)$.

Remark 2.6. The general case when the line bundle L is allowed to vary, as well as the lift of the G-action from X to the line bundle, can be reduced to the case above by a trick due to Thaddeus [Th]. Suppose that a given action of G on X lifts to ample line bundles $L_0, \ldots, L_m$ giving linearisations $\mathcal{L}_0, \ldots, \mathcal{L}_m$ over X, and consider the projective variety

$$Y = \mathbf{P}(L_0 \oplus \cdots \oplus L_m).$$

Then the induced action of G on Y has a natural linearisation, and the complex torus $T = \mathbf{C}^{m+1}$ also acts on Y, commuting with the action of G, with a natural linearisation which can be modified using any character χ of T. Taking χ to be the jth projection $\chi_j : T \to \mathbf{C}^*$ and using the fact that the GIT quotient operations with respect to G and T commute, we find that

$$X/\!/_{\mathcal{L}_j} G \cong (Y/\!/G)/\!/_{\chi_j} T$$

and hence the general question of variation of GIT quotients with linearisations reduces to the special case when the variation is by multiplication by a character of the group. The conclusion [DH, Res, Th] is that, roughly speaking, the space of all possible ample fractional linearisations of a given G-action on a projective variety X is divided into finitely many polyhedral chambers within which the GIT quotient is constant. Moreover, when a wall between two chambers is crossed, the quotient undergoes a transformation which typically can be thought of as a blow-up followed by a blow-down. If $\mathcal{L}_+$ and $\mathcal{L}_-$ represent fractional linearisations in the interiors of two adjoining chambers, and $\mathcal{L}_0$ represents a fractional linearisation in the interior of the wall between them, then we have inclusions

$$X^{s,\mathcal{L}_0} \subseteq X^{s,\mathcal{L}_+} \cap X^{s,\mathcal{L}_-} \quad \text{and} \quad X^{ss,\mathcal{L}_+} \cup X^{ss,\mathcal{L}_-} \subseteq X^{ss,\mathcal{L}_0}$$

inducing morphisms

$$X/\!/_{\mathcal{L}_{\pm}} G \to X/\!/_{\mathcal{L}_0} G$$

which are isomorphisms over $X^{s,\mathcal{L}_0}/G$. In addition, under mild conditions there are sheaves of ideals on the two quotients $X/\!/_{\mathcal{L}_{\pm}} G$ whose blow-ups are both isomorphic to a component of the fibred product of the two quotients over the quotient $X/\!/_{\mathcal{L}_0} G$ on the wall.

Remark 2.7. If $\varphi : X \rightarrow Y$ is a categorical quotient for a G-action on a variety X then its restriction to a G-invariant open subset of X is not necessarily a categorical quotient for the action of G on U. In the situation above, as in Example 2.5, we have $X^{ss,\mathcal{L}_+} \subseteq X^{ss,\mathcal{L}_0}$ but the restriction of the categorical quotient $X^{ss,\mathcal{L}_0} \rightarrow X /\!/_{\mathcal{L}_0} G$ to $X^{ss,\mathcal{L}_+}$ is not a categorical quotient for the G action on $X^{ss,\mathcal{L}_+}$.

3 Quotients by non-reductive actions

Translation actions appear all over geometry, so it is not surprising that there are many cases of moduli problems which involve non-reductive group actions, where Mumford's GIT does not apply. One example is that of hypersurfaces in a toric variety Y. The case we shall consider in detail in this paper is when Y is the weighted projective plane $\mathbf{P}(1,1,2)$ (cf. [Fa3]), with homogeneous coordinates x, y, z (that is, Y is the quotient of $\mathbf{C}^3 \setminus \{0\}$ by the action of $\mathbf{C}^*$ with weights 1,1 and 2, and x, y and z are coordinates on $\mathbf{C}^3 \setminus \{0\}$). Let H be the automorphism group of $Y = \mathbf{P}(1,1,2)$, which is the quotient by $\mathbf{C}^*$ of a semidirect product of the unipotent group $U = (\mathbf{C}^+)^3$ acting on Y via

$$[x : y : z] \mapsto [x : y : z + \lambda x^2 + \mu xy + \nu y^2] \quad \text{for } (\lambda, \mu, \nu) \in (\mathbf{C}^+)^3$$

and the reductive group $GL(2; \mathbf{C}) \times GL(1; \mathbf{C})$ acting on the (x, y) co-ordinates and the z coordinate. H acts linearly on the projective space X_d of weighted degree d polynomials in x, y, z.

Example 3.1. When $d = 4$, a basis for the weighted degree d polynomials is

$$\{x^4, x^3 y, x^2 y^2, xy^3, y^4, x^2 z, xyz, y^2 z, z^2\},$$

and with respect to this basis, the U-action is given by

$$\begin{pmatrix} 1 & 0 & 0 & 0 & 0 & \lambda & 0 & 0 & \lambda^2 \\ 0 & 1 & 0 & 0 & 0 & \mu & \lambda & 0 & 2\lambda\mu \\ 0 & 0 & 1 & 0 & 0 & \nu & \mu & \lambda & 2\lambda\nu + \mu^2 \\ 0 & 0 & 0 & 1 & 0 & 0 & \nu & \mu & 2\mu\nu \\ 0 & 0 & 0 & 0 & 1 & 0 & 0 & \nu & \nu^2 \\ 0 & 0 & 0 & 0 & 0 & 1 & 0 & 0 & 2\lambda \\ 0 & 0 & 0 & 0 & 0 & 0 & 1 & 0 & 2\mu \\ 0 & 0 & 0 & 0 & 0 & 0 & 0 & 1 & 2\nu \\ 0 & 0 & 0 & 0 & 0 & 0 & 0 & 0 & 1 \end{pmatrix}.$$

The tautological family $\mathcal{H}^{(d)}$ parametrised by X_d of hypersurfaces in Y has the following two properties:

(i) the hypersurfaces $\mathcal{H}_s^{(d)}$ and $\mathcal{H}_t^{(d)}$ parametrised by weighted degree d polynomials s and t are isomorphic as hypersurfaces in Y if and only if s and t lie in the same orbit of the natural action of $H \cong U \rtimes GL(2; \mathbf{C})$ on X_d, and

(ii) (local universal property) any family of hypersurfaces in Y is locally equivalent to the pullback of $\mathcal{H}^{(d)}$ along a morphism to X_d.

This means that the construction of a (coarse) moduli space of weighted degree d hypersurfaces in Y is equivalent to constructing an orbit space for the action of H on X_d ([Ne] Proposition 2.13).

Now let H be any affine algebraic group, with unipotent radical U, acting linearly on a complex projective variety X with respect to an ample line bundle L. Of course the most immediate difficulty when trying to generalise Mumford's GIT to a non-reductive situation is that the ring of invariants

$$\hat{\mathcal{O}}_L(X)^H = \bigoplus_{k \geq 0} H^0(X, L^{\otimes k})^H$$

is not necessarily finitely generated as a graded complex algebra, so that $\mathrm{Proj}\,(\hat{\mathcal{O}}_L(X)^H)$ is not well-defined as a projective variety. However $\mathrm{Proj}\,(\hat{\mathcal{O}}_L(X)^H)$ does make sense as a scheme, and the inclusion of $\hat{\mathcal{O}}_L(X)^H$ in $\hat{\mathcal{O}}_L(X)$ gives us a rational map of schemes q from X to $\mathrm{Proj}\,(\hat{\mathcal{O}}_L(X)^H)$, whose image is a constructible subset of $\mathrm{Proj}\,(\hat{\mathcal{O}}_L(X)^H)$ (that is, a finite union of locally closed subschemes).

The action on X of the unipotent radical U of H is studied in [DK1] (building on earlier work such as [Fa, Fa2, GP, GP2, W]), where the following definitions are made and results proved.

Definition 3.2 Let $I = \bigcup_{m>0} H^0(X, L^{\otimes m})^U$ and for $f \in I$ let X_f be the U-invariant affine open subset of X where f does not vanish, with $\mathcal{O}(X_f)$ its coordinate ring. A point $x \in X$ is *naively semistable* if the rational map q from X to $\mathrm{Proj}\,(\hat{\mathcal{O}}_L(X)^U)$ is well-defined at x; that is, if there exists some $f \in I$ which does not vanish at x. The set of naively semistable points is $X^{nss} = \bigcup_{f \in I} X_f$. The *finitely generated semistable set* of X is

$$X^{ss,fg} = \bigcup_{f \in I^{fg}} X_f$$

where

$$I^{fg} = \{f \in I \mid \mathcal{O}(X_f)^U \text{ is finitely generated}\}.$$

The set of *naively stable* points of X is

$$X^{ns} = \bigcup_{f \in I^{ns}} X_f$$

where

$$I^{ns} = \{f \in I \mid \mathcal{O}(X_f)^U \text{ is finitely generated, and}$$

$$q : X_f \longrightarrow \operatorname{Spec}(\mathcal{O}(X_f)^U) \text{ is a geometric quotient}\}.$$

The set of *locally trivial stable* points is

$$X^{lts} = \bigcup_{f \in I^{lts}} X_f$$

where

$$I^{lts} = \{f \in I \mid \mathcal{O}(X_f)^U \text{ is finitely generated, and}$$

$$q : X_f \longrightarrow \operatorname{Spec}(\mathcal{O}(X_f)^U) \text{ is a locally trivial geometric quotient}\}.$$

$\triangle$

Definition 3.3 Let $q : X^{ss,fg} \to \operatorname{Proj}(\hat{\mathcal{O}}_L(X)^U)$ be the natural morphism of schemes. The *enveloped quotient* of $X^{ss,fg}$ is $q : X^{ss,fg} \to q(X^{ss,fg})$, where $q(X^{ss,fg})$ is a dense constructible subset of the *enveloping quotient*

$$X//U = \bigcup_{f \in I^{ss,fg}} \operatorname{Spec}(\mathcal{O}(X_f)^U)$$

of $X^{ss,fg}$. Note that $q(X^{ss,fg})$ is not necessarily a subvariety of $X//U$, as is demonstrated by the example studied in [DK1] §6 of $U = \mathbf{C}^+$ acting on $X = \mathbf{P}^n$ via the nth symmetric product of its standard representation on $\mathbf{C}^2$ when n is even. $\triangle$

Proposition 3.4. ([DK1] 4.2.9 and 4.2.10). *The enveloping quotient $X//U$ is a quasi-projective variety with an ample line bundle $L_H \to X//U$ which pulls back to a positive tensor power of L under the natural map $q : X^{ss,fg} \to X//U$. If $\hat{\mathcal{O}}_L(X)^U$ is finitely generated then $X//U$ is the projective variety $\operatorname{Proj}(\hat{\mathcal{O}}_L(X)^U)$.*

Now suppose that G is a complex reductive group with U as a closed subgroup. Let $G \times_U X$ denote the quotient of $G \times X$ by the free action of U defined by $u(g, x) = (gu^{-1}, ux)$ for $u \in U$, which is a quasi-projective variety by [PV] Theorem 4.19. There is an induced G-action on $G \times_U X$ given by left multiplication of G on itself. If the action of U on X extends to an action of G there is an isomorphism of G-varieties

$$G \times_U X \cong (G/U) \times X$$

given by

$$[g, x] \mapsto (gU, gx). \tag{3.1}$$

When U acts linearly on X with respect to a very ample line bundle L inducing an embedding of X in $\mathbf{P}^n$, and G is a subgroup of $SL(n+1; \mathbf{C})$, then we get a very ample G-linearisation (which by abuse of notation we will also denote by L) on $G \times_U X$ as follows:

$$G \times_U X \hookrightarrow G \times_U \mathbf{P}^n \cong (G/U) \times \mathbf{P}^n,$$

by taking the trivial bundle on the quasi-affine variety G/U. If we choose a G-equivariant embedding of G/U in an affine space $\mathbf{A}^m$ with a linear G-action we get a G-equivariant embedding of $G \times_U X$ in

$$\mathbf{A}^m \times \mathbf{P}^n \subset \mathbf{P}^m \times \mathbf{P}^n \subset \mathbf{P}^{nm+m+n}$$

and the G-invariants on $G \times_U X$ are given by

$$\bigoplus_{m \geq 0} H^0(G \times_U X, L^{\otimes m})^G \cong \bigoplus_{m \geq 0} H^0(X, L^{\otimes m})^U = \hat{\mathcal{O}}_L(X)^U. \tag{3.2}$$

Definition 3.5 The sets of *Mumford stable points* and *Mumford semistable points* in X are $X^{ms} = i^{-1}((G \times_U X)^s)$ and $X^{mss} = i^{-1}((G \times_U X)^{ss})$ where $i : X \to G \times_U X$ is the inclusion given by $x \mapsto [e, x]$ for e the identity element of G. Here $(G \times_U X)^s$ and $(G \times_U X)^{ss}$ are defined as in Definition 2.3 for the induced linear action of G on the quasi-projective variety $G \times_U X$. $\triangle$

In fact it follows from Theorem 3.10 below that X^{ms} and X^{mss} are equal and are independent of the choice of G.

Definition 3.6 A *finite separating set of invariants* for the linear action of U on X is a collection of invariant sections $\{f_1, \ldots, f_n\}$ of positive tensor powers of L such that, if x, y are any two points of X then $f(x) = f(y)$ for all invariant sections f of $L^{\otimes k}$ and all $k > 0$ if and only if

$$f_i(x) = f_i(y) \qquad \forall i = 1, \ldots, n.$$

If G is any reductive group containing U, a finite separating set S of invariant sections of positive tensor powers of L is a *finite fully separating set of invariants* for the linear U-action on X if

(i) for every $x \in X^{ms}$ there exists $f \in S$ with associated G-invariant F over $G \times_U X$ (under the isomorphism (3.2)) such that $x \in (G \times_U X)_F$ and $(G \times_U X)_F$ is affine; and

(ii) for every $x \in X^{ss,fg}$ there exists $f \in S$ such that $x \in X_f$ and S is a generating set for $\mathcal{O}(X_f)^U$. $\triangle$

This definition is in fact independent of the choice of G (see [DK1] Remark 5.2.3).

Definition 3.7 Let X be a quasi-projective variety with a linear U-action with respect to an ample line bundle L on X, and let G be a complex reductive group containing U as a closed subgroup. A G-equivariant projective completion $\overline{G \times_U X}$ of $G \times_U X$, together with a G-linearisation with respect to a line bundle L which restricts to the given U-linearisation on X, is a *reductive envelope* of the linear U-action on X if every U-invariant f in some finite fully separating set of invariants S for the U-action on X extends to a G-invariant section of a tensor power of L over $\overline{G \times_U X}$.

If moreover there exists such an S for which every $f \in S$ extends to a G-invariant section F over $\overline{G \times_U X}$ such that $(\overline{G \times_U X})_F$ is affine, then $(\overline{G \times_U X}, L')$ is a *fine reductive envelope*, and if L is ample (in which case $(\overline{G \times_U X})_F$ is always affine) it is an *ample reductive envelope*.

If every $f \in S$ extends to a G-invariant F over $\overline{G \times_U X}$ which vanishes on each codimension 1 component of the boundary of $G \times_U X$ in $\overline{G \times_U X}$, then a reductive envelope for the linear U-action on X is called a *strong reductive envelope*. $\triangle$

It will be useful to add an extra definition which does not appear in [DK1].

Definition 3.8 In the notation of Definitions 3.6 and 3.7 above, a reductive envelope is called *stably fine* if for every $x \in X^{ms}$ there exists a U-invariant f which extends to a G-invariant section F over $\overline{G \times_U X}$ such that $x \in (G \times_U X)_F$ and both $(G \times_U X)_F$ and $(\overline{G \times_U X})_F$ are affine. $\triangle$

Definition 3.9 Let X be a projective variety with a linear U-action and a reductive envelope $\overline{G \times_U X}$. The set of *completely stable points* of

X with respect to the reductive envelope is

$$X^{\overline{s}} = (j \circ i)^{-1}(\overline{G \times_U X}^s)$$

and the set of *completely semistable points* is

$$X^{\overline{ss}} = (j \circ i)^{-1}(\overline{G \times_U X}^{ss}),$$

where $i : X \hookrightarrow G \times_U X$ and $j : G \times_U X \hookrightarrow \overline{G \times_U X}$ are the inclusions, and $\overline{G \times_U X}^s$ and $\overline{G \times_U X}^{ss}$ are the stable and semistable sets for the linear G-action on $\overline{G \times_U X}$. Following Remark 2.4 we also define

$$X^{\overline{nss}} = (j \circ i)^{-1}(\overline{G \times_U X}^{nss});$$

then $X^{\overline{nss}} = X^{\overline{ss}}$ when the reductive envelope is ample, but not in general otherwise. $\triangle$

Theorem 3.10. ([DK1] 5.3.1 and 5.3.5). *Let X be a normal projective variety with a linear U-action, for U a connected unipotent group, and let $(\overline{G \times_U X}, L)$ be any fine reductive envelope. Then*

$$X^{\overline{s}} \subseteq X^{lts} = X^{ms} = X^{mss} \subseteq X^{ns} \subseteq X^{ss,fg} \subseteq X^{\overline{ss}} = X^{nss}.$$

The stable sets $X^{\overline{s}}$, $X^{lts} = X^{ms} = X^{mss}$ and X^{ns} admit quasi-projective geometric quotients, given by restrictions of the quotient map $q = \pi \circ j \circ i$ where

$$\pi : (\overline{G \times_U X})^{ss} \to \overline{G \times_U X}//G$$

is the classical GIT quotient map for the reductive envelope and i, j are as in Definition 3.9. The quotient map q restricted to the open subvariety $X^{ss,fg}$ is an enveloped quotient with $q : X^{ss,fg} \to X//U$ an enveloping quotient. Moreover $X//U$ is an open subvariety of $\overline{G \times_U X}//G$ and there is an ample line bundle L_U on $X//U$ which pulls back to a tensor power $L^{\otimes k}$ of the line bundle L for some $k > 0$ and extends to an ample line bundle on $\overline{G \times_U X}//G$.

If furthermore $\overline{G \times_U X}$ is normal and provides a fine strong reductive envelope for the linear U-action on X, then $X^{\overline{s}} = X^{lts}$ and $X^{ss,fg} = X^{nss}$.

Definition 3.11 ([DK1] 5.3.7). Let X be a projective variety equipped with a linear U-action. A point $x \in X$ is called *stable* for the linear U-action if $x \in X^{lts}$ and *semistable* if $x \in X^{ss,fg}$, so from now on we will write X^s (or $X^{s,U}$) for X^{lts} and X^{ss} (or $X^{ss,U}$) for $X^{ss,fg}$. $\triangle$

Thus in the situation of Theorem 3.10 we have a diagram of quasi-projective varieties

$$
\begin{array}{ccccccccc}
X^{\bar{s}} & \subseteq & X^{s} & \subseteq & X^{ns} & \subseteq & X^{ss} & \subseteq & X^{\overline{ss}} = X^{nss} \\
\downarrow & & \downarrow & & \downarrow & & \downarrow & & \downarrow \\
X^{\bar{s}}/U & \subseteq & X^{s}/U & \subseteq & X^{ns}/U & \subseteq & X/\!/U & \subseteq & \overline{G \times_{U} X}/\!/G
\end{array}
$$

where all the inclusions are open and all the vertical morphisms are restrictions of $\pi : (\overline{G \times_U X})^{ss} \to \overline{G \times_U X}/\!/G$, and each except the last is a restriction of the map of schemes $q : X^{nss} \to \mathrm{Proj}\,(\hat{\mathcal{O}}_L(X)^U))$ associated to the inclusion $\hat{\mathcal{O}}_L(X)^U \subseteq \hat{\mathcal{O}}_L(X)$. In particular we have

$$
\begin{array}{ccccc}
X^{s} & \subseteq & X^{ss} & \subseteq & X \\
\downarrow & & \downarrow & & \\
X^{s}/U & \subseteq & X/\!/U & &
\end{array} \tag{3.3}
$$

which looks very similar to the situation for reductive actions (see diagram (2.1) above), with the major differences that

(i) $X/\!/U$ is not always projective,
and (even if the ring of invariants $\hat{\mathcal{O}}_L(X)^U$ is finitely generated and $X/\!/U = \mathrm{Proj}\,(\hat{\mathcal{O}}_L(X)^U)$ is projective)

(ii) the morphism $X^{ss} \to X/\!/U$ is not in general surjective.

Remark 3.12. The proofs of [DK1] Theorem 5.3.1 and Theorem 5.3.5 show that if X is a normal projective variety with a linear U-action and $(\overline{G \times_U X}, L)$ is any stably fine reductive envelope in the sense of Definition 3.8, then

$$
X^{\bar{s}} \subseteq X^{s} \subseteq X^{ns} \subseteq X^{ss}
$$

and if furthermore $\overline{G \times_U X}$ is normal and provides a reductive envelope which is both strong and stably fine, then $X^{\bar{s}} = X^{s}$. Indeed, this is still true even if $(\overline{G \times_U X}, L)$ is not a reductive envelope at all, provided that it satisfies all the conditions except for omitting (ii) in Definition 3.6.

There always exists an ample, and hence fine, but not necessarily strong, reductive envelope for any linear U-action on a projective variety X, at least if we replace the line bundle L with a suitable positive tensor power of itself, by [DK1] Proposition 5.2.8. By Theorem 3.10 above a choice of fine reductive envelope $\overline{G \times_U X}$ provides a projective completion

$$
\overline{X/\!/U} = \overline{G \times_U X}/\!/G
$$

of the enveloping quotient $X/\!/U$. This projective completion in general depends on the choice of reductive envelope, but when $\hat{\mathcal{O}}_L(X)^U$

is finitely generated then $X//U = \text{Proj}\,(\hat{\mathcal{O}}_L(X)^U)$ is itself projective, which implies that $X//U = \overline{G \times_U X}//G$ for any fine reductive envelope $\overline{G \times_U X}$.

The proof of [DK1] Theorem 5.3.1 also gives us the following result for any reductive envelope, not necessarily fine or strong.

Proposition 3.13. *Let X be a normal projective variety with a linear U-action, for U a connected unipotent group, and let $(\overline{G \times_U X}, L)$ be any reductive envelope. Then*

$$X^{\bar{s}} \subseteq X^s \subseteq X^{ss} \subseteq X^{\overline{nss}},$$

and if the graded algebra $\bigoplus_{k \geq 0} H^0(\overline{G \times_U X}, L^{\otimes k})$ is finitely generated then the projective completion

$$\overline{\overline{G \times_U X}//G} = \text{Proj}\,(\bigoplus_{k \geq 0} H^0(\overline{G \times_U X}, L^{\otimes k}))$$

of $\overline{G \times_U X}//G$ (cf. Remark 2.4) is a projective completion of $X//U$ with a commutative diagram

$$
\begin{array}{ccccccc}
X^{\bar{s}} & \subseteq & X^s & \subseteq & X^{ss} & \subseteq & X^{\overline{nss}} \\
\downarrow & & \downarrow & & \downarrow & & \downarrow \\
X^{\bar{s}}/U & \subseteq & X^s/U & \subseteq & X//U & \subseteq & \overline{X//U} = \overline{\overline{G \times_U X}//G}.
\end{array}
\tag{3.4}
$$

4 Choosing reductive envelopes

Let H be a connected affine algebraic group over $\mathbf{C}$. Then H has a unipotent radical U, which is a normal subgroup of H with reductive quotient group $R = H/U$. We can hope to quotient first by the action of U, and then by the induced action of the reductive group H/U, provided that the unipotent quotient is sufficiently canonical to inherit an induced linear action of the reductive group R. Moreover U has canonical series of normal subgroups $\{1\} = U_0 \leq U_1 \leq \cdots \leq U_s = U$ such that each successive subquotient is isomorphic to $(\mathbf{C}^+)^r$ for some r (for example the descending central series of U), so we can hope to quotient successively by unipotent groups of the form $(\mathbf{C}^+)^r$, and then finally by the reductive group R. Therefore we will concentrate on the case when $U \cong (\mathbf{C}^+)^r$ for some r; of course this is the situation in our example concerning hypersurfaces in the weighted projective plane $\mathbf{P}(1,1,2)$, when H is the automorphism group of $\mathbf{P}(1,1,2)$ and U is its unipotent radical.

More generally, let us assume first that U is a unipotent group with a one-parameter group of automorphisms $\lambda : \mathbf{C}^* \to \text{Aut}(U)$ such that the weights of the induced $\mathbf{C}^*$ action on the Lie algebra $\mathbf{u}$ of U are all nonzero. When $U = (\mathbf{C}^+)^r$ we can take λ to be the inclusion of the central $\mathbf{C}^*$ in $\mathrm{Aut}(U) \cong GL(r; \mathbf{C})$. Then we can form the semidirect product

$$\hat{U} = \mathbf{C}^* \ltimes U$$

given by $\mathbf{C}^* \times U$ with group multiplication

$$(z_1, u_1).(z_2, u_2) = (z_1 z_2, (\lambda(z_2^{-1})(u_1))u_2).$$

U meets the centre of $\hat{U}$ trivially, so we have an inclusion

$$U \hookrightarrow \hat{U} \to \mathrm{Aut}(\hat{U}) \to GL(\mathrm{Lie}\hat{U}) = GL(\mathbf{C} \oplus \mathbf{u})$$

where $\hat{U}$ maps to its group of inner automorphisms. Thus U is isomorphic to a closed subgroup of the reductive group $G = SL(\mathbf{C} \oplus \mathbf{u})$.

In particular when $U = (\mathbf{C}^+)^r$ we have $U \leq G = SL(r+1; \mathbf{C})$, and then

$$G/U \cong \{\alpha \in (\mathbf{C}^r)^* \otimes \mathbf{C}^{r+1} \mid \alpha : \mathbf{C}^r \to \mathbf{C}^{r+1} \text{ is injective}\}$$

with the natural G-action $g\alpha = g \circ \alpha$. Since the injective linear maps from $\mathbf{C}^r$ to $\mathbf{C}^{r+1}$ form an open subset in the affine space $(\mathbf{C}^r)^* \otimes \mathbf{C}^{r+1}$ whose complement has codimension two, it follows that $U = (\mathbf{C}^+)^r$ is a Grosshans subgroup of $G = SL(r+1; \mathbf{C})$ and

$$\mathcal{O}(G)^U \cong \mathcal{O}(G/U) \cong \mathcal{O}((\mathbf{C}^r)^* \otimes \mathbf{C}^{r+1})$$

is finitely generated [Gr].

4.1 Actions of $(\mathbf{C}^+)^r$ which extend to $SL(r+1; \mathbf{C})$

Let X be a normal projective variety with a linear action of $U = (\mathbf{C}^+)^r$ with respect to an ample line bundle L. Suppose first that the linear action of $U = (\mathbf{C}^+)^r$ on X extends to a linear action of $G = SL(r+1; \mathbf{C})$, giving us an identification of G-spaces

$$G \times_U X \cong (G/U) \times X$$

as at (3.1) via $[g, x] \mapsto (gU, gx)$. Then (as in the Borel transfer theorem [Do, Lemma 4.1])

$$\hat{\mathcal{O}}_L(X)^U \cong \hat{\mathcal{O}}_L(G \times_U X)^G \cong [\mathcal{O}(G/U) \otimes \hat{\mathcal{O}}_L(X)]^G$$

is finitely generated [Gr2] and we have a reductive envelope

$$\overline{G \times_U X} = \mathbf{P}^{r(r+1)} \times X,$$

where $\mathbf{P}^{r(r+1)} = \mathbf{P}(\mathbf{C} \oplus ((\mathbf{C}^r)^* \otimes \mathbf{C}^{r+1}))$, with

$$\overline{G \times_U X} //G \cong X//U = \mathrm{Proj}\,(\hat{\mathcal{O}}_L(X)^U).$$

More precisely, if we choose for our linearisation on $\overline{G \times_U X}$ the line bundle

$$L^{(N)} = \mathcal{O}_{\mathbf{P}^{r(r+1)}}(N) \otimes L$$

with $N > 0$ sufficiently large, then by [DK1] Lemma 5.3.14 we obtain a reductive envelope which is strong as well as ample, and so by Theorem 3.10 we have

$$X^{\bar{s}} = X^s \text{ and } X^{\bar{s}s} = X^{ss}. \tag{4.1}$$

Remark 4.1. Even if X is nonsingular, this quotient

$$X//U = (\mathbf{P}^{r(r+1)} \times X)//G$$

may have serious singularities if there are semistable points which are not stable. However provided that $X^s \neq \emptyset$ we can construct a partial desingularisation

$$\widetilde{X//U}^{(G)} = (\widetilde{\mathbf{P}^{r(r+1)} \times X})//G$$

as in §2.2 by blowing $\mathbf{P}^{r(r+1)} \times X$ up successively along G-invariant closed subvarieties, all disjoint from $(\mathbf{P}^{r(r+1)} \times X)^s$ and hence from $X^{\bar{s}} = X^s$, to get a linear G-action on the resulting blow-up $\widetilde{\mathbf{P}^{r(r+1)} \times X}$ for which all semistable points are stable. This construction is determined by the linear G-action, and if X is nonsingular the resulting quotient is an orbifold. Since $X^{\bar{s}} = X^s$ and the morphism

$$\widetilde{X//U}^{(G)} = (\widetilde{\mathbf{P}^{r(r+1)} \times X})//G \to (\mathbf{P}^{r(r+1)} \times X)//G = X//U$$

is an isomorphism over $(\mathbf{P}^{r(r+1)} \times X)^s$, it follows that $\widetilde{X//U}^{(G)} \to X//U$ is an isomorphism over X^s/U, and hence we have two compactifications of the geometric quotient X^s/U:

$$\begin{array}{ccc} X^s/U & \subseteq & \widetilde{X//U}^{(G)} \\ \| & & \downarrow \\ X^s/U & \subseteq & X//U \end{array}$$

where $\widetilde{X//U}^{(G)}$ is an orbifold.

$U = (\mathbf{C}^+)^r$ is the unipotent radical of a parabolic subgroup

$$P = U \rtimes GL(r; \mathbf{C}) \tag{4.2}$$

in $SL(r+1; \mathbf{C})$ with Levi subgroup $GL(r; \mathbf{C})$ embedded in $SL(r+1; \mathbf{C})$ as

$$g \mapsto \begin{pmatrix} g & 0 \\ 0 & \det g^{-1} \end{pmatrix}.$$

We have

$$G \times_U X \cong G \times_P (P \times_U X)$$

where $P/U \cong GL(r; \mathbf{C})$ and $G/P \cong \mathbf{P}^r$ is projective. If $\overline{P \times_U X}$ is a P-equivariant projective completion of $P \times_U X$ then $G \times_P (\overline{P \times_U X})$ is a projective completion of $G \times_U X$. When the action of U on X extends to a G-action as above, we can choose $\overline{P \times_U X}$ to be the closure of $P \times_U X$ in

$$\overline{G \times_U X} = \overline{G/U} \times X = \mathbf{P}^{r(r+1)} \times X;$$

that is,

$$\overline{P \times_U X} = \mathbf{P}^{r^2} \times X \subseteq \overline{G \times_U X}$$

where $\mathbf{P}^{r^2} = \mathbf{P}(\mathbf{C} \oplus ((\mathbf{C}^r)^* \otimes \mathbf{C}^r))$. There is then a birational morphism

$$G \times_P (\overline{P \times_U X}) \to \overline{G \times_U X}$$

given by $[g, y] \mapsto gy$ which is an isomorphism over $G \times_U X$. The resulting pullback $\hat{L} = \hat{L}^{(N)}$ to $G \times_P (\overline{P \times_U X})$ of $\mathcal{O}_{\mathbf{P}^{r(r+1)}}(N) \otimes L$ is isomorphic to the induced line bundle

$$G \times_P (\mathcal{O}_{\mathbf{P}^{r^2}}(N) \otimes L)$$

on $G \times_P (\overline{P \times_U X})$, where the P-action on $\mathcal{O}_{\mathbf{P}^{r^2}}(N) \otimes L$ is the restriction of the G-action on $\mathcal{O}_{\mathbf{P}^{r(r+1)}}(N) \otimes L$. If we regard $G \times_P (\overline{P \times_U X})$ as a subvariety in the obvious way of

$$G \times_P (\overline{G \times_U X}) = G \times_P (\mathbf{P}^{r(r+1)} \times X)$$

$$\cong (G/P) \times \mathbf{P}^{r(r+1)} \times X \cong \mathbf{P}^r \times \mathbf{P}^{r(r+1)} \times X$$

then the birational morphism

$$G \times_P (\overline{P \times_U X}) \to \overline{G \times_U X} \cong \mathbf{P}^{r^2} \times X$$

given by $[g, y] \mapsto gy$ extends to the projection

$$\mathbf{P}^r \times \mathbf{P}^{r(r+1)} \times X \to \mathbf{P}^{r(r+1)} \times X$$

and so $\hat{L}^{(N)}$ is the restriction to $G \times_P (\overline{P \times_U X})$ of $\mathcal{O}_{\mathbf{P}^{r(r+1)}}(N) \otimes L$. Thus this line bundle $\hat{L} = \hat{L}^{(N)}$ is not ample, but its tensor product

$\hat{L}_\epsilon = \hat{L}_\epsilon^{(N)}$ with the pullback via the morphism

$$G \times_P (\overline{P \times_U X}) \to G/P \cong \mathbf{P}^r,$$

of the fractional line bundle $\mathcal{O}_{\mathbf{P}^r}(\epsilon)$, where $\epsilon \in \mathbb{Q} \cap (0, \infty)$, provides an ample fractional linearisation for the action of G on $G \times_P (\overline{P \times_U X})$ with, when ϵ is sufficiently small, an induced surjective birational morphism

$$\widehat{X/\!/U} =_{df} G \times_P (\overline{P \times_U X})/\!/_{\hat{L}_\epsilon} G \to \overline{G \times_U X}/\!/G = X/\!/U \qquad (4.3)$$

(cf. [Ki2, Rei]) which is an isomorphism over

$$(G \times_U X^{\bar{s}})/G \cong X^{\bar{s}}/U = X^s/U.$$

Note that $\hat{L}_\epsilon$ can be thought of as the bundle $G \times_P (\mathcal{O}_{\mathbf{P}^{r^2}}(N) \otimes L)$ on $G \times_P (\overline{P \times_U X})$, where now the P-action on $\mathcal{O}_{\mathbf{P}^{r^2}}(N) \otimes L$ is no longer the restriction of the G-action on $\mathcal{O}_{\mathbf{P}^{r(r+1)}}(N) \otimes L$ but has been twisted by ϵ times the character of P which restricts to the determinant on $GL(r; \mathbf{C})$.

Remark 4.2. It follows from variation of GIT [Res] for the G-action on $G \times_P (\overline{P \times_U X})$ that $\widehat{X/\!/U} = \widehat{X/\!/U}^{N,\epsilon}$ is independent of N and ϵ, provided that N is sufficiently large and $\epsilon > 0$ is sufficiently small, depending on N.

Remark 4.3. When $\epsilon > 0$ the projective completion $G \times_P (\overline{P \times_U X})$ equipped with the induced ample fractional linearisation on $\hat{L}_\epsilon$ is not in general a reductive envelope for the U-action on X, though it satisfies all the remaining conditions when (ii) is omitted from Definition 3.6 (cf. Remark 3.12). If we use the linearisation on $\hat{L}_0 = \hat{L}$ instead, then we do obtain a reductive envelope, but it is not ample; nonetheless the conditions of Proposition 3.13 are satisfied and we have

$$\overline{G \times_P (\overline{P \times_U X})/\!/_{\hat{L}_0} G} = \overline{G \times_U X}/\!/G = X/\!/U.$$

Example 4.4. Let $U = \mathbf{C}^+$ act linearly on a projective space $\mathbf{P}^n$. Then we can choose coordinates so that $1 \in \mathrm{Lie}(\mathbf{C}^+) = \mathbf{C}$ has Jordan normal form with blocks

$$\begin{pmatrix} 0 & 1 & 0 & 0 & \cdots & 0 \\ 0 & 0 & 1 & 0 & \cdots & 0 \\ & & & \cdots & & \\ 0 & 0 & \cdots & 0 & 0 & 1 \\ 0 & 0 & \cdots & 0 & 0 & 0 \end{pmatrix}$$

of sizes $k_1+1, \ldots, k_s+1$ where $\sum_{j=1}^s (k_j+1) = n+1$. Then the $\mathbf{C}^+$ action extends to an action of $G = SL(2; \mathbf{C})$ via the identifications

$$\mathbf{C}^+ \cong \{ \begin{pmatrix} 1 & a \\ 0 & 1 \end{pmatrix} : a \in \mathbf{C} \} \leq G$$

and

$$\mathbf{C}^{n+1} \cong \bigoplus_{j=1}^{s} \mathrm{Sym}^{k_j}(\mathbf{C}^2)$$

where $\mathrm{Sym}^k(\mathbf{C}^2)$ is the kth symmetric power of the standard representation $\mathbf{C}^2$ of $G = SL(2; \mathbf{C})$. Moreover

$$G/\mathbf{C}^+ \cong \mathbf{C}^2 \smallsetminus \{0\} \subseteq \mathbf{C}^2 \subseteq \mathbf{P}^2 = \overline{G/\mathbf{C}^+}$$

and thus we have

$$\mathbf{P}^n /\!/ \mathbf{C}^+ \cong \mathrm{Proj}(\mathbf{C}[x_0, \dots, x_n]^{\mathbf{C}^+}) \cong (\mathbf{P}^2 \times \mathbf{P}^n) /\!/ G$$

with respect to the linearisation $\mathcal{O}_{\mathbf{P}^2}(N) \otimes \mathcal{O}_{\mathbf{P}^n}(1)$ on $\mathbf{P}^2 \times \mathbf{P}^n$ for N a sufficiently large positive integer. When $G = SL(2; \mathbf{C})$ acts on $\mathbf{P}^2$ we have $(\mathbf{P}^2)^{ss,G} = \mathbf{C}^2$ (and $(\mathbf{P}^2)^{s,G} = \emptyset$), so since N is large we have

$$(\mathbf{P}^2 \times \mathbf{P}^n)^{ss,G} \subseteq \mathbf{C}^2 \times \mathbf{P}^n = (G \times_{\mathbf{C}^+} \mathbf{P}^n) \sqcup (\{0\} \times \mathbf{P}^n)$$

and if semistability implies stability then

$$\mathbf{P}^n /\!/ \mathbf{C}^+ = (\mathbf{P}^n)^{s,U}/\mathbf{C}^+ \sqcup (\{0\} \times \mathbf{P}^n)/\!/ SL(2; \mathbf{C}).$$

In this example the parabolic subgroup P of $G = SL(2; \mathbf{C})$ is its Borel subgroup

$$B = \{ \begin{pmatrix} a & b \\ 0 & a^{-1} \end{pmatrix} : a \in \mathbf{C}^*, b \in \mathbf{C} \}$$

with $\overline{B/\mathbf{C}^+} = \overline{\mathbf{C}^*} = \mathbf{P}^1$ and

$$\overline{B \times_{\mathbf{C}^+} \mathbf{P}^n} = \mathbf{P}^1 \times \mathbf{P}^n,$$

while $G \times_B \overline{B/\mathbf{C}^+} = G \times_B \mathbf{P}^1$ is the blow-up of $\mathbf{P}^2$ at the origin $0 \in \mathbf{C}^2 \subseteq \mathbf{P}^2$. Similarly $G \times_B (\overline{B \times_{\mathbf{C}^+} \mathbf{P}^n})$ is the blow-up of $\overline{G \times_{\mathbf{C}^+} \mathbf{P}^n} \cong \mathbf{P}^2 \times \mathbf{P}^n$ along $\{0\} \times \mathbf{P}^n$, and its quotient $\widehat{X /\!/ U}$ is the blow-up of $\mathbf{P}^n /\!/ \mathbf{C}^+$ along its 'boundary'

$$\begin{aligned} \mathbf{P}^n /\!/ SL(2; \mathbf{C}) &\cong (\{0\} \times \mathbf{P}^n) /\!/ SL(2; \mathbf{C}) \subseteq (\mathbf{P}^2 \times \mathbf{P}^n) /\!/ SL(2; \mathbf{C}) \\ &= \mathbf{P}^n /\!/ \mathbf{C}^+. \end{aligned}$$

Let us continue to assume that $U = (\mathbf{C}^+)^r$ acts linearly on X and that the action extends to $G = SL(r + 1; \mathbf{C})$. Notice that there are surjections

$$\overline{P \times_U X}^{ss,P,\epsilon} \to \widehat{X//U} \to X//U \qquad (4.4)$$

where $\overline{P \times_U X}^{ss,P,\epsilon}$ is the intersection of $\overline{P \times_U X}$ with the G-semistable set in $G \times_P \overline{P \times_U X}$ with respect to the linearisation $\hat{L}_\epsilon$, and y_1, $y_2 \in \overline{P \times_U X}^{ss,P,\epsilon}$ map to the same point in $\widehat{X//U}$ if and only if the closures of their P-orbits Py_1 and Py_2 meet in $\overline{P \times_U X}^{ss,P,\epsilon}$.

Consider the linear action of the Levi subgroup $GL(r; \mathbf{C}) \leq P$ on $\overline{P \times_U X} = \mathbf{P}^{r^2} \times X$. It follows from the Hilbert-Mumford criteria (Proposition 2.2 above) that

$$\overline{P \times_U X}^{ss,P,\epsilon} \subseteq \overline{P \times_U X}^{ss,GL(r;\mathbf{C}),\epsilon} \subseteq \overline{P \times_U X}^{ss,SL(r;\mathbf{C})} \qquad (4.5)$$

where $\overline{P \times_U X}^{ss,GL(r;\mathbf{C}),\epsilon}$ and $\overline{P \times_U X}^{ss,SL(r;\mathbf{C})}$ (independent of ϵ) denote the $GL(r; \mathbf{C})$ and $SL(r; \mathbf{C})$-semistable sets of $\overline{P \times_U X}$ after twisting the linearisation by ϵ times the character det of $GL(r; \mathbf{C})$; this character is of course trivial on $SL(r; \mathbf{C})$.

It is not hard to check that if the action of $GL(r; \mathbf{C})$ on $\overline{P/U} = \mathbf{P}^{r^2}$ is linearised with respect to $\mathcal{O}_{\mathbf{P}^{r^2}}(1)$ by twisting by the fractional character $\frac{1}{2}$ det then

$$\overline{P/U}^{ss,GL(r;\mathbf{C}),1/2} = \overline{P/U}^{s,GL(r;\mathbf{C}),1/2} = GL(r; \mathbf{C}) \subseteq (\mathbf{C}^r)^* \otimes \mathbf{C}^r \subseteq \mathbf{P}^{r^2}. \qquad (4.6)$$

Thus, if instead of choosing ϵ close to 0 we choose ϵ to be approximately $N/2$, where N is the sufficiently large positive integer chosen above, then we see from the Hilbert-Mumford criteria (Proposition 2.2) that

$$\overline{P \times_U X}^{ss,GL(r;\mathbf{C}),\epsilon} = (\mathbf{P}^{r^2} \times X)^{ss,GL(r;\mathbf{C}),\epsilon} = GL(r; \mathbf{C}) \times X$$

and so quotienting we get

$$\overline{P \times_U X} //_{\hat{L}_{N/2}^{(N)}} GL(r; \mathbf{C}) = X.$$

A GIT quotient of a nonsingular complex projective variety Y by a linear action of $GL(r; \mathbf{C})$ can always be constructed by first quotienting by $SL(r; \mathbf{C})$ and then quotienting by the induced linear action of $\mathbf{C}^* = GL(r; \mathbf{C})/SL(r; \mathbf{C})$: we have

$$Y//GL(r; \mathbf{C}) = (Y//SL(r; \mathbf{C})) // \mathbf{C}^*.$$

Therefore if we set

$$\mathcal{X} = \overline{P \times_U X} /\!\!/_{\hat{L}(N)} SL(r; \mathbf{C}) = (\mathbf{P}^{r^2} \times X) /\!\!/_{\hat{L}(N)} SL(r; \mathbf{C}) \qquad (4.7)$$

for $N > 0$ sufficiently large, then $\mathcal{X}$ is a projective variety with a linear action of $\mathbf{C}^*$ which we can twist by ϵ times the standard character of $\mathbf{C}^*$, such that when $\epsilon = N/2$ we get

$$\mathcal{X} /\!\!/_{N/2} \mathbf{C}^* \cong X \qquad (4.8)$$

while for $\epsilon > 0$ sufficiently small it follows from (4.4) and (4.5) that we have a surjection from an open subset of $\mathcal{X} /\!\!/_\epsilon \mathbf{C}^*$ onto $\widehat{X /\!\!/ U}$, and hence onto $X /\!\!/ U$. More precisely, the inclusion

$$(\hat{\mathcal{O}}_{\hat{L}_\epsilon}(G \times_P (\overline{P \times_U X})))^G = (\hat{\mathcal{O}}_{\hat{L}_\epsilon}(\overline{P \times_U X}))^P \subseteq (\hat{\mathcal{O}}_{\hat{L}_\epsilon}(\overline{P \times_U X}))^{GL(r;\mathbf{C})}$$

induces a rational map

$$\mathcal{X} /\!\!/_\epsilon \mathbf{C}^* = \overline{P \times_U X} /\!\!/_{\hat{L}_\epsilon} GL(r; \mathbf{C}) - - \to G \times_P \overline{P \times_U X} /\!\!/_{\hat{L}_\epsilon} G = \widehat{X /\!\!/ U}$$
$$(4.9)$$

whose composition with the surjection

$$\overline{P \times_U X}^{ss, GL(r;\mathbf{C}), \epsilon} \to \mathcal{X} /\!\!/_\epsilon \mathbf{C}^*$$

induced by the inclusion

$$(\hat{\mathcal{O}}_{\hat{L}_\epsilon}(\overline{P \times_U X}))^{GL(r;\mathbf{C})} \subseteq (\hat{\mathcal{O}}_{\hat{L}_\epsilon}(\overline{P \times_U X}))^{SL(r;\mathbf{C})}$$

is the rational map

$$\overline{P \times_U X}^{ss, GL(r;\mathbf{C}), \epsilon} - - \to G \times_P \overline{P \times_U X} /\!\!/_{\hat{L}_\epsilon} G = \widehat{X /\!\!/ U}$$

which restricts to a surjection

$$\overline{P \times_U X}^{ss, P, \epsilon} \to \widehat{X /\!\!/ U}.$$

Hence the restriction of $\mathcal{X} /\!\!/_\epsilon \mathbf{C}^* - - \to \widehat{X /\!\!/ U}$ to its domain of definition is surjective.

Definition 4.5 Let $(\mathcal{X} /\!\!/_\epsilon \mathbf{C}^*)^{\hat{s}s}$ denote the open subset of $\mathcal{X} /\!\!/_\epsilon \mathbf{C}^*$ which is the domain of definition of the rational map (4.9) from $\mathcal{X} /\!\!/_\epsilon \mathbf{C}^*$ to $\widehat{X /\!\!/ U}$, where as above $\mathcal{X}$ is the projective variety

$$\mathcal{X} = (\overline{P \times_U X}) /\!\!/ SL(r; \mathbf{C})$$

with the induced linear $\mathbf{C}^*$-action, and $0 < \epsilon \ll 1$. Let $(\mathcal{X} /\!\!/_\epsilon \mathbf{C}^*)^{\hat{s}}$ be the open subset $\overline{P \times_U X}^{s,P,\epsilon}/GL(r;\mathbf{C})$ of

$$\overline{P \times_U X}^{s,GL(r;\mathbf{C}),\epsilon}/GL(r;\mathbf{C}) = (\overline{P \times_U X}^{s,GL(r;\mathbf{C}),\epsilon}/SL(r;\mathbf{C}))/\mathbf{C}^*$$

$$= \mathcal{X}^{s,\epsilon}/\mathbf{C}^* \subseteq \mathcal{X}//_\epsilon \mathbf{C}^*.$$

Let $\mathcal{X}^{\hat{s}s,\epsilon} = \pi^{-1}((\mathcal{X}//_\epsilon\mathbf{C}^*)^{\hat{s}s})$ and $\mathcal{X}^{\hat{s},\epsilon} = \pi^{-1}((\mathcal{X}//_\epsilon\mathbf{C}^*)^{\hat{s}})$ where $\pi :$
$\mathcal{X}^{ss,\epsilon} \to \mathcal{X}//_\epsilon\mathbf{C}^*$ is the quotient map, so that

$$(\mathcal{X}//_\epsilon\mathbf{C}^*)^{\hat{s}} = \mathcal{X}^{\hat{s},\epsilon}/\mathbf{C}^*.$$

△

Then we have

Proposition 4.6. *If $\epsilon > 0$ is sufficiently small, the rational map from* $\mathcal{X}//_\epsilon\mathbf{C}^*$ *to* $\widehat{X//U}$ *induced by the inclusion of* $(\hat{\mathcal{O}}_{\hat{L}_\epsilon}(\overline{P \times_U X}))^P$ *in* $(\hat{\mathcal{O}}_{\hat{L}_\epsilon}(\overline{P \times_U X}))^{GL(r;\mathbf{C})}$ *restricts to surjective morphisms*

$$(\mathcal{X}//_\epsilon\mathbf{C}^*)^{\hat{s}s} \to \widehat{X//U} \to X//U$$

and

$$(\mathcal{X}//_\epsilon\mathbf{C}^*)^{\hat{s}} \to X^s/U.$$

Remark 4.7. Using the theory of variation of GIT [DH, Res, Th], as described in Remark 2.6, we can relate the quotient $\mathcal{X}//_\epsilon\mathbf{C}^*$ which appears in Proposition 4.6 above to $\mathcal{X}//_{N/2}\mathbf{C}^* \cong X$ via a sequence of flips which occur as walls are crossed between the linearisations corresponding to ϵ and to $N/2$. Thus we have a diagram

$$
\begin{array}{ccccccc}
(\mathcal{X}//_\epsilon\mathbf{C}^*)^{\hat{s}} & \subseteq & (\mathcal{X}//_\epsilon\mathbf{C}^*)^{\hat{s}s} & \subseteq & \mathcal{X}//_\epsilon\mathbf{C}^* & \longleftarrow\text{--}\to & X = \mathcal{X}//_{N/2}\mathbf{C}^* \\
\downarrow & & \downarrow & & & \text{flips} & \\
X^s/U & \subseteq & \widehat{X//U} & & & & \\
\| & & \downarrow & & & & \\
X^s/U & \subseteq & X//U & & & &
\end{array}
$$

where the vertical maps are all *surjective*, in contrast to (3.3), and the inclusions are all open.

Note also that by variation of GIT if $0 < \epsilon \ll 1$ there is a birational surjective morphism

$$\mathcal{X}//_\epsilon\mathbf{C}^* \to \mathcal{X}//_0\mathbf{C}^*.$$

When $\epsilon = 0$ the inclusion

$$(\hat{\mathcal{O}}_{\hat{L}_0}(G \times_P (\overline{P \times_U X})))^G = (\hat{\mathcal{O}}_{\hat{L}_0}(\overline{P \times_U X}))^P \subseteq (\hat{\mathcal{O}}_{\hat{L}_0}(\overline{P \times_U X}))^{GL(r;\mathbf{C})}$$

induces a rational map

$$\mathcal{X}/\!/_0 \mathbf{C}^* = \overline{P \times_U X}/\!/_{\hat{L}_0} GL(r; \mathbf{C}) - - \to G \times_P \overline{P \times_U X}/\!/_{\hat{L}_0} G = X/\!/U \tag{4.10}$$

whose composition with the surjective morphism

$$\mathcal{X}/\!/_\epsilon \mathbf{C}^* \to \mathcal{X}/\!/_0 \mathbf{C}^*$$

is the composition of (4.9) with the surjective morphism $\widetilde{X/\!/U} \to X/\!/U$. Thus the restriction of the rational map (4.10) from $\mathcal{X}/\!/_0 \mathbf{C}^*$ to $X/\!/U$ to its domain of definition is surjective.

Remark 4.8. Note that the GIT quotient $\mathbf{P}^{r^2}/\!/SL(r; \mathbf{C})$ is isomorphic to $\mathbf{P}^1$, but we do not have $(\mathbf{P}^{r^2})^{ss} = (\mathbf{P}^{r^2})^s$ for this action of $SL(r; \mathbf{C})$. It is therefore convenient to replace the compactification $\mathbf{P}^{r^2}$ of $GL(r; \mathbf{C})$ by its *wonderful compactification* $\widetilde{\mathbf{P}^{r^2}}$ given by blowing up $\mathbf{P}^{r^2} = \{[z : (z_{ij})_{i,j=1}^r]\}$ along the (proper transforms of the) subvarieties defined by

$$z = 0 \text{ and } \mathrm{rank}(z_{ij}) \le \ell$$

for $\ell = 0, 1, \ldots, r$ and by

$$\mathrm{rank}(z_{ij}) \le \ell$$

for $\ell = 0, 1, \ldots, r - 1$ [Ka]. The action of $SL(r; \mathbf{C})$ on $\widetilde{\mathbf{P}^{r^2}}$, linearised with respect to a small perturbation $\mathcal{O}_{\widetilde{\mathbf{P}^{r^2}}}(1)$ of the pullback of $\mathcal{O}_{\mathbf{P}^{r^2}}(1)$, satisfies

$$\widetilde{\mathbf{P}^{r^2}}^{ss} = \widetilde{\mathbf{P}^{r^2}}^{s} \text{ and } \widetilde{\mathbf{P}^{r^2}}/\!/SL(r; \mathbf{C}) \cong \mathbf{P}^1.$$

Thus if we replace $\overline{P \times_U X} = \mathbf{P}^{r^2} \times X$ with

$$\widetilde{P \times_U X} = \widetilde{\mathbf{P}^{r^2}} \times X$$

and define $\widetilde{X/\!/U} = G \times_P (\widetilde{P \times_U X})/\!/_{\hat{L}_\epsilon} G$ and

$$\widetilde{\mathcal{X}} = \widetilde{P \times_U X}/\!/_{\hat{L}(N)} SL(r; \mathbf{C}) = (\widetilde{\mathbf{P}^{r^2}} \times X)/\!/_{\hat{L}(N)} SL(r; \mathbf{C}) \tag{4.11}$$

for $N \gg 0$, then all the properties of $\mathcal{X}$ given above still hold for $\widetilde{\mathcal{X}}$, and in addition $\widetilde{\mathcal{X}}$ fibres over $\mathbf{P}^1$ as

$$\widetilde{\mathcal{X}} = (\widetilde{\mathbf{P}^{r^2}} \times X)/\!/_{\hat{L}(N)} SL(r; \mathbf{C}) = (\widetilde{\mathbf{P}^{r^2}}^{ss} \times X)/SL(r; \mathbf{C})$$

$$\to \widetilde{\mathbf{P}^{r^2}}^{ss}/SL(r; \mathbf{C}) = \mathbf{P}^1$$

with fibres isomorphic to the quotient of X by the finite centre of $SL(r; \mathbf{C})$. If X is nonsingular then it turns out that $\widetilde{\mathcal{X}}$ and $\widetilde{\mathcal{X}}/\!/_\epsilon \mathbf{C}^*$

(for $0 < \epsilon \ll 1$) and $\widetilde{X/\!/U}$ are orbifolds, so that $\widetilde{X/\!/U}$ is a projective completion of X^s/U which is a partial desingularisation of $X/\!/U$ (cf. Remark 4.1).

4.2 General $(\mathbf{C}^+)^r$ actions

Of course the constructions described in §4.1 only work if the action of $U = (\mathbf{C}^+)^r$ on X extends to an action of $G = SL(\mathbf{C} \oplus \mathbf{u})$, which is a rather special situation when the ring of invariants $\hat{\mathcal{O}}_L(X)^U$ is always finitely generated. Moreover at least *a priori* these constructions may depend on the choice of this extension, although $\overline{G \times_U X}/\!/G = X/\!/U = \mathrm{Proj}(\hat{\mathcal{O}}_L(X)^U)$ depends only on the linearisation of the U-action on X. So next we need to consider what happens if the linear U-action on X does not extend to a linear action of G. Suppose that we can associate to the linear U-action on X a normal projective variety Y containing X, with an action of $G = SL(\mathbf{C} \oplus \mathbf{u})$ and a G-linearisation on a line bundle L_Y, which restricts to the given linearisation of the U-action on X and is such that every U-invariant in a finite fully separating set of U-invariants on X extends to a U-invariant on Y. Then we can embed X in the G-variety

$$\mathbf{P}^{r(r+1)} \times Y$$

as $\{\iota\} \times X$ where $\iota \in (\mathbf{C}^r)^* \otimes \mathbf{C}^{r+1} \subseteq \mathbf{P}^{r(r+1)}$ is the standard embedding of $\mathbf{C}^r$ in $\mathbf{C}^{r+1}$, and the closure of $GX \cong G \times_U X$ in $\mathbf{P}^{r(r+1)} \times Y$ will provide us with a reductive envelope $\overline{G \times_U X}$. Therefore we will next consider how, given any linearised U-action on X, we can choose a G-variety Y with these properties. We will find that for any sufficiently divisible positive integer m we can choose such a variety Y_m in a canonical way, depending only on m and the linear action of U on X, giving us a reductive envelope $\overline{G \times_U X}^m$.

Let S be any finite fully separating set of invariants (in the sense of Definition 3.6) on X. By replacing the elements of S with suitable powers of themselves, we can assume that $S \subseteq H^0(X, L^{\otimes m})^U$ for some $m > 0$ for which $L^{\otimes m}$ is very ample. Then $X \subseteq \mathbf{P}(H^0(X, L^{\otimes m})^*)$ and every $\sigma \in S$ extends to a U-invariant section of $\mathcal{O}(1)$ on $\mathbf{P}(H^0(X, L^{\otimes m})^*)$.

Now consider the linear action of U on $V_m = H^0(X, L^{\otimes m})^*$, and let P be the parabolic subgroup of $G = SL(r+1; \mathbf{C})$ with unipotent radical U, as at (4.2) above. Since P is a semi-direct product

$$P = U \rtimes GL(r; \mathbf{C})$$

we have

$$P \times_U V_m \cong GL(r; \mathbf{C}) \times V_m$$

with the P-action on $GL(r; \mathbf{C}) \times V_m$ given for $(h, v) \in GL(r; \mathbf{C}) \times V_m$ by

$$p.(h, v) = (gh, (h^{-1}uh).v)$$

where $p = gu$ with $g \in GL(r; \mathbf{C})$ and $u \in U$, and $h^{-1}uh$ acts on $v \in V_m$ via the given U-action. Of course $GL(r; \mathbf{C}) \times V_m$ is an affine variety with

$$\mathcal{O}(GL(r; \mathbf{C}) \times V_m) \cong \mathbf{C}[h_{ij}, (\det h)^{-1}, v_k]$$

where $\det h$ is the determinant of the $r \times r$ matrix $(h_{ij})_{i,j=1}^r$ and (v_k) are coordinates on V_m. Let

$$\varphi_{V_m} : \mathbf{C}^r = \mathrm{Lie}\, U \to \mathrm{Lie}(GL(V_m)) \tag{4.12}$$

be the infinitesimal action of U on V_m and let U_{V_m} be its image in $\mathrm{Lie}(GL(V_m))$. Since U is unipotent we have

$$V_m \supseteq U_{V_m}(V_m) \supseteq (U_{V_m})^2(V_m) \supseteq \cdots \supseteq (U_{V_m})^{\dim V_m}(V_m) = 0$$

where

$$(U_{V_m})^j(V_m) = \{u_1 u_2 \cdots u_j(v) \ : \ u_1, \ldots, u_j \in U_{V_m}, \ v \in V_m\}$$

$$= \{\varphi_{V_m}(\tilde{u}_1)\varphi_{V_m}(\tilde{u}_2) \cdots \varphi_{V_m}(\tilde{u}_j)(v) \ : \ \tilde{u}_1, \ldots, \tilde{u}_j \in \mathrm{Lie}\, U, \ v \in V_m\}.$$

For $0 \le j \le \dim V_m - 1$ let $\Theta_{j,m}$ be the complex vector space consisting of all polynomial functions

$$\theta : (\mathbf{C}^r)^j \times ((\mathbf{C}^r)^* \otimes \mathbf{C}^r)^{\dim V_m - 1} \to \mathbf{C}$$

$$(u_1, \ldots, u_j, h_1, \ldots, h_{\dim V_m - 1}) \mapsto \theta(u_1, \ldots, u_j, h_1, \ldots, h_{\dim V_m - 1})$$

which are simultaneously homogeneous of degree 1 in the coordinates of each $u_i \in \mathbf{C}^r$ separately, for $1 \le i \le j$, and homogeneous of total degree $r(\dim V_m - 1) - (r - 1)j$ in the coordinates of all the $h_k \in (\mathbf{C}^r)^* \otimes \mathbf{C}^r$

together, for $1 \leq k \leq \dim V_m - 1$. Let

$$W_m = \bigoplus_{j=0}^{\dim V_m - 1} \Theta_{j,m} \otimes (U_{V_m})^j (V_m). \tag{4.13}$$

Then we can embed V_m linearly into W_m via

$$v \mapsto \psi(v) = \sum_{j=0}^{\dim V_m - 1} \psi_j(v) \tag{4.14}$$

where $\psi_j(v) : (\mathbf{C}^r)^j \times ((\mathbf{C}^r)^* \otimes \mathbf{C}^r)^{(\dim V_m - 1)} \to (U_{V_m})^j(V_m)$ for $0 \leq j \leq \dim V_m - 1$ and $v \in V_m$ sends $(u_1, \ldots, u_j, h_1, \ldots, h_{\dim V_m - 1})$ to the product of

$$\det(h_1) \det(h_2) \cdots \det(h_{\dim V_m - j - 1})$$

with

$$\varphi_{V_m}(h_{\dim V_m - 1} u_1) \cdots \varphi_{V_m}(h_{\dim V_m - j} u_j)(v).$$

In particular $\psi_0(v) : ((\mathbf{C}^r)^* \otimes \mathbf{C}^r)^{(\dim V_m - 1)}) \to (U_{V_m})^0(V_m) = V_m$ sends $(h_1, \ldots, h_{\dim V_m - 1})$ to

$$(\det(h_1) \det(h_2) \cdots \det(h_{\dim V_m - 1}))v \in V_m.$$

This embedding of V_m into W_m is U-equivariant with respect to the linear U-action on W_m for which the infinitesimal action of $u \in \mathrm{Lie}\, U = \mathbf{C}^r$ is given by

$$u \left(\sum_{j=0}^{\dim V_m - 1} \alpha_j \right) = \sum_{j=0}^{\dim V_m - 1} u \cdot \alpha_{j+1} \tag{4.15}$$

with $u \cdot \alpha_{j+1} : (\mathbf{C}^r)^j \times ((\mathbf{C}^r)^* \otimes \mathbf{C}^r)^{\dim V_m - 1} \to (U_{V_m})^j(V_m)$ defined to be 0 for $j = \dim V_m$ and for any $0 \leq j < \dim V_m$ defined by

$$u \cdot \alpha_{j+1}(u_1, \ldots, u_j, h_1, \ldots, h_{\dim V_m - 1})$$

$$= \alpha_{j+1}(u_1, \ldots, u_j, \mathrm{adj}(h_{\dim V_m - 1 - j})(u), h_1, \ldots, h_{\dim V_m - 1}).$$

Here $\mathrm{adj}(h)$ is the adjoint matrix of h (so that $h\, \mathrm{adj}(h) = \mathrm{adj}(h)\, h$ is $\det(h)$ times the identity matrix ι), which is homogeneous of degree $r - 1$ in the coefficients of h. Moreover this linear action of U on W_m extends to a linear action of $P = U \rtimes GL(r; \mathbf{C})$ on W_m where $g \in GL(r; \mathbf{C})$ acts as

$$g \left(\sum_{j=0}^{\dim V_m - 1} \alpha_j \right) = \sum_{j=0}^{\dim V_m - 1} (\det g)^j \, g \cdot \alpha_j \qquad (4.16)$$

with $g \cdot \alpha_j : (\mathbf{C}^r)^j \times ((\mathbf{C}^r)^* \otimes \mathbf{C}^r)^{\dim V_m - 1} \to (U_{V_m})^j (V_m)$ defined for any $0 \le j < \dim V_m$ and any $\alpha_j : (\mathbf{C}^r)^j \times ((\mathbf{C}^r)^* \otimes \mathbf{C}^r)^{\dim V_m - 1} \to (U_{V_m})^j (V_m)$ by

$$g \cdot \alpha_j (u_1, \ldots, u_j, h_1, \ldots, h_{\dim V_m - 1})$$

$$= \alpha_j (gu_1, \ldots, gu_j, gh_1 g^{-1}, \ldots, gh_{\dim V_m - 1} g^{-1}).$$

Since the U-action on $\mathbf{P}(W_m)$ extends to a linear P-action, we can construct the projective variety

$$Y_m = G \times_P \mathbf{P}(W_m)$$

and we can equip Y_m with the line bundles $L^\epsilon_{Y_m} = G \times_P \mathcal{O}_{\mathbf{P}(W_m)}(1)$ for $\epsilon \in \mathbb{Q}$ where the P-action on $\mathcal{O}_{\mathbf{P}(W_m)}(1)$ is induced from the given P-action on W_m twisted by ϵ times the pullback to P of the character $\det$ of $P/U \cong GL(r; \mathbf{C})$. Equivalently $L^\epsilon_{Y_m}$ is the tensor product of $L^0_{Y_m}$ with the pullback via

$$G \times_P \mathbf{P}(W_m) \to G/P \cong \mathbf{P}^r$$

of $\mathcal{O}_{\mathbf{P}^r}(\epsilon)$.

Remark 4.9. The action of P on W_m extends to a linear action of $G = SL(r + 1; \mathbf{C})$ on $\mathcal{W}_m \supseteq W_m$, where

$$\mathcal{W}_m = H^0 (G \times_P \mathbb{P}(W_m \otimes \bigoplus_{\lambda \in \Delta} \mathbb{C}_\lambda); G \times_P \mathcal{O}(1))$$

for a sufficiently large finite set Δ of weights of P, with $\mathbb{C}_\lambda$ denoting a copy of $\mathbb{C}$ on which P acts with weight λ.

Thus $G \times_P \mathbf{P}(W_m) \subseteq G \times_P \mathbf{P}(\mathcal{W}_m) \cong G/P \times \mathbf{P}(\mathcal{W}_m) \cong \mathbf{P}^r \times \mathbf{P}(\mathcal{W}_m)$, and with respect to this identification we have

$$L^0_{Y_m} = \mathcal{O}_{\mathbf{P}(\mathcal{W}_m)}(1)|_{Y_m}.$$

Hence if $\epsilon > 0$ it follows that $L^\epsilon_{Y_m}$ is the restriction of the line bundle $\mathcal{O}_{\mathbf{P}^r}(\epsilon) \otimes \mathcal{O}_{\mathbf{P}(\mathcal{W}_m)}(1)$ on $\mathbf{P}^r \times \mathbf{P}(\mathcal{W}_m)$, and so $L^\epsilon_{Y_m}$ is ample.

This gives us for every sufficiently divisible positive integer m a U-equivariant embedding

$$X \subseteq \mathbf{P}(H^0(X, L^{\otimes m})^*) = \mathbf{P}(V_m) \subseteq \mathbf{P}(W_m)$$

in a projective variety with a linear P-action, such that every σ in the finite fully separating set of invariants S extends to a U-invariant linear functional on V_m, and hence extends to a U-invariant (in fact P-invariant) linear functional on W_m defined by

$$\sum_{j=0}^{\dim V_m} \alpha_j \mapsto \sigma(\alpha_0(\iota, \ldots, \iota)).$$

As each $\sigma \in S$ extends to a P-invariant linear functional on W_m, it extends to a G-invariant section of $L^0_{Y_m}$. Thus from §4.1 we have

Proposition 4.10. *Let X be embedded in $Y_m = G \times_P \mathbf{P}(W_m)$ as above, for a sufficiently divisible positive integer m, and let $\iota \in (\mathbf{C}^r)^* \otimes \mathbf{C}^{r+1} \subseteq \mathbf{P}^{r(r+1)}$ be the standard embedding of $\mathbf{C}^r$ in $\mathbf{C}^{r+1}$. If N is sufficiently large (depending on m), then the linear action of $U = (\mathbf{C}^+)^r$ on X has a reductive envelope given by the closure $\overline{G \times_U X}^m$ of $G \times_U X$ embedded in $\mathbf{P}^{r(r+1)} \times Y_m$ as $G(\{\iota\} \times X)$, equipped with the restriction of the G-linearisation on $\mathcal{O}_{\mathbf{P}^{r(r+1)}}(N) \otimes L^0_{Y_m}$.*

Note that by Remark 4.9 the line bundle $L^0_{Y_m}$ is not in general ample (although $L^\epsilon_{Y_m}$ is ample for any $\epsilon > 0$), so we do not necessarily have an ample reductive envelope here. Nonetheless by Proposition 3.13 if $X^{\bar{s}}$ and $X^{\overline{nss}}$ are defined using this reductive envelope we have

Corollary 4.11. $X^{\bar{s}} \subseteq X^s \subseteq X^{ss} \subseteq X^{\overline{nss}}$.

If moreover the ring of invariants $\hat{\mathcal{O}}_L(X)^U$ is finitely generated and m is sufficiently divisible that $\hat{\mathcal{O}}_{L^{\otimes m}}(X)^U$ is generated by $H^0(X, L^{\otimes m})^U$, then for $N \gg 0$ the restriction map

$$\rho_m : \bigoplus_{k \geq 0} H^0(\overline{G \times_U X}^m, (\mathcal{O}_{\mathbf{P}^{r(r+1)}}(N) \otimes L^0_{Y_m})^{\otimes k})^G \to \hat{\mathcal{O}}_{L^{\otimes m}}(X)^U$$

is an isomorphism and $X//U = \mathrm{Proj}\, \hat{\mathcal{O}}_L(X)^U$ is the canonical projective completion of $\overline{G \times_U X}^m //G$ (see Proposition 3.13 again). Even when the ring of invariants $\hat{\mathcal{O}}_L(X)^U$ is not finitely generated, if m is sufficiently divisible that $H^0(X, L^{\otimes m})$ contains a finite fully separating set of invariants, then for any multiple $m' = k'm$ of m the subalgebra $\hat{\mathcal{O}}_L^{m'}(X)^U$ of

$\hat{\mathcal{O}}_L(X)^U$ generated by $H^0(X, L^{\otimes m'})^U$ is finitely generated and provides a projective completion

$$\overline{X /\!/ U}^m = \text{Proj } \hat{\mathcal{O}}_L^{m'}(X)^U$$

of $X /\!/ U$, while the restriction of ρ_m to the subalgebra of

$$\bigoplus_{k \geq 0} H^0(\overline{G \times_U X}^m, (\mathcal{O}_{\mathbf{P}^{r(r+1)}}(N) \otimes L_{Y_m}^0)^{\otimes k})^G$$

generated by $H^0(\overline{G \times_U X}^m, (\mathcal{O}_{\mathbf{P}^{r(r+1)}}(N) \otimes L_{Y_m}^0)^{\otimes k'})^G$ gives an isomorphism onto $\hat{\mathcal{O}}_L^{m'}(X)^U$.

By analogy with the construction of $\widehat{X /\!/ U}$ and $\widetilde{X /\!/ U}$ in §4.1, let us also consider the closures $\widehat{G \times_U X}^m$ and $\widetilde{G \times_U X}^m$ of $G \times_U X = G(\{\iota\} \times X)$ in $G \times_P (\mathbf{P}^{r^2} \times Y_m)$ and $G \times_P (\widetilde{\mathbf{P}^{r^2}} \times Y_m)$ respectively. Since $Y_m = G \times_P \mathbf{P}(W_m)$ we have

$$\widehat{G \times_U X}^m \cong G \times_P (\overline{P \times_U X}^m) \cong \overline{G \times_U X}^m$$

and

$$\widetilde{G \times_U X}^m \cong G \times_P (\widetilde{P \times_U X}^m)$$

where $\overline{P \times_U X}^m$ and $\widetilde{P \times_U X}^m$ are the closures of $P \times_U X = P(\{\iota\} \times X)$ in $\mathbf{P}^{r^2} \times \mathbf{P}(W_m)$ and $\widetilde{\mathbf{P}^{r^2}} \times \mathbf{P}(W_m)$ respectively.

Definition 4.12 Let $\hat{L}_\epsilon = \hat{L}_\epsilon^{(N)}$ be the tensor product of the pullback via

$$\widehat{G \times_U X}^m \cong G \times_P (\overline{P \times_U X}^m) \to G/P \cong \mathbf{P}^r$$

of $\mathcal{O}_{\mathbf{P}^r}(\epsilon)$ with the line bundle $G \times_P \hat{L}^{(N)}$ on $G \times_P (\overline{P \times_U X}^m)$ where

$$\hat{L}^{(N)} = \mathcal{O}_{\mathbf{P}^{r^2}}(N) \otimes \mathcal{O}_{\mathbf{P}(W_m)}(1)|_{\overline{P \times_U X}^m} \, ;$$

equivalently $\hat{L}_\epsilon^{(N)} = G \times_P \hat{L}^{(N)}$ where the action of P on $\hat{L}^{(N)}$ is twisted by ϵ times the character det of $P/U \cong GL(r; \mathbf{C})$. Similarly let $\tilde{L}_\epsilon = \tilde{L}_\epsilon^{(N)}$ be the tensor product of the pullback via

$$\widetilde{G \times_U X}^m \cong G \times_P (\widetilde{P \times_U X}^m) \to G/P \cong \mathbf{P}^r$$

of $\mathcal{O}_{\mathbf{P}^r}(\epsilon)$ with the line bundle $G \times_P \tilde{L}^{(N)}$ on $G \times_P (\widetilde{P \times_U X}^m)$, where $\tilde{L}^{(N)} = \mathcal{O}_{\widetilde{\mathbf{P}^{r^2}}}(N) \otimes \mathcal{O}_{\mathbf{P}(W_m)}(1)|_{\widetilde{P \times_U X}^m}$ and as in Remark 4.8 $\mathcal{O}_{\widetilde{\mathbf{P}^{r^2}}}(1)$ is a small perturbation of the pullback of $\mathcal{O}_{\mathbf{P}^{r^2}}(1)$ along $\widetilde{\mathbf{P}^{r^2}} \to \mathbf{P}^{r^2}$ such that the $SL(r; \mathbf{C})$-action lifts to $\mathcal{O}_{\widetilde{\mathbf{P}^{r^2}}}(1)$ and satisfies $(\widetilde{\mathbf{P}^{r^2}})^{ss} =$

$(\widetilde{\mathbf{P}^{r^2}})^s$ and $\widetilde{\mathbf{P}^{r^2}}//SL(r;\mathbf{C}) \cong \mathbf{P}^1$. Note that the line bundles $\hat{L}_\epsilon^{(N)}$ on $\widetilde{G \times_U X}^m \cong G \times_P (\overline{P \times_U X}^m)$ and $\mathcal{O}_{\mathbf{P}^{r(r+1)}}(N) \otimes L_{Y_m}^\epsilon$ on

$$\widetilde{G \times_U X}^m \cong G \times_P (\overline{P \times_U X}^m) \cong \overline{G \times_U X}^m$$

are both G-invariant and both restrict to $\mathcal{O}_{\mathbf{P}^{r^2}}(N) \otimes \mathcal{O}_{\mathbf{P}(W_m)}(1)$ on $\overline{P \times_U X}^m$ with the same P-action, so they are isomorphic to each other. $\triangle$

The line bundles $\hat{L}_\epsilon = \hat{L}_\epsilon^{(N)}$ and $\widetilde{L}_\epsilon = \widetilde{L}_\epsilon^{(N)}$ on $\overline{G \times_U X}^m$ and $\widetilde{G \times_U X}^m$ are ample for $\epsilon > 0$, and the G-actions on $\overline{G \times_U X}^m$ and $\widetilde{G \times_U X}^m$ lift to linear actions on $\hat{L}_\epsilon$ and $\widetilde{L}_\epsilon$.

Definition 4.13 For a positive integer N sufficiently large (depending on m) let

$$\mathcal{X}_m = \overline{P \times_U X}^m //_{\hat{L}^{(N)}} SL(r;\mathbf{C}) \quad \text{and} \quad \widetilde{\mathcal{X}}_m = \widetilde{P \times_U X}^m //_{\widetilde{L}^{(N)}} SL(r;\mathbf{C})$$

and for $\epsilon > 0$ sufficiently small (depending on m and N) let

$$\overline{X//U}^m = \overline{G \times_U X}^m //_{\hat{L}_\epsilon^{(N)}} G \quad \text{and} \quad \widetilde{X//U}^m = \widetilde{G \times_U X}^m //_{\widetilde{L}_\epsilon^{(N)}} G.$$
                                                                  $\triangle$

Remark 4.14. The line bundles $\hat{L}^{(N)}$ and $\widetilde{L}^{(N)}$ are ample on $\overline{P \times_U X}^m$ and $\widetilde{P \times_U X}^m$, and for $\epsilon > 0$ the line bundles $\hat{L}_\epsilon^{(N)}$ and $\widetilde{L}_\epsilon^{(N)}$ are ample on $\overline{G \times_U X}^m$ and $\widetilde{G \times_U X}^m$. Thus it follows from variation of GIT [Res] for the $SL(r;\mathbf{C})$-actions on $\overline{P \times_U X}^m$ and $\widetilde{P \times_U X}^m$ and the $G = SL(r+1;\mathbf{C})$-actions on $\overline{G \times_U X}^m$ and $\widetilde{G \times_U X}^m$ that $\mathcal{X}_m$ and $\widetilde{\mathcal{X}}_m$ and $\overline{X//U}^m$ and $\widetilde{X//U}^m$ are independent of N and ϵ, provided that m is fixed and N is sufficiently large, depending on m, and ϵ is sufficiently small, depending on N and ϵ.

Remark 4.15. Recall that $\hat{L}_\epsilon^{(N)}$ can be identified with the line bundle $G \times_P \hat{L}^{(N)}$ on $G \times_P (\overline{P \times_U X}^m)$ when the action of P on $\hat{L}^{(N)}$ is twisted by ϵ times the character det of $P/U \cong GL(r;\mathbf{C})$. The character det extends to a section of the line bundle $\mathcal{O}_{\mathbf{P}^{r^2}}(r)$ over the projective completion $\mathbf{P}^{r^2}$ of $GL(r;\mathbf{C})$ on which P acts by multiplication by the character det, and thus to a section of the line bundle $G \times_P \mathcal{O}_{\mathbf{P}^{r^2}}$ over $G \times_P (\overline{P \times_U X}^m)$ when the action of P on $\mathcal{O}_{\mathbf{P}^{r^2}}$ is twisted by this character. Tensoring with this section gives us an injection

$$H^0(\overline{G \times_U X}^m, (\hat{L}_0^{(N)})^{\otimes k})^G \to H^0(\overline{G \times_U X}^m, (\hat{L}_\epsilon^{(N+r\epsilon)})^{\otimes k})^G \quad (4.17)$$

where $\epsilon = 1/k$, whose composition with the injection given by the restriction map

$$\rho_m^\epsilon : \bigoplus_{k \geq 0} H^0(\overline{G \times_U X}^m, (\hat{L}_\epsilon^{(N+r\epsilon)})^{\otimes k})^G \to \hat{\mathcal{O}}_{L^{\otimes m}}(X)^U$$

is the restriction map $\rho_m = \rho_m^0$. If the ring of invariants $\hat{\mathcal{O}}_L(X)^U$ is finitely generated and m is sufficiently divisible that ρ_m is an isomorphism, it follows that (4.17) is an isomorphism and thus that

$$(\overline{G \times_U X}^m)^{ss,\epsilon,G} \subseteq (\overline{G \times_U X}^m)^{ss,0,G}$$

and that the inclusion of $(\overline{G \times_U X}^m)^{ss,\epsilon,G}$ in $(\overline{G \times_U X}^m)^{ss,0,G}$ induces a birational surjective morphism

$$\widetilde{X//U}^m \to X//U = \mathrm{Proj}\,(\hat{\mathcal{O}}_L(X)^U).$$

Even when $\hat{\mathcal{O}}_L(X)^U$ is not finitely generated, if m and k are sufficiently divisible that $H^0(X, L^{\otimes m})^U$ contains a fully separating set of invariants and $\mathcal{O}_{(\hat{L}_\epsilon^{(N+r\epsilon)})^{\otimes k}}(\overline{G \times_U X}^m)$ is generated by

$$H^0(\overline{G \times_U X}^m, (\hat{L}_\epsilon^{(N+r\epsilon)})^{\otimes k})^G,$$

then we have a birational surjective morphism

$$\widetilde{X//U}^m \to \mathrm{Proj}\,(\hat{\mathcal{O}}_L^{m'}(X)^U) = \overline{X//U}^{m'}$$

where $m' = km$. The same is also true when $\widetilde{X//U}^m$ is replaced with $\widetilde{X//U}^m$.

As in Proposition 4.6 and Remark 4.8 we obtain

Theorem 4.16. *If m is a sufficiently divisible positive integer and $N \gg 0$ then $\mathcal{X}_m$ and $\widetilde{\mathcal{X}}_m$ are projective varieties with linear actions of $\mathbf{C}^*$ which we can twist by ϵ times the standard character of $\mathbf{C}^*$, such that when $\epsilon = N/2$ we have*

$$\mathcal{X}_m//{}_{N/2}\mathbf{C}^* \cong X \quad and \quad \widetilde{\mathcal{X}}_m//{}_{N/2}\mathbf{C}^* \cong X,$$

while if $\epsilon > 0$ is sufficiently small then the rational maps from $\mathcal{X}_m//{}_\epsilon\mathbf{C}^*$ to $\overline{X//U}^m$ and from $\widetilde{\mathcal{X}}_m//{}_\epsilon\mathbf{C}^*$ to $\widetilde{X//U}^m$ induced by the inclusions of $(\hat{\mathcal{O}}_{\hat{L}_\epsilon}(\overline{P \times_U X}^m))^P$ in $(\hat{\mathcal{O}}_{\hat{L}_\epsilon}(\overline{P \times_U X}^m))^{GL(r;\mathbf{C})}$ and $(\hat{\mathcal{O}}_{\hat{L}_\epsilon}(\widetilde{P \times_U X}^m))^P$ in $(\hat{\mathcal{O}}_{\hat{L}_\epsilon}(\widetilde{P \times_U X}^m))^{GL(r;\mathbf{C})}$ restrict to surjective morphisms

$$(\mathcal{X}_m//{}_\epsilon\mathbf{C}^*)^{\hat{s}s} \to \widetilde{X//U}^m \quad and \quad (\widetilde{\mathcal{X}}_m//{}_\epsilon\mathbf{C}^*)^{\hat{s}s} \to \widetilde{X//U}^m$$

where $(\mathcal{X}_m /\!\!/_\epsilon \mathbf{C}^*)^{\acute{s}s}$ and $(\widetilde{\mathcal{X}}_m /\!\!/_\epsilon \mathbf{C}^*)^{\acute{s}s}$ are open subsets of $\mathcal{X}_m /\!\!/_\epsilon \mathbf{C}^*$ and $\widetilde{\mathcal{X}}_m /\!\!/_\epsilon \mathbf{C}^*$ defined as in Definition 4.5.

Thus when m is sufficiently divisible we have the following diagrams (cf. Remark 4.7):

$$
\begin{array}{ccccc}
(\mathcal{X}_m /\!\!/_\epsilon \mathbf{C}^*)^{\acute{s}s} & \subseteq & \mathcal{X}_m /\!\!/_\epsilon \mathbf{C}^* & \leftarrow\!-\!\rightarrow & X = \mathcal{X}_m /\!\!/_{N/2} \mathbf{C}^* \\
\downarrow & & & \text{flips} & \\
\widehat{X /\!\!/ U}^m & & & &
\end{array}
\qquad (4.18)
$$

and

$$
\begin{array}{ccccc}
(\widetilde{\mathcal{X}}_m /\!\!/_\epsilon \mathbf{C}^*)^{\acute{s}s} & \subseteq & \widetilde{\mathcal{X}}_m /\!\!/_\epsilon \mathbf{C}^* & \leftarrow\!-\!\rightarrow & X = \widetilde{\mathcal{X}}_m /\!\!/_{N/2} \mathbf{C}^* \\
\downarrow & & & \text{flips} & \\
\widetilde{X /\!\!/ U}^m & & & &
\end{array}
\qquad (4.19)
$$

where the vertical maps are surjective and the inclusions are open. When the ring of invariants $\hat{\mathcal{O}}_L(X)^U$ is finitely generated these can be extended by Remark 4.15 to

$$
\begin{array}{ccccccc}
(\mathcal{X}_m /\!\!/_\epsilon \mathbf{C}^*)^{\hat{s}} & \subseteq & (\mathcal{X}_m /\!\!/_\epsilon \mathbf{C}^*)^{\acute{s}s} & \subseteq & \mathcal{X}_m /\!\!/_\epsilon \mathbf{C}^* & \underset{\text{flips}}{\leftarrow -\rightarrow} & X = \mathcal{X}_m /\!\!/_{N/2} \mathbf{C}^* \\
\downarrow & & & \downarrow & & & \\
& & & \widehat{X /\!\!/ U}^m & & & \\
\downarrow & & & \downarrow & & & \\
X^s /U & \subseteq & & X /\!\!/ U & & &
\end{array}
$$

$$ (4.20) $$

(and a similar diagram involving $\widetilde{X /\!\!/ U}^m$), where $(\mathcal{X}_m /\!\!/_\epsilon \mathbf{C}^*)^{\hat{s}}$ is the inverse image of X^s /U in $(\mathcal{X}_m /\!\!/_\epsilon \mathbf{C}^*)^{\acute{s}s}$. In general for sufficiently divisible m and k we have

$$
\begin{array}{ccccccc}
(\mathcal{X}_m /\!\!/_\epsilon \mathbf{C}^*)^{\hat{s}} & \subseteq & (\mathcal{X}_m /\!\!/_\epsilon \mathbf{C}^*)^{\acute{s}s} & \subseteq & \mathcal{X}_m /\!\!/_\epsilon \mathbf{C}^* & \underset{\text{flips}}{\leftarrow -\rightarrow} & X = \mathcal{X}_m /\!\!/_{N/2} \mathbf{C}^* \\
\downarrow & & & \downarrow & & & \\
& & & \widehat{X /\!\!/ U}^m & & & \\
\downarrow & & & \downarrow & & & \\
X^s /U & \subseteq X /\!\!/ U \subseteq & \widehat{X /\!\!/ U}^{m'} & & & &
\end{array}
$$

$$ (4.21) $$

where $m' = km$.

4.3 Naturality properties

Given a linear action of $U = (\mathbf{C}^+)^r$ on a projective variety X, we have embedded X in a normal projective variety Y_m such that the linear action of U on X extends to a linear action of $G = SL(r+1; \mathbf{C}) \geq U$ on Y_m and (if m is sufficiently divisible) every U-invariant in a finite fully separating set of U-invariants on X extends to a U-invariant on Y_m. We have then constructed the GIT quotients

$$\widehat{X/\!/U}^m = G \times_P (\widehat{P \times_U X}^m)/\!/_{\widehat{\tilde{L}_t^{(N)}}} G,$$

$$\widetilde{X/\!/U}^m = G \times_P (\widetilde{P \times_U X}^m)/\!/_{\widetilde{\tilde{L}_t^{(N)}}} G$$

and

$$\mathcal{X}_m = \overline{P \times_U X}^m /\!/_{\hat{L}^{(N)}} SL(r; \mathbf{C}) \quad \text{and} \quad \widetilde{\mathcal{X}}_m = \widetilde{P \times_U X}^m /\!/_{\widetilde{L}^{(N)}} SL(r; \mathbf{C})$$

where $\overline{P \times_U X}^m$ and $\widetilde{P \times_U X}^m$ are the closures of $P \times_U X$ in the projective completions $\mathbf{P}^{r^2} \times Y_m$ and $\widetilde{\mathbf{P}^{r^2}} \times Y_m$ of $P \times_U Y_m \cong P/U \times Y_m$. One difficulty with using this construction in practice is that it is not easy to tell how divisible m has to be for there to be a finite fully separating set of U-invariants on X extending to U-invariants on Y_m. However the construction does have the following nice property, which enables us to study the families $\widehat{X/\!/U}^m$ and $\widetilde{X/\!/U}^m$ by embedding X in other projective varieties Y.

Proposition 4.17. *Let Y be a projective variety with a very ample line bundle L and a linear action of $U = (\mathbf{C}^+)^r$ on L, and let X be a U-invariant projective subvariety of Y with the inherited linear action of U. Then the inclusion of X in Y induces inclusions of projective varieties*

$$\widehat{X/\!/U}^m \subseteq \widehat{Y/\!/U}^m \quad \text{and} \quad \widetilde{X/\!/U}^m \subseteq \widetilde{Y/\!/U}^m$$

for all $m > 0$, as well as $\mathcal{X}_m \subseteq \mathcal{Y}_m$ and $\widetilde{\mathcal{X}}_m \subseteq \widetilde{\mathcal{Y}}_m$ when $\mathcal{X}_m$ and $\widetilde{\mathcal{X}}_m$ are defined as in Definition 4.13 and $\mathcal{Y}_m$ and $\widetilde{\mathcal{Y}}_m$ are defined similarly with Y replacing X.

Proof The construction of $\widehat{X/\!/U}^m$ and $\widetilde{X/\!/U}^m$ starts by embedding X into $\mathbf{P}(V_m^X)$ where $V_m^X = H^0(X, L^{\otimes m})^*$, and then embedding $\mathbf{P}(V_m^X)$ into $\mathbf{P}(W_m^X)$ where

$$W_m^X = \bigoplus_{j=0}^{\dim V_m^X - 1} \Theta_{j,m}^X \otimes (U_{V_m^X})^j (V_m^X)$$

and $U_{V_m^X}$ represents the infinitesimal action of $\mathbf{u} = \mathrm{Lie}\,(U) = \mathbf{C}^r$ on V_m^X. Here $\Theta_{j,m}^X$ is the space of complex valued polynomial functions on

$$\{(u_1, \ldots, u_j, h_1, \ldots, h_{\dim V_m^X - 1}) \in (\mathbf{C}^r)^j \times ((\mathbf{C}^r)^* \otimes \mathbf{C}^r)^{\dim V_m^X - 1}\}$$

which are homogeneous of degree 1 in the coordinates of each $u \in \mathbf{C}^r$ separately, and homogeneous of total degree $r(\dim V_m^X - 1) - (r-1)j$ in the coordinates of all the $h_k \in (\mathbf{C}^r)^* \otimes \mathbf{C}^r$ together. The surjection $H^0(Y, L^{\otimes m}) \to H^0(X, L^{\otimes m})$ given by restriction gives us an inclusion of V_m^X into V_m^Y and of $(U_{V_m^X})^j(V_m^X)$ into $(U_{V_m^Y})^j(V_m^Y)$ for all $j \geq 0$. We then get a map

$$\Theta_{j,m}^X \otimes (U_{V_m^X})^j(V_m^X) \to \Theta_{j,m}^Y \otimes (U_{V_m^Y})^j(V_m^Y)$$

$$\alpha \mapsto \alpha^Y$$

where $\alpha^Y(u_1, \ldots, u_j, h_1, \ldots h_{\dim V_m^Y - 1})$ is given by

$$\det(h_1) \ldots \det(h_{\dim V_m^Y - \dim V_m^X})$$

times

$$\alpha(u_1, \ldots, u_j, h_{\dim V_m^Y - \dim V_m^X + 1}, \ldots, h_{\dim V_m^Y - 1}).$$

This gives us a commutative diagram of U-equivariant embeddings

$$
\begin{array}{ccccc}
X & \to & \mathbf{P}(V_m^X) & \to & \mathbf{P}(W_m^X) \\
\downarrow & & \downarrow & & \downarrow \\
Y & \to & \mathbf{P}(V_m^Y) & \to & \mathbf{P}(W_m^Y)
\end{array}
$$

where the righthand vertical map sends

$$\sum_{j=0}^{\dim V_m^X - 1} \alpha_j \mapsto \sum_{j=0}^{\dim V_m^X - 1} \alpha_j^Y$$

and is equivariant with respect to the linear P-actions on W_k^X and W_k^Y. We thus get a commutative diagram of embeddings

$$
\begin{array}{ccccccc}
X & \to & \mathbf{P}(V_m^X) & \to & \mathbf{P}(W_m^X) & \to & G \times_P \mathbf{P}(W_m^X) \\
\downarrow & & \downarrow & & \downarrow & & \downarrow \\
Y & \to & \mathbf{P}(V_m^Y) & \to & \mathbf{P}(W_m^Y) & \to & G \times_P \mathbf{P}(W_m^Y).
\end{array}
$$

$\widetilde{X /\!/ U}^m$ is the GIT quotient $G \times_U \overline{X}^m /\!/_{\widetilde{L}_\epsilon^{(N)}} G$ where $G \times_U \overline{X}^m$ is the closure of $G(\{\iota\} \times X) \cong G \times_U X$ in $G \times_P (\overline{\mathbf{P}^{r^2}} \times (G \times_P \mathbf{P}(W_m^X)))$, and $\widetilde{Y /\!/ U}^m$ is constructed in the same way. Since the line bundle $\widetilde{L}_\epsilon^{(N)}$ is ample for $\epsilon > 0$, this gives us an inclusion of $\widetilde{X /\!/ U}^m$ into $\widetilde{Y /\!/ U}^m$, and the other inclusions follow similarly.

Suppose now that $U \cong (\mathbf{C}^+)^r$ is a normal subgroup of an algebraic group H acting linearly on X with respect to the line bundle L. This linear action induces an action of H on $V_m = H^0(X, L^{\otimes m})^*$ for each $m > 0$, and thus H acts by conjugation on $GL(V_m)$ and its Lie algebra. H also acts by conjugation on U and $\mathbf{C}^r = \mathrm{Lie}\, U$ as U is a normal subgroup of H, and the Lie algebra homomorphism $\varphi_{V_m} : \mathrm{Lie}\, U \to \mathrm{Lie}(GL(V_m))$ defined at (4.12) is H-equivariant with respect to these actions. Hence the action of H on V_m preserves the subspaces $(U_{V_m})^j(V_m)$ of V_m, and from this action and the action of H on $\mathbf{C}^r = \mathrm{Lie}\, U$ we get induced actions of H on W_m. If $h \in H$ and $p = gu \in P = GL(r; \mathbf{C}) \ltimes U$ with $g \in GL(r; \mathbf{C})$ and $u \in U$, then the actions of H and P are related by

$$h(pw) = h(guw) = ((\psi(h)g\psi(h)^{-1})(huh^{-1}))(hw) = \Psi_h(p)(hw)$$

for any w in W_m, where $\psi : H \to GL(r; \mathbf{C})$ is the group homomorphism defining the action of H on $\mathrm{Lie}\, U = \mathbf{C}^r$ by conjugation and

$$\Psi_h(p) = (\psi(h)g\psi(h)^{-1})(huh^{-1}) \in GL(r; \mathbf{C}) \ltimes U = P.$$

Thus we get actions of a semidirect product $P \rtimes H$ on W_m. (Note that the action of U as a subgroup of P defined at (4.15) is different from the action of U as a subgroup of H defined above.) In fact the subgroup $P \rtimes U$ of $P \rtimes H$ is a direct product $P \times U$, since $U \cong (\mathbf{C}^+)^r$ acts trivially on itself by conjugation, so if $p \in P$ and $h \in U \le H$ then $\psi(h)$ is the identity element of $GL(r; \mathbf{C})$ and $\Psi_h(p) = p$.

H also acts on $\mathbf{P}^{r^2}$ and on $\widetilde{\mathbf{P}^{r^2}}$ via the homomorphism $\psi : H \to GL(r; \mathbf{C})$, giving us an action of $P \rtimes H$ on $\widetilde{\mathbf{P}^{r^2}} \times \mathbf{P}(W_m)$. Since X is H-invariant it follows that the closure $\widetilde{P \times_U X}^m$ of $P \times_U X = P(\{\iota\} \times X)$ in $\widetilde{\mathbf{P}^{r^2}} \times Y_m$ is also H-invariant. Since H normalises $SL(r; \mathbf{C})$ and commutes with the central $\mathbf{C}^*$ subgroup of $GL(r; \mathbf{C})$, we get an induced linear action of $H \times \mathbf{C}^*$ on

$$\widetilde{\mathcal{X}}_m = \widetilde{P \times_U X}^m /\!/_{L_{N/2}^{(N)}} SL(r; \mathbf{C}),$$

preserving the open subset $(\widetilde{\mathcal{X}}_m /\!/_\epsilon \mathbf{C}^*)^{\hat{s}s}$, and of H/U on

$$\widetilde{X/\!/U}^m = \widetilde{P \times_U X}^m /\!/_{L_\epsilon^{(N)}} P.$$

Remark 4.18. Suppose that $(\mathbf{C}^+)^r = U \trianglelefteq H$ and H acts linearly on X as above. Suppose also that H contains a one-parameter subgroup $\lambda : \mathbf{C}^* \to H$ whose weights for the induced (conjugation) action on $\mathbf{u} = \mathbf{C}^r$ are all strictly positive. Then the subgroup $\hat{U}$ of H generated by $\lambda(\mathbf{C}^*)$ and U is a semidirect product

$$\hat{U} \cong U \rtimes \mathbf{C}^*.$$

Moreover this $\mathbf{C}^*$ acts on $\mathbf{C}^{r(r+1)} \subseteq \mathbf{P}^{r(r+1)}$ with all weights strictly positive, and we have $GL(r;\mathbf{C}) = \psi(\lambda(\mathbf{C}^*))SL(r;\mathbf{C})$ with $\psi(\lambda(\mathbf{C}^*)) \cap SL(r;\mathbf{C})$ finite. Recall from Remark 4.8 that if $\overline{P/U} = \mathbf{P}^{r^2}$ then

$$\overline{P/U}^{ss,SL(r;\mathbf{C})} = \overline{P/U}^{s,SL(r;\mathbf{C})} \quad \text{and} \quad \overline{P/U}//SL(r;\mathbf{C}) \cong \mathbf{P}^1$$

with the induced action of $\lambda(\mathbf{C}^*)$ on $\mathbf{P}^1$ a positive power of the standard action on $\mathbf{C}^*$ on $\mathbf{P}^1$. Thus using variation of GIT (as in Remark 2.7) there are rational numbers $\delta_- < \delta_+$ such that the induced action of $\lambda(\mathbf{C}^*)$ on $\mathbf{P}^1 = \overline{P/U}//SL(r;\mathbf{C})$ twisted by δ times the standard character of $\mathbf{C}^*$ satisfies

$$(\mathbf{P}^1)^{ss,\delta} = (\mathbf{P}^1)^{s,\delta} = \begin{cases} \mathbf{C}^* & \text{if } \delta \in (\delta_-,\delta_+) \\ \emptyset & \text{if } \delta \notin [\delta_-,\delta_+] \end{cases}$$

and hence

$$\overline{P/U}^{ss,GL(r;\mathbf{C}),\delta} =$$

$$\overline{P/U}^{s,GL(r;\mathbf{C}),\delta} = \begin{cases} GL(r;\mathbf{C}) = P/U & \text{if } \delta \in (\delta_-,\delta_+) \\ \emptyset & \text{if } \delta \notin [\delta_-,\delta_+] \end{cases}.$$

It follows that if the linearisation of the H-action is twisted by δ times the standard character of $\lambda(\mathbf{C}^*)$ for $\delta \neq \delta_-,\delta_+$ (which is possible to arrange if, for example, $\lambda(\mathbf{C}^*)$ centralises H/U), then (for sufficiently large N) all the points of $\widetilde{\mathcal{X}}_m = \widetilde{P \times_U X}^m //_{L(N)} SL(r;\mathbf{C})$ which are semistable for the induced action of $\lambda(\mathbf{C}^*)$ are contained in the image of $P \times_U X \cong GL(r;\mathbf{C}) \times X$. Hence the inverse image of $(\widetilde{X//U}^m)^{ss,\lambda(\mathbf{C}^*)}$ under the surjection

$$(\widetilde{\mathcal{X}}_m //_\epsilon \mathbf{C}^*)^{\hat{s}s} \to \widetilde{X//U}^m$$

in Theorem 4.16 is an open subset of $\widetilde{\mathcal{X}}_m //_\epsilon \mathbf{C}^*$ which is unaffected by the flips $\widetilde{\mathcal{X}}_m //_\epsilon \mathbf{C}^* \leftarrow - \to \widetilde{\mathcal{X}}_m //_{N/2}\mathbf{C}^* = X$ and thus can be identified canonically with an open subset $X^{ss,\hat{U}}$ of X. Similarly the inverse image of $(\widetilde{X//U}^m)^{s,\lambda(\mathbf{C}^*)}$ under the restriction of this surjection to $(\mathcal{X}_m //_\epsilon \mathbf{C}^*)^{\hat{s}}$ is an open subset $X^{s,\hat{U}}$ of $X^{ss,\hat{U}}$ which, like $X^{ss,\hat{U}}$, is independent of m when m is sufficiently divisible. Indeed it turns out that $x \in X^{ss,\hat{U}}$ (respectively $x \in X^{s,\hat{U}}$) if and only if x is semistable (respectively stable) for every conjugate of $\lambda : \mathbf{C}^* \to \hat{U}$ in $\hat{U}$, or equivalently for every one-parameter subgroup $\hat{\lambda} : \mathbf{C}^* \to \hat{U}$ of $\hat{U}$. If the linear U-action on X extends to G then the same is true when $\widetilde{X//U}^m$ is replaced with $\overline{X//U}$ defined as in §4.1.

In particular this means that when, in addition, H/U is reductive and is centralised by $\lambda(\mathbf{C}^*)$, then if the linearisation of the H-action is twisted by a suitable character of H/U the induced GIT quotients

$$\widetilde{X/\!/U}^{\,m}/\!/(H/U)$$

for m sufficiently divisible are independent of m (and are isomorphic to $\widetilde{X/\!/U}/\!/(H/U)$ when the linear U-action on X extends to G). In fact the proof of [DK1] Theorem 5.3.18 shows that in this situation, with the linearisation of the H-action suitably twisted, the ring of invariants $\hat{\mathcal{O}}_L(X)^H$ is finitely generated, with associated projective variety

$$X/\!/H = \mathrm{Proj}(\hat{\mathcal{O}}_L(X)^H),$$

and we have

$$X/\!/H \cong \widetilde{X/\!/U}^{\,m}/\!/(H/U).$$

Indeed, it turns out that in this situation, with the linearisation of the H-action suitably twisted, we have essentially the same situation as for classical GIT for reductive group actions: there is a diagram

$$
\begin{array}{ccccc}
X^{s,H}/H & \subseteq & X^{ss,H} & \subseteq & X \\
\downarrow & & \downarrow & & \\
X^{s,H}/H & \subseteq & X/\!/H & &
\end{array}
\qquad (4.22)
$$

where the vertical maps are surjective and the inclusions are open, and in addition the Hilbert-Mumford criteria for stability and semistability hold as in the reductive case (Proposition 2.3 above), and two semistable orbits in X represent the same point of $X/\!/H$ if and only if their closures meet in $X^{ss,H}$. We will see an example of this phenomenon for hypersurfaces in $\mathbf{P}(1,1,2)$ in §5.3 below. Moreover the partial desingularisation (defined as in §2.2)

$$\widetilde{\widetilde{X/\!/U}^{\,m}}/\!/(H/U)$$

of the GIT quotient of $\widetilde{X/\!/U}^{\,m}$ by the action of the reductive group H/U is independent of m and provides a partial desingularisation $\widetilde{X/\!/H}$ of $X/\!/H$.

Remark 4.19. In practice it is not difficult to find the range of characters of $\lambda(\mathbf{C}^*)$ with which the linearisation can be twisted in order to achieve the nice situation described in Remark 4.18. The picture described in Remark 4.18 is valid for all $\delta \in \mathbb{Q} \smallsetminus \{\delta_-, \delta_+\}$, and if $\delta \notin [\delta_-, \delta_+]$ then $X^{ss,\hat{U},\delta} = \emptyset$ and hence $X/\!/_\delta H = \emptyset$. If we attempt to use the

Hilbert-Mumford criteria to calculate $X^{ss,\hat{U},\delta}$ for all $\delta \in \mathbb{Q}$, we will find finitely many rational numbers $a_0 < a_1 < \ldots < a_q$ such that, when calculated according to the Hilbert-Mumford criteria, $X^{ss,\hat{U},\delta}$ is empty for $\delta < a_0$ and for $\delta > a_q$, and is nonempty but constant for $\delta \in (a_{j-1}, a_j)$ when $j = 1, \ldots, q$. Then we must have $\delta_- \leq a_0$ and $\delta_+ \geq a_q$, and moreover $X^{ss,\hat{U},\delta}$ and $X^{ss,H,\delta}$ are as predicted by the Hilbert-Mumford criteria for any $\delta \neq a_0, a_q$. Thus $X/\!/_\delta H = \emptyset$ if $\delta \notin [a_0, a_q]$ and the situation described in Remark 4.18 holds for every $\delta \in (a_0, a_q)$.

5 Hypersurfaces in $\mathbf{P}(1,1,2)$

Recall from §3 that the moduli problem of hypersurfaces of weighted degree d in the weighted projective plane $\mathbf{P}(1,1,2)$ is essentially equivalent to constructing a quotient for the action of

$$H = (\mathbf{C}^+)^3 \rtimes GL(2; \mathbf{C})$$

on (an open subset of) the projective space X_d of weighted degree d polynomials in the three weighted homogeneous coordinates x, y, z on $\mathbf{P}(1,1,2)$. Here H is the automorphism group of $\mathbf{P}(1,1,2)$, where $(\alpha, \beta, \gamma) \in U = (\mathbf{C}^+)^3$ acts on $\mathbf{P}(1,1,2)$ via

$$[x:y:z] \mapsto [x:y:z+\alpha x^2 + \beta xy + \gamma y^2]$$

and $g \in GL(2; \mathbf{C})$ acts in the standard fashion on $(x,y) \in \mathbf{C}^2$ and as scalar multiplication by $(\det g)^{-1}$ on z. Thus $g \in GL(2; \mathbf{C})$ acts by conjugation on U as the standard action of $GL(2; \mathbf{C})$ on $\mathrm{Sym}^2(\mathbf{C}^2) \cong \mathbf{C}^3$ twisted by the character det.

Remark 5.1. Notice that the central one-parameter subgroup $\lambda : \mathbf{C}^* \to GL(2; \mathbf{C})$ of $GL(2; \mathbf{C})$ satisfies the conditions of Remark 4.18 above: the weights of its action (by conjugation) on $\mathbf{u} = \mathbf{C}^3$ are all strictly positive, as they are all equal to 4.

We wish to study the action of H on the projective space

$$X_d = \mathbf{P}(\mathbf{C}_{(d)}[x,y,z])$$

where $\mathbf{C}_{(d)}[x,y,z]$ is the linear subspace of the polynomial ring $\mathbf{C}[x,y,z]$ consisting of polynomials p of the form

$$p(x,y,z) = \sum_{\substack{i,j,k \geq 0 \\ i+j+2k = d}} a_{ijk} x^i y^j z^k$$

for some $a_{ijk} \in \mathbf{C}$ [Co, CK]. Here $h \in H$ acts as $p \mapsto h \cdot p$ with $h \cdot p(x, y, z) = p(h^{-1}x, h^{-1}y, h^{-1}z)$. This representation

$$H \to GL(\mathbf{C}_d[x, y, z])$$

gives us a linearisation of the action of H on X_d, which we can twist by any multiple $\epsilon \in \mathbb{Q}$ of the character det of $GL(2; \mathbf{C})$ to get a fractional linearisation $\mathcal{L}_\epsilon$.

Remark 5.2. If $m \geq 1$ then $H^0(X_d, \mathcal{O}_{X_d}(m))^* \cong \mathbf{C}_{(md)}[x, y, z]$ and the natural embedding of X_d in the projective space $\mathbf{P}(H^0(X_d, \mathcal{O}_{X_d}(m))^*)$ is given by $p(x, y, z) \mapsto (p(x, y, z))^m$.

5.1 *The action of* $U = (\mathbf{C}^+)^3$

First let us consider the action of the unipotent radical $U = (\mathbf{C}^+)^3$ of H on X_d. Consider

$$Y_d = \mathbf{P}(\mathbf{C}_{\lceil d/2 \rceil}[X, Y, W, z])$$

where $\lceil d/2 \rceil$ denotes the least integer $n \geq d/2$, and $\mathbf{C}_{\lceil d/2 \rceil}[X, Y, W, z]$ is the space of homogeneous polynomials of degree $\lceil d/2 \rceil$ in X, Y, W, z. By multiplying by x if d is odd, we can identify $\mathbf{C}_{(d)}[x, y, z]$ with the set of polynomials of the form

$$p(x, y, z) = \sum_{\substack{i \geq 2\lceil d/2 \rceil - d, \ j, k \geq 0 \\ i + j + 2k = 2\lceil d/2 \rceil}} a_{ijk} x^i y^j z^k.$$

Then we can embed X_d in Y_d via $p \mapsto \hat{p}$ where $\hat{p}(X, Y, W, z)$ equals the sum over $i \geq 2\lceil d/2 \rceil - d$ and $j, k \geq 0$ satisfying $i + j + 2k = 2\lceil d/2 \rceil$ of

$$a_{ijk} X^{(i - M_{ij})/2 - \lceil (m_{ij} - M_{ij})/2 \rceil} W^{M_{ij} + 2\lceil (m_{ij} - M_{ij})/2 \rceil} Y^{(j - M_{ij})/2 - \lceil (m_{ij} - M_{ij})/2 \rceil} z^k$$

for $m_{ij} = \min\{i, j\}$ and $M_{ij} = \max\{i, j\}$. Thus $\hat{p}(x^2, y^2, xy, z) = p(x, y, z)$ if d is even and $\hat{p}(x^2, y^2, xy, z) = xp(x, y, z)$ if d is odd. For simplicity we will assume from now on that d is even; by Remark 5.2 this involves very little loss of generality.

The action of U on X_d extends to an action on Y_d such that $(\alpha, \beta, \gamma) \in U$ acts via

$$p(X, Y, W, z) \mapsto p(X, Y, W, z + \alpha X + \beta W + \gamma Y).$$

This extends to the standard action of $G = SL(4; \mathbf{C})$ on $\mathbf{C}_{d/2}[X, Y, W, z]$. Thus

$$Y_d /\!/ U = (\mathbf{P}^{12} \times Y_d) /\!/ G \quad \text{and} \quad \widetilde{Y_d /\!/ U} = (G \times_P (\mathbf{P}^9 \times Y_d)) /\!/ G$$

where $\mathbf{P}^{12} = \mathbf{P}(\mathbf{C} \oplus ((\mathbf{C}^3)^* \otimes \mathbf{C}^4))$ and $\mathbf{P}^9 = \mathbf{P}(\mathbf{C} \oplus ((\mathbf{C}^3)^* \otimes \mathbf{C}^3))$. Here the linearisation on $\mathbf{P}^{12} \times Y_d$ is $\mathcal{O}_{\mathbf{P}^{12}}(N) \otimes \mathcal{O}_{Y_d}(1)$ for $N >> 0$ and the linearisation on $\mathbf{P}^9 \times Y_d$ is $\mathcal{O}_{\mathbf{P}^9}(N) \otimes \mathcal{O}_{Y_d}(1)$.

The weights of the action of the standard maximal torus T_c of $G = SL(4; \mathbf{C})$ on $\mathbf{P}^{12} = \mathbf{P}(\mathbf{C} \oplus ((\mathbf{C}^3)^* \otimes \mathbf{C}^4))$ with respect to $\mathcal{O}_{\mathbf{P}^{12}}(1)$ are 0 (with multiplicity 1) and $\chi_1, \chi_2, \chi_3, \chi_4$ (each with multiplicity 3) where $\chi_1, \chi_2, \chi_3, \chi_4 = -\chi_1 - \chi_2 - \chi_3$ are the weights of the standard representation of $SL(4; \mathbf{C})$ on $\mathbf{C}^4$. The weights of the action of T_c on $Y_d = \mathbf{P}(\mathbf{C}_{d/2}[X, Y, W, z])$ with respect to $\mathcal{O}_{Y_d}(1)$ are

$$\{0\} \cup \{i\chi_1 + j\chi_2 + k\chi_3 + \ell\chi_4 : i, j, k, \ell \geq 0 \text{ and } i + j + k + \ell = d/2\}.$$

A point $a = [a_0 : a_{11} : a_{12} : a_{13} : a_{14} : a_{21} : a_{22} : a_{23} : a_{24} : a_{31} : a_{32} : a_{33} : a_{34}] \in \mathbf{P}^{12}$ is semistable for this action of $SL(4; \mathbf{C})$ if and only if $a_0 \neq 0$. Therefore if $N >> 0$ we have $a_0 \neq 0$ whenever $(a, y) \in (\mathbf{P}^{12} \times Y_d)^{ss, G}$ for any $y \in Y_d$. Moreover if $a = [1 : a_{ij}] \in \mathbf{P}^{12}$ and $y \in Y_d$ is represented by

$$p(X, Y, W, z) = \sum_{\substack{i, j, k, \ell \geq 0 \\ i + j + k + \ell = d/2}} b_{ijk\ell} X^i Y^j W^k z^\ell \in \mathbf{C}_{d/2}[X, Y, W, z]$$

then by the Hilbert-Mumford criteria $(a, y) \in (\mathbf{P}^{12} \times Y_d)^{ss, G}$ if and only if $(ga, gy) \in (\mathbf{P}^{12} \times Y_d)^{ss, T_c}$ for every $g \in G$, and $(a, y) \in (\mathbf{P}^{12} \times Y_d)^{ss, T_c}$ if and only if 0 lies in the convex hull of the set of weights

$$\{i\chi_1 + j\chi_2 + k\chi_3 + \ell\chi_4 : b_{ijk\ell} \neq 0\} \cup S_1 \cup S_2 \cup S_3 \cup S_4$$

where

$$S_1 = \left\{ \begin{array}{ll} \{N\chi_1 + i\chi_1 + j\chi_2 + k\chi_3 + \ell\chi_4 : b_{ijk\ell} \neq 0\} & \text{if } (a_{11}, a_{21}, a_{31}) \neq 0 \\ \emptyset & \text{if } (a_{11}, a_{21}, a_{31}) = 0 \end{array} \right.$$

and S_2, S_3, S_4 are defined similarly. Let us write

$$(\mathbf{P}^{12} \times Y_d)^{ss, G} =$$

$$(\mathbf{P}^{12} \times Y_d)_0^{ss, G} \sqcup (\mathbf{P}^{12} \times Y_d)_1^{ss, G} \sqcup (\mathbf{P}^{12} \times Y_d)_2^{ss, G} \sqcup (\mathbf{P}^{12} \times Y_d)_3^{ss, G} \quad (5.1)$$

where $(\mathbf{P}^{12} \times Y_d)_q^{ss, G} = \{(a, y) \in (\mathbf{P}^{12} \times Y_d)^{ss, G} : \text{rank}((a_{ij})) = q\}$. Then

$$(\mathbf{P}^{12} \times Y_d)_0^{ss, G} = \{[1 : 0 : \ldots : 0]\} \times Y_d^{ss, G}$$

and

$$(\mathbf{P}^{12} \times Y_d)_1^{ss, G} = G \times_{U_1} (\{[1 : \iota_1]\} \times Y_d^{ss, 1})$$

where

$$\iota_1 = \begin{pmatrix} 1 & 0 & 0 & 0 \\ 0 & 0 & 0 & 0 \\ 0 & 0 & 0 & 0 \end{pmatrix},$$

U_1 is its stabiliser $\{(g_{ij} \in G : g_{11} = 1 \text{ and } g_{21} = g_{31} = g_{41} = 0\}$ in $G = SL(4; \mathbf{C})$ and

$$Y_d^{ss,1} = \{y \in Y_d : uy \in Y_d^{ss,T_c} \text{ for all } u \in U_1\}.$$

Similarly

$$(\mathbf{P}^{12} \times Y_d)_2^{ss,G} = G \times_{U_2} (\{[1 : \iota_2]\} \times Y_d^{ss,2})$$

where

$$\iota_2 = \begin{pmatrix} 1 & 0 & 0 & 0 \\ 0 & 1 & 0 & 0 \\ 0 & 0 & 0 & 0 \end{pmatrix},$$

U_2 is its stabiliser $\{(g_{ij} \in G : g_{11} = g_{22} = 1 \text{ and } g_{12} = g_{21} = g_{31} = g_{32} = g_{41} = g_{42} = 0\}$ in G and

$$Y_d^{ss,2} = \{y \in Y_d : uy \in Y_d^{ss,T_c^2} \text{ for all } u \in U_2\}$$

for $T_c^2 = \{(g_{ij}) \in T_c : g_{11} = g_{22}\}$, while

$$(\mathbf{P}^{12} \times Y_d)_3^{ss,G} = G \times_U (\{[1 : \iota]\} \times Y_d^{ss,3})$$

where

$$Y_d^{ss,3} = \{y \in Y_d : uy \in Y_d^{ss,T_c^3} \text{ for all } u \in U\}$$

for $T_c^3 = \{(g_{ij}) \in T_c : g_{11} = g_{22} = g_{33}\}$.

5.2 The action of $\hat{U} = \mathbf{C}^3 \rtimes \mathbf{C}^*$

Now let us consider the action of the subgroup $\hat{U} = \mathbf{C}^* \ltimes U$ of H on X_d, where $\mathbf{C}^*$ is the centre of $GL(2; \mathbf{C})$ and acts by conjugation on $\mathrm{Lie}\,(U) = \mathbf{C}^3$ with weights all equal to 4. The action of $t \in \mathbf{C}^*$ on $p(x, y, z) \in \mathbf{C}_d[x, y, z]$ is given by

$$tp(x, y, z) = p(tx, ty, t^{-2}z).$$

This action extends to the action on $\mathbf{C}_{d/2}[X, Y, W, z]$ given by

$$tP(X, Y, W, z) = P(t^2 X, t^2 Y, t^2 W, t^{-2} z)$$

and thus the action of $\hat{U}$ on X_d extends to a linear action on Y_d, which is the restriction of the $GL(4;\mathbf{C})$-action on Y_d via the embedding of $\hat{U}$ in $GL(4;\mathbf{C})$ such that

$$t \mapsto \begin{pmatrix} t^2 & 0 & 0 & 0 \\ 0 & t^2 & 0 & 0 \\ 0 & 0 & t^2 & 0 \\ 0 & 0 & 0 & t^{-2} \end{pmatrix} \text{ for } t \in \mathbf{C}^*. \qquad (5.2)$$

If we twist this action by 2δ times the standard character of $\mathbf{C}^*$ then we get a fractional linearisation of the action of $\hat{U}$ on Y_d which extends the fractional linearisation $\mathcal{L}_\delta$ on X_d. We also get an action of $\mathbf{C}^* = \hat{U}/U$ on $\mathbf{P}^{12} = \overline{G/U}$ via

$$t[a_0 : a_{ij}] = [a_0 : t^4 a_{ij}].$$

Note that, since $(t_1, t_2, t_3, t_4) = (t^2 \tau_1, t^2 \tau_2, t^2 \tau_3, t^{-2}(\tau_1 \tau_2 \tau_3)^{-1})$ if and only if

$$t^4 = t_1 t_2 t_3 t_4 \text{ and } \tau_1 = t^{-2} t_1,\ \tau_2 = t^{-2} t_2,\ \tau_3 = t^{-2} t_3,$$

$\mathbf{C}^* T_c \cong (\mathbf{C}^* \times T_c)/(\mathbf{Z}/4\mathbf{Z})$ is the maximal torus of $GL(4;\mathbf{C})$, acting on $\mathbf{P}^{12}$ with weights 0, $2\chi_1 + \chi_2 + \chi_3 + \chi_4$, $\chi_1 + 2\chi_2 + \chi_3 + \chi_4$, $\chi_1 + \chi_2 + 2\chi_3 + \chi_4$ and $\chi_1 + \chi_2 + \chi_3 + 2\chi_4$, and acting on Y_d with fractional weights

$$\{ i\chi_1 + j\chi_2 + k\chi_3 + \ell\chi_4 + \frac{\epsilon}{2}(\chi_1 + \chi_2 + \chi_3 - \chi_4) : \ i, j, k, \ell \geq 0$$

$$\text{and } i + j + k + \ell = \frac{d}{2} \}$$

where $\chi_1, \chi_2, \chi_3, \chi_4$ are now the weights of the standard representation of $GL(4;\mathbf{C})$ on $\mathbf{C}^4$. Let us break up $(\mathbf{P}^{12} \times Y_d)^{ss, GL(4;\mathbf{C}), \delta}$ as at (5.1) as

$$(\mathbf{P}^{12} \times Y_d)_0^{ss, GL(4;\mathbf{C}), \delta} \sqcup (\mathbf{P}^{12} \times Y_d)_1^{ss, GL(4;\mathbf{C}), \delta}$$

$$\sqcup (\mathbf{P}^{12} \times Y_d)_2^{ss, GL(4;\mathbf{C}), \delta} \sqcup (\mathbf{P}^{12} \times Y_d)_3^{ss, GL(4;\mathbf{C}), \delta} \qquad (5.3)$$

where $(\mathbf{P}^{12} \times Y_d)_q^{ss, GL(4;\mathbf{C}), \delta}$ equals

$$\{(a, y) \in (\mathbf{P}^{12} \times Y_d)^{ss, GL(4;\mathbf{C}), \delta} : \operatorname{rank}((a_{ij})) = q\}.$$

We find by considering the central $\mathbf{C}^*$ in $GL(4;\mathbf{C})$ that

$$(\mathbf{P}^{12} \times Y_d)_0^{ss, GL(4;\mathbf{C}), \delta} = (\mathbf{P}^{12} \times Y_d)_1^{ss, GL(4;\mathbf{C}), \delta} = (\mathbf{P}^{12} \times Y_d)_2^{ss, GL(4;\mathbf{C}), \delta} = \emptyset$$

unless $\delta = -d/2$, while

$$(\mathbf{P}^{12} \times Y_d)^{ss, GL(4;\mathbf{C}), \delta} = (\mathbf{P}^{12} \times Y_d)_3^{ss, GL(4;\mathbf{C}), \delta} \cong GL(4;\mathbf{C}) \times_{\hat{U}} Y_d^{ss, \hat{U}, \delta}$$

where
$$Y_d^{ss,\hat{U},\delta} = \{y \in Y_d : uy \in Y_d^{ss,\mathbf{C}^*,\delta} \text{ for all } u \in U\}.$$

Similarly if $\delta \neq -d/2$ then
$$(G \times_P (\widetilde{\mathbf{P}^9} \times Y_d))^{ss,GL(4;\mathbf{C}),\delta} = (G \times_P (\widetilde{\mathbf{P}^9} \times Y_d))_3^{ss,GL(4;\mathbf{C}),\delta}$$

$$\cong GL(4;\mathbf{C}) \times_{\hat{U}} Y_d^{ss,\hat{U},\delta}$$

and if m is sufficiently divisible
$$(G \times_P (\widetilde{P \times_U Y_d}^m))^{ss,GL(4;\mathbf{C}),\delta} \cong GL(4;\mathbf{C}) \times_{\hat{U}} Y_d^{ss,\hat{U},\delta},$$

while
$$(\mathbf{P}^{12} \times Y_d)^{s,GL(4;\mathbf{C}),\delta} \cong (G \times_P (\widetilde{\mathbf{P}^9} \times Y_d))^{s,GL(4;\mathbf{C}),\delta}$$

$$\cong (G \times_P (\widetilde{P \times_U Y_d}^m))^{s,GL(4;\mathbf{C}),\delta} \cong GL(4;\mathbf{C}) \times_{\hat{U}} Y_d^{s,\hat{U},\delta}$$

where
$$Y_d^{s,\hat{U},\delta} = \{y \in Y_d : uy \in Y_d^{s,\mathbf{C}^*,\delta} \text{ for all } u \in U\}.$$

Thus if $\delta \neq -d/2$, for sufficiently divisible m we have
$$\widetilde{Y_d /\!/ U}^m /\!/_\delta \mathbf{C}^* = Y_d^{ss,\hat{U},\delta} / \sim_{\hat{U}}$$

and
$$(\widetilde{Y_d /\!/ U}^m)^{s,\mathbf{C}^*,\delta} / \mathbf{C}^* = Y_d^{s,\hat{U},\delta} /\hat{U},$$

and so by Proposition 4.14 for sufficiently divisible m we have
$$\widetilde{X_d /\!/ U}^m /\!/_\delta \mathbf{C}^* = X_d^{ss,\hat{U},\delta} / \sim_{\hat{U}}$$

and
$$(\widetilde{X_d /\!/ U}^m)^{s,\mathbf{C}^*,\delta} / \mathbf{C}^* = X_d^{s,\hat{U},\delta} /\hat{U}$$

where
$$X_d^{ss,\hat{U},\delta} = \{y \in X_d : uy \in X_d^{ss,\mathbf{C}^*,\delta} \text{ for all } u \in U\}$$

and
$$X_d^{s,\hat{U},\delta} = \{y \in X_d : uy \in X_d^{s,\mathbf{C}^*,\delta} \text{ for all } u \in U\}$$

and $x \sim_{\hat{U}} y$ if and only if $\hat{U}x \cap \hat{U}y \cap X_d^{ss,\hat{U},\delta} \neq \emptyset$. Using this, the proof of [DK1] Theorem 5.3.18 shows that in fact, for the linearisation $\mathcal{L}_\delta$ when $\delta \neq -d/2$, the ring of invariants $\hat{\mathcal{O}}_{\mathcal{L}_\delta}(X_d)^{\hat{U}}$ is finitely generated, and

$$X_d /\!/_\delta \hat{U} = \text{Proj}(\hat{\mathcal{O}}_{\mathcal{L}_\delta}(X_d)^{\hat{U}}) = \widetilde{X_d /\!/ U}^m /\!/_\delta \mathbf{C}^*$$

for sufficiently divisible m. Thus for all $\delta \neq -d/2$

$$X_d /\!/_\delta \hat{U} = X_d^{ss,\hat{U},\delta} / \sim_{\hat{U}}$$

is a projective completion of $X_d^{s,\hat{U},\delta}/\hat{U}$ (cf. Remark 4.18).

5.3 The action of H

Let $T_c(GL(2;\mathbf{C}))$ be the standard maximal torus of $GL(2;\mathbf{C}) = H/U$. It now follows immediately that when $\delta \neq -d/2$, the ring of invariants $\hat{\mathcal{O}}_{\mathcal{L}_\delta}(X_d)^H = (\hat{\mathcal{O}}_{\mathcal{L}_\delta}(X_d)^{\hat{U}})^{SL(2;\mathbf{C})}$ is finitely generated, and

$$X_d /\!/_\delta H = \mathrm{Proj}(\hat{\mathcal{O}}_{\mathcal{L}_\delta}(X_d)^H) = X_d^{ss,H,\delta} / \sim_H \tag{5.4}$$

is a projective completion of $X_d^{s,H,\delta}/H$, where

$$X_d^{ss,H,\delta} = \{y \in X_d : uy \in X_d^{ss,GL(2;\mathbf{C}),\delta} \text{ for all } u \in U\}$$

$$= \{y \in X_d : hy \in X_d^{ss,T_c(GL(2;\mathbf{C})),\delta} \text{ for all } h \in H\}$$

and

$$X_d^{s,H,\delta} = \{y \in X_d : uy \in X_d^{s,GL(2;\mathbf{C}),\delta} \text{ for all } u \in U\}$$

$$= \{y \in X_d : hy \in X_d^{s,T_c(GL(2;\mathbf{C})),\delta} \text{ for all } h \in H\}$$

and $x \sim_H y$ if and only if $Hx \cap Hy \cap X_d^{ss,H,\delta} \neq \emptyset$.

The weights of the action on X_d of

$$T_c(GL(2;\mathbf{C})) = \left\{ \begin{pmatrix} t_1 & 0 \\ 0 & t_2 \end{pmatrix} : t_1, t_2 \in \mathbf{C}^* \right\}$$

with respect to the linearisation $\mathcal{L}_\delta$ are given by

$(t_1, t_2) \mapsto t_1^{i-k+\delta} t_2^{j-k+\delta}$ for integers $i, j, k \geq 0$ such that $i + j + 2k = d$.

Thus $p(x, y, z) = \sum_{i+j+2k=d} a_{ijk} x^i y^j z^k \in \mathbf{C}_{(d)}[x, y, z]$ represents a point of $X_d^{ss,T_c(GL(2;\mathbf{C})),\delta}$ (respectively a point of $X_d^{s,T_c(GL(2;\mathbf{C})),\delta}$) if and only if 0 lies in the convex hull (respectively 0 lies in the interior of the convex hull) in $\mathbf{R}^2$ of the subset

$$\{(i - k + \delta, j - k + \delta) : i, j, k \geq 0, \ i + j + 2k = d \text{ and } a_{ijk} \neq 0\}$$

of the set of weights

$$\{(i - k + \delta, j - k + \delta) : i, j, k \geq 0, \ i + j + 2k = d\}$$

whose convex hull is the triangle in $\mathbf{R}^2$ with vertices $(d + \delta, \delta)$, $(\delta, d + \delta)$

and $(\delta - d/2, \delta - d/2)$. Notice that the bad case $\delta = -d/2$ occurs precisely when the origin lies on the edge of this triangle joining the vertices $(d + \delta, \delta)$ and $(\delta, d + \delta)$, and that we have $X_d^{ss, H, \delta} = \emptyset$ if $\delta \notin (-d/2, d/2)$ (cf. Remark 4.19).

Combining this with (5.4) gives an explicit description of $X_d /\!/_\delta H$ and $X_d^{s, H, \delta}/H$ whenever $\delta \neq -d/2$.

Remark 5.3. For small d this description can be expressed in terms of singularities of hypersurfaces (cf. the description in [MFK] Chapter 4 §2 of stability and semistability for hypersurfaces in the ordinary projective plane $\mathbf{P}^2$).

5.4 Symplectic descriptions

Classical GIT quotients in complex algebraic geometry are closely related to the process of reduction in symplectic geometry. Suppose that a compact, connected Lie group K with Lie algebra $\mathbf{k}$ acts smoothly on a symplectic manifold X and preserves the symplectic form ω. A moment map for the action of K on X is a smooth map $\mu : X \to \mathbf{k}^*$ which is equivariant with respect to the given action of K on X and the coadjoint action of K on $\mathbf{k}^*$, and satisfies

$$d\mu(x)(\xi).a = \omega_x(\xi, a_x)$$

for all $x \in X$, $\xi \in T_x X$ and $a \in \mathbf{k}$, where $x \mapsto a_x$ is the vector field on X defined by the infinitesimal action of $a \in \mathbf{k}$. The quotient $\mu^{-1}(0)/K$ then inherits a symplectic structure and is the symplectic reduction at 0, or symplectic quotient, of X by the action of K.

Now let X be a nonsingular complex projective variety embedded in complex projective space $\mathbf{P}^n$, and let G be a complex reductive group acting on X via a complex linear representation $\rho : G \to GL(n + 1; \mathbf{C})$. If K is a maximal compact subgroup of G, we can choose coordinates on $\mathbf{P}^n$ so that the action of K preserves the Fubini-Study form ω on $\mathbf{P}^n$, which restricts to a symplectic form on X. There is a moment map $\mu : X \to \mathbf{k}^*$ defined by

$$\mu(x).a = \frac{\overline{\hat{x}}^t \rho_*(a) \hat{x}}{2\pi i \|\hat{x}\|^2} \tag{5.5}$$

for all $a \in \mathbf{k}$, where $\mu(x).a$ denotes the natural pairing between $\mu(x) \in \mathbf{k}^*$ and $a \in \mathbf{k}$, while $\hat{x} \in \mathbf{C}^{n+1} - \{0\}$ is a representative vector for $x \in \mathbf{P}^n$ and the representation $\rho : K \to U(n + 1)$ induces $\rho_* : \mathbf{k} \to \mathbf{u}(n + 1)$

and dually $\rho^* : \mathbf{u}(n+1)^* \to \mathbf{k}^*$. Note that we can think of μ as a map $\mu : X \to \mathbf{g}^*$ defined by

$$\mu(x).a = \mathrm{re}\left(\frac{\overline{\hat{x}}^t \rho_*(a)\hat{x}}{2\pi i \|\hat{x}\|^2}\right)$$

for $a \in \mathbf{g} = \mathbf{k} \otimes_{\mathbf{R}} \mathbf{C}$; then μ satisfies $\mu(x).a = 0$ for all $a \in i\mathbf{k}$.

In this situation the GIT quotient $X/\!/G$ can be canonically identified with the symplectic quotient $\mu^{-1}(0)/K$. More precisely [Ki], any $x \in X$ is semistable if and only if the closure of its G-orbit meets $\mu^{-1}(0)$, while x is stable if and only if its G-orbit meets

$$\mu^{-1}(0)_{\mathrm{reg}} = \{x \in \mu^{-1}(0) \mid d\mu(x) : T_x X \to \mathbf{k}^* \text{ is surjective}\},$$

and the inclusions of $\mu^{-1}(0)$ into X^{ss} and of $\mu^{-1}(0)_{\mathrm{reg}}$ into X^s induce homeomorphisms

$$\mu^{-1}(0)/K \to X/\!/G \text{ and } \mu^{-1}(0)_{\mathrm{reg}} \to X^s/G.$$

Thus the moment map picks out a unique K-orbit in each stable G-orbit, and also in each equivalence class of strictly semistable G-orbits, where x and y in X^{ss} are equivalent if the closures of their G-orbits meet in X^{ss}; that is, if their images under the natural surjection $q : X^{ss} \to X/\!/G$ agree.

Remark 5.4. It follows from the formula (5.5) that if we change the linearisation of the G-action of X by multiplying by a character $\chi : G \to \mathbf{C}^*$ of G, then the moment map is modified by the addition of a central constant c_χ in $\mathbf{k}^*$, which we can identify with the restriction to $\mathbf{k}$ of the derivative of χ.

When a non-reductive affine algebraic group H with unipotent radical U acts linearly on a projective variety X there are 'moment-map-like' descriptions of suitable projectivised quotients $\widetilde{X/\!/U} = G \times_U X/\!/G$ and the resulting quotients $(\widetilde{X/\!/U})/\!/(H/U)$, which are analogous to the description of a reductive GIT quotient $Y/\!/G$ as a symplectic quotient $\mu^{-1}(0)/K$, and can be obtained from the symplectic quotient description of the reductive GIT quotient $\widetilde{G \times_U X/\!/G}$ (see [Ki5] and [Ki6] for more details). This is very closely related to the 'symplectic implosion' construction of Guillemin, Jeffrey and Sjamaar [GJS].

The case of the automorphism group H of $\mathbf{P}(1,1,2)$ acting on $X_d = \mathbf{P}(\mathbf{C}_{(d)}[x,y,z])$ as above with respect to the linearisation $\mathcal{L}_\epsilon$ for any $\epsilon \neq -d/2$ is particularly simple. We have seen that then $X_d/\!/_\epsilon H =$

$(X_d/\!/_\epsilon \hat{U})/\!/SL(2;\mathbf{C})$ where $X_d/\!/_\epsilon \hat{U}$ is the image of $X_d^{ss,\hat{U},\epsilon}$ in $Y_d/\!/_\epsilon \hat{U} = (\mathbf{P}^{12} \times Y_d)/\!/_\epsilon GL(4;\mathbf{C})$. There is a moment map

$$\mu_{U(4)} : \mathbf{P}^{12} \times Y_d \to \text{Lie } U(4)^*$$

for the action of the maximal compact subgroup $U(4)$ of $GL(4;\mathbf{C})$ on $\mathbf{P}^{12} \times Y_d$ associated to the linearisation $\mathcal{O}_{\mathbf{P}^{12}}(N) \times \mathcal{O}_{Y_d}(1)$ for $N \gg 0$, given by

$$\mu_{U(4)}(a,y) = N\mu_{U(4)}^{\mathbf{P}^{12}}(a) + \mu_{U(4)}^{Y_d}(y) \qquad (5.6)$$

for $(a,y) \in \mathbf{P}^{12} \times Y_d$. Here $\mu_{U(4)}^{\mathbf{P}^{12}} : \mathbf{P}^{12} \to \text{Lie } U(4)^*$ and $\mu_{U(4)}^{Y_d} : Y_d \to \text{Lie } U(4)^*$ are the moment maps given by formula (5.5) for the actions of $U(4)$ on $\mathbf{P}^{12}$ and on $Y_d = \mathbf{P}(\mathbf{C}_{d/2}[X,Y,W,z])$. We can identify $Y_d/\!/_\epsilon \hat{U} = (\mathbf{P}^{12} \times Y_d)/\!/_\epsilon GL(4;\mathbf{C})$ with $\mu_{U(4)}^{-1}(-\epsilon)/U(4) = (\mu_{SU(4)}^{-1}(0) \cap \mu_{S^1}^{-1}(-\epsilon))/S^1 SU(4)$, where S^1 is the maximal compact subgroup of the subgroup $\mathbf{C}^*$ of $GL(4;\mathbf{C})$ given at (5.2). We can use the standard invariant inner product on the Lie algebra of $U(4)$ to identify Lie $U(4)$ with Lie $U(4)^*$ and with $\text{Lie}(U(3) \times U(1)) \oplus (\text{Lie}(U(3) \times U(1)))^\perp$, and thus with

$$\text{Lie} S^1 \ \oplus \ \text{Lie } S(U(3) \times U(1)) \ \oplus \ (\text{Lie}(U(3) \times U(1)))^\perp$$

where $(\text{Lie}(U(3) \times U(1)))^\perp$ is the orthogonal complement to $\text{Lie}(U(3) \times U(1))$ in $\text{Lie} U(4)$, and $S(U(3) \times U(1)) = (U(3) \times U(1)) \cap SU(4)$ so that $U(3) \times U(1) = S^1 S(U(3) \times U(1))$. With respect to this decomposition we can write $\mu_{U(4)} = \mu_{S^1} \oplus \mu_{S(U(3) \times U(1))} \oplus \mu_\perp$ where $\mu_\perp$ is the orthogonal projection of $\mu_{U(4)}$ onto $(\text{Lie }(U(3) \times U(1)))^\perp$. We find that if $\epsilon \neq -d/2$ and N is sufficiently large then

$$Y_d/\!/_\epsilon \hat{U} \cong \mu_{U(4)}^{-1}(-\epsilon)/U(4) \cong (\mu_{S^1}^{-1}(-\epsilon) \cap \mu_\perp^{-1}(0))/S^1,$$

and restricting to X_d we get an identification

$$X_d/\!/_\epsilon \hat{U} \cong \mu_{\hat{U}}^{-1}(-\epsilon)/S^1$$

where $\mu_{\hat{U}} : X_d \to \text{Lie}(\hat{U})^* \cong (\text{Lie} S^1 \otimes_{\mathbf{R}} \mathbf{C})^* \oplus (\text{Lie}(U(3) \times U(1)))^\perp$ is a 'moment map' for the action of $\hat{U}$ on X_d (which takes into account the Kähler structure on X_d, not just its symplectic structure), defined by

$$\mu_{\hat{U}}(x).a = \text{re} \left(\frac{\overline{\hat{x}}^t \rho_*(a)\hat{x}}{2\pi i \|\hat{x}\|^2} \right)$$

for $a \in \text{Lie}(\hat{U})$. Moreover if $\epsilon \neq -d/2$ then $X_d/\!/_\epsilon H = (X_d/\!/_\epsilon \hat{U})/\!/SL(2;\mathbf{C})$

can be identified with

$$\frac{\mu_{\hat{U}}^{-1}(-\epsilon) \cap \mu_{SU(2)}^{-1}(0)}{S^1 \; SU(2)} = \frac{\mu_H^{-1}(-\epsilon)}{U(2)}$$

where the 'moment map' $\mu_H : X_d \to \mathrm{Lie}(H)^*$ is defined by

$$\mu(x).a = \mathrm{re}\left(\frac{\overline{\hat{x}}^t \rho_*(a)\hat{x}}{2\pi i \|\hat{x}\|^2}\right)$$

for $a \in \mathrm{Lie}(H)$, and $U(2)$ is a maximal compact subgroup of H.

Bibliography

[AD] A. Asok and B. Doran, *On unipotent quotients and some $\mathbf{A}^1$-contractible smooth schemes*, Int. Math. Research Papers **5** (2007), article ID rpm005.

[AD2] A. Asok and B. Doran, *Vector bundles on contractible smooth schemes*, Duke Mathematical Journal **143** (2008), 513–530.

[AB] M.F. Atiyah and R. Bott, *The Yang-Mills equations over Riemann surfaces*, Phil. Trans. Roy. Soc. London **308** (1982), 523–615.

[BBDG] A. Beilinson, J. Bernstein and P. Deligne, *Faisceaux pervers*, Analysis and topology on singular spaces I (Luminy, 1981), Astrisque **100**, Soc. Math. France, Paris, 1982, 5–171.

[BP] M. Brion and C. Procesi, *Action d'un tore dans une variété projective*, Progress in Mathematics **192** (1990), 509–539.

[Co] D Cox, *The homogeneous coordinate ring of a toric variety*, J. Algebraic Geom. **4** (1995), 17–50.

[CK] D Cox and S Katz, *Mirror symmetry and algebraic geometry*, Mathematical Surveys and Monographs **68**, American Mathematical Society, Providence, RI, 1999.

[Do] I. Dolgachev, *Lectures on invariant theory*, London Mathematical Society Lecture Note Series **296**, Cambridge University Press, 2003.

[DH] I. Dolgachev and Y. Hu, *Variation of Geometric Invariant Theory quotients*, Publ. Math. I.H.E.S. **87** (1998) (with an appendix by N. Ressayre).

[DK1] B. Doran and F. Kirwan, *Towards non-reductive geometric invariant theory*, Pure Appl. Math. Quarterly **3** (2007), 61–105.

[DK2] B. Doran and F. Kirwan, *Effective non-reductive geometric invariant theory*, in preparation.

[Fa] A. Fauntleroy, *Categorical quotients of certain algebraic group actions*, Illinois Journal Math. **27** (1983), 115–124.

[Fa2] A. Fauntleroy, *Geometric invariant theory for general algebraic groups*, Compositio Mathematica **55** (1985), 63–87.

[Fa3] A. Fauntleroy, *On the moduli of curves on rational ruled surfaces*, Amer. J. Math. **109** (1987), 417–452.

[Gi] D. Gieseker, *Geometric invariant theory and applications to moduli problems*, Invariant theory (Montecatini, 1982) Lecture Notes in Math. **996**, Springer (1983), 45–73.

[GM] M. Goresky and R. MacPherson, *On the topology of algebraic torus actions*, Algebraic groups, Utrecht 1986, 73–90, Lecture Notes in Mathematics, 1271.

[GP] G.-M. Greuel and G. Pfister, *Geometric quotients of unipotent group actions*, Proc. London Math. Soc. (3) **67** (1993) 75–105.

[GP2] G.-M. Greuel and G. Pfister, *Geometric quotients of unipotent group actions II*, Singularities (Oberwolfach 1996), 27–36, Progress in Math. 162, Birkhauser, Basel 1998.

[Gr] F. Grosshans, *Algebraic homogeneous spaces and invariant theory*, Lecture Notes in Math. 1673, Springer-Verlag, Berlin, 1997.

[Gr2] F. Grosshans, *The invariants of unipotent radicals of parabolic subgroups*, Invent. Math. **73** (1983), 1–9.

[GJS] V. Guillemin, L. Jeffrey and R. Sjamaar, *Symplectic implosion*, Transformation Groups **7** (2002), 155–184.

[JK] L. Jeffrey and F. Kirwan, *Localization for nonabelian group actions*, Topology **34** (1995), 291–327.

[JKKW] L. Jeffrey, Y.-H. Kiem, F. Kirwan, and J. Woolf, *Cohomology pairings on singular quotients in geometric invariant theory*, Transform. Groups **8** (2003), 217–259.

[Ka] I. Kausz, *A modular compactification of the general linear group*, Doc. Math. **5** (2000), 553–594.

[KM] S. Keel and S. Mori, *Quotients by groupoids*, Annals of Math. (2) **145** (1997), 193–213.

[Ki] F. Kirwan, *Cohomology of quotients in symplectic and algebraic geometry*, Mathematical Notes **31** Princeton University Press, Princeton, NJ, 1984.

[Ki2] F. Kirwan, *Partial desingularisations of quotients of nonsingular varieties and their Betti numbers*, Annals of Math. (2) **122** (1985), 41–85.

[Ki3] F. Kirwan, *Rational intersection cohomology of quotient varieties*, Invent. Math. **86** (1986), 471–505.

[Ki4] F. Kirwan, *Rational intersection cohomology of quotient varieties II*, Invent. Math. **90** (1987), 153–167.

[Ki5] F. Kirwan, *Symplectic implosion and non-reductive quotients*, to appear in the proceedings of the 65th birthday conference for Hans Duistermaat, "Geometric Aspects of Analysis and Mechanics", Utrecht, 2007.

[Ki6] F. Kirwan, *Generalised symplectic implosion*, in preparation.

[Muk] S. Mukai, *An introduction to invariants and moduli*, Cambridge University Press 2003.

[Muk2] S. Mukai, *Geometric realization of T-shaped root systems and counterexamples to Hilbert's fourteenth problem*, Algebraic transformation groups and algebraic varieties, 123–129, Encyclopaedia Math. Sci. **132**, Springer, Berlin, 2004.

[MFK] D. Mumford, J. Fogarty and F. Kirwan, *Geometric invariant theory*, 3rd edition, Springer, 1994.

[Na] M. Nagata, *On the 14-th problem of Hilbert*, Amer. J. Math. 81, 1959, 766–772.

[Ne] P.E. Newstead, *Introduction to moduli problems and orbit spaces*, Tata Institute Lecture Notes, Springer, 1978.

[Ne2] P.E. Newstead, *Geometric invariant theory*, Moduli spaces and vector bundles, ed. L. Brambila-Paz, S.Bradlow, O. Garcia-Prada, and S. Ramanan, Cambridge University Press (to appear).

[Po] V. Popov, *On Hilbert's theorem on invariants*, Dokl. Akad. Nauk SSSR 249 (1979), 551–555. English translation: Soviet Math. Dokl. 20 (1979), 1318–1322 (1980).

[PV] V. Popov and E. Vinberg, *Invariant theory*, Algebraic geometry IV, Encyclopaedia of Mathematical Sciences v. 55, 1994.

[Rei] Z. Reichstein, *Stability and equivariant maps*, Invent. Math. **96** (1989), 349–383.

[Res] N. Ressayre, *The GIT-equivalence for G-line bundles*, Geom. Dedicata **81** (2000), 295–324.

[Th] M. Thaddeus, *Geometric invariant theory and flips*, Journal of Amer. Math. Soc. **9** (1996), 691–723.

[W] J. Winkelmann, *Invariant rings and quasiaffine quotients*, Math. Z. 244 (2003), 163–174.

11

Dualities on $T^*\mathcal{SU}_X(2,\mathcal{O}_X)$

E. Previato

Department of Mathematics and Statistics
Boston University
Boston, MA 02215-2411 USA
e-mail: ep@bu.edu

To P.E. Newstead with best wishes,
in gratitude, celebration of the past
and anticipation of his future leadership.

Abstract. The notion of Algebraic Complete Integrability (ACI) of certain mechanical systems, introduced in the early 1980s, has given great impetus to the study of moduli spaces of holomorphic vector bundles over an algebraic curve (or a higher-dimensional variety, still at a much less developed stage). Several notions of 'duality' have been the object of much interest in both theories. There is one example, however, that appears to be a beautiful isolated feature of genus-2 curves. In this note such example of duality, which belongs to a 'universal' class of ACIs, namely (generalized) Hitchin systems, is interpreted in the setting of the classical geometry of Klein's quadratic complex, following the Newstead and Narasimhan-Ramanan programme of studying moduli spaces through explicit projective models.

§0. Introduction

In this volume's conference, dedicated to Peter Newstead and his work, one of the prominent objects was $\mathcal{SU}_X(2,\xi)$, the moduli space of (semi)stable, rank-2 vector bundles over a Riemann surface X of genus $g \geq 2$, with fixed determinant ξ. The cases of degree(ξ) even, odd respectively, give rise to isomorphic varieties (by tensoring with a line bundle, since $\mathrm{Jac}(X)$ is a divisible group), usually denoted by $\mathcal{SU}_X(2,0)$, $\mathcal{SU}_X(2,1)$ respectively, when ξ is not important. When the rank is coprime with the degree, a semistable bundle must be stable and the variety is nonsingular. Newstead contributed much to the knowledge of these spaces, by now treated in texts such as [M], where proofs

367

of the basic properties we quote can be found; in [New] he gave the first projective model of $\mathcal{SU}_X(2,1)$ when X has genus two and Weierstrass points $\lambda_1, \ldots, \lambda_6$, namely Klein's quadratic complex

$$\{(x_1, \ldots, x_6) | \sum_{i=1}^{6} x_i^2 = 0\} \cap \{(x_1, \ldots, x_6) | \sum_{i=1}^{6} \lambda_i x_i^2 = 0\}$$

(cf. also [DR], [R]). The relevance of moduli spaces of bundles to mathematical physics as well was enhanced by the theory of "spectral curves" originated (in the context of integrable systems, to my knowledge) by Hitchin [H1]: when the spectrum of an $r \times r$ matrix L that depends on a parameter $\lambda \in \mathbb{C}$ is fixed, the spectral equation $\det(L(\lambda) - \mu) = 0$ can be viewed as that of a plane curve, the eigenspaces as fibres of line bundles over the curve if they are generically one-dimensional. This idea connected bundles over a curve to the "algebraically completely integrable" Hamiltonian systems (ACI) that enjoyed much activity since the 1970s. The first examples were given by equations of "Lax-pair type",

$$\dot{L}(\lambda) = [B(\lambda), L(\lambda)]$$

for a pair of matrices that depend on positions and momenta (q_i, p_i); Hitchin again [H2] provided the 'next level', which enables us to view λ no longer as a number but as a point on a curve of higher genus, and the matrix as a section of a twisted endomorphism of a rank-r vector bundle over the curve. As a particular case, Hitchin showed that the cotangent bundle $T^*\mathcal{SU}_X(2, \xi)^s$ to the set of stable points is the phase space of an ACI. The question then arose of writing explicit equations of motion for the system, and this was achieved in [vGP] and [GT-N-B1] for the case of $\mathcal{SU}_X(2, \mathcal{O}_X)$ in genus 2, not surprisingly perhaps since it is the only case in which Torelli fails: for any curve, the moduli space is isomorphic to $\mathbb{P}^3$, while the memory of the curve is retained by the locus of strictly semistable points, the classical Kummer quartic surface $\mathcal{K}$.

The goal of this short note is to give classical context to a duality discovered by Gawędzki and Tran-Ngoc-Bich interchanging position and momentum of the integrable system. Of the many dualities that Hitchin(-type) systems have been shown to enjoy (mainly based on a Fourier-Mukai transform of the integral manifolds which are abelian varieties and have natural duals, cf. e.g. [HT1-2] for connections to mirror symmetry), this one seems to be relatively unexplored, and it may appeal to Newstead in that one of his many talents has been to give a contemporary interpretation to the beautiful classical constructions on the Grassmannian of lines in $\mathbb{P}^3$.

In Section 1 I recall the definition of ACI, pointing out issues that are not yet understood, and present the Hitchin Hamiltonians. In Section 2 I treat the dualities and in Section 3 I gather other aspects of the system, some related to the geometry of Klein's complex, whose understanding, in my opinion, is worth pursuing.

Acknowledgements†. I am immensely indebted to Peter Newstead in many ways, as a mathematician and, to Ann Newstead as well, as generous friends. I am too fortunate to be able with this small note to celebrate his birthday and to wish many happy returns, and for this fortune I am indebted to the conference organizers, S. Bradlow, L. Brambila-Paz, D. Ellwood, O. García-Prada and S. Ramanan, as I am also for their creating a meeting of great warmth, presenting admirable contributions that spanned several generations of mathematical research. Lastly I would like to thank Nigel Hitchin for the question, "What is this duality?", which is the origin of this note, although I am not presuming that it was a momentuous question to him (in fact it was asked as the organizers were leading us to explore the beautiful San Miguel, Gto.)

§1. ACI

As recalled in the Introduction, Hitchin showed as a special case of his theory [H2] that $T^*\mathcal{SU}_X(r, \xi)^s$ supports $(r^2 - 1)(g - 1)$ functionally independent holomorphic functions that Poisson commute with respect to the cotangent-bundle natural symplectic structure (also holomorphic).

1.1 Aspects of integrability. The aforementioned property defines a "completely integrable" Hamiltonian system. Classically, the "phase space" of such a system is a real $2n$-dimensional manifold M; if n independent commuting Hamiltonians $H_1, \ldots, H_n$ have compact level sets, the Arnold-Liouville theorem guarantees that the invariant manifolds $I_c = \{p \in M | H_i(p) = c_i \in \mathbb{R}\}$ are tori $\mathbb{T}^n$; more generally, if solutions to Hamilton equations exist for all time, $I_c \cong \mathbb{T}^k \times \mathbb{R}^{n-k}$ for some $0 \leq k \leq n$. It may therefore seem sufficient to require that the Hamiltonians be complex-holomorphic, and the symplectic structure local-holomorphic, $\omega = dp \wedge dq$ (with p_i, q_i complex coordinates), for the system to be ACI, namely "the generic fibre is an open set in an abelian variety and the vector fields are linear" [H2].

However, the ACI property is more special; as Flaschka observes [F], a simple example such as $H = p^2 - q^3$, whose fibres are "circles×lines"

---

† Many thanks to the Referees for a very close reading which weeded out several inaccuracies.

over $\mathbb{R}^2$, when complexified as $q = x_1 + iy_1$, $p = x_2 + iy_2$, gives rise to

$$I_1 = x_1^2 - y_1^2 - x_2^3 + 3x_2 y_2^2, \qquad I_2 = 2y_1 x_1 - 3x_2^2 y_2 + y_2^3$$

(the real and imaginary parts of H) which by the Cauchy-Riemann equations are in involution with respect to the canonical Poisson bracket, $\{y_i, x_j\} = \delta_{ij}$, $\{y_1, y_2\} = \{x_1, x_2\} = 0$. However, there are no action-angle variables, although each fibre (except the singular curve $p^2 - q^3 = 0$) is now an elliptic curve with some missing point: the (real) Liouville tori are not the real part of a complex abelian torus; Flaschka infers, "Evidently (...) one can choose initial conditions such that [the real flow] passes through any desired number of [period] parallellograms before hitting a vertex". The missing requirement is that the solution exist for all time. In such absence of action-angle variable, one may ask the following

Question 1. [F] Find canonical models for the behavior of integrable Hamiltonian systems of two degrees of freedom whose level surfaces are punctured Riemann surfaces.

Mumford [Mum, IIIa, §4] points out that even if the Liouville tori are real ovals in abelian varieties, the ACI property may fail, and gives the following simplest non-example (that is, an example which fails to be ACI). His (equivalent) requirement for ACI is the following. He defines a completely integrable system on a symplectic manifold M of (real) dimension $2n$ to be such that there exists n functions in involution whose differentials are linearly independent (possibly, generically on M), and in addition he requires compactness of the fibres I_c. When M is a component of the set of real points of an algebraic variety $M_\mathbb{C}$ and the symplectic form ω and Hamiltonian function H are rational without poles on M, then the tori I_c are the real points of complex algebraic varieties. The system is ACI if the vector fields corresponding to the Hamiltonians in involution still have no poles on a compactification of the fibres on $M_\mathbb{C}$.

His non-example is

$$M = \mathbb{R}^2, \quad \omega = dx \wedge dy, \quad H = x^4 + y^4$$

Here a compactification of the fibre, the affine curve $x^4 + y^4 = c$, is the projective curve $X^4 + Y^4 = cZ^4$, which is smooth (provided $c \neq 0$) and has 4 points at infinity. The vector field X_H defined by H, $X_H \lrcorner \omega = -dH$, is tangent to the fibre in the affine plane, i.e. a multiple of $\chi_*(\frac{\partial}{\partial t})$, where $\chi : t \mapsto (x(t), y(t))$ is a map to the curve. Indeed, differentiating $x^4 + y^4 = c$ gives $\frac{\partial x}{\partial t} x^3 + \frac{\partial y}{\partial t} y^3 = 0$, so $(\frac{\partial x}{\partial t}, \frac{\partial y}{\partial t}) \propto (y^3, -x^3)$.

At infinity $Z = 0$, $X, Y \neq 0$, the vector field $X_H = 4y^3 \frac{\partial}{\partial x} - 4x^3 \frac{\partial}{\partial y}$ has a pole, since coordinates at infinity are $(Z/X = 1/x, Y/X = y/x)$ and $\frac{\partial}{\partial x} = -\frac{1}{x^2} \frac{\partial}{\partial(1/x)}$, and since $4y^3/(-x^2) = -4Y/Z$.

Note: In the algebraically completely integrable situation, the fibres are abelian varieties or extensions of such by $\mathbb{C}^{*k}$ for some power k. Indeed, the holomorphic tangent bundle of an abelian variety is trivial.

In particular, the Hamiltonian $H = x^3 + y^3$ would be ACI; now the compactification of the Fermat curves $x^3 + y^3 = cz^3$ $(c \neq 0)$ have three points at infinity where $1/x$ is a local parameter, $z/x = 0$ is an inflectionary tangent, and the vector field X_H has no poles. By general theory a regular vector field on an abelian variety is translation invariant, a fact that does not seem easy to check directly, under the geometric definition of addition on the curve (using lines through two points and their third intersection with the cubic curve). Notice also that there is no contradiction even though two 'orbits' of the motion meet. Indeed, any two Fermat curves only meet at the three points at infinity $1/x = 0, y/x = -e^{k2\pi i/3}$, $k = 0, 1, 2$, each of the intersections has multiplicity three because each curve has the same inflectionary tangent at the point, but there is no global symplectic structure on $\mathbb{P}^2$, which doesn't have trivial canonical bundle.

In view of Mumford's non-example, it makes sense to extend the definition of ACI to Hamiltonians that are (complex-)completely integrable but whose flows are not (locally) linear on abelian varieties. This was done by Vanhaecke [V] and exemplified by Abenda and Fedorov [AF1-2] who defined "deficient integrability" as the case of Liouville integrability (on real tori) when the complexification has fibres that are "$n \leq g$-dimensional $(\dots)$ non-linear subvarieties of [g-dimensional] abelian tori, and $(\dots)$ on universal coverings of the abelian tori there exist coordinates $\varphi_1, \dots, \varphi_g$ such that under the Hamiltonian flow $\varphi_1, \dots, \varphi_n$ change linearly in time, whereas the rest of the coordinates are analytic functions of the former."

Question 2. There seems to be no Lax-pair characterization of "deficient integrability".

1.2 Hitchin Hamiltonians in genus two. We specialize the Hitchin system to the (stable) principal bundles with structure group $SL(2, \mathbb{C}) \cong Sp(2, \mathbb{C})$, thus getting the moduli space of (stable) vector bundles with fixed determinant of even degree. The case genus$(X) = 1$ is void because there are no stable rank-2 bundles of even determinant over an elliptic curve. We consider the case $g = 2$ and let the distinct complex numbers $\lambda_1, \dots, \lambda_6$ determine the Weierstrass points of the curve. In this case, the moduli space is smooth, while for $g > 2$ the locus of strictly semistable bundles equals the singular locus, and we can extend Hitchin's construction over $\mathcal{SU}_X(2, \mathcal{O}_X)$, which we denote henceforth by $\mathcal{M}$. The points of the space $T^*\mathcal{M}$ are pairs (E, ψ), where ψ is a holomorphic $(1,0)$ form with values in the bundle of traceless endomorphisms of E with poles bounded by K, the canonical bundle of X. The Hitchin map:

$$H : T^*\mathcal{M} \longrightarrow H^0(X, 2K), \quad \psi \mapsto \det(\psi), \tag{1}$$

gives three Hamiltonian functions, generically independent holomorphic functions with values in the 3-dimensional vector space of quadratic differentials. Since det is homogeneous (of degree two) in the variable ψ, we look instead for a map

$$H : \mathbb{P}T^*\mathcal{M} \longrightarrow \mathbb{P}H^0(X, 2K) = |2K|.$$

In our situation, this projectivized tangent bundle is the incidence correspondence:

$$\mathbb{P}T^*\mathbb{P}V \cong I := \{\ (x, h) \in \mathbb{P}V \times \mathbb{P}V^* : x \in h\ \},$$

where $V = H^0(\mathrm{Pic}^1(X), 2\Theta)$ and $\Theta \subset \mathrm{Pic}^{g-1}$ is the canonical theta divisor. Denoting by $q = (q_1, \ldots, q_4)$ coordinates in $\mathbb{P}V$ and by $p = (p_1, \ldots, p_4)$ coordinates in $\mathbb{P}V^*$, we consider the following sections of the bundle projection: $\mathbb{P}T^*\mathbb{P}^3 \longrightarrow \mathbb{P}^3$,

$$\Phi_i : \mathbb{P}^3 \longrightarrow \mathbb{P}T^*\mathbb{P}^3 = I \subset \mathbb{P}^3 \times (\mathbb{P}^3)^*, \quad q \mapsto (q; \epsilon_i(q)) := (q; X_i(q, -)),$$

where $(q; p)$ is the natural pairing and the linear forms X_i are the Klein coordinates of a line $l = \langle (Z_0 : \ldots : Z_3), (W_0 : \ldots : W_3) \rangle \subset (\mathbb{P}^3)^*$, $p_{ij} := Z_i W_j - W_i Z_j$, namely:

$$X_1 = p_{01} + p_{23}, \qquad X_3 = \imath(p_{02} + p_{13}), \qquad X_5 = p_{03} + p_{12}$$
$$X_2 = \imath(p_{01} - p_{23}), \qquad X_4 = p_{02} - p_{13}, \qquad X_6 = \imath(p_{03} - p_{12}).$$

Theorem [vGP] *Any three among the following functions Poisson commute and can be taken as a complete set of Hamiltonians for the Hitchin system, possibly up to multiplication by functions from the base, an open set in $\mathbb{P}^3$,*

$$H_i(p, q) := \sum_{j \neq i} \frac{x_{ij}^2}{\lambda_i - \lambda_j}, \quad \text{with} \quad x_{ij} := X_j(\langle \epsilon_i(q), p \rangle),$$

the Klein coordinates of the line $\langle \epsilon_i(q), p \rangle \subset (\mathbb{P}^3)^*$.

In [GT-N-B1], the authors showed that these are actually Hitchin Hamiltonians by constructing a self-duality on the Hitchin system in genus 2, and this will be the topic of the next section.

§2. What is this duality?

In [GT-N-B1] the authors provide a parametrization of the phase space in terms of second-order theta functions and reveal that this case of the Hitchin system is self-dual under interchange of positions and momenta. The aim of this section is to identify this duality in geometric terms. In rephrasing results from [GT-N-B1] we shall be necessarily sketchy, referring to the original paper for technical provisos.

2.1 Wirtinger duality. Points of the phase space of the system are pairs (E, ψ), where the bundles E can be given as an extension (class)

$$0 \to \ell^{-1} \to E \to \ell \to 0,$$

ℓ a line bundle of degree $g - 1 = 1$, and ψ is a holomorphic $(1, 0)$ form with values in the bundle of traceless endomorphisms of E. It is useful to fix $\det(E) \cong \mathcal{O}_X$ because in this case E is self dual. By associating to E the subvariety C_E of $\operatorname{Pic}^{g-1}(X)$ consisting of the line bundles ℓ such that $H^0(X, \ell \otimes E) \neq 0$ (equivalently, E is an extension of ℓ), M.S. Narasimhan and S. Ramanan showed that $\mathcal{SU}_X(2, \mathcal{O}_X) \xrightarrow{\cong} \mathbb{P}^3 = \mathbb{P} H^0(\operatorname{Pic}^1, 2\Theta)$, $\Theta := \{\xi \in \operatorname{Pic}^{g-1} : H^0(X, \xi) \neq 0\}$, $E \mapsto \mathbb{C}^* \theta_E$ where θ_E is a second-order theta function that vanishes precisely on the curve C_E (twisted by suitable line bundle of degree -1). There is a natural theta divisor $\Theta \subset \operatorname{Pic}^{g-1}(X)$, whereas a symmetric theta divisor $\Theta_0 \subset \operatorname{Pic}^0(X) = \operatorname{Jac}(X)$ is only determined up to a point of order 2 in $\operatorname{Jac}(X)$, so that the linear system $|2\Theta_0|$ is canonical. The natural map that identifies (for any genus) $|2\Theta_0|$ and $|2\Theta|^*$ ("Wirtinger duality") allows us to use second-order theta functions to coordinatize the projective image of $\mathcal{M}$, as the following diagram commutes (A. Beauville):

$$
\begin{array}{ccc}
|2\Theta| & = \mathbb{P}^{2^g - 1} & \\
\nearrow \quad \theta \nearrow & & \\
\operatorname{Pic}^0 X \quad \to \quad \mathcal{SU}_X(2, \mathcal{O}) & \quad \| & \text{(Wirtinger)} \\
\searrow \quad \searrow & & \\
|2\Theta_0|^* & = |\mathcal{L}|^* & \text{(Beauville)}
\end{array}
$$

where $\Theta \subset \operatorname{Pic}^{g-1} X$ is the natural theta divisor, $\mathcal{L}$ the ample generator of $\operatorname{Pic}(\mathcal{SU}_X(2, \mathcal{O}_X))$, and for $\alpha \in \operatorname{Pic}^0 X$, $\alpha \mapsto \Theta_\alpha + \Theta_{\alpha^{-1}} \in |2\Theta|$, $\alpha \mapsto \alpha \oplus \alpha^{-1} \in \mathcal{SU}_X(2, \mathcal{O}_X)$ and $\theta(E) = C_E = \{\xi \in \operatorname{Pic}^{g-1} X | H^0(X, E \otimes \xi) \neq$

0}. By a choice of theta characteristic, $\mathrm{Pic}^{g-1}(X)$ is identified with the Jacobian of X, $\mathbb{C}^2/\Lambda$; let Δ be the vector that corresponds to the chosen theta characteristic.

2.1.1 Two versions. The classical duality is intrinsic. Indeed, the theta divisor in Pic^{g-1} is canonically defined, as is the morphism $\mathrm{Pic}^0 \to |2\Theta|$, $\eta \mapsto \Theta_\eta + \Theta_{\eta^{-1}}$, and this identifies $|2\Theta|^*$, the linear forms on $|2\Theta|$, with $|2\Theta_0| = H^0(\mathrm{Pic}^0, 2\Theta_0)$, because the hyperplane class pulls back to $2\Theta_0$. In [GT-N-B1], this is presented in an analytic way using extrinsic, explicit coordinates that depend on the choice of suitable bases. Upon choosing a (standard) homology basis on X, the associated theta function $\vartheta(u)$ gives rise to a second-order theta function (in both u and u') $\vartheta(u'-u)\vartheta(u'+u)$, obeying Riemann's addition theorem:

$$\vartheta(u'-u)\vartheta(u'+u) = \sum_{e \in \mathbb{Z}^g} \theta_{2,e}(u')\theta_{2,e}(u), \text{where} :$$

$$\theta_{k,e}(u) = \sum_{n \in \mathbb{Z}^g} e^{\pi i k^t (n+e/k)\tau(n+e/k)+2\pi i k^t (n+e/k)\cdot u} .$$

The duality in coordinates. The linear isomorphism

$$\iota : H^0(\mathrm{Pic}^0, 2\Theta_0)^* \to H^0(\mathrm{Pic}^0, 2\Theta_0), \quad \iota(\varphi)(u) = \langle \vartheta(\cdot - u)\vartheta(\cdot + u), \varphi \rangle,$$

where ϑ depends on the choice of a theta characteristic, is such that by Riemann's addition theorem ι interchanges the basis $(\theta_{2,e})$ of $H^0(\mathrm{Pic}^0, 2\Theta_0)$ with the dual basis $(\theta_{2,e}^*)$ of $H^0(\mathrm{Pic}^0, 2\Theta_0)^*$. As noted in [GT-N-B1], this definition relies on the choice of a basis of the vector space underlying $\mathbb{P}^3$ up to a $(\mathbb{Z}/2\mathbb{Z})^4$-action.

Remark. This duality coincides with the Wirtinger duality, as can be seen by noting that to an element $\varphi \in |2\Theta_0|^*$ there should correspond an element $\theta \in |2\Theta_0| = H^0(\mathrm{Pic}^0, 2\Theta_0)$ according to the intrinsic duality, and since $\iota(\varphi)(u) = \langle \vartheta(\cdot - u)\vartheta(\cdot + u), \varphi \rangle$, this is the divisor $\iota(\varphi) = \{u : \varphi((\Theta_0)_u) + ((\Theta_0)_{-u}) = 0\}$. In this setup, Lemma 3.4 of [vGP] says that a point ξ in the Kummer surface $\varphi_{|2\Theta|}(\mathrm{Pic}^1 X)$ belongs to the curve associated to a bundle E if and only if the second-order theta function θ_E associated to E is perpendicular to the dual hyperplane $\Theta_\xi + \Theta_{K\xi^{-1}}$.

Definition. We consider the following duality on $T^*\mathcal{M}$: $(\theta, \varphi) \mapsto \theta' := \iota(\varphi), \varphi' := \iota^{-1}(\theta)$.

This duality interchanges positions $p_1, \ldots, p_4$ and momenta $q_1, \ldots, q_4$ (given by a dual basis) for the standard symplectic form $dp \wedge dq$, which is the canonical one on the cotangent bundle, used by Hitchin. It is by this device that the authors are able to show that $H(p,q) = H(q,p)$ for a Hitchin Hamiltonian H, a property "far from obvious in the original formulation of the Hitchin system" [GT-N-B2, §5.4].

2.1.2 Equations for the Kummer surface. A notable advantage of the explicit approach is the following. It is known that the classical Kummer surface is self-dual: the surface in $\mathbb{P}^{3*}$ consisting of tangent planes of $\mathcal{K} \subset \mathbb{P}^3$ is isomorphic to $\mathcal{K}$. This was proved classically using properties of the associated quadratic complex. It can be shown [GT-N-B1, Appendix 3] that the two surfaces are isomorphic by expressing Wirtinger's duality in extrinsic coordinates, since the Kummer equation is the only quartic invariant under the action of the Heisenberg group (up to rescaling) and the (transpose) Heisenberg action is computed explicitly on the dual basis. Moreover, by differentiating in the abelian variables and evaluating at the branch points, Gawędzki and Tran-Ngoc-Bich express the coefficients of the Kummer quartic, which are given by thetanulls, in terms of the branchpoints of the curve: as remarked in [GT-N-B1], this calculation differs essentially from the standard geometric argument (the projective geometry of the complex). However, we note that the expression of the thetanulls in terms of the branchpoints of the curve is also provided by the classical Thomae's formula [Mum IIIa §8], and the expression for the projection λ to $\mathbb{P}^1$ of a point x of the curve, involving derivatives: $\lambda(x) = -\dfrac{\partial_1 \vartheta(\int_{x_0}^x \omega - \Delta)}{\partial_2 \vartheta(\int_{x_0}^x \omega - \Delta)}$ had also been proved in [G] and [Jo]. On the other hand, the authors' method might be useful in investigating analogous self-duality properties of the higher-genus Heisenberg-invariant quartics, which enter the equations of the projective model of $\mathcal{SU}_X(2, \mathcal{O}_X)$ in $|2\Theta|$.

2.2 Fay lines and planes. The interchange $(\theta, \varphi) \mapsto (\iota(\varphi), \iota^{-1}(\theta))$ takes place on a projective variety, so a natural geometric question is, "what is this duality?" In [GT-N-B1] it is presented in a strictly coordinate-dependent way. We interpret it below as a transformation on the point-plane incidence correspondence by which we described $T^*\mathcal{M}$.

Let's recall how [GT-N-B1] traces the bundle to its projective coordinates $(p_1, \ldots, p_4) \in \mathbb{P}^3$. In the Atiyah-Bott construction of vector bundles via connections, the (semistable) trivial-determinant bundles E, $0 \to \ell^{-1} \to E \to \ell \to 0$, are regarded as the bundle $\ell^{-1} \oplus \ell$ with complex structure ($\bar\partial$-operator) $\bar\partial_E = \bar\partial_{\ell^{-1} \oplus \ell} + B$, $B = \begin{bmatrix} 0 & b \\ 0 & 0 \end{bmatrix}$ (this defines the smooth, $SL(2)$-valued $(0,1)$-form B and b corresponds to the extension $[b] \in H^1(X, \ell^{-2})$), and a Higgs field is represented by a matrix $\psi = \begin{bmatrix} -\mu & \nu \\ \eta & \mu \end{bmatrix}$ with $\mu \in A^{(1,0)}$ (denoting a smooth $(1,0)$ form),

$\nu \in A^{(1,0)}(\ell^{-2}), \eta \in \Omega^1(\ell^2)$ (Ω^1 denotes holomorphic 1-forms). The Hitchin Hamiltonians, by definition of the Hitchin map H in (1), correspond to the determinant of this matrix. The question is to write explicitly the η, μ, ν associated to a point $\langle \theta, \varphi \rangle$ in the cotangent space. In [vGP, Prop. 2.6 and 3.6], this was done for $\eta = 0$, in which case the determinant is the square of a holomorphic differential, by showing that the Hitchin map coincides with the Veronese of the Gauss map on the tangent plane of the Kummer surface. In other words, the Hitchin map was computed on the special points of $\mathbb{P}\mathcal{T}^*\mathcal{M}$ given by (ξ, π), where ξ is a point of the curve in $|2\Theta|$ associated to E, and π is the tangent plane to the Kummer surface that contains ξ. These functions, however, are only determined on a Zariski-open set of the $\xi \in \mathbb{P}^3$. To show that they are (up to constant) Hitchin's Hamiltonians, in [GT-N-B1] the determinant is computed on the remaining points. By deforming the complex structure with connection given above and computing $\delta\theta$ for the corresponding second-order theta function in terms of δb, the holomorphic form η with values in ℓ^2 (a specific line bundle $\ell = \ell_{u_1}$ will be chosen below) corresponding to a point (θ, φ) of the cotangent bundle is shown to be:

$$\eta(x) = \langle \vartheta(\int_{x_0}^x \omega - u - \cdot - \Delta)\vartheta(\int_{x_0}^x \omega - u + \cdot - \Delta), \varphi \rangle,$$

where the left-hand side is a second-order theta function in the variable $(\cdot)$ restricted to the Abel image of the curve. We now want to exclude the case $\eta = 0$. As an element of $H^0(X, \ell^2 K)$, therefore, η has four zeros, $x_1, \ldots, x_4 \in X$, and for a suitable complex number a_2 it can be written (as can be checked by matching zeros and poles) as $\eta = a_2\eta_{\varphi_{u_2}}$ for

$$u_1 + u_2 = \int_{x_0}^{x_1} \omega + \int_{x_0}^{x_2} \omega - 2\Delta, \quad u_1 - u_2 = \int_{x_0}^{x_3} \omega + \int_{x_0}^{x_4} \omega - 2\Delta, \quad (2)$$

where the point u_1 corresponds to the given $\ell = \ell_{u_1}$ (recall that the points ℓ_{u_1} of the curve C_E associated to E are characterized by $[b] \in H^1(X, \ell_{u_1}^{-2})$), and where φ_u is the form that computes θ at the point u, so that $\varphi = a_1\varphi_{u_1} + a_2\varphi_{u_2}$ for another suitable constant a_1. By observing that there is a 2-dimensional choice of theta functions vanishing at u_1, u_2 ($2u_1, 2u_2, u_1 \pm u_2 \notin \Lambda$), and the four zeroes $x_1, \ldots, x_4$ of

$$\vartheta\left(\int_{x_0}^x \omega - u_1 \pm u_2 - \Delta\right)$$

satisfy the equations in (2), the effect of the choice of a quotient bundle ℓ_{u_1} of E is clarified.

The Hitchin quadratic differential becomes:

$$H(\theta, \varphi)(x_i) = -\frac{1}{16\pi^2}(a_1 \partial_a \theta(u_1)\omega^a(x_i) \pm a_2 \partial_a \theta(u_2)\omega^a(x_i))^2, \qquad (3)$$

with the $+$ sign for $i = 1, 2$ and the $-$ sign for $i = 3, 4$. This is proved in [GT-N-B1] by setting up a system of linear equations involving the entries of the Higgs field ψ, that hold on spaces of forms with values in the bundles in question, and evaluating at points. Also by evaluation, the numbers a_1, a_2 are determined below. To construct the Hamiltonians in dual coordinates, the authors compute $H(\theta' = \iota(\varphi), \varphi' = \iota^{-1}(\theta))$. This duality by definition interchanges the bundle E that corresponds to the curve of zeroes of its associated second-order θ_E, with the bundle that corresponds to the curve of zeroes of $\theta' = \iota(\varphi)$, $\{u : \langle \vartheta(\cdot + u)\vartheta(\cdot - u), \varphi \rangle = 0\}$. We wish to interpret this interchange more geometrically.

Notice that since φ is a linear form on the $\mathbb{P}^3$ of vector bundles, $\iota(\varphi)$ is a point in the same $\mathbb{P}^3$. Since we wrote $\varphi = a_1\varphi_{u_1} + a_2\varphi_{u_2}$, the $u_i' := \int_{x_0}^{x_i} \omega - u_1 - \Delta$ are the four intersections of the line $\mathbb{P}\langle \iota(\varphi_{u_1}), \iota(\varphi_{u_2}) \rangle$ with the Kummer surface.

Notice also that in this coordinate choice, the hyperelliptic involution $x \mapsto x'$ lifted to the Jacobian changes the sign of the abelian coordinate u, $\int_{x_0}^{x} \omega - \Delta = -\int_{x_0}^{x'} \omega + \Delta$. Calculation shows that the duality has the following effect:

$$H(\theta', \varphi')(x_i) = -\frac{1}{16\pi^2}(a_1' \partial_a \theta'(u_1')\omega^a(x_i) \mp a_2' \partial_a \theta'(u_2')\omega^a(x_i))^2, \qquad (3)'$$

with $-$ sign if $i = 1, 2$ and $+$ sign if $i = 3, 4$ and

$$u_1' - u_2' = \int_{x_0}^{x_1} \omega + \int_{x_0}^{x_2'} \omega - 2\Delta, \quad u_1' + u_2' = \int_{x_0}^{x_3'} \omega + \int_{x_0}^{x_4'} \omega - 2\Delta \qquad (4)$$

(there is complete symmetry among $x_1, \ldots, x_4$ achieved by regrouping them.) Notice that the four points $\ell_i' := \ell_{u_i'}$ correspond to four points y_i which, upon suitably arranging the coordinate choices made, are images of the x_i under the hyperelliptic involution.

By differentiating in the abelian coordinates and evaluating, we have

$$a_1 = \frac{\partial_k \iota(\varphi)(u_2')\omega^k(x_1)}{\vartheta(w_1 + w_3 + w_4)\partial_k \vartheta(w_2)\omega^k(x_1)},$$

summing over $k = 1, 2$ and having set $w_i = \int_{x_0}^{x_i} \omega - \Delta$ (a similar formula holds for a_2). This is the final calculation that shows the Hitchin

Hamiltonians to be self-dual, in the sense that $(3) = (3)'$ (the differential forms take the same value on $x_1, \ldots, x_4$ as $y_1, \ldots, y_4$).

To understand this geometrically, we have to keep track of the choices we made. The points $x_1, \ldots, x_4$ are the four zeroes of the entry η in the Higgs field (which corresponds to φ), and by choosing a point ℓ_1 in the divisor associated to E (i.e., a zero of θ_E), we determined the pair u_1, u_2: the divisors of $\ell_1 \ell_2$ and $\ell_1 \ell_2^{-1} K$ are $x_1 + x_2$, $x_3 + x_4$, respectively. Now tracing the choices backwards, a generic pair (θ, φ) can be written as $\varphi = a_1 \varphi_{u_1} + a_2 \varphi_{u_2}$ where θ vanishes at u_1 and u_2.

In other words, given E and an associated θ_E corresponding to a curve C_E, with a line bundle ℓ_1 corresponding to a point on it, one chooses a line through the point of $|2\Theta_0|$ that corresponds to θ_E and a φ in its orthogonal (projectively, a line); the orthogonal line meets the Kummer dual in four points φ_{ℓ_i}, $i = 1, \ldots, 4$. For the dual pair (θ', φ'), the four points $\iota(\varphi_{\ell_i'})$ correspond to four points y_i of $\mathcal{K} = \iota^{-1} \mathcal{K}^*$.

To conclude, recall that, as was classically known, each line in $\mathbb{P}^3$ is a Fay quadrisecant of the Kummer surface; in [OPP] this is rephrased in terms of rank-2 bundles since each line is also a Hecke line, $\mathbb{P}\mathrm{Hom}(F, \mathbb{C}_x)$ made up of the kernels $E \colon 0 \to E \to F \to \mathbb{C}_x \to 0$, where F is a rank-2 bundle with determinant $\mathcal{O}_x$. In genus greater than 2, the Hecke lines that are Fay lines are precisely those which are trisecant to the Kummer variety [OPP, 2.1].

2.2.1 Proposition. *The duality that interchanges positions and momenta has the following geometric interpretation. Given a pair $(\theta, \varphi) \in \mathbb{P}T^*\mathcal{M}$, choose a Hecke line corresponding to the bundle E, and consider its orthogonal line in the dual space. If the Hecke line meets the Kummer surface in four points $x_1, \ldots, x_4$, then the orthogonal line corresponding to the dual pair meets the dual Kummer $\mathcal{K}^*$ in four points whose dual images $y_1, \ldots, y_4$ are the images of the x_i under the hyperelliptic involution. Since every Fay line is a Hecke line, the bundle E' in the dual pair corresponds to one of the points in the quadrisecant through $y_1, \ldots, y_4$.*

Proof. Given a bundle $E \in \mathcal{M}$, there is a 2-dimensional variety of Hecke lines through the point $\langle \theta_E \rangle$, but note that their intersection with the Kummer surface $\mathcal{K}$ is one-dimensional (there is a curve C_E of line bundles ℓ that realize E as an extension of ℓ by ℓ^{-1}). Choosing one of these lines gives us the four points $x_1, \ldots, x_4$, which constitute its intersection with the Kummer surface $\mathcal{K}$, by equations (2). These, up to a finite number of choices (see **Remark 1** following), give the line bundles ℓ_1, ℓ_2. Choosing a point in the orthogonal line gives one more dimension, to account for the projectivized tangent space to $\mathcal{M}$ at E, and the pair (θ, φ). Since under the duality ι the bundles ℓ_1', ℓ_2' are computed as in (4), the ι-images $y_1, \ldots, y_4$ of the points of $\mathcal{K}^*$ of intersection with the orthogonal line corresponding to the pair $(\iota(\varphi), \iota^{-1}(\theta))$ are seen to be hyperelliptic conjugate to $x_1, \ldots, x_4$ (corresponding to their negatives on the Jacobian). QED

Remarks. 1. If the dual line to the chosen Hecke line is spanned by $\varphi_{\ell_1}, \varphi_{\ell_2}$ with $\ell_1 \ell_2 = \mathcal{O}(x_1 + x_2)$ and $\ell_1 \ell_2^{-1} K = \mathcal{O}(x_3 + x_4)$, then it is also spanned by $\varphi_{\ell_3}, \varphi_{\ell_4}$ with $\ell_1 \ell_3 = \mathcal{O}(x_1 + x_3)$, $\ell_1 \ell_3^{-1} K = \mathcal{O}(x_2 + x_4)$, $\ell_1 \ell_4 = \mathcal{O}(x_1 + x_4)$ and $\ell_1 \ell_4^{-1} K = \mathcal{O}(x_2 + x_3)$, thus, as noted above regarding equations (2) and (4), seemingly different choices give the same formulas.

2. The duality naturally extends over the non-generic case in which the x_i are not distinct, in particular the six pairs Z_s of bitangents to $\mathcal{K}^*$: $\{p | q \cdot p = 0, H(q, p)(\lambda_s) = 0\}$, where $(\lambda_s, 0)$ are the six Weierstrass points of the curve $X : \mu^2 = \prod_{s=1}^{6} (\lambda - \lambda_s)$, on which the Hitchin Hamiltonians are computed in [vGP].

§3. Lax-pair representations; other dualities?

3.1 Hitchin's spectral Prym. The approach to integrability in [vGP] is as it were 'orthogonal' to Hitchin's, since we fix a bundle E and determine the image of the fibre $T^* \mathcal{M} \to \mathcal{M}$ under the Hitchin map, proving that the Hitchin map is the same as the Gauss map on the $|2\Theta|$-divisor associated to E, and that its image in $H^0(X, 2K_X)$ is a quadratic system of conics on which we define certain natural functions, the Hitchin Hamiltonians. However, Hitchin had shown that the fibres of $H : T^* \mathcal{M} \to H^0(X, 2K)$ are Prym varieties associated with a "spectral" double cover $X_\eta \to X$, for a generic quadratic differential η. Thus, finding the Hamiltonians of the system leaves unanswered the question of explicit linearization. We make the following observation:

Let $X : y^2 = (x - \lambda_1) \ldots (x - \lambda_6)$ be a genus two curve. Choose two more points λ_7, λ_8 on the projective line: the inverse image of these two points gives a degree 4 divisor on X which lies in $|2K|$. The corresponding double cover of X is a spectral curve Γ, since the Hitchin map is onto the space of quadratic differentials. Define a genus three curve by

$$C_z : y^2 = (z - \lambda_1) \ldots (z - \lambda_8), \qquad \text{let} \quad p_i := (\lambda_i, 0) \ (\in C_z).$$

The Prym P of the covering $\Gamma \to X$ is isomorphic to the Jacobian of C_z modulo the subgroup generated by the point of order two $p_7 - p_8$:

$$P = JC_z / \langle p_7 - p_8 \rangle.$$

To integrate the system by quadratures, one would need to find the angle variables on P. In [GT-N-B2], the authors succeed by representing the Hitchin system as a Neumann-type system with 6×6 Lax-pair matrix. Notice that this is not the Lax pair with 'constraints' identified by Krichever [Kr] for the Hitchin system in Tyurin coordinates, for the size of Krichever's matrix is the rank of the bundle (two, in this case).

3.2 The Lax pair. Let us recall the notation in which six Hitchin Hamiltonians (three among which are functionally independent) are written as quartic polynomials in [vGP] and also [GT-N-B1]: by choosing a suitable basis of second-order theta functions and their dual to coordinatize $\mathbb{P}^3$ and the dual by q and p,

$$\theta = q_1\theta_{2,(0,0)} + q_2\theta_{2,(1,0)} + q_3\theta_{2,(0,1)} + q_4\theta_{2,(1,1)},$$

$$\varphi = p_1\theta^*_{2,(0,0)} + p_2\theta^*_{2,(1,0)} + p_3\theta^*_{2,(0,1)} + p_4\theta^*_{2,(1,1)}.$$

Now

$$H(q,p)(x_s) = h_s \sum_{t \neq s} \frac{r_{st}(q,p)}{\lambda_s - \lambda_t},$$

with $s, t = 1, \ldots, 6$, and h_s constants such that

$$H(q,p) = -\frac{1}{128\pi^2} \sum_{s \neq t} \frac{r_{st}(q,p)}{(\lambda - \lambda_s)(\lambda - \lambda_t)}(d\lambda)^2,$$

$$r_{12} = (q_1p_1 + q_2p_2 - q_3p_3 - q_4p_4)^2 \qquad r_{13} = (q_1p_4 - q_2p_3 - q_3p_2 + q_4p_1)^2$$
$$r_{14} = -(q_1p_4 + q_2p_3 - q_3p_2 - q_4p_1)^2 \qquad r_{15} = -(q_1p_3 - q_2p_4 - q_3p_1 + q_4p_2)^2$$
$$r_{16} = (q_1p_3 + q_2p_4 + q_3p_1 + q_4p_2)^2 \qquad r_{23} = -(q_1p_4 - q_2p_3 + q_3p_2 - q_4p_1)^2$$
$$r_{24} = (q_1p_4 + q_2p_3 + q_3p_2 + q_4p_1)^2 \qquad r_{25} = (q_1p_3 - q_2p_4 + q_3p_1 - q_4p_2)^2$$
$$r_{26} = -(q_1p_3 + q_2p_4 - q_3p_1 - q_4p_2)^2 \qquad r_{34} = (q_1p_1 - q_2p_2 + q_3p_3 - q_4p_4)^2$$
$$r_{35} = (q_1p_2 + q_2p_1 + q_3p_4 + q_4p_3)^2 \qquad r_{36} = -(q_1p_2 - q_2p_1 - q_3p_4 + q_4p_3)^2$$
$$r_{45} = -(q_1p_2 - q_2p_1 + q_3p_4 - q_4p_3)^2 \qquad r_{46} = (q_1p_2 + q_2p_1 - q_3p_4 - q_4p_3)^2$$
$$r_{56} = (q_1p_1 - q_2p_2 - q_3p_3 + q_4p_4)^2$$

where $r_{st} = r_{ts}$; notably, $r_{st}(q,p) = r_{st}(p,q)$.

The fact that the Hamiltonians

$$H_s = \sum_{t \neq s} \frac{r_{st}}{\lambda_s - \lambda_t}$$

commute in the canonical Poisson structure, is equivalent to the classical Yang-Baxter equations:

$$\left\{ \frac{r_{st}}{\lambda_s - \lambda_t}, \frac{r_{sv}}{\lambda_s - \lambda_v} \right\} + \left\{ \frac{r_{st}}{\lambda_s - \lambda_t}, \frac{r_{tv}}{\lambda_t - \lambda_v} \right\} + \left\{ \frac{r_{sv}}{\lambda_s - \lambda_v}, \frac{r_{tv}}{\lambda_t - \lambda_v} \right\} = 0,$$

as well as the equations obtained by cyclically permuting the indices, and

$$\left\{ \frac{r_{st}}{\lambda_s - \lambda_t}, \frac{r_{vw}}{\lambda_v - \lambda_w} \right\} = 0 \quad \text{if } \{s, t\} \cap \{v, w\} = 0.$$

M. Olshanetsky observed that the Hamiltonians of the Neumann system have a very similar form, namely:

$$T = \sum_{t \neq s} \frac{J_{st}^2}{\lambda_s - \lambda_t},$$

$J_{st} = q_s p_t - p_t q_s$ the functions on $T^*\mathbb{C}^6$ that generate the infinitesimal action of SO_6,

$$\{J_{st}, J_{tv}\} = -J_{sv} \quad s, t, v \text{ distinct}; \quad \{J_{st}, J_{vw}\} = 0 \quad s, t, v, w \text{ distinct}.$$

This prompted the authors of [GT-N-B2] to construct a symplectomorphism: the Hitchin phase space $T^*\mathcal{M}$ identified with $\{(q, p) | q \cdot p = 0\}/\mathbb{C}^*$, where $\mathbb{C}^*$ acts by $(q, p) \mapsto (tq, t^{-1}p)$, may be identified with the coadjoint orbit of $SL(4)$ consisting of the traceless rank-1 4×4 matrices $(p_1, .., p_4)^T(q_1, \ldots, q_4)$. Using the isomorphism $\mathfrak{sl}(4) \cong \mathfrak{so}(6)$, and setting $J_{st} = -J_{ts}$ for the functions on the given $SL(4)$ orbit which generate the action of $\mathfrak{so}(6)$, one sees that $r_{st} = -4J_{st}^2$ and the Poisson brackets are as above. Thanks to the explicit formulas, the authors were then able to construct an explicit 6×6 Lax-pair representation for the Hamiltonians, with the advantage of having the genus-3 curve as its spectrum, and were able to give the explicit linearization of the flows on the Jacobian by differential equations of the zeroes and poles of the eigenline-bundle, as customary in ACI [GT-N-B2]:

$$L_{mn}(\zeta) = \zeta J_{mn} + \lambda_n \delta_{mn},$$

with spectral curve

$$0 = \det(L(\zeta) - z)$$

$$= \prod_{n=1}^{6} (z - \lambda_n) \left(1 + \frac{1}{2} \zeta^2 \sum_{n \neq m \in \{1, \ldots, 6\}} \frac{J_{nm}^2}{(z - \lambda_n)(z - \lambda_m)} \right),$$

which upon substituting $\sigma := (i\zeta)^{-1} \prod_n (z - \lambda_n)$ becomes more clearly hyperelliptic of genus 3:

$$\sigma^2 = \prod_{n=1}^{6} (z - \lambda_n)^2 \sum_{n \neq m \in \{1, \ldots, 6\}} \frac{J_{nm}^2}{(z - \lambda_n)(z - \lambda_m)},$$

a polynomial of degree 8. In Hitchin's approach, the flows linearize on the Prym of the covering curve Γ:

$$\xi^2 = -4 \prod_{n=1}^{6} (z - \lambda_n) \sum_{n \neq m \in \{1,\ldots,6\}} \frac{J_{nm}^2}{(z - \lambda_n)(z - \lambda_m)},$$

where now z is a function on $X : y^2 = \prod_{1 \leq n \leq 6}(z - \lambda_6)$. As outlined in **3.1**, Γ is a cover of both X (by forgetting ξ) and C_z (by letting $\sigma = \imath(\sqrt{2}/4\xi y)$).

In particular, the symplectomorphism equivariant under the group actions is of the type that Adams, Harnad and Hurtubise used to go from $n \times n$ to $r \times r$ Lax pair representations of the same system, Moser's "rank-r perturbations of an $n \times n$ matrix" [AHH]. To indicate the corresponding group actions, which give rise to the ACI orbits under symplectic reduction, I only display the following commutative diagram, found in [AHH]. Here a tilde over a Lie algebra denotes the loop algebra of formal power series on it (the minus signs denote truncations to retain only negative exponents) J's are the moment maps for the infinitesimal action corresponding to the natural action† of $r \times r$ or $n \times n$ matrix groups on pairs of $n \times r$ matrices which make up the phase space M, for g an element of the group:

$$GL(n) : M \to M, \ (F, G) \mapsto (gF, (g^{-1})^T G), \ J_r(F, G) = -G^T(A - \lambda)^{-1} F,$$

$$GL(r) : M \to M \ (F, G) \mapsto (Fg^{-1}, Gg^T), \ J_n(F, G) = F(Y - z)^{-1} G^T,$$

and the subscripts A, Y denote the stabilizers under the action of the indicated matrix by conjugation:

$$\widetilde{gl(n)}^{-} \quad \overset{J_n}{\longleftarrow} \quad M \quad \overset{J_r}{\longrightarrow} \quad \widetilde{gl(r)}^{-}$$

$$\overset{J_n^Y}{\nwarrow} \qquad \swarrow \quad \searrow \qquad \overset{J_r^A}{\nearrow}$$

$$\downarrow \quad M/GL(r)_Y \qquad \qquad M/GL(n)_A \quad \downarrow$$

$$\searrow \qquad \swarrow$$

$$\widetilde{gl(n)}^{-}/GL(n)_A \longleftarrow M/(GL(r)_Y \times GL(n)_A) \longrightarrow \widetilde{gl(r)}^{-}/GL(r)_Y.$$

3.3 Other dualities? It appears that the authors have not compared the duality they constructed (see §2) in [GT-N-B1] with other dualities known for Neumann systems.

---

† The diagram is understood on appropriate open dense subsets where the action is free.

3.3.1 The geodesic and Neumann systems. Knörrer in [K] identified a correspondence between the Neumann system,

$$\ddot{\xi} = -A\xi + u\xi, \ \xi_1^2 + \ldots + \xi_n^2 = 1, \ u = \langle \xi, A\xi \rangle - \langle \dot{\xi}, \dot{\xi} \rangle$$

(assume for simplicity that A is a diagonal matrix with distinct non-zero eigenvalues) and the geodesic motion on the quadric

$$2q(x) = \langle x, Bx \rangle - 1, \ B = \text{diag}\left(\frac{1}{a_1}, \ldots, \frac{1}{a_n}\right).$$

He gave a beautiful geometric interpretation (rooted in Riemannian geometry) of the following:

If $x(t)$ is a geodesic on the quadric, then the zeroes of the polynomial in z:

$$\left(\sum_{i=1}^n \frac{\dot{x}_i^2}{a_i - z} - \frac{1}{2}\sum_{i,j=1}^n \frac{(x_i\dot{x}_j - x_j\dot{x}_i)^2}{(a_i - z)(a_j - z)}\right) \cdot \prod_{i=1}^n (a_i - z)$$

are independent of t. For these zeroes $0, \lambda_1, \ldots, \lambda_{n-2}$, then $0, \dfrac{1}{\lambda_1}, \ldots,$

$\dfrac{1}{\lambda_{n-2}}$ *are the zeroes of the polynomial:*

$$\left(\sum_{i=1}^n \frac{\xi_i^2}{\alpha_i - z} - \frac{1}{2}\sum_{i,j=1}^n \frac{(\dot{\xi}_i\xi_j - \dot{\xi}_j\xi_i)^2}{(\alpha_i - z)(\alpha_j - z)}\right) \cdot \prod_{i=1}^n (\alpha_i - z), \ \alpha_i := \frac{1}{a_i},$$

and are independent integrals of the Neumann system.

This being an interchange of positions and momenta, it is again a symplectic transformation (the system is in fact a rank-2 perturbation recalled above, but the groups $GL(r)$ do not enter Knörrer's construction). It seems unlikely that performing this duality on the 6×6 Lax pair should correspond to the interchange of q and p coordinates of §2, but it may be interesting to compute the transformed Hitchin system.

3.3.2 AC and AA dualities. In [FGNR] for the first time (to my knowledge) the authors explored two kinds of dualities related to Hitchin-type systems, and subsequently a partial connection with the S-duality of quantum field theory was developed. I recall the definition and propose to compare the duality in [GT-N-B1] with these, by taking the limit of the curve X to a punctured sphere or a genus-1 curve (this is a joint project with S. Abenda): the Hitchin system for those degenerate cases has been shown to be related to the Calogero-Moser or Gaudin system [ER, N], which have been shown to be self-dual in the following sense:

In the differentiable category, on (open sets) of a manifold of dimension $2m$, a completely integrable system admits action-angle variables (I_i, φ^i), $\omega = \sum_{i=1}^{m} dI_i \wedge d\varphi^i$ being the symplectic form, given in Darboux coordinates as $\sum_{i=1}^{m} dp_i \wedge dq^i$, $h_j = f_j(\underline{I})$ the Hamiltonians (functions of the action variables only), and $\underline{I}(b_2) - \underline{I}(b_1) = \int_{\Gamma} \omega$, where b_1, b_2 are two points in the base joined by a path γ and $\Gamma \in H_2(\underline{h}^{-1}(\gamma), \underline{h}^{-1}(b_1 \cup b_2); \mathbb{Z})$, $\underline{h}$ is the Hamiltonian map and the lift of γ to $H_1(\underline{h}^{-1}(\gamma), \mathbb{Z})$ is transported through the Gauss-Manin connection.

Definition. [FGNR] Two Hamiltonian systems are dual to each other in the sense of action-coordinate (AC) duality if the action variables I_i of the first system coincide with the coordinates q^i of the second and vice-versa.

For the holomorphic-symplectic case, the extra structure enables the authors to give another definition of duality, which they call AA (Action-Angle):

Definition. [FGNR] AA duality is obtained by choosing a symplectic basis (A, B) of $H_1(\underline{h}^{-1}(b_1), \mathbb{Z})$, which is also transported via the Gauss-Manin connection along a path γ connecting the points b_1, b_2 of the base, giving a decomposition of

$$\Gamma \in H_2(\underline{h}^{-1}(\gamma), \underline{h}^{-1}(b_1 \cup b_2), \mathbb{Z}), \quad \Gamma = \Gamma_A + \Gamma_B.$$

The dual action variables are:

$$\underline{I}(b_2) - \underline{I}(b_1) = \int_{\Gamma_A} \omega, \quad \underline{I}^D(b_2) - \underline{I}^D(b_1) = \int_{\Gamma_B} \omega.$$

3.3.3 Involutions on Sato's Grassmannian.

There are many geometrically relevant involutions on Sato's Grassmann manifold, which is a space where the phase space of an ACI can be mapped, up to additional choices, ranging from the classical "adjoint" of differential algebra in the operator setup, to the "bispectrality" transformation for the Baker function. However, there seems to be little information on how dualities of ACIs reflect on dualities on the Grassmannian for the PDE problem that corresponds, certainly in a non-trivial way, to the Hamiltonian system (for instance, KdV correspond to the Neumann system, cf. [Mum]), with the following exception. In [C], Sutherland quantized potentials (in several variables (x_i)),

$$-\Delta + \sum_{1 \le i < j \le n} \frac{1}{2} m(m+1) \sinh^{-2}\left(\frac{x_i - x_j}{2}\right),$$

are shown to be bispectral to a commutative ring of difference operators, the rational Ruijsenaars operators, cf. [Rui] in which Ruijsenaars first detected a duality. In [GT-N-B2], upon finding Hamiltonians for the Hitchin system in genus 2, the authors propose an explicit quantization; in this operator model, it would be important to interpret their duality on the level of the Grassmannian; the missing link with bispectrality is an explicit wave/Baker function, which Chalykh is able to produce [C].

3.3.4 Lie's sphere geometry. The quadratic complex is key in the integration of the Hitchin system. Sophus Lie devised a way for parametrizing lines in real projective space alternative to Plücker's Grassmannian (Lie's line-sphere correspondence [J, Art. 215]). He then gave a characterization of linear complexes in terms of differential equations. Certain "special" quadratic complexes were also characterized, where special means that they consist of all the lines tangent to a given surface. The equation is $\sum_{i=0}^{5}(\frac{\partial \varphi}{\partial x_i})^2 = 0$, $\varphi = \sum_{i=0}^{5} \lambda_i x_i^2$ being the equation of the complex in Klein coordinates (i.e., the Grassmannian has equation: $\sum x_i^2 = 0$). There are 4 co-singular ("confocal" would be the word for quadrics) complexes, $\sum_{i=0}^{5} \frac{x_i^2}{\lambda_i + \mu} = 0$, that go through any given line $y := [y_0, \dots, y_5] \in \mathbb{P}^5$, given by the values $\mu_1, \dots, \mu_4$, say, of the parameter. Then, y can be described by the "elliptic coordinates" $[\mu_1, \dots, \mu_4]$ in such a way that

$$[y_i] \sim \left[\frac{(\lambda_i + \mu_1)(\lambda_i + \mu_2)(\lambda_i + \mu_3)(\lambda_i + \mu_4)}{f'(-\lambda_i)}\right], \quad f(\mu) = \prod_{i=0}^{5}(\lambda_i + \mu)$$

and the PDE for the complex becomes:

$$\sum_{i=1}^{4} \left(\frac{\partial \varphi}{\partial \mu_i}\right)^2 \frac{f(\mu_i)}{\prod_{j \ne i}(\mu_i - \mu_j)} = 0.$$

Jacobi had already integrated this equation, by genus-2 integrals:

$$\varphi = \sum_{i=1}^{4} \left(\int d\mu_i \frac{\sqrt{(\mu_i - a)(\mu_i - b)}}{\sqrt{f(\mu_i)}}\right) + C, \ a, b \in \mathbb{C} \text{ arbitrary.}$$

386   *E. Previato*

We see that this system is a transformed Neumann/geodesic system, by letting the parameters of the base (the fixed values of the Hamiltonians) correspond to moving curves $\mu = a, \mu = b$, the family of hyperelliptic curves given by fixing $g + 1(= 3)$ Weierstrass points (plus a 'normalized' one, such as ∞ in the Neumann notation, cf. [Mum], e.g.) and varying the other g. It seems worthwhile to ask what is the effect of duality, both Knörrer's and the 'Wirtinger' duality of [GT-N-B1], on this differential equation.

Bibliography

[AF1] S. Abenda and Yu. Fedorov, On the weak Kowalevski-Painlevé property for hyperelliptically separable systems, *Acta Appl. Math.* **60** (2000), no. 2, 137–178.

[AF2] S. Abenda and Yu. Fedorov, Complex angle variables for constrained integrable Hamiltonian systems, in "Nonlinear evolution equations and dynamical systems (Kolimbary, 1999)", *J. Nonlinear Math. Phys.* **8** (2001), suppl., 1–4.

[AHH] M.R. Adams, J. Harnad and J. Hurtubise, Dual moment maps into loop algebras, *Lett. Math. Phys.* **20** (1990), 299–308.

[C] O.A. Chalykh, Duality of the generalized Calogero and Ruijsenaars problems, *Uspekhi Mat. Nauk* **52** (1997), no. 6(318), 191–192; translation in *Russian Math. Surveys* **52** (1997), no. 6, 1289–1291.

[DR] U. Desale and S. Ramanan, Classification of vector bundles of rank 2 over hyperelliptic curves, *Invent. Math.* **38** (1977), 161–186.

[ER] B. Enriquez and V. Rubtsov, Hitchin systems, higher Gaudin operators and R-matrices, *Math. Res. Lett.* **3** (1996), no. 3, 343–357.

[F] H. Flaschka, A remark on integrable Hamiltonian systems, *Phys.Lett. A* **131**, no 9 (1988), 505–508.

[FGNR] V. Fock, A. Gorsky, N. Nekrasov and V. Rubtsov, Duality in integrable systems and gauge theories, *J. High Energy Phys.* **2000**, no. 7, Paper 28, 40 pp.

[GT-N-B1] K. Gawędzki and P. Tran-Ngoc-Bich, Self-duality of the SL_2 Hitchin integrable system at genus 2, *Comm. Math. Phys.* **196** (1998), no. 3, 641–670.

[GT-N-B2] K. Gawędzki and P. Tran-Ngoc-Bich, Hitchin systems at low genera, *J. Math. Phys.* **41** (2000), no. 7, 4695–4712.

[vGP] B. van Geemen and E. Previato, On the Hitchin System, *Duke Math. J.* **85** (1996), 659–683.

[G] D. Grant, A generalization of Jacobi's derivative formula to dimension two, *J. Reine Angew. Math.* **392** (1988), 125–136.

[HT1] T. Hausel and M. Thaddeus, Mirror symmetry, Langlands duality, and the Hitchin system, *Invent. Math.* **153** (2003), no. 1, 197–229.

[HT2] T. Hausel and M. Thaddeus, Examples of mirror partners arising from integrable systems, *C. R. Acad. Sci. Paris Sér. I Math.* **333** (2001), no. 4, 313–318.

[H1] N.J. Hitchin, The self-duality equations on a Riemann surface, *Proc. London Math. Soc.* (3) **55** (1987), no. 1, 59–126.

[H2] N. Hitchin, Stable bundles and integrable systems, *Duke Math. J.* **54** (1987), no. 1, 91–114.

[J] C.M. Jessop, *A treatise on the line complex*, Cambridge Univ. Press, Cambridge, 1903.

[Jo] J. Jorgenson, On directional derivatives of the theta function along its divisor, *Israel J. Math.* **77** (1992), 273–284.

[K] H. Knörrer, Geodesics on quadrics and a mechanical problem of C. Neumann, *J. Reine Angew. Math.* **334** (1982), 69–78.

[Kr] I.M. Krichever, Vector bundles and Lax equations on algebraic curves, *Comm. Math. Phys.* **229** (2002), no. 2, 229–269.

[M] S. Mukai, *An introduction to invariants and moduli*, translated from the 1998 and 2000 Japanese editions by W. M. Oxbury. Cambridge Studies in Advanced Mathematics, 81. Cambridge University Press, Cambridge, 2003.

[Mum] D. Mumford, Tata lectures on theta II, *Progr. Math.* **43**, Birkhäuser, Boston, 1984.

[N] N. Nekrasov, Holomorphic bundles and many-body systems, *Comm. Math. Phys.* **180** (1996), no. 3, 587–603.

[New] P.E. Newstead, Stable bundles of rank 2 and odd degree over a curve of genus 2, *Topology* **7** (1968), 205–215.

[OPP] W.M. Oxbury, C. Pauly, E. Previato, Subvarieties of $\mathcal{SU}_C(2)$ and 2θ-divisors in the Jacobian, *Trans. Amer. Math. Soc.* **350** (1998), 3587–3614.

[R] M. Reid, The complete intersection of two or more quadrics, Thesis, Cambridge Univ. 1972.

[Rui] S.N.M. Ruijsenaars, Complete integrability of relativistic Calogero-Moser systems and elliptic function identities, *Comm. Math. Phys.* **110** (1987), no. 2, 191–213.

[V] P. Vanhaecke, Integrable systems and symmetric products of curves, *Math. Z.* **227** (1998), no. 1, 93–127.

12

Moduli Spaces for Principal Bundles

Alexander H.W. Schmitt

*Freie Universität Berlin, Institut für Mathematik,
Arnimallee 3, D-14195 Berlin, Germany
e-mail: schmitta@mi.fu-berlin.de*

Dedicated to Peter Newstead on occasion of his 65th birthday

Abstract

In this note, we will survey recent developments in the construction of
moduli spaces of principal bundles with reductive structure group on
algebraic varieties with special emphasis on smooth projective curves.
An application of the theory of moduli spaces to the theory of moduli
stacks will also be mentioned.

1 Introduction

Let k be an algebraically closed field, X a smooth, connected, and projective curve over k, and G a connected reductive linear algebraic group over k. This survey article is devoted to various aspects of the classification of principal G-bundles on X.

To begin with, let us look at the topological side. If $k = \mathbb{C}$, then any principal G-bundle may be viewed as a holomorphic principal G-bundle over X (viewed as a compact Riemann surface). As such, it defines a topological principal G-bundle on X. It is not hard to see that there is a bijection between the set of isomorphy classes of topological principal G-bundles on X and the fundamental group $\pi_1(G)$ ([Ram01], Proposition 5.1 see also Section 5.1). The fundamental group of G may be defined in terms of the root data of G. In particular, it is defined for reductive linear algebraic groups over any field k. Then, one can show that, over any field k, $\pi_1(G)$ parameterizes the algebraic equivalence classes of principal G-bundles on X ([DS], Proposition 5, [Holla], Proposition 3.15).

388

Now, we turn to the classification of principal G-bundles on X up to isomorphy. If $X = \mathbb{P}_1$, then Grothendieck's splitting theorem [Groth] (see also [Ba02]) gives the classification: Let $T \subset G$ be a maximal torus. Then, any principal G-bundle $\mathscr{P}$ on X possesses a reduction of the structure group to T. Fixing an isomorphism $T \cong \mathbb{G}_m(k)^{\times n}$, we see that the classification of principal T-bundles is equivalent to the classification of tuples $(\mathscr{L}_1, \ldots, \mathscr{L}_n)$ of line bundles on X. Thus, isomorphy classes of principal T-bundles are in bijection to tuples $(d_1, \ldots, d_n)$ of integers. The Weyl group $W := W(T, G) :=$ (Normalizer of T in G)/T acts on T and therefore on the set of principal T-bundles. It is easy to see that two principal T-bundles give isomorphic principal G-bundles, if and only if they belong to the same W-orbit. Hence, there is a bijection between the set of isomorphy classes of principal G-bundles on $\mathbb{P}_1$ and the set $\mathbf{Z}^{\oplus n}/W$.

If $g \geq 1$, then one can still construct principal G-bundles from principal T-bundles. In this way, we see that there are positive dimensional families of pairwise non-isomorphic principal G-bundles. Moreover, we see that the family of isomorphy classes of principal G-bundles on X is not bounded, i.e., cannot be (over-)parameterized by a scheme of finite type over k, even if we fix the topological type $\vartheta \in \pi_1(G)$. In order to get a meaningful solution to the classification problem, we have to admit **moduli spaces** as the answer. Due to the unboundedness phenomenon, the presence of non-trivial automorphisms, and some bad deformation properties of principal bundles, the answer can only be an **algebraic stack**. Since the theory of stacks was also conceived to solve moduli problems in a very general context, it is no wonder that the moduli stack $\mathscr{B}un_G(X)$ does exist. It contains the closed substack $\mathscr{B}un_G^{\vartheta}(X)$ that parameterizes the principal G-bundles of topological type $\vartheta \in \pi_1(G)$. By the aforementioned result on algebraic equivalence of principal G-bundles,

$$\mathscr{B}un_G(X) = \bigsqcup_{\vartheta \in \pi_1(G)} \mathscr{B}un_G^{\vartheta}(X)$$

is the decomposition of $\mathscr{B}un_G(X)$ into its connected components. A benefit of the approach via stacks is that there is a universal family on $\mathscr{B}un_G(X) \times X$. Furthermore, $\mathscr{B}un_G(X)$ is smooth, because X is a curve. We refer the reader to Sorger's lecture notes [S] for a more detailed introduction to the moduli stack of principal G-bundles. The reader may also consult [Beh02].

If we are more traditionally minded and would like to have a scheme or a variety as the moduli space, we first have to get rid of the unboundedness phenomenon. This means that we have to exclude many principal G-bundles from our considerations and restrict to a suitable subclass of objects. In the setting of the structure group $G = \mathrm{GL}_r(\mathbb{C})$, i.e., vector bundles of rank r on X, one knows that this subclass is the class of semistable vector bundles. Ramanathan introduced in [Ram02] a general notion of semistability for principal G-bundles (see also [Ba02] and Section 2.1). As for vector bundles, it serves the purpose of granting boundedness. Furthermore, deformations within the class of semistable objects behave more nicely, so that one may expect the existence of a moduli space $\mathscr{M}_G^\vartheta(X)$ for Ramanathan-semistable principal G-bundles of topological type $\vartheta \in \pi_1(G)$. Ramanathan managed in [Ram02] to construct $\mathscr{M}_G^\vartheta(X)$ as a **projective variety**, if $k = \mathbb{C}$. Note that there is typically no universal family on $\mathscr{M}_G^\vartheta(X) \times X$, because 1) the phenomenon of S-equivalence identifies some non-isomorphic semistable but not stable principal G-bundles (which usually exist by [Ram01], Proposition 7.8) and 2) there might be stable principal G-bundles with non-central automorphisms ([Ram01], Remark 4.1). (The paper [BBNN] gives more precise information in the case of the trivial topological type ϑ, and [BHof] will settle the case of arbitrary topological type.) Restricting to semistable bundles is not an artificial process: 1) The Harder-Narasimhan reduction ([Beh01], Proposition 8.2) tells us that any principal G-bundle is related in a canonical way to a semistable principal L-bundle, L being a Levi component of some parabolic subgroup Q of G, and 2) (assuming $g \geq 2$ and G to be semisimple) the theorem of Narasimhan-Seshadri-Ramanathan ([NS] and [Ram01]) shows, over $k = \mathbb{C}$, that any S-equivalence class of semistable principal G-bundles contains a unique isomorphy class of a principal G-bundle which is constructed from the universal covering $\widetilde{X} \longrightarrow X$ by means of a representation $\varrho \colon \pi_1(X) \longrightarrow K$ of the fundamental group of X in a compact real form K of G. This makes $\mathscr{M}_G^\vartheta(X)$ into a valuable model for a component of the real analytic representation space for $\pi_1(X)$ in K.

Regardless whether one considers $\mathscr{B}un_G^\vartheta(X)$ or $\mathscr{M}_G^\vartheta(X)$ to be the answer to the classification result, the answer is at first a mere existence result. It thus becomes part of the classification problem to study the moduli spaces further, for example, to determine their cohomology (with or without its ring structure). We recall the strategy of Newstead from his paper [N] for finding generators of the cohomology ring of a moduli space: Let $\mathscr{U}$ be a moduli space for principal G-bundles (a variety or a stack) and assume that there is a universal family $\mathscr{P}_{\mathscr{U}}$ on $\mathscr{U} \times X$. The characteristic classes of $\mathscr{P}_{\mathscr{U}}$ give certain cohomology classes $c_1, \ldots, c_r \in H^\star(\mathscr{U} \times X, \mathbb{C})$. Then, we may perform the Künneth decomposition of these classes in order to define cohomology classes $\gamma_1, \ldots, \gamma_N \in H^\star(\mathscr{U}, \mathbb{C})$. In the case that $\mathscr{U} = \mathscr{U}_X(2,1)$ is the moduli space of stable vector bundles of rank 2 and degree 1, Newstead proves that the classes $\gamma_1, \ldots, \gamma_N$ do generate the cohomology ring. (Note that Mumford and Newstead had already used Künneth components in [MN] to find (additive) generators for $H^3(\mathscr{U}_X(2,1), \mathbb{C})$ and that Ramanan [Raman] independently computed the cohomology ring $H^\star(\mathscr{U}_X(2,1), \mathbb{C})$ over a curve of genus 3, also using Künneth components.) In 1983, Atiyah and Bott [AB] showed that the classes $\gamma_1, \ldots, \gamma_N \in H^\star(\mathscr{B}un_G^\vartheta(X), \mathbb{C})$ are **free generators** for the cohomology ring $H^\star(\mathscr{B}un_G^\vartheta(X), \mathbb{C})$. Their approach uses Morse theory.

In this article, we will remind the reader of the existing constructions of the moduli spaces $\mathscr{M}_G^\vartheta(X)$ in characteristic zero and positive characteristic. In the case of arbitrary positive characteristic, this construction was completed only recently in [GLSS01]. (The projectivity of these moduli spaces in low characteristic is still open, see Section 2.3.) It is based on techniques developed by the author in [Sch01] and [Sch03]. We will briefly present these techniques, including applications to other base varieties, such as higher dimensional projective manifolds and nodal curves. At the end, we would like to sketch the purely algebraic proof of the theorem of Atiyah and Bott from [HeSc] which also brings it to positive characteristic. Though this is a result concerning the moduli stack, the results on moduli schemes are a central tool in the proof.

In the theory of moduli spaces of principal G-bundles on curves, loop groups, affine Graßmannians, and stacks play a very important role. For this reason, we conclude our article by some references for these topics.

Conventions

We will always work over a ground field k, mostly algebraically closed. A *scheme* is a scheme of finite type over k, and a *variety* is a reduced and irreducible scheme.

If H is a linear algebraic group, then a *one parameter subgroup* λ will always be a non-constant one (in order to avoid writing down trivial conditions in the definition of semistability). Likewise, a *parabolic subgroup of H* will be a proper one.

Acknowledgments

My thanks go to the organizers of the "Newstead Fest" for the opportunity to participate in that unique and pleasant event and present the

results of this article. During the preparation of this article, the author
was supported by the SFB/TR 45 "Perioden, Modulräume und Arith-
metik algebraischer Varietäten" and a Heisenberg fellowship of the DFG.
I thank the referees for their comments and suggestions.

2 Moduli spaces of principal bundles over curves

In this section, we will assume that X is a smooth projective curve
over the (algebraically closed) field k and try to discuss elements of the
history of moduli spaces of semistable principal bundles on X.

2.1 Preliminaries

Ramanathan's definition of (semi)stability ([Ram01], Definition 1.1 and
Lemma 2.1, [Ram02], Definition 2.13 and Lemma 2.15) reads as follows:
A principal G-bundle $\mathscr{P}$ on X is said to be *(semi)stable*, if for every
parabolic subgroup $Q \subsetneq G$, every section $\sigma \colon X \longrightarrow \mathscr{P}/Q$, leading by
means of the cartesian diagram

$$
\begin{array}{ccc}
\mathscr{Q} & \longrightarrow & \mathscr{P} \\
\downarrow & & \downarrow \\
X & \overset{\sigma}{\longrightarrow} & \mathscr{P}/Q
\end{array}
$$

to the principal Q-bundle $\mathscr{Q}$ over X, and every anti-dominant charac-
ter $\chi \colon Q \longrightarrow \mathbb{G}_m(k)$, the line bundle $\mathscr{L}(\mathscr{Q}, \chi)$ associated to $\mathscr{Q}$ and χ
satisfies

$$
\deg\bigl(\mathscr{L}(\mathscr{Q}, \chi)\bigr)(\geq)0.
$$

(An anti-dominant character is the inverse of a dominant character as
defined in [Ram01], §2, or [Ba02].)

If the structure group is $G = \mathrm{GL}_r(k)$, then the groupoid of principal G-bundles with isomorphisms is equivalent to the groupoid of vector
bundles of rank r with isomorphisms. Moreover, one checks that the
above definition of (semi)stability of principal G-bundles agrees with
the familiar notion of (semi)stability for vector bundles (see [Ram01],
Lemma 3.3, [HM], Corollary 1, or Example 3.4.4). Therefore, the moduli space of semistable vector bundles of rank r and fixed degree d is
also the moduli space of semistable principal G-bundles of the topological type specified by the integer d. Now, the construction of the moduli
space of semistable vector bundles of fixed rank and degree is one of
the first applications of Mumford's Geometric Invariant Theory to the

construction of moduli spaces (see [Mum01] and [Ses01]). Furthermore, there is also a well-developed theory of moduli spaces for vector bundles with additional structures (see [Sch02] and [Sch06]). Therefore, it seems a natural idea to fix a representation $\varrho \colon G \longrightarrow \mathrm{GL}_r(k)$ in order to describe principal G-bundles as vector bundles with additional structures.

Example 2.1.1. i) Let $G := \mathrm{Sp}_{2n}(k)$ be the symplectic group and $\varrho \colon \mathrm{Sp}_{2n}(k) \longrightarrow \mathrm{SL}_{2n}(k)$ the inclusion. The groupoid of principal G-bundles with isomorphisms is equivalent to the groupoid whose objects are pairs (E, φ) which consist of a vector bundle E of rank $2n$ and a bilinear map $\varphi \colon E \otimes E \longrightarrow \mathscr{O}_X$ which is fiberwise non-degenerate and anti-symmetric and whose isomorphisms are isomorphisms between the vector bundles which respect the bilinear form. Such pairs (E, φ) are examples of decorated vector bundles as studied in [Sch02] and [GS01].

ii) Let $\mathfrak{g}$ be a semisimple Lie algebra and $G := \mathrm{Aut}(\mathfrak{g}) \subset \mathrm{GL}(\mathfrak{g})$ its automorphism group (as a Lie algebra). Let us assume $\mathrm{Char}(k) = 0$. The Lie-bracket on $\mathfrak{g}$ defines an element $L \in \mathrm{Hom}(\mathfrak{g} \otimes \mathfrak{g}, \mathfrak{g})$. There is an obvious left action of $\mathrm{GL}(\mathfrak{g})$ on $\mathrm{Hom}(\mathfrak{g} \otimes \mathfrak{g}, \mathfrak{g})$. By the assumption on the characteristic of k, the orbit map $G \longrightarrow \mathrm{Hom}(\mathfrak{g} \otimes \mathfrak{g}, \mathfrak{g})$, $g \longmapsto g \cdot L$, is separable and induces an isomorphism $\mathrm{GL}(\mathfrak{g})/G \cong G \cdot L$. Using this isomorphism, one may establish an equivalence between the groupoid of principal G-bundles with isomorphisms and the groupoid whose objects are pairs (E, ψ), consisting of a vector bundle E with typical fiber $\mathfrak{g}$ and a bilinear map $\psi \colon E \otimes E \longrightarrow E$ which is fiberwise isomorphic to the Lie bracket L on $\mathfrak{g}$, and whose isomorphisms are isomorphisms between vector bundles, respecting the Lie bracket. Again, pairs (E, ψ) as above are special instances of decorated vector bundles.

2.2 Moduli spaces over the field of complex numbers

In this section, we review the existing constructions of moduli spaces over the field of complex numbers.

2.2.1 The work of Ramanathan

The first construction of $\mathscr{M}_G^\vartheta(X)$ is due to Ramanathan. It is the content of his PhD thesis at the Tata institute finished in 1976. It was posthumously published in 1996 [Ram02]. Ramanathan used the following chain of homomorphisms

$$G \xrightarrow{\;\mathrm{Ad}\;} \mathrm{Aut}^0\big(\mathrm{Lie}(G)\big) \subset \mathrm{Aut}\big(\mathrm{Lie}(G)\big).$$

Here, Ad is the adjoint representation of G on its Lie algebra. It maps G surjectively onto the connected component of the identity of $\mathrm{Aut}(\mathrm{Lie}(G))$ and has the center of G as the kernel.

As we have explained in Example 2.1.1, ii), we may describe principal $\mathrm{Aut}(\mathrm{Lie}(G))$-bundles as *Lie vector bundles*, i.e., as vector bundles together with a Lie bracket which is fiberwise isomorphic to $\mathrm{Lie}(G)$. With a grain of salt, we may say that Ramanathan first solves the moduli problem of Lie vector bundles with GIT and then conceives additional techniques in order to get from Lie vector bundles to principal $\mathrm{Aut}^0(\mathrm{Lie}(G))$-bundles and then to principal G-bundles. We emphasize the following: 1) Let $\mathscr{P}$ be a principal G-bundle with associated Lie bundle (E, ψ). Ramanathan needs that $\mathscr{P}$ is semistable if and only if the vector bundle E is semistable. This follows from the correspondence between polystable principal G-bundles and conjugacy classes of representations of the fundamental group $\pi_1(X)$ in a compact real form K of G (see [Ram02], Corollary 3.18). A purely algebraic proof was later given by Ramanan and Ramanathan ([RR], Theorem 3.18). The corresponding property in positive characteristic is only true, if the adjoint representation is of **low height** (see [IMP], Theorem 3.1 and Theorem 3.6); 2) In analyzing semistability of Lie vector bundles, one has to study the invariant theory on $\mathrm{Hom}(\mathfrak{g} \otimes \mathfrak{g}, \mathfrak{g})$: One needs that the point $L \in \mathrm{Hom}(\mathfrak{g} \otimes \mathfrak{g}, \mathfrak{g})$ is polystable. This results from the rigidity of semisimple Lie algebras ([NR], Theorem 7.2). Probably, rigidity holds in large characteristic as well, but there seems to be no reference.

2.2.2 The work of Faltings

Faltings gives in his paper [Fal01] from 1993 a construction of moduli spaces of semistable principal G-Higgs bundles which covers the case of semistable principal G-bundles as well. Let us describe his work without mentioning the Higgs fields. He first gives a new construction of the moduli space of semistable vector bundles, by using theta functions and a new characterization of semistability rather than the Hilbert-Mumford criterion. Then, he also invokes the fact that a principal G-bundle $\mathscr{P}$ is semistable, if and only if the associated vector bundle E with typical fiber $\mathrm{Lie}(G)$ is semistable. The existence of the moduli space of semistable vector bundles thus implies the existence of $\mathscr{M}_G^\vartheta(X)$ as a **quasi-projective variety**. In order to prove projectivity, one may establish:

THEOREM (Semistable reduction). *Let R be a discrete valuation ring with quotient field K. If $\mathscr{P}_K$ is a family of semistable principal G-bundles on $\mathrm{Spec}(K) \times X$, then there are a finite extension K'/K and an extension of the family $\mathscr{P}_{K'} := \mathscr{P}_K \times_{\mathrm{Spec}(K)} \mathrm{Spec}(K')$ to a family $\mathscr{P}_{R'}$ of semistable principal G-bundles on $\mathrm{Spec}(R') \times X$, R' being the integral closure of R in K'.*

Faltings proves the semistable reduction theorem with the help of deformation theory of Lie algebras. Again, it is not clear under which hypotheses these techniques work in positive characteristic.

2.2.3 The work of Balaji and Seshadri

In the paper [BS] from 2002, Balaji and Seshadri give an alternative construction of the moduli space of semistable principal bundles with **semisimple structure group** with an eye on a possible extension to positive characteristic (see Section 2.3.1). They choose a faithful representation $\varrho \colon G \longrightarrow \mathrm{SL}(V)$. Any principal G-bundle $\mathscr{P}$ then gives rise to a vector bundle $E = \mathscr{P}(V)$ with typical fiber V. As before, the characteristic zero hypothesis implies that $\mathscr{P}$ is semistable if and only if E is a semistable vector bundle. This fact together with the existence of the moduli space of semistable vector bundles gives, as in Faltings's work, the existence of $\mathscr{M}_G^\vartheta(X)$ as a quasi-projective variety. To prove properness of $\mathscr{M}_G^\vartheta(X)$, Balaji and Seshadri demonstrate the semistable reduction theorem with the help of Bruhat-Tits theory. A more detailed discussion of this approach is contained in Balaji's lecture notes [Ba02].

2.2.4 The work of Gómez and Sols

In [GS02], Gómez and Sols construct moduli spaces of principal G-sheaves on a projective manifold X of arbitrary dimension. First, one needs the notion of a *Lie sheaf*. This is a pair $(\mathcal{E}, \psi)$ which consists of a torsion free sheaf $\mathcal{E}$ on X and a "Lie bracket" $\psi \colon \mathcal{E} \otimes \mathcal{E} \longrightarrow \mathcal{E}^{\vee\vee}$ which is fiberwise isomorphic to the Lie algebra of G on the open subset $U_\mathcal{E}$ where $\mathcal{E}$ is locally free. Using the theory of decorated sheaves developed in [Sch02] and [GS01], one finds a notion of semistability for these Lie sheaves and a GIT construction of the **projective moduli space** of semistable Lie sheaves. The projectivity rests again on the fact that semisimple Lie algebras are rigid in characteristic zero. A *principal G-sheaf* is now a triple $(\mathcal{E}, \psi, \mathscr{P})$ which consists of a Lie sheaf $(\mathcal{E}, \psi)$ and a principal G-bundle $\mathscr{P}$ on $U_\mathcal{E}$ whose associated Lie bundle agrees with $(\mathcal{E}_{|U_\mathcal{E}}, \psi_{|U_\mathcal{E}})$. The triple $(\mathcal{E}, \psi, \mathscr{P})$ is *(semi)stable*, if the pair $(\mathcal{E}, \psi)$ is.

The core of the work of Gómez and Sols is to show that the lifting techniques of Ramanathan can be extended to higher dimensions. This is a bit surprising, because the principal G-bundle $\mathscr{P}$ is just defined on an open subset of X and not on the whole of X. (This difficulty apparently disappears on curves.)

If one specializes these results to the case of a curve, one gets another construction of $\mathscr{M}_G^\vartheta(X)$ as a projective variety, following the strategy of Ramanathan but being simpler in the details. Note that Gómez and Sols do not have to use the fact that a principal G-bundle is semistable if and only if the associated vector bundle with fiber $\mathrm{Lie}(G)$ is so.

2.2.5 The work of Schmitt

In [Sch01] and [Sch03], the author developed the moduli theory of pseudo G-bundles and singular principal G-bundles for semisimple structure groups on smooth projective varieties of arbitrary dimension. We will outline this approach in Section 3. This method uses only techniques from GIT and immediately yields projective moduli spaces, i.e., no separate proof of the semistable reduction theorem is required. These results were extended by Gómez, Langer, Schmitt, and Sols to arbitrary reductive groups in the paper [GLSS02].

2.3 Moduli spaces over fields of positive characteristic

If the characteristic of the ground field is positive and the group G is not a classical one, then the construction of $\mathscr{M}_G^\vartheta(X)$ gets much harder and was completed only recently. The projectivity of $\mathscr{M}_G^\vartheta(X)$ in very low characteristic is still an open problem (see below).

2.3.1 The work of Balaji and Parameswaran

The first general construction of moduli spaces of principal G-bundles on curves over fields of positive characteristic is given in [BP]. The authors fix again a faithful representation $\varrho\colon G \longrightarrow \mathrm{SL}(V)$. To such a representation, one can attach its height $\mathrm{ht}(\varrho)$ (see [Serre], Lecture 2, [IMP], Definition 2.1) and its separable index $\mathrm{si}(\varrho)$ ([BP], Definition 6). If ϱ is of low height, i.e., if $\mathrm{Char}(p) > \mathrm{ht}(\varrho)$, then a principal G-bundle is semistable if and only if its associated vector bundle with fiber V is ([IMP], Theorem 3.1). Therefore, $\mathscr{M}_G^\vartheta(X)$ exists as a quasi-projective variety.

Now, assume that there are a representation $\kappa\colon \mathrm{SL}(V) \longrightarrow \mathrm{GL}(W)$ and a G-equivariant closed embedding $\iota\colon \mathrm{SL}(V)/G \hookrightarrow \mathrm{GL}(W)$ and that

κ is of low separable index, i.e., $\text{Char}(k) > \text{si}(\kappa)$. Then, the semistable reduction argument relying on Bruhat-Tits theory can be adapted. Note that one may suppose that G, ϱ, and κ are defined over the integers. The result of Balaji and Parameswaran therefore asserts that $\mathscr{M}_G^\vartheta(X)$ exists as a projective variety provided $\text{Char}(p) \gg 0$.

2.3.2 The work of Gómez, Langer, Schmitt, and Sols

In [GLSS01], the authors investigate to which extent the approach from [Sch01] and [Sch03] works in positive characteristic. As a main result, the authors show that the moduli space $\mathscr{M}_G^\vartheta(X)$ **exists in any characteristic** as a **quasi-projective variety** and prove projectivity under mild and **explicit** assumptions on the characteristic of the base field k (see Theorem 3.4.10). Over higher dimensional base manifolds in positive characteristic, the recent advances in the theory of torsion free sheaves ([Lan01], [Lan02]) play a crucial role.

We will present in Section 3.4.2 more information on the issue of the characteristic of k in the approach of [Sch01], [Sch03], and [GLSS01].

2.3.3 The work of Heinloth

The semistable reduction theorem for principal G-bundles with structure group $G = \text{GL}_r(k)$ on curves (and more generally, semistable torsion free sheaves on projective manifolds) is due to Langton [Langt]. The current status of the semistable reduction theorem for principal G-bundles on curves over fields of positive characteristic is the following (see [Hei02] and [Hei03]):

THEOREM 2.3.1 (Heinloth). *Let G be a connected reductive group. Under the following assumptions on the characteristic of k, the semistable reduction theorem for principal G-bundles on the curve X holds true:*

- $\text{Char}(k) = 0$.
- $\text{Char}(k) \geq 3$, *if G contains a simple factor of type G_2.*
- $\text{Char}(k) \geq 11$, *if G contains a simple factor of type F_4 or E_6.*
- $\text{Char}(k) \geq 17$, *if G contains a simple factor of type E_7.*
- $\text{Char}(k) \geq 37$, *if G contains a simple factor of type E_8.*

Heinloth uses the theory of affine Graßmannians ([S], §8) in his proof. These are ind-projective schemes. Given a discrete valuation ring R with field of fractions K and a family $\mathscr{P}_K$ of semistable principal G-bundles on $X \times \text{Spec}(K)$, the projectivity of the affine Graßmannian allows to find some extension $\mathscr{P}_R$ of this family to $X \times \text{Spec}(R)$. Furthermore, one can do quite explicit cocycle computations, using the description of the affine Graßmannian as a coset space. This enables Heinloth to mimic Langton's original algorithm for making the principal bundle on the special fiber less unstable. A technical tool which enters the proof is Behrend's conjecture on the rationality of the Harder-Narasimhan reduction ([Beh01], Conjecture 7.6). This conjecture was verified under the low height assumption by Biswas and Holla [BHol] and under the above hypotheses by Heinloth [Hei03]. It is interesting to note that Heinloth found in [Hei03] a counter-example to Behrend's conjecture (for G_2 in characteristic 2).

3 Pseudo G-bundles

From now on, we will assume that G is a **connected semisimple** linear algebraic group and X an arbitrary scheme.

3.1 Principal bundles as decorated vector bundles

In Example 2.1.1, we have illustrated how the choice of a suitable representation $\varrho\colon G \longrightarrow \mathrm{SL}(V)$ may be used to describe principal G-bundles as vector bundles with additional structures. (Note that a connected semisimple group possesses only the trivial character, therefore any representation $\varrho\colon G \longrightarrow \mathrm{GL}(V)$ actually factorizes over $\mathrm{SL}(V)$.) Here, we will explain how one can do a similar thing based on an arbitrary **faithful** representation ϱ.

As explained in [Ba02], any principal G-bundle gives rise to a vector bundle $E = \mathscr{P}(V)$ and a principal $\mathrm{GL}(V)$-bundle $\varrho_\star(\mathscr{P}) := \mathscr{P}(\mathrm{GL}(V))$. These two objects are related by a canonical isomorphism

$$\varrho_\star(\mathscr{P}) \cong \mathscr{I}som(V \otimes \mathscr{O}_X, E).$$

Moreover, the representation $\varrho\colon G \longrightarrow \mathrm{GL}(V)$ gives rise to the G-equivariant embedding

$$\iota\colon \mathscr{P} = \mathscr{P}(G) \hookrightarrow \mathscr{P}(\mathrm{GL}(V)) = \mathscr{I}som(V \otimes \mathscr{O}_X, E).$$

If we take the G-quotient, we get the section

$$\sigma := \iota/G\colon X = \mathscr{P}/G \longrightarrow \mathscr{I}som(V \otimes \mathscr{O}_X, E)/G.$$

Conversely, given a vector bundle E with typical fiber V and a section $\sigma\colon X \longrightarrow \mathscr{I}som(V \otimes \mathscr{O}_X, E)/G$, we define the principal G-bundle $\mathscr{P}$ on X via the cartesian diagram

$$
\begin{array}{ccc}
\mathscr{P} & \longrightarrow & \mathscr{I}som(V \otimes \mathscr{O}_X, E) \\
\downarrow & & {\scriptstyle G\text{-}}\Big\downarrow{\scriptstyle \text{bundle}} \\
X & \xrightarrow{\ \sigma\ } & \mathscr{I}som(V \otimes \mathscr{O}_X, E)/G.
\end{array}
$$

There is the obvious notion of an *isomorphism between pairs* (E, σ) *and* (E', σ'). The above assignments give rise to an equivalence between the groupoid of principal G-bundles with isomorphisms and the groupoid of pairs (E, σ) with isomorphisms.

The fact $\varrho(G) \subset \mathrm{SL}(V)$ grants the existence of the commutative diagram

$$
\begin{array}{ccc}
\mathscr{I}som(V \otimes \mathscr{O}_X, E) & \rightarrowtail & \mathscr{H}om(V \otimes \mathscr{O}_X, E) \\
\downarrow & & \downarrow \\
\mathscr{I}som(V \otimes \mathscr{O}_X, E)/G & \rightarrowtail & \mathscr{H}om(V \otimes \mathscr{O}_X, E)/\!/G
\end{array}
$$

in which the horizontal arrows are open embeddings. Next. note that

$$
\mathscr{H}om(V \otimes \mathscr{O}_X, E)/\!/G = \mathscr{S}pec\big(\mathscr{S}ym^{\star}(V \otimes E^{\vee})^G\big).
$$

Altogether, we see that a principal G-bundle is determined by a pair (E, τ) which consists of a vector bundle E with typical fiber V and a homomorphism $\tau\colon \mathscr{S}ym^{\star}(V \otimes E^{\vee})^G \longrightarrow \mathscr{O}_X$ of $\mathscr{O}_X$-algebras.

3.2 The notion of a pseudo bundle

Let X be a scheme. Motivated by the discussion of the last section, we define a *pseudo G-bundle on X* as a pair $(\mathscr{A}, \tau)$ which consists of a coherent $\mathscr{O}_X$-module $\mathscr{A}$ and a homomorphism $\tau\colon \mathscr{S}ym^{\star}(V \otimes \mathscr{A})^G \longrightarrow \mathscr{O}_X$ of $\mathscr{O}_X$-algebras.

Remark 3.2.1. i) The sheaf $\mathscr{S}ym^{\star}(V \otimes \mathscr{A})$ is a sheaf of finitely generated and graded $\mathscr{O}_X$-algebras. Therefore, it defines a scheme together with an affine morphism onto X. We write this object as

$$
\mathscr{H}om(V \otimes \mathscr{O}_X, \mathscr{A}^{\vee}) := \mathscr{S}pec\big(\mathscr{S}ym^{\star}(V \otimes \mathscr{A})\big) \longrightarrow X.
$$

By the Hilbert-Nagata theorem on the finite generation of invariant rings, $\mathscr{S}ym^{\star}(V \otimes \mathscr{A})^G$ is also a sheaf of finitely generated and graded

$\mathscr{O}_X$-algebras, and

$$\mathscr{H}om(V \otimes \mathscr{O}_X, \mathscr{A}^\vee)/\!/G := \mathscr{S}pec(\mathscr{S}ym^\star(V \otimes \mathscr{A})^G) \longrightarrow X$$

is the categorical quotient of $\mathscr{H}om(V \otimes \mathscr{O}_X, \mathscr{A}^\vee)$ by the G-action coming from the representation ϱ.

ii) The datum of a homomorphism $\tau \colon \mathscr{S}ym^\star(V \otimes \mathscr{A})^G \longrightarrow \mathscr{O}_X$ is the same as the datum of a section

$$\sigma \colon X \longrightarrow \mathscr{H}om(V \otimes \mathscr{O}_X, \mathscr{A}^\vee)/\!/G.$$

Any pseudo G-bundle $(\mathscr{A}, \tau)$ therefore defines a geometric object $\mathscr{P}(\mathscr{A}, \tau) \longrightarrow X$ by means of the cartesian diagram

$$
\begin{array}{ccc}
\mathscr{P}(\mathscr{A},\tau) & \longrightarrow & \mathscr{H}om(V \otimes \mathscr{O}_X, \mathscr{A}^\vee) \\
\downarrow & & \downarrow \\
X & \stackrel{\sigma}{\longrightarrow} & \mathscr{H}om(V \otimes \mathscr{O}_X, \mathscr{A}^\vee)/\!/G.
\end{array}
$$

Note that $\mathscr{H}om(V \otimes \mathscr{O}_X, \mathscr{A}^\vee)/\!/G$ has a natural zero section. It corresponds to the homomorphism $\tau^0 \colon \mathscr{S}ym^\star(V \otimes \mathscr{A})^G \longrightarrow \mathscr{O}_X$ which projects everything onto $\mathscr{O}_X = \mathscr{S}ym^0(V \otimes \mathscr{A})^G$. The pair $(\mathscr{A}, \tau^0)$ will be called the *trivial pseudo G-bundle*.

Example 3.2.2. Suppose X is a smooth projective variety and $\varrho \colon \mathrm{Sp}_{2n}(k) \subset \mathrm{SL}_{2n}(k)$ is the standard representation. Then, a pseudo G-bundle $(\mathscr{A}, \tau)$ in which $\mathscr{A}$ is torsion free is the same as a torsion-free sheaf $\mathscr{A}$ endowed with an anti-symmetric bilinear map $\mathscr{A} \otimes \mathscr{A} \longrightarrow \mathscr{O}_X$, i.e., there is no condition on non-degeneracy whatsoever. We see that pseudo G-bundles are much more general objects than principal G-bundles.

3.3 Semistability

In this section, we discuss the notion of δ-(semi)stability which origins from our paper [Sch01].

3.3.1 Some GIT-considerations

First, we look at the action of G on the vector space $\mathrm{Hom}(V, k^r)$ and its projectivization

$$P(\mathrm{Hom}(V, k^r)) := (\mathrm{Hom}(V, k^r) \smallsetminus \{0\})/\mathbb{G}_m(k).$$

The coordinate algebra of the affine variety $\mathrm{Hom}(V, k^r)$ is the symmetric algebra $\mathrm{Sym}^\star(V \otimes k^r)$. Hence, the GIT-process gives the categorical quotients

$$\mathrm{Hom}(V, k^r) /\!\!/ G = \mathrm{Spec}\big(\mathrm{Sym}^\star(V \otimes k^r)^G\big)$$
$$\text{and} \quad P\big(\mathrm{Hom}(V, k^r)\big) /\!\!/ G = \mathrm{Proj}\big(\mathrm{Sym}^\star(V \otimes k^r)^G\big).$$

Note that we have on $\mathrm{Hom}(V, k^r)$ a natural $(\mathrm{GL}_r(k) \times G)$-action, so that $\mathrm{GL}_r(k)$ acts on those two quotients. We will have to study the $\mathrm{SL}_r(k)$-action on $P(\mathrm{Hom}(V, k^r)) /\!\!/ G$. Observe that

$$\mathrm{Sym}^\star(V \otimes k^r)^G = \bigoplus_{d \geq 0} \mathrm{Sym}^d(V \otimes k^r)^G$$

is a graded algebra. For $s > 0$, we set

$$\mathrm{Sym}^{(s)}(V \otimes k^r)^G := \bigoplus_{d \geq 0} \mathrm{Sym}^{d \cdot s}(V \otimes k^r)^G.$$

It is well-known that

$$\mathrm{Proj}\big(\mathrm{Sym}^\star(V \otimes k^r)^G\big) \cong \mathrm{Proj}\big(\mathrm{Sym}^{(s)}(V \otimes k^r)^G\big),$$

for all $s > 0$. Suppose that the algebra $\mathrm{Sym}^{(s)}(V \otimes k^r)^G$ is generated in degree one, i.e., by the elements of $\mathrm{Sym}^s(V \otimes k^r)^G$. Then, the surjection

$$\mathrm{Sym}^\star(\mathbb{V}_s) \longrightarrow \mathrm{Sym}^{(s)}(V \otimes k^r)^G, \quad \mathbb{V}_s := \mathrm{Sym}^s(V \otimes k^r)^G,$$

gives rise to the $\mathrm{GL}_r(k)$-equivariant embedding

$$v_s \colon P\big(\mathrm{Hom}(V, k^r)\big) /\!\!/ G \cong \mathrm{Proj}\big(\mathrm{Sym}^{(s)}(V \otimes k^r)^G\big) \hookrightarrow \mathbb{P}(\mathbb{V}_s).$$

It is easy to check that one can always find an $s > 0$, such that $\mathrm{Sym}^{(s)}(V \otimes k^r)^G$ is generated in degree one ([Mum02], III.8, Lemma).

Next, we point out that

$$P\big(\mathrm{Hom}(V, k^r)\big) /\!\!/ G \cong \Big(\big(\mathrm{Hom}(V, k^r) /\!\!/ G\big) \smallsetminus \{0\}\Big) /\!\!/ \mathbb{G}_m(k).$$

A point $0 \neq h \in \mathrm{Hom}(V, k^r) /\!\!/ G$ therefore gives, for $s > 0$, the point $v_s[h] \in \mathbb{P}(\mathbb{V}_s)$. For any one parameter subgroup $\lambda \colon \mathbb{G}_m(k) \longrightarrow \mathrm{SL}_r(k)$, we have the number $\mu(\lambda, v_s[h])$ (cf. appendix), and we set

$$\overline{\mu}(\lambda, h) := \frac{1}{s} \cdot \mu(\lambda, v_s[h]).$$

The latter number is well-defined, i.e., it does not depend on the choice of s. Finally, we point out that the formation of invariant rings and GIT-quotients does, in **any characteristic**, commute with **flat** base change.

All the above considerations therefore remain valid over an extension K/k, algebraically closed or not.

3.3.2 Definition of semistability and moduli spaces

Now, assume that $(\mathscr{A}, \tau)$ is a pseudo G-bundle on the projective **variety** X, that $\mathscr{A}$ is a torsion free sheaf, and that τ is non-trivial in the sense of Remark 3.2.1, ii). Recall that a subsheaf $\{0\} \subsetneq \mathscr{B} \subsetneq \mathscr{A}$ is called *saturated*, if the quotient $\mathscr{A}/\mathscr{B}$ remains torsion free. The test objects for the (semi)stability of $(\mathscr{A}, \tau)$ are weighted filtrations: A *weighted filtration of $\mathscr{A}$* is a pair $(\mathscr{A}_\bullet, \alpha_\bullet)$ which is composed of a filtration

$$\mathscr{A}_\bullet : \{0\} \subsetneq \mathscr{A}_1 \subsetneq \cdots \subsetneq \mathscr{A}_s \subsetneq \mathscr{A}$$

of $\mathscr{A}$ by saturated subsheaves and a tuple $\alpha_\bullet = (\alpha_1, \ldots, \alpha_s)$ of positive rational numbers. Fix once and for all a polarization $\mathscr{O}_X(1)$ on X, so that Hilbert polynomials and degrees of coherent $\mathscr{O}_X$-modules are defined. Then, we associate to a weighted filtration $(\mathscr{A}_\bullet, \alpha_\bullet)$ the polynomial

$$M(\mathscr{A}_\bullet, \alpha_\bullet) := \sum_{i=1}^{s} \alpha_i \cdot \big(P(\mathscr{A}) \cdot \mathrm{rk}(\mathscr{A}_i) - P(\mathscr{A}_i) \cdot \mathrm{rk}(\mathscr{A})\big)$$

and the rational number

$$L(\mathscr{A}_\bullet, \alpha_\bullet) := \sum_{i=1}^{s} \alpha_i \cdot \big(\deg(\mathscr{A}) \cdot \mathrm{rk}(\mathscr{A}_i) - \deg(\mathscr{A}_i) \cdot \mathrm{rk}(\mathscr{A})\big).$$

Note that the degree of the polynomial $M(\mathscr{A}_\bullet, \alpha_\bullet)$ is at most $\dim(X) - 1$.

Next, we proceed to the definition of the number $\mu(\mathscr{A}_\bullet, \alpha_\bullet, \tau)$. For this, let $K := k(X)$ be the function field of X and $\eta \in X$ the generic point. As we observed before, we may use the constructions and definitions from Section 3.3.1 for K as well. Let $\mathbb{A}$ be the restriction of $\mathscr{A}$ to the generic point and

$$\sigma_K \in \big(\mathrm{Hom}(V \otimes_k K, \mathbb{A}) /\!/ G\big) \smallsetminus \{0\}$$

the point defined by τ. If all the numbers α_i lie in $\mathbb{Z}[1/r]$, there exists a one parameter subgroup $\lambda \colon \mathbb{G}_m(K) \longrightarrow \mathrm{SL}(\mathbb{A})$ whose weighted flag (see appendix) in $\mathbb{A}$ is $(\mathbb{A}_\bullet := \mathscr{A}_{\bullet|\{\eta\}}, \alpha_\bullet)$. We define

$$\mu(\mathscr{A}_\bullet, \alpha_\bullet, \tau) := \overline{\mu}(\lambda, \sigma_K).$$

By Corollary 6.3.2, this number depends only on the weighted flag $(\mathbb{A}_\bullet, \alpha_\bullet)$ and not on the concrete choice of the one parameter subgroup λ which induces that weighted flag.

In general, we may choose $t \gg 0$, such that $t \cdot \alpha_i \in \mathbb{Z}[1/r]$, $i = 1, \ldots, s$, and define

$$\mu(\mathscr{A}_\bullet, \alpha_\bullet, \tau) := \frac{1}{t} \cdot \mu(\mathscr{A}_\bullet, t \cdot \alpha_\bullet, \tau).$$

It is straightforward to check that the choice of t does not influence the definition.

The notion of (semi)stability for pseudo G-bundles will depend on a parameter δ which is a positive polynomial over the rationals of degree at most $\dim(X) - 1$. Fix such a parameter $\delta \in \mathbb{Q}[x]_+$. A pseudo G-bundle $(\mathscr{A}, \tau)$ is called δ-*(semi)stable*, if 1) $\mathscr{A}$ is torsion free, 2) $\tau \neq \tau^0$, and 3) the inequality

$$M(\mathscr{A}_\bullet, \alpha_\bullet) + \delta \cdot \mu(\mathscr{A}_\bullet, \alpha_\bullet, \tau)(\succeq)0$$

of polynomials in the lexicographic ordering holds for any weighted filtration $(\mathscr{A}_\bullet, \alpha_\bullet)$ of $\mathscr{A}$.

THEOREM 3.3.1. *Let $(X, \mathscr{O}_X(1))$ be a polarized projective variety. Fix a Hilbert polynomial P and a stability parameter δ and assume that X is smooth or that $\mathrm{Char}(k) = 0$. Then, the moduli space $\mathscr{M}^P_{\varrho,\delta}(X)$ for δ-semistable pseudo G-bundles $(\mathscr{A}, \tau)$ on X, such that $\mathscr{A}$ has Hilbert polynomial P with respect to $\mathscr{O}_X(1)$, exists as a projective scheme.*

Proof The theorem appeared for smooth projective varieties over the complex numbers in [Sch01]. Later Bhosle [Bho02] proved it for singular varieties under the following:

Assumption 3.3.2. Given a smooth curve C, a closed point $c_0 \in C$, an open subset $U \subset X \times C$, containing $X \times (C \smallsetminus \{c_0\})$ and meeting $X \times \{c_0\}$, and a locally free $\mathscr{O}_{X \times C}$-module $\mathcal{E}$, the natural homomorphism

$$\mathcal{E} \longrightarrow j_\star\big(j^\star(\mathcal{E})\big), \quad j \colon U \longrightarrow X \times C \text{ being the inclusion,}$$

is an isomorphism.

Bhosle checked that semi-normal varieties and varieties verifying Serre's condition S_2 satisfy the above assumption. Recently, Langer [Lan04] found an easy argument that the assumption is always satisfied.

In [GLSS01], the theorem was established for smooth varieties of arbitrary dimension over algebraically closed fields of positive characteristic. The extension to non-semisimple reductive structure group may be found in [GLSS02]. $\qquad \square$

By Theorem 3.3.1, pseudo G-bundles are very flexible objects which admit projective moduli spaces in very general situations. Example 3.2.2 shows, however, that pseudo G-bundles may be very strange objects and be very far from principal G-bundles. We will illustrate in the next sections that improvements are possible.

3.4 Singular principal G-bundles on smooth varieties

Here, we assume that X is a smooth projective variety. We add the conditions 1) $\mathscr{A}$ is torsion free and 2) $\det(\mathscr{A}) \cong \mathscr{O}_X$ to the definition of a pseudo G-bundle $(\mathscr{A}, \tau)$. A *singular principal G-bundle on X* is a pseudo G-bundle $(\mathscr{A}, \tau)$, such that

$$\sigma(U_{\mathscr{A}}) \subset \mathscr{I}som(V \otimes \mathscr{O}_{U_{\mathscr{A}}}, \mathscr{A}^{\vee}_{U_{\mathscr{A}}})/G.$$

Here, $U_{\mathscr{A}}$ is again the maximal open subset were $\mathscr{A}$ is locally free.

Remark 3.4.1. Since $\mathscr{A}$ is torsion free, the open subset $U_{\mathscr{A}}$ is **big**, i.e., its complement has codimension at least two. The definition means that $\mathscr{P}(\mathscr{A}, \tau)_{|U_{\mathscr{A}}}$ is a principal G-bundle. Therefore, $(U_{\mathscr{A}}, \mathscr{P}(\mathscr{A}, \tau)_{|U_{\mathscr{A}}})$ is a rational principal G-bundle in the sense of [RR], Definition 4.6. Hence, singular principal G-bundles appear to be valid analogs to torsion free sheaves for a semisimple group G.

LEMMA 3.4.2. *Let $(\mathscr{A}, \tau)$ be a pseudo G-bundle Then, $(\mathscr{A}, \tau)$ is a singular principal G-bundle, if and only if there exists a single point $x \in U_{\mathscr{A}}$, such that $\sigma(x) \in \mathscr{I}som(V \otimes \mathscr{O}_{U_{\mathscr{A}}}, \mathscr{A}^{\vee}_{|U_{\mathscr{A}}})/G$.*

Proof (see [Sch01], Remark 3.3 and Corollary 3.4) A pseudo G-bundle induces a homomorphism $\det(\tau)\colon \mathscr{O}_X \longrightarrow \det(\mathscr{A})$. Since this is a homomorphism between trivial invertible $\mathscr{O}_X$-modules, it is an isomorphism if and only if it is an isomorphism in a single point of $U_{\mathscr{A}}$. The latter condition is equivalent to the one in the lemma. $\square$

This lemma characterizes singular principal G-bundles among pseudo G-bundles and demonstrates that, in any family of pseudo G-bundles, the locus in the parameterizing scheme that belongs to the singular principal G-bundles is open.

3.4.1 Semistability for singular principal G-bundles
(see [Sch03], Introduction)

Let $(\mathscr{A}, \tau)$ be a singular principal G-bundle and $\lambda\colon \mathbb{G}_m(k) \longrightarrow G$ a one parameter subgroup. A *reduction of $(\mathscr{A}, \tau)$ to λ* is a pair (U, β) which consists of a big open subset $U \subseteq U_{\mathscr{A}}$ and a section

$$\beta : U \longrightarrow \mathscr{P}(\mathscr{A}, \tau)_{|U} / Q_G(\lambda).$$

Given a reduction (U, β) of $(\mathscr{A}, \tau)$ to λ, we form the composite section

$$\beta' : U \xrightarrow{\ \beta\ } \mathscr{P}(\mathscr{A}, \tau)_{|U} / Q_G(\lambda) \lhook\joinrel\longrightarrow \mathscr{I}som(V \otimes \mathscr{O}_U, \mathscr{A}_{|U}^{\vee}) / Q_{\mathrm{GL}(V)}(\lambda).$$

Let $(V_{\bullet}(\lambda), \alpha_{\bullet}(\lambda))$ be the weighted flag of λ in the vector space V (cf. appendix). Since $\mathscr{I}som(V \otimes \mathscr{O}_U, \mathscr{A}_{|U}^{\vee}) / Q_{\mathrm{GL}(V)}(\lambda)$ is the bundle of flags in the fibers of $\mathscr{A}_{|U}^{\vee}$ of the same "dimension-type" as $V_{\bullet}(\lambda)$, the section β' corresponds to a filtration

$$\mathscr{A}_{\bullet}' : \{0\} \subsetneq \mathscr{A}_1' \subsetneq \cdots \subsetneq \mathscr{A}_s' \subsetneq \mathscr{A}_{|U}^{\vee}$$

of $\mathscr{A}_{|U}^{\vee}$ by subbundles in which the rank of $\mathscr{A}_i'$ is the dimension of the corresponding subvectorspace in the flag $V_{\bullet}(\lambda)$, $i = 1, \ldots, s$. By dualizing, we get a filtration of $\mathscr{A}_{|U}$ by subbundles which extends to a filtration $\mathscr{A}_{\bullet}(\beta)$ of $\mathscr{A}$ by saturated subsheaves. Assume $\alpha_{\bullet}(\lambda) = (\alpha_1', \ldots, \alpha_s')$. Then, we set $\alpha_{\bullet}(\beta) := (\alpha_1, \ldots, \alpha_s) := (\alpha_s', \ldots, \alpha_1')$. Altogether, we have associated to the reduction (U, β) of $(\mathscr{A}, \tau)$ to λ a weighted filtration $(\mathscr{A}_{\bullet}(\beta), \alpha_{\bullet}(\beta))$ of $\mathscr{A}$. We call $(\mathscr{A}, \tau)$ *(semi)stable*, if

$$M\big(\mathscr{A}_{\bullet}(\beta), \alpha_{\bullet}(\beta)\big)(\succeq)0$$

holds for every one parameter subgroup λ of G and every reduction (U, β) of $(\mathscr{A}, \tau)$ to λ.
   If

$$L\big(\mathscr{A}_{\bullet}(\beta), \alpha_{\bullet}(\beta)\big)(\geq)0$$

is satisfied for every one parameter subgroup λ of G and every reduction (U, β) of $(\mathscr{A}, \tau)$ to λ, we say that $(\mathscr{A}, \tau)$ is a *slope (semi)stable singular principal G-bundle*.

Remark 3.4.3. Let $(\mathscr{A}, \tau)$ be a singular principal G-bundle. Recall that it defines the rational principal G-bundle $(U_{\mathscr{A}}, \mathscr{P}(\mathscr{A}, \tau)_{|U_{\mathscr{A}}})$. In [Sch03], Proposition 5.4 (see also [GLSS01], Section 3.2), the author verified that $(\mathscr{A}, \tau)$ is slope (semi)stable, if and only if $(U_{\mathscr{A}}, \mathscr{P}(\mathscr{A}, \tau)_{|U_{\mathscr{A}}})$ is (semi)stable in the sense of Ramanathan ([RR], Definition 4.7).

Example 3.4.4. Assume $G = \mathrm{SL}(V)$ and $\varrho = \mathrm{id}$. Then, one immediately sees that a singular principal $\mathrm{SL}(V)$-bundle $(\mathscr{A}, \tau)$ is (semi)stable, if and only if $\mathscr{A}$ is a Gieseker (semi)stable torsion free sheaf. The analogous statement holds for slope (semi)stability.

LEMMA 3.4.5 ([Sch03], Lemma 4.4 and Proposition 4.5, [GLSS01], Lemma 5.4.2). *Let* $(\mathscr{A}, \tau)$ *be a singular principal G-bundle. Then, the following conditions on a weighted filtration* $(\mathscr{A}_\bullet, \alpha_\bullet)$ *of* $\mathscr{A}$ *are equivalent:*

1. $\mu(\mathscr{A}_\bullet, \alpha_\bullet, \tau) = 0$.
2. *There are a one parameter subgroup* λ *of* G *and a reduction* (U, β) *of* $(\mathscr{A}, \tau)$ *to* λ, *such that* $(\mathscr{A}_\bullet, \alpha_\bullet) = (\mathscr{A}_\bullet(\beta), \alpha_\bullet(\beta))$.

In particular, we see that a singular principal G-bundle $(\mathscr{A}, \tau)$ which is δ-(semi)stable for some stability parameter δ is also (semi)stable in the new sense. Let P be a fixed Hilbert polynomial. Then, the family of isomorphy classes of slope semistable singular principal G-bundles $(\mathscr{A}, \tau)$ with $P(\mathscr{A}) = P$ is bounded: In characteristic zero, this follows from the fact that $\mathscr{A}$ is a slope semistable torsion free sheaf (see [RR], Theorem 3.18), and in positive characteristic from the results in [Lan04]. Using this, one can easily prove (see [Sch03], Proposition 5.3, [GLSS01], Theorem 5.4.1):

PROPOSITION 3.4.6. *Fix the Hilbert polynomial P. Then, there is a positive polynomial* δ_0, *such that for every other positive polynomial* $\delta \succ \delta_0$, *and every singular principal G-bundle* $(\mathscr{A}, \tau)$, *such that* $P(\mathscr{A}) = P$, *the following conditions are equivalent:*

1. $(\mathscr{A}, \tau)$ *is* δ-*(semi)stable.*
2. $(\mathscr{A}, \tau)$ *is (semi)stable (in the sense of the present section).*

For a polynomial $\delta \succ \delta_0$, the family of semistable singular principal G-bundles with Hilbert polynomial P is closed under S-equivalence within the family of δ-semistable pseudo G-bundles with Hilbert polynomial P ([GLSS01], Remark 5.4.3). Together with Lemma 3.4.2, this implies:

COROLLARY 3.4.7. *For fixed Hilbert polynomial P and* $\delta \succ \delta_0$ *as in Proposition 3.4.6, the moduli space* $\mathscr{M}_\varrho^P(X)$ *of semistable singular principal G-bundles with Hilbert polynomial P exists as an open subscheme of the projective moduli scheme* $\mathscr{M}_{\varrho, \delta}^P(X)$ *for* δ-*semistable pseudo G-bundles with Hilbert polynomial P.*

If X is a curve, then $\mathscr{M}_G^\vartheta(X)$ is a connected component of $\mathscr{M}_\varrho^P(X)$ for the Hilbert polynomial $P = \dim(V) \cdot (n + 1 - g(X))$. The corollary therefore gives the existence of $\mathscr{M}_G^\vartheta(X)$ in any characteristic.

On higher dimensional varieties, we expect that $\mathscr{M}_\varrho^P(X)$ does depend on the choice of the representation ϱ.

3.4.2 The semistable reduction theorem

In order to settle the projectivity of $\mathscr{M}_\varrho^P(X)$, and thereby the semistable reduction theorem (see [GLSS01], Section 5.5), it suffices to show that, for large stability parameters δ, a δ-semistable pseudo G-bundle is a singular principal G-bundle.

We will use the notation of Section 3.3.2. Recall that any pseudo G-bundle $(\mathscr{A}, \tau)$ defines a point

$$\sigma_K \in \mathbb{H} \smallsetminus \{0\}, \quad \mathbb{H} := \mathrm{Hom}(V \otimes_k K, \mathbb{A})/\!/G.$$

The coordinate algebra $K[\mathbb{H}]$ is graded. Thus, we may use Mumford's definition of semistability: A point $h \in \mathbb{H}$ is *semistable*, if there are an integer $d > 0$ and a homogeneous function $f \in K[\mathbb{H}]^{\mathrm{SL}(\mathbb{A})}$ of degree d with $f(h) \neq 0$.

LEMMA 3.4.8 ([Sch03], Lemma 4.1, [GLSS01], Lemma 2.1.3). *The point σ is semistable under the action of* $\mathrm{SL}(\mathbb{A})$ *if and only if* $\sigma_K \in \mathrm{Isom}(V \otimes_k K, \mathbb{A})/G$.

Together with Lemma 3.4.2, this gives a GIT characterization of singular principal G-bundles among pseudo G-bundles. If K were algebraically closed, we could test this condition with the Hilbert-Mumford criterion (using the quantity $\overline{\mu}$ from Section 3.3.1).

PROPOSITION 3.4.9. *Fix the Hilbert polynomial P. Then, there is a polynomial δ_∞, such that, for every polynomial $\delta \succ \delta_\infty$ and every pseudo G-bundle $(\mathscr{A}, \tau)$ with $P(\mathscr{A}) = P$, the following conditions are equivalent:*

1. *The pseudo G-bundle $(\mathscr{A}, \tau)$ is δ-semistable.*
2. *For every weighted filtration $(\mathscr{A}_\bullet, \alpha_\bullet)$ of $\mathscr{A}$, one has*

$$\mu(\mathscr{A}_\bullet, \alpha_\bullet, \tau) \geq 0,$$

and, if equality holds at this stage, also

$$M(\mathscr{A}_\bullet, \alpha_\bullet)(\succeq)0.$$

In particular, $(\mathscr{A}, \tau)$ is δ-unstable with respect to any positive polynomial $\delta \succ \delta_\infty$, if there is a one parameter subgroup $\lambda_K : \mathbb{G}_m(K) \longrightarrow \mathrm{SL}(\mathbb{A})$ with $\overline{\mu}(\lambda_K, \sigma_K) < 0$.

Proof In [GLSS01], Theorem 4.2.1, it is shown that the family of isomorphy classes of torsion free sheaves $\mathscr{A}$ with Hilbert polynomial P for which there exists **some** polynomial $\delta \succ 0$ of degree at most $\dim(X) - 1$ and a δ-semistable pseudo G-bundle $(\mathscr{A}, \tau)$ is bounded. This fact makes it easy to characterize the notion of δ-(semi)stability for large values of δ (loc. cit., Corollary 4.2.2). $\square$

This proposition tells us where the difference between characteristic zero and positive characteristic lies in our approach: In characteristic zero, the Hilbert-Mumford criterion holds also over non-algebraically closed fields, by the theory of the instability flag (see [Hesse], [Kempf], and [RR]). In positive characteristic, this is not so. An easy counter-example is given in [Hesse], Example 5.6.

Nevertheless, one may hope that the approach may be saved under some extra conditions on the characteristic of k depending on ϱ. Indeed, if ϱ is of low separable index or ϱ is the adjoint representation and of low height, then the existence of λ_K as in Proposition 3.4.9 was established in [GLSS01], Corollary 2.2.8 and 2.2.10. Over curves, this and some extra arguments for classical root systems yield the following result on semistable reduction which is a bit weaker than Heinloth's theorem 2.3.1:

THEOREM 3.4.10 ([GLSS01], Introduction). *Let G be a connected reductive group. Under the following assumptions on the characteristic of k, the semistable reduction theorem for principal G-bundles on the curve X holds true:*

- Char$(k) = 0$.
- Char$(k) \geq 3$, *if G contains a simple factor of type B, C, or D.*
- Char$(k) \geq 11$, *if G contains a simple factor of type G_2.*
- Char$(k) \geq 23$, *if G contains a simple factor of type F_4 or E_6.*
- Char$(k) \geq 37$, *if G contains a simple factor of type E_7.*
- Char$(k) \geq 59$, *if G contains a simple factor of type E_8.*

Unlike Heinloth's result, this theorem also generalizes to quasi-parabolic principal bundles (see Section 4.1.2).

3.4.3 The work of Balaji

Here, we let k be a field of characteristic zero. Balaji proved in [Ba01] that the semistable reduction theorem is also true for slope semistable singular principal G-bundles. The proof uses his work with Seshadri [BS] and the restriction theorem of Mehta and Ramanathan ([HL], Section 7.2). As a very interesting application, he constructs the **Uhlenbeck compactification** $\mathscr{U}_\varrho^P(X)$ on a (smooth, projective) **surface** X.

It is a kind of moduli space for slope semistable singular principal G-bundles. (The reader may consult [HL], Section 8.2, for the respective result on vector bundles.) A remarkable fact is that the topological space underlying $\mathscr{U}_\varrho^P(X)$ **does not depend on the choice of the representation** ϱ, if G is a **simple** linear algebraic group ([Ba01], Theorem 1.2, 3). Last but not least, Balaji also proves the existence of stable principal G-bundles ([Ba01], Theorem 7.2).

3.5 Singular principal G-bundles on nodal curves

The ground field is now of characteristic zero. In this section, X will be a **nodal curve**, i.e., an irreducible curve with exactly one singular point x_0 which is an ordinary double point. Set $U := X \smallsetminus \{x_0\}$. We will only admit pseudo G-bundles $(\mathscr{A}, \tau)$ with $\deg(\mathscr{A}) = 0$. A *singular principal G-bundle on X* is a pseudo G-bundle $(\mathscr{A}, \tau)$, such that

$$\sigma(U) \subset \mathscr{I}som(V \otimes \mathcal{O}_U, \mathscr{A}_{|U}^\vee)/G.$$

A singular principal G-bundle defines the principal G-bundle $\mathscr{P}(\mathscr{A}, \tau)_{|U}$ on U. (Beware of the different meanings of "pseudo G-bundle" and "singular G-bundle" in [Sch05]!)

LEMMA 3.5.1 ([Sch05], Proposition 3.5). *Let (V, φ) be a finite dimensional k-vector space together with a non-degenerate symmetric or anti-symmetric bilinear form $\varphi: V \otimes V \longrightarrow k$. Assume that $\varrho: G \longrightarrow \mathrm{SL}(V)$ is a faithful representation with image in the isotropy group of (V, φ). Then, a pseudo G-bundle $(\mathscr{A}, \tau)$ is a singular principal G-bundle, if and only if there exists a point $x \in U$ with $\sigma(x) \in \mathscr{I}som(V \otimes \mathcal{O}_U, \mathscr{A}_{|U}^\vee)/G$.*

This lemma is the analog to Lemma 3.4.2. Note that we can easily find a representation ϱ which satisfies the assumption of the lemma: Start with any faithful representation $\widetilde{\varrho}: G \longrightarrow \mathrm{SL}(W)$, set $V := W \oplus W^\vee$ and $\varrho := \widetilde{\varrho} \oplus \widetilde{\varrho}^\vee: G \longrightarrow \mathrm{SL}(V)$. The image of ϱ respects the bilinear forms $\varphi_\pm: V \otimes V \longrightarrow k$, $(w, l) \otimes (w', l') \longmapsto l(w') \pm l'(w)$.

3.5.1 Semistability ([Sch05], Section 1.1)

Let $(\mathscr{A}, \tau)$ be a singular principal G-bundle on X and $\lambda: \mathbb{G}_m(k) \longrightarrow G$ a one parameter subgroup. A *reduction of $(\mathscr{A}, \tau)$ to λ* is a section

$$\beta: U \longrightarrow \mathscr{P}(\mathscr{A}, \tau)_{|U}/Q_G(\lambda).$$

As before, such a reduction gives rise to a weighted filtration

$$(\mathscr{A}_\bullet(\beta), \alpha_\bullet(\beta))$$

of $\mathscr{A}$. We say that $(\mathscr{A}, \tau)$ is *(semi)stable*, if the inequality

$$L\big(\mathscr{A}_\bullet(\beta), \alpha_\bullet(\beta)\big)(\geq)0$$

holds for every one parameter subgroup λ of G and every reduction β of $(\mathscr{A}, \tau)$ to λ. For the following, note that the stability parameter δ for pseudo G-bundles is just a positive rational number and that the Hilbert polynomial is always $P = \dim(V) \cdot (n + 1 - g(X))$.

THEOREM 3.5.2 ([Sch05], Corollary 3.3). *Let $\varrho\colon G \longrightarrow \mathrm{SL}(V)$ be as in Lemma 3.5.1. Then, there is a positive rational number δ_0, such that for $\delta > \delta_0$, the following assertions are true:*

1. *A δ-semistable pseudo G-bundle is a singular principal G-bundle.*
2. *A singular principal G-bundle is δ-(semi)stable, if and only if it is (semi)stable (in the sense of this section).*

In particular, the moduli space $\mathscr{M}_\varrho(X)$ for semistable singular principal G-bundles exists as a projective scheme.

The idea of the proof is the following: Let $\nu\colon \widetilde{X} \longrightarrow X$ be the normalization of X. Then, by work of Bhosle [Bho01], a torsion free sheaf on X may be described as a **generalized parabolic bundle** on $\widetilde{X}$. In a similar vein, the author checked that a singular principal G-bundle on X can be described as a **descending principal G-bundle** on $\widetilde{X}$ (see [Sch04], Section 1.2, and [Sch05], Section 2). Thus, we have to study principal G-bundles on the smooth curve $\widetilde{X}$, so that we are in the setting of Section 3.4 and may apply the respective results to conclude.

Adrian Langer has informed me that he expects that the theory will work to a large extent on any singular variety and also over fields of sufficiently large positive characteristic. The details will appear in his paper [Lan04].

3.5.2 Degenerations

Moduli spaces for semistable singular principal G-bundles might become interesting for studying moduli spaces for principal G-bundles on smooth curves: Let $R := k[\![t]\!]$, $K := k(\!(t)\!)$, and $\mathscr{C} \longrightarrow \mathrm{Spec}(R)$ a flat family of curves, such that $\mathscr{C}$ is smooth over $\mathrm{Spec}(K)$ and the special fiber is a nodal curve. Then, one can construct a moduli space $f\colon \mathscr{M} \longrightarrow \mathrm{Spec}(R)$, such that the generic fiber is the moduli space of Ramanathan-semistable principal G-bundles and the special fiber is the moduli space of semistable singular principal G-bundles. This might be useful for setting up a degeneration argument. It is, however, not clear that f will be flat.

For degenerations to a nodal curve from the stack point of view, the reader may consult [Fal03].

3.6 Decorated principal bundles

Here, X will be a smooth projective curve over a field of characteristic zero. A large class of interesting moduli problems is of the following shape: Fix a representation $\kappa\colon G \longrightarrow \mathrm{GL}(W)$ and study the problem of classifying pairs $(\mathscr{P}, \sigma)$ which consist of a principal G-bundle $\mathscr{P}$ and a section $\sigma\colon X \longrightarrow \mathscr{P}_\kappa := \mathscr{P}(W)$ with respect to the natural notion of isomorphy. A famous special case is the one of the adjoint representation which leads to the theory of Higgs bundles (see, e.g., [Fal01]).

A first general treatise of this moduli problem was given in [Sch02] for $G = \mathrm{GL}_r(\mathbb{C})$ and ϱ a homogeneous representation. With our concept of a pseudo G-bundle, we may use this special case to find the general solution: Let $\varrho\colon G \longrightarrow \mathrm{SL}(V)$ be a faithful representation. Then, there is a representation $\widetilde{\kappa}\colon \mathrm{GL}(V) \longrightarrow \mathrm{GL}(\widetilde{W})$, such that κ is a direct summand of $\widetilde{\kappa} \circ \varrho$. First, we look at the classification of triples $(\mathscr{A}, \tau, \sigma)$ where $(\mathscr{A}, \tau)$ is a pseudo G-bundle and $\sigma\colon X \longrightarrow \mathscr{A}_{\widetilde{\varrho}} := \mathscr{P}(\widetilde{W})$, $\mathscr{P} := \mathscr{I}som(V \otimes \mathscr{O}_X, \mathscr{A})$, is a section. The solution of this moduli problem is, as for pseudo G-bundles, obtained from the one in the case $G = \mathrm{GL}_r(\mathbb{C})$. Invoking the techniques from Section 3.3.1 and 3.3.2, one gets a nice theory of moduli spaces for decorated principal G-bundles. The complete construction will be carried out in the book [Sch06].

4 Quasi-parabolic principal bundles and the theorem of Atiyah and Bott

Let X be a smooth projective curve over the field of complex numbers. In their paper [MS], Mehta and Seshadri introduced quasi-parabolic and parabolic vector bundles which are holomorphic objects on X associated to representations of the fundamental group of the punctured Riemann surface $X \smallsetminus \{x_1, \ldots, x_n\}$ in $\mathrm{U}_r(\mathbb{C})$ and extended the theorem of Narasimhan and Seshadri [NS]. Here, we will consider the naive generalizations of these notions to the setting of principal bundles over fields of arbitrary characteristic. Over $\mathbb{C}$, such results are discussed in [BR] and [TW]. In [BBN01] and [BBN02], the authors discuss the Tannakian approach to parabolic principal bundles. (Note that this approach is confined to characteristic zero, because semistability of parabolic principal bundles is not preserved under extensions of the structure group.) Our aim is to establish certain properties of the cohomology of the moduli stack $\mathscr{B}un_G(X)$.

4.1 Quasi-parabolic principal bundles

Let X be a smooth projective curve over an algebraically closed field k of arbitrary characteristic. We fix a tuple $\underline{x} = \{x_1, \ldots, x_n\}$ of n distinct closed points on X. Moreover, we select a tuple $\underline{P} = (P_1, \ldots, P_n)$ of parabolic subgroups of G. A *quasi-parabolic principal G-bundle of type* $(\underline{x}, \underline{P})$ is a tuple $(\mathscr{P}, \underline{s})$ which consists of a principal G-bundle $\mathscr{P}$ on X and a tuple $\underline{s} = (s_1, \ldots, s_n)$ of points $s_i \in (\mathscr{P}/P_i)_{|\{x_i\}}$, $i = 1, \ldots, n$.

4.1.1 Moduli stacks for quasi-parabolic principal G-bundles

Given the data $\underline{x}$, $\underline{P}$, and $\vartheta \in \pi_1(G)$, we have the moduli stacks

$$f : \mathscr{B}un_{G,\underline{x},\underline{P}}(X) \longrightarrow \mathscr{B}un_G(X)$$

of quasi-parabolic principal G-bundles of type $(\underline{x}, \underline{P})$ and

$$f^\vartheta : \mathscr{B}un_{G,\underline{x},\underline{P}}^\vartheta(X) \longrightarrow \mathscr{B}un_G^\vartheta(X)$$

of quasi-parabolic principal G-bundles $(\mathscr{P}, \underline{s})$ of type $(\underline{x}, \underline{P})$ where the topological type of $\mathscr{P}$ is ϑ together with their forgetful maps onto the respective moduli stack for principal G-bundles. Note that f and f^ϑ are bundle maps (in the Zariski topology) with typical fiber $\mathsf{X}_{i=1}^n(G/P_i)$. Hence, the cohomology of $\mathscr{B}un_G(X)$ and $\mathscr{B}un_G^\vartheta(X)$ appears as a direct summand of the cohomology of $\mathscr{B}un_{G,\underline{x},\underline{P}}(X)$ and $\mathscr{B}un_{G,\underline{x},\underline{P}}^\vartheta(X)$, respectively.

4.1.2 Semistability

Let H be a linear algebraic group and $T \subset G$ a maximal torus. We have the abelian group $X_\star(T)$ of one parameter subgroups of T and the $\mathbb{Q}$-vector space $X_{\star,\mathbb{Q}}(T) := X_\star(T) \otimes_{\mathbb{Z}} \mathbb{Q}$. Note that the Weyl group $W := W(T, H)$ acts on both objects. The *set of rational one parameter subgroups of H* is $\bigsqcup_{T \subset G \text{ max. torus}} X_{\star,\mathbb{Q}}(T)$. To a rational one parameter subgroup λ of H, we may still associate the parabolic subgroup $Q_H(\lambda)$ as in the appendix. The group H acts by conjugation on the set of rational one parameter subgroups. The set of conjugacy classes identifies with $X_{\star,\mathbf{Q}}(T)/W$. If G is a reductive algebraic group and $P \subset G$ is a parabolic subgroup, the set of conjugacy classes of rational one parameter subgroups of P equals also $X_{\mathbf{Q}}(P)^{\vee}$, $X_{\mathbf{Q}}(P) := X(P) \otimes_{\mathbf{Z}} \mathbf{Q}$, and we write $X_{\mathbf{Q}}(P)^{\vee}_{+}$ for the cone of rational one parameter subgroups λ of P with the property $Q_G(-\lambda) = P$.

The stability parameters for quasi-parabolic principal G-bundles are tuples $\underline{a} = (\lambda_1, \ldots, \lambda_n)$ with $\lambda_i \in X_{\mathbf{Q}}(P_i)^{\vee}_{+}$, $i = 1, \ldots, n$. If $(\mathscr{P}, \underline{s})$ is a quasi-parabolic principal G-bundle of type $(\underline{x}, \underline{P})$, $Q \subset G$ is a parabolic subgroup, $\chi \in X(Q)$, and $\beta \colon X \longrightarrow \mathscr{P}/Q$ is a reduction of the structure group of $\mathscr{P}$ to Q, then one may use the natural pairing between one parameter subgroups and characters to define

$$\langle \lambda_i, \chi \rangle_{\beta}, \ i = 1, \ldots, n, \quad \text{and} \quad \langle \underline{a}, \chi \rangle_{\beta} := \sum_{i=1}^{n} \langle \lambda_i, \chi \rangle_{\beta}.$$

This is explained in Lemma 4.1.3 in [HeSc]. We now say that $(\mathscr{P}, \underline{s})$ is $\underline{a}$-*(semi)stable*, if

$$\deg(\mathscr{L}(\mathscr{Q}, \chi)) + \langle \underline{a}, \chi \rangle_{\beta}(\geq)0$$

holds for every parabolic subgroup $Q \subset G$, every anti-dominant character χ on Q, and every reduction $\beta \colon X \longrightarrow \mathscr{P}/Q$ of the structure group.

Let $P \subset G$ be a parabolic subgroup and $T \subset P$ a maximal torus. A rational one parameter subgroup $\lambda \in X_{\star,\mathbf{Q}}(T)$ of P is said to be *admissible*, if $|\langle \lambda, \alpha \rangle| \leq 1/2$ holds for every root α of T. We say that the stability parameter $\underline{a} = (\lambda_1, \ldots, \lambda_n)$ is *admissible*, if λ_i is an admissible rational one parameter subgroup of P_i, $i = 1, \ldots, n$. A stability parameter $\underline{a}$ is said to be of *coprime type*, if the notion of $\underline{a}$-semistability for quasi-parabolic principal G-bundles of type $(\underline{x}, \underline{P})$ is equivalent to the notion of $\underline{a}$-stability.

Remark 4.1.1. Let B be a Borel subgroup of G and $\underline{P} = (B, \ldots, B)$. Then, it is not hard to see that a stability parameter outside a certain finite union of hyperplanes in $\bigoplus_{i=1}^{n} X_{\mathbf{Q}}(B)^{\vee}$ is of coprime type ([HeSc], Lemma 4.2.1).

THEOREM 4.1.2 (Heinloth/Schmitt [HeSc]). *Fix the type $(\underline{x}, \underline{P})$, the topological type $\vartheta \in \pi_1(G)$, and the admissible stability parameter $\underline{a}$. Then, the moduli space $\mathscr{M}^{\underline{a}}_{G,\vartheta,\underline{x},\underline{P}}(X)$ for $\underline{a}$-semistable quasi-parabolic principal G-bundles $(\mathscr{P}, \underline{s})$ of type $(\underline{x}, \underline{P})$ with $\mathscr{P}$ of topological type ϑ exists as a quasi-projective variety. It is projective, if the characteristic of the ground field k satisfies the assumptions in Theorem 3.4.10.*

The method of proof follows the strategy explained in Section 3.4 with one little difficulty: If one fixes a faithful representation ϱ, one needs a notion of admissibility which depends on ϱ and is usually stronger than the one presented above. (In fact, it equals the above notion in the case of the adjoint representation.) Thus, one first finds a weaker result (which still suffices for the applications we have in mind). To prove the theorem in its full strength, one uses its version for adjoint groups with the adjoint representation and involves the techniques from Section 5 in [GLSS02].

4.1.3 Moduli stacks of semistable quasi-parabolic principal G-bundles

Let the data $\underline{x}$, $\underline{P}$, and ϑ as above be given. For an open substack $\mathscr{U} \subset \mathscr{B}un_G^\vartheta(X)$, we set

$$\mathscr{U}_{\underline{x},\underline{P}} := \mathscr{B}un_{G,\underline{x},\underline{P}}^\vartheta(X) \times_{\mathscr{B}un_G^\vartheta(X)} \mathscr{U}$$

and let $\mathscr{U}_{\underline{x},\underline{P}}^{\underline{a}} \subset \mathscr{U}_{\underline{x},\underline{P}}$ be the open substack of those quasi-parabolic principal G-bundles that satisfy the condition of $\underline{a}$-semistability.

PROPOSITION 4.1.3 ([HeSc], Proposition 4.2.2). *Let $\mathscr{U} \subset \mathscr{B}un_G^\vartheta(X)$ be an open substack of **finite type over** k, N a positive integer, and $B \subset G$ a Borel subgroup. For every $n \gg 0$ and every collection $\underline{x} = \{ x_1, \ldots, x_n \}$ of n distinct points on X, there is an admissible stability parameter $\underline{a}$ of coprime type, such that the map*

$$\mathscr{U}_{\underline{x},\underline{P}}^{\underline{a}} \subset \mathscr{U}_{\underline{x},\underline{P}} \longrightarrow \mathscr{U}, \quad \underline{P} := (B, \ldots, B),$$

is surjective and the codimension of $\mathscr{U}_{\underline{x},\underline{P}} \setminus \mathscr{U}_{\underline{x},\underline{P}}^{\underline{a}}$ in $\mathscr{U}_{\underline{x},\underline{P}}$ is at least N.

Remark 4.1.4. i) The morphism $f^\vartheta \colon \mathscr{B}un_{G,\underline{x},\underline{P}}^\vartheta(X) \longrightarrow \mathscr{B}un_G^\vartheta(X)$ is proper. Let $\underline{a}$ be an admissible stability parameter, $\mathscr{B}un_{G,\vartheta,\underline{x},\underline{P}}^{\underline{a}}(X)$ the open substack of $\underline{a}$-semistable quasi-parabolic principal G-bundles, and $\mathscr{Z} := \mathscr{Z}_{G,\vartheta,\underline{x},\underline{P}}^{\underline{a}}(X)$ its complement in $\mathscr{B}un_{G,\underline{x},\underline{P}}^\vartheta(X)$. There is the open substack $\mathscr{U} := \mathscr{U}(G,\vartheta,\underline{x},\underline{P},\underline{a}) \subset \mathscr{B}un_G^\vartheta(X)$ that is characterized by the property that a point $x \in \mathscr{B}un_G^\vartheta(X)$ lies in $\mathscr{U}$, if and only if the fiber of f^ϑ over x does not meet $\mathscr{Z}$. By Proposition 4.1.3, we may restrict to those open substacks of $\mathscr{B}un_G^\vartheta(X)$ that are of the form $\mathscr{U}(G,\vartheta,\underline{x},\underline{P},\underline{a})$ for suitable $\underline{x}$, $\underline{P}$, and $\underline{a}$.

ii) Let $\underline{a}$ be a stability parameter of the coprime type and assume that the characteristic of k fulfills the requirements of Theorem 3.4.10. Then, the cohomology of the moduli stack $\mathscr{B}un_{G,\vartheta,\underline{x},\underline{P}}^{\underline{a}}(X)$ equals the cohomology of its projective moduli space $\mathscr{M}^a_{G,\vartheta,\underline{x},\underline{P}}(X)$ ([HeSc], proof of Corollary 3.3.2).

4.2 The purity of the cohomology of $\mathscr{B}un^a_G(X)$

Let $q = p^s$ be a power of the prime p, $\mathbb{F}_q$ the corresponding finite field, and k an algebraic closure of $\mathbb{F}_q$. We let X be a smooth projective curve over k which is defined over $\mathbb{F}_q$, i.e., there is a curve Y over $\mathbb{F}_q$, such that $X = Y \times_{\mathrm{Spec}(\mathbb{F}_q)} \mathrm{Spec}(k)$. The Frobenius $x \longmapsto x^q$ on k gives a morphism $\mathrm{Spec}(k) \longrightarrow \mathrm{Spec}(k)$ over $\mathrm{Spec}(\mathbb{F}_q)$ and therefore a morphism $F \colon X \longrightarrow X$. The operation of pulling back a principal bundle on X by means of F yields a map $F^\# \colon \mathscr{B}^\vartheta \longrightarrow \mathscr{B}^\vartheta$, $\mathscr{B}^\vartheta := \mathscr{B}un^\vartheta_G(X)$. The resulting $\overline{\mathbb{Q}}_\ell$-linear automorphism $H^\star(\mathscr{B}^\vartheta, \overline{\mathbb{Q}}_\ell) \longrightarrow H^\star(\mathscr{B}^\vartheta, \overline{\mathbb{Q}}_\ell)$ on the ℓ-adic cohomology of $\mathscr{B}^\vartheta$ is called the *arithmetic Frobenius*. We say that the cohomology of $\mathscr{B}^\vartheta$ is *pure*, if

$$|\lambda| = q^{-\frac{i}{2}}$$

holds for every eigenvalue λ of the restriction of the arithmetic Frobenius to $H^i(\mathscr{B}^\vartheta, \overline{\mathbb{Q}}_\ell)$ and all $i \geq 0$.

THEOREM 4.2.1 (Heinloth/Schmitt [HeSc]). *Under the assumption on the prime number p from Theorem 3.4.10, the cohomology of $\mathscr{B}^\vartheta$ is pure.*

Proof Suppose that $i \geq 0$ and that λ is an eigenvalue of the arithmetic Frobenius on $H^i(\mathscr{B}^\vartheta, \overline{\mathbb{Q}}_\ell)$. Then, there is a substack of $\mathscr{U} \subset \mathscr{B}^\vartheta$ of finite type over k with $H^i(\mathscr{B}^\vartheta, \overline{\mathbb{Q}}_\ell) = H^i(\mathscr{U}, \overline{\mathbb{Q}}_\ell)$. The substack $\mathscr{U}$ may be obtained as the (stack) quotient $[U/\mathrm{GL}_R(k)]$ of a smooth quasi-projective variety U by an appropriate $\mathrm{GL}_R(k)$-action. Behrend and Dhillon ([BD], Theorem 5.21) show that this implies $|\lambda| \leq q^{-i/2}$.

By Remark 4.1.4, i), we may assume $\mathscr{U} = \mathscr{U}(G, \vartheta, \underline{x}, \underline{P}, \underline{a})$ for suitable data $\underline{x}$, $\underline{P}$, and $\underline{a}$, such that Proposition 4.1.3 holds for some $N > i$. By that proposition, Section 4.1.1, and Remark 4.1.4, ii), the cohomology group $H^i(\mathscr{B}^\vartheta, \overline{\mathbb{Q}}_\ell)$ is a direct summand of $H^i(\mathscr{M}^a_{G,\vartheta,\underline{x},\underline{P}}(X), \overline{\mathbb{Q}}_\ell)$. By our assumption on p, $\mathscr{M}^a_{G,\vartheta,\underline{x},\underline{P}}(X)$ is a projective variety, so that the inequality $|\lambda| \geq q^{-i/2}$ follows from Deligne's work on the Weil conjectures ([D], Théorème I). $\qquad\square$

4.3 The theorem of Atiyah and Bott

Set $\mathscr{B} := \bigsqcup_{\vartheta \in \pi_1(G)} \mathscr{B}^\vartheta = \mathscr{B}un_G(X)$. Let us first show how we can prove that the theorem of Atiyah and Bott holds for the algebraic closure k

of a finite field. As in the introduction, the Künneth components of the characteristic classes of the universal principal G-bundle on $\mathscr{B} \times X$ give cohomology classes $\gamma_1, \ldots, \gamma_N \in H^*(\mathscr{B}, \overline{\mathbb{Q}}_\ell)$.

PROPOSITION 4.3.1 ([HeSc], Proposition 3.1.1). *The classes* $\gamma_1, \ldots, \gamma_N$ *generate a free subalgebra* Can* *of* $H^*(\mathscr{B}, \overline{\mathbb{Q}}_\ell)$.

Proof The idea of proof is to look at a maximal torus $T \subset G$ and the restriction map $H^*(\mathscr{B}, \overline{\mathbb{Q}}_\ell) \longrightarrow H^*(\mathscr{B}un_T(X), \overline{\mathbb{Q}}_\ell)$. $\qquad \square$

Since Can* is a free algebra, its Poincaré series is readily computed. In order to conclude, we have to determine the Poincaré series of $H^*(\mathscr{B}, \overline{\mathbb{Q}}_\ell)$. By Behrend's trace formula ([Beh02], [Beh03]), one can compute the trace of the arithmetic Frobenius (and all its powers) on $H^*(\mathscr{B}, \overline{\mathbb{Q}}_\ell)$. By the purity of the cohomology (Theorem 4.2.1), we may derive the Poincaré series. It turns out that the Poincaré series of Can* and $H^*(\mathscr{B}, \overline{\mathbb{Q}}_\ell)$ do agree, if and only if the so-called Tamagawa number $\tau(G)$ (see, e.g., [KR], Section 2) equals the number of connected components of $\mathscr{B}$, i.e., $\#\pi_1(G)$. This is true, by a theorem of Harder [Har02] and Ono ([O], [BD], Theorem 6.1). It is obvious that the analogous result holds for $\mathscr{B}^\vartheta$, $\vartheta \in \pi_1(G)$, too.

Next, we would like to prove the theorem also over the complex numbers. Let X be a smooth projective curve over $\mathbb{C}$. It is defined over a finite extension K of $\mathbb{Q}$. We note that there are a certain one dimensional ring R with fraction field K and finite residue fields and a smooth curve $\mathscr{X} \longrightarrow \operatorname{Spec}(R)$, such that the generic fiber is X. The theory of base change (see [HeSc], Corollary 3.3.4) now allows to deduce the theorem for the generic fiber from the theorem on a special fiber. This gives a purely algebraic proof of the theorem of Atiyah and Bott.

Remark 4.3.2. i) We can also argue backwards: If we assume the theorem of Atiyah and Bott (over $\mathbb{C}$), we also get it in positive characteristic (under the provisos of Theorem 3.4.10). This has the benefit that we may **conclude** $\tau(G) = \#\pi_1(G)$ along the way, i.e., we find a geometric proof of the theorem of Harder and Ono on the Tamagawa number.

ii) The $GL_r(k)$-case was treated by Bifet, Ghione, and Letizia in [BGL] with different methods. In his diploma thesis [Hei01], Heinloth presented an alternative proof for $GL_r(k)$, using an approach similar to the one we outlined here.

iii) Note that with respect to purity and base change, $\mathscr{B}un_G(X)$ behaves like a smooth projective variety, i.e., it inherits smoothness from its stacky existence and projectivity from the coarse projective moduli spaces of quasi-parabolic principal G-bundles.

5 Some references on moduli stacks, affine Graßmannians, and loop groups

For the sake of completeness, we add here a few references regarding other areas of research in the theory of principal bundles on curves. As a conceptual introduction to these techniques, we recommend Sorger's lecture notes [S].

5.1 Loop groups

If one wants to show that the set of topological principal G-bundles with connected reductive structure group G over a closed and oriented topological surface X is in bijection to the elements of the fundamental group $\pi_1(G)$, one picks a point $x_0 \in X$ and a disc $D \subset X$ with center x_0. Since $U := X \smallsetminus \{x_0\}$ is homotopy equivalent to a wedge product of circles, any topological principal G-bundle on U is trivial. The same goes for a topological principal G-bundle on D. Hence, only the gluing datum matters. This is a map $D^\star := D \smallsetminus \{x_0\} \longrightarrow G$. It defines an element of $\pi_1(G)$, because $D^\star$ is homotopy equivalent to S^1.

If we assume that G is semi-simple and that X is a smooth projective curve over the algebraically closed ground field k, a similar reasoning applies: By work of Harder [Har01] or Drinfel'd and Simpson [DS], an algebraic principal G-bundle is trivial on $X \smallsetminus \{x_0\}$, $x_0 \in X$ a previously fixed point. The gluing datum is now a morphism $\mathrm{Spec}(k((t))) \longrightarrow G$. To equip the set of such morphisms with an algebraic structure, we study the functor which assigns to a k-algebra R the set $G(R((t)))$ of $R((t))$-valued points of G. This functor is representable by an **ind-scheme** (a—usually infinite dimensional—inductive limit of schemes nested by closed embeddings) LG, called the *loop group*. In case that $k = \mathbb{C}$, such loop groups were studied in [BL], [BLS], and [Fal02]. Further information about the significance of loop groups is also contained in the book [Kumar]. In [Fal03], the analogous object is introduced and studied over nodal curves and, in [LS], in connection with quasi-parabolic principal bundles. Faltings investigated loop groups over ground fields of positive characteristic [Fal04]. Twisted versions of loops groups appear in relation with torsors under non-constant group schemes over the base curve ([PR], [Hei04]).

5.2 Affine Graßmannians

The loop group LG contains the *positive loop group* L^+G that is associated to the functor $R \longmapsto G(R[[t]])$ on k-algebras. Clearly, $G(k[[t]])$ parameterizes the possible trivializations of a principal G-bundle over the disc $D = \mathrm{Spec}(R[[t]])$. The quotient LG/L^+G carries the structure of an **ind-projective scheme**. It is called the *affine* or *infinite Graß-mannian*. By its construction, it is the (infinite dimensional) moduli space for principal G-bundles on X together with a trivialization on $X \smallsetminus \{x_0\}$.

References for affine Graßmannians are the papers of the last section. In addition, we mention [KNR]. In that paper, it is shown inter alia that the moduli space $\mathscr{M}_G(X)$ is unirational, if G is a simple and simply connected affine algebraic group. (Under this hypothesis, there is only one topological type of principal G-bundles on X, so that we omitted it from the notation.) Also note that the affine Graßmannian is a projective moduli space for principal G-bundles on X. This fact permits to extend any family of principal G-bundles on X parameterized by the spectrum of the quotient field K of a discrete valuation ring R to a family of principal G-bundles on X parameterized by the spectrum of R. This observation is the first step in Heinloth's proof of the semistable reduction theorem [Hei02].

5.3 Picard groups and the Verlinde formula

Assume that G is a semi-simple affine algebraic group. The above techniques were used to determine the Picard groups of the moduli spaces $\mathscr{M}_G^\vartheta(X)$, $\vartheta \in \pi_1(G)$ ([LS], [BLS], [KN]). The complete picture for all reductive groups will be presented in the forthcoming paper [BHof]. If G is simple and simply connected, then $\mathrm{Pic}(\mathscr{M}_G(X)) \cong \mathbb{Z}$ (see [BK] for the explicit description of the ample generator).

Let us stay in the setting of a simple and simply connected affine algebraic group G. Write $\mathscr{L}$ for the ample generator of the Picard group of $\mathscr{M}_G(X)$. It is a natural task to compute the dimensions of the vector spaces $H^0(\mathscr{M}_G(X), \mathscr{L}^{\otimes m})$, $m \in \mathbb{N}$. It turns out that these vector spaces are isomorphic to spaces of so-called *conformal blocks* of theoretical physics. The dimensions of these vector spaces were predicted by the famous Verlinde formula. A main stimulus for work on principal G-bundles was to obtain a rigorous mathematical proof for the Verlinde formula. It was finally obtained in the papers [BL], [Fal02], and [KNR].

Appendix: One parameter subgroups

Let G be a reductive linear algebraic group and $\kappa\colon G \longrightarrow \mathrm{GL}(W)$ a representation. This gives the action

$$\begin{aligned}
\alpha\colon G \times \mathbb{P}(W) &\longrightarrow \mathbb{P}(W) = \big(W^{\vee} \smallsetminus \{0\}\big)/\mathbb{G}_m(k) \\
(g, [l]) &\longmapsto \big[\varrho^{\vee}(l)\big].
\end{aligned}$$

For a one parameter subgroup $\lambda\colon \mathbb{G}_m(k) \longrightarrow G$, there are the integral weights $\gamma_1 < \cdots < \gamma_{s+1}$ and the decomposition

$$W = W^1 \oplus \cdots \oplus W^{s+1},$$

$$W^i := \big\{\, w \in W \mid \varrho\big(\lambda(z)\big)(w) = z^{\gamma_i} \cdot w \ \forall z \in \mathbb{G}_m(k) \,\big\}, \ i = 1, \dots, s+1,$$

of W into non-trivial eigenspaces. We construct the flag

$$W_{\bullet}(\lambda) : \{0\} \subsetneqq W_1 \subsetneqq W_2 \subsetneqq \cdots \subsetneqq W_s \subsetneqq W_{s+1} = W,$$

$$W_i := \bigoplus_{j=1}^{i} W^j, \ i = 1, \dots, s+1,$$

and the vector

$$\alpha_{\bullet}(\lambda) = (\alpha_1, \dots, \alpha_s), \quad \alpha_i := \frac{\gamma_{i+1} - \gamma_i}{\dim(W)}, \quad i = 1, \dots, s,$$

of positive rational numbers. The pair $(W_{\bullet}(\lambda), \alpha_{\bullet}(\lambda))$ is called the *weighted flag of λ in W*.

If λ is a one parameter subgroup of G and $[l] \in \mathbb{P}(W)$ is the class of the linear form $l\colon W \longrightarrow k$, then

$$\mu\big(\lambda, [l]\big) := -\min\big\{\, \gamma_i, \ i = 1, \dots, s \mid l_{|W_i} \neq 0 \,\big\}.$$

A one parameter subgroup λ of G also defines the **parabolic subgroup**

$$Q_G(\lambda) := \big\{\, g \in G \mid \lim_{z \to \infty} \lambda(z) \cdot g \cdot \lambda(z)^{-1} \text{ exists in } G \,\big\}$$

of G. If $G = \mathrm{GL}(W)$, one readily verifies that $Q_G(\lambda)$ is the stabilizer of the flag $W_{\bullet}(\lambda)$.

PROPOSITION 6.3.1 ([Mum02], Proposition 2.7, p. 57). *In the above setting, let λ be a one parameter subgroup of G and $g \in G$. Then,*

$$\mu\big(\lambda, [l]\big) = \mu\big(g \cdot \lambda \cdot g^{-1}, [l]\big), \quad \forall [l] \in \mathbb{P}(W).$$

As an immediate consequence, we note:

COROLLARY 6.3.2. *Let ϱ: $\mathrm{SL}(V) \longrightarrow \mathrm{GL}(W)$ be a representation and λ, λ' two one parameter subgroups of $\mathrm{SL}(V)$, such that $(V_\bullet(\lambda), \alpha_\bullet(\lambda)) = (V_\bullet(\lambda'), \alpha_\bullet(\lambda'))$. Then,*

$$\mu(\lambda, [l]) = \mu(\lambda', [l]), \quad \forall [l] \in \mathbb{P}(W).$$

We stress that we take the weighted flags inside V and not inside W. The corollary means that weighted flags in V are the genuine test objects for the Hilbert-Mumford criterion for actions of $\mathrm{SL}(V)$.

Added in proof: Jochen Heinloth managed to prove the semistable reduction argument without using Behrend's conjecture. Thus, theorem 2.3.1 is valid in any characteristic. His preprint *Addendum to "semistable reduction for G-bundles on curves."* is available at http://staff. science.uva.nl/%7Eheinloth/. Most likely the restrictions on the characteristic in theorem 4.1.2 can be removed, too.

Bibliography

[AB] M.F. Atiyah, R. Bott, *The Yang-Mills equations over Riemann surfaces*, Philos. Trans. Roy. Soc. London Ser. A **308** (1983), 523–615.

[Ba01] V. Balaji, *Principal bundles on projective varieties and the Donaldson-Uhlenbeck compactification*, J. Differential Geom. **76** (2007), 351–98.

[Ba02] V. Balaji, *Lectures on principal bundles*, Moduli spaces and vector bundles, ed. L. Brambila-Paz, S.Bradlow, O. Garcia-Prada, and S. Ramanan, London Math. Soc. Lecture Notes Series 359, Cambridge University Press.

[BBN01] V. Balaji, I. Biswas, D.S. Nagaraj, *Principal bundles over projective manifolds with parabolic structure over a divisor*, Tohoku Math. J. (2) **53** (2001), 337–67.

[BBN02] V. Balaji, I. Biswas, D.S. Nagaraj, *Ramified G-bundles as parabolic bundles*, J. Ramanujan Math. Soc. **18** (2003), 123–38.

[BBNN] V. Balaji, I. Biswas, D.S. Nagaraj, P.E. Newstead, *Universal families on moduli spaces of principal bundles on curves*, Int. Math. Res. Not. **2006**, Art. ID 80641, 16 pp.

[BP] V. Balaji, A.J. Parameswaran, *Semistable principal bundles. II. Positive characteristics*, Transform. Groups **8** (2003), 3–36.

[BS] V. Balaji, C.S. Seshadri, *Semistable principal bundles. I. Characteristic zero*, Special issue in celebration of Claudio Procesi's 60th birthday, J. Algebra **258** (2002), 321–47.

[BL] A. Beauville, Y. Laszlo, *Conformal blocks and generalized theta functions*, Comm. Math. Phys. **164** (1994), 385–419.

[BLS] A. Beauville, Y. Laszlo, C. Sorger, *The Picard group of the moduli of G-bundles on a curve*, Compositio Math. **112** (1998), 183–216.

[Beh01] K. Behrend, *Semi-stability of reductive group schemes over curves*, Math. Ann. **301** (1995), 281–305.

[Beh02] K. Behrend, *The Lefschetz trace formula for the moduli stack of principal bundles*, http://www.math.ubc.ca/~behrend/thesis.ps, 96 pp.

[Beh03] K. Behrend, *The Lefschetz trace formula for algebraic stacks*, Invent. Math. **112** (1993), 127–49.

[BD] K. Behrend, A. Dhillon, *Connected components of moduli stacks of torsors via Tamagawa numbers*, arXiv:math/0503383, 34 pp.

[Bho01] U.N. Bhosle, *Generalised parabolic bundles and applications to torsionfree sheaves on nodal curves*, Ark. Mat. **30** (1992), 187–215.

[Bho02] U.N. Bhosle, *Tensor fields and singular principal bundles*, Int. Math. Res. Not. **2004:57** (2004), 3057–77.

[BR] U.N. Bhosle, A. Ramanathan, *Moduli of parabolic G-bundles on curves*, Math. Z. **202** (1989), 161–80.

[BGL] E. Bifet, F. Ghione, M. Letizia, *On the Abel-Jacobi map for divisors of higher rank on a curve*, Math. Ann. **299** (1994), 641–72.

[BHof] I. Biswas, N. Hoffmann, *Line bundles and Poincaré families on moduli of principal bundles on curves*, in preparation.

[BHol] I. Biswas, Y.I. Holla, *Harder-Narasimhan reduction of a principal bundle*, Nagoya Math. J. **174** (2004), 201–23.

[BK] A. Boysal, S. Kumar, *Explicit determination of the Picard group of moduli spaces of semistable G-bundles on curves*, Math. Ann. **332** (2005), 823–42.

[D] P. Deligne, *La conjecture de Weil. II.*, Inst. Hautes Études Sci. Publ. Math. **52** (1980), 137–252.

[DS] V.G. Drinfel'd, C. Simpson, *B-structures on G-bundles and local triviality*, Math. Res. Lett. **2** (1995), 823–9.

[Fal01] G. Faltings, *Stable G-bundles and projective connections*, J. Algebraic Geom. **2** (1993), 507–68.

[Fal02] G. Faltings, *A proof for the Verlinde formula*, J. Algebraic Geom. **3** (1994), 347–74.

[Fal03] G. Faltings, *Moduli-stacks for bundles on semistable curves*, Math. Ann. **304** (1996), 489–515.

[Fal04] G. Faltings, *Algebraic loop groups and moduli spaces of bundles*, J. Eur. Math. Soc. (JEMS) **5** (2003), 41–68.

[GLSS01] T. L. Gómez, A. Langer, A. Schmitt, I. Sols, *Moduli spaces for principal bundles in arbitrary characteristic*, Ad. Math. **219** (2008), 177–245.

[GLSS02] T.L. Gómez, A. Langer, A.H.W. Schmitt, I. Sols, *Moduli spaces for principal bundles in large characteristic*, 77 pp., to appear in the proceedings of the Allahabad International Workshop on Teichmüller Theory and Moduli Problems, 2006.

[GS01] T.L. Gómez, I. Sols, *Stable tensors and moduli space of orthogonal sheaves*, math.AG/0103150, 36 pp.

[GS02] T.L. Gómez, I. Sols, *Moduli space of principal sheaves over projective varieties*, Ann. of Math. **161** (2005), 1033–88.

[Groth] A. Grothendieck, *Sur la classification des fibrés holomorphes sur la sphère de Riemann*, Amer. J. Math. **79** (1957), 121–38.

[Har01] G. Harder, *Halbeinfache Gruppenschemata über Dedekindringen*, Invent. Math. **4** (1967), 165–91.

[Har02] G. Harder, *Chevalley groups over function fields and automorphic forms*, Ann. of Math. **100** (1974), 249–306.

[Hei01] J. Heinloth, *Über den Modulstack der Vektorbündel auf Kurven*, http://www.uni-due.de/~hm0002/dalklein.dvi, 64 pp.

[Hei02] J. Heinloth, *Semistable reduction for G-bundles on curves*, J. Algebraic Geom. **17** (2008), 167–83.

[Hei03] J. Heinloth, *Bounds for Behrend's conjecture on the canonical reduction*, Int. Math. Res. Notices 2008 article JD rnn045-17, 17 pp.

[Hei04] J. Heinloth, *Uniformization of G-bundles*, http://staff.science. uva.nl/~heinloth/Uniformization.pdf, 19 pp.

[HeSc] J. Heinloth, A.H.W. Schmitt, *The cohomology ring of moduli stacks of principal bundles over curves*, http://www.uni-due.de/~hm0002/HS_v1.pdf, 53 pp.

[Hesse] W.H. Hesselink, *Uniform instability in reductive groups*, J. reine angew. Math. **303/304** (1978), 74–96.

[Holla] Y.I. Holla, *Parabolic reductions of principal bundles*, math.AG/0204219, 38 pp.

[HL] D. Huybrechts, M. Lehn, *The geometry of moduli spaces of sheaves*, Aspects of Mathematics, E31, Friedr. Vieweg & Sohn, Braunschweig, 1997, xiv+269 pp.

[HM] D. Hyeon, D. Murphy, *Note on the stability of principal bundles*, Proc. Amer. Math. Soc. **132** (2004), 2205–13.

[IMP] S. Ilangovan, V.B. Mehta, A.J. Parameswaran, *Semistability and semisimplicity in representations of low height in positive characteristic* in *A tribute to C.S. Seshadri* (Chennai, 2002), 271–82, Trends Math., Birkhäuser, Basel, 2003.

[KR] C. Kaiser, J.-E. Riedel, *Tamagawazahlen und die Poincaréreihen affiner Weylgruppen*, J. reine angew. Math. **519** (2000), 31–9.

[Kempf] G.R. Kempf, *Instability in invariant theory*, Ann. of Math. **108** (1978), 299–316.

[Kumar] S. Kumar, *Kac-Moody groups, their flag varieties and representation theory*, Progress in Mathematics, 204, Birkhäuser Boston, Inc., Boston, MA, 2002, xvi+606 pp.

[KN] S. Kumar, M.S. Narasimhan, *Picard group of the moduli spaces of G-bundles*, Math. Ann. **308** (1997), 155–73.

[KNR] S. Kumar, M.S. Narasimhan, A. Ramanathan, *Infinite Grassmannians and moduli spaces of G-bundles*, Math. Ann. **300** (1994), 41–75.

[Lan01] A. Langer, *Semistable sheaves in positive characteristic*, Ann. of Math. **159** (2004), 251–76: Addendum, Ann. of Math. **160** (2004), 1211–3.

[Lan02] A. Langer, *Moduli spaces of sheaves in mixed characteristic*, Duke Math. J. **124** (2004), 571–86.

[Lan03] A. Langer, *Semistable principal G-bundles in positive characteristic*, Duke Math. J. **128** (2005), 511–40.

[Lan04] A. Langer, *Moduli spaces of principal bundles on singular varieties*, in preparation.

[Langt] S.G. Langton, *Valuative criteria for families of vector bundles on algebraic varieties*, Ann. of Math. **101** (1975), 88–110.

[LS] Y. Laszlo, C. Sorger, *The line bundles on the moduli of parabolic G-bundles over curves and their sections*, Ann. Sci. École Norm. Sup. (4) **30** (1997), 499–525.

[MS] V.B. Mehta, C.S. Seshadri, *Moduli of vector bundles on curves with parabolic structures*, Math. Ann. **248** (1980), 205–39.

[Mum01] D. Mumford, *Projective invariants of projective structures and applications*, Proc. Internat. Congr. Mathematicians (Stockholm, 1962), pp. 526–30, Inst. Mittag-Leffler, Djursholm, 1963.

[Mum02] D. Mumford, *The red book of varieties and schemes*, second, expanded edition, includes the Michigan lectures (1974) on curves and their Jacobians, with contributions by Enrico Arbarello, Lecture Notes in Mathematics, 1358, Springer-Verlag, Berlin, 1999, x+306 pp.

[Mum03] D. Mumford et al., *Geometric Invariant Theory*, third edition, Ergebnisse der Mathematik und ihrer Grenzgebiete (2), 34, Springer-Verlag, Berlin, 1994, xiv+292 pp.

[MN] D. Mumford, P.E. Newstead, *Periods of a moduli space of bundles on curves*, Amer. J. Math. **90** (1968), 1200–8.

[NS] M.S. Narasimhan, C.S. Seshadri, *Stable and unitary vector bundles on a compact Riemann surface*, Ann. of Math. **82** (1965), 540–67.

[N] P.E. Newstead, *Characteristic classes of stable bundles of rank 2 over an algebraic curve*, Trans. Amer. Math. Soc. **169** (1972), 337–45.

[NR] A. Nijenhuis, R.W. Richardson, Jr, *Deformations of Lie algebra structures*, J. Math. Mech. **17** (1967), 89–105.

[O] T. Ono, *On the relative theory of Tamagawa numbers*, Ann. of Math. **82** (1965), 88–111.

[PR] G. Pappas, M. Rapoport, *Twisted loop groups and their affine flag varieties*, Adv. Math. **219** (2008), 118–98.

[Raman] S. Ramaman, *The moduli spaces of vector bundles over an algebraic curve*, Math. Ann. **200** (1973), 69–84.

[RR] S. Ramanan, A. Ramanathan, *Some remarks on the instability flag*, Tohoku Math. J. **36** (1984), 269–291.

[Ram01] A. Ramanathan, *Stable principal bundles on a compact Riemann surface*, Math. Ann. **213** (1975), 129–52.

[Ram02] A. Ramanathan, *Moduli for principal bundles over algebraic curves* I-II, Proc. Indian Acad. Sci. Math. Sci. **106** (1996), 301–28, 421–49.

[Sch01] A.H.W. Schmitt, *Singular principal bundles over higher-dimensional manifolds and their moduli spaces*, Int. Math. Res. Not. **2002:23** (2002), 1183–209.

[Sch02] A.H.W. Schmitt, *A universal construction for moduli spaces of decorated vector bundles over curves*, Transform. Groups **9** (2004), 167–209.

[Sch03] A.H.W. Schmitt, *A closer look at semistability for singular principal bundles*, Int. Math. Res. Not. **2004:62** (2004), 3327–66.

[Sch04] A.H.W. Schmitt, *Singular principal G-bundles on nodal curves*, J. Eur. Math. Soc. (JEMS) **7** (2005), 215–51.

[Sch05] A.H.W. Schmitt, *Moduli spaces for semistable honest singular principal bundles on a nodal curve which are compatible with degeneration. A remark on U.N. Bhosle's paper: "Tensor fields and singular principal bundles"*, Int. Math. Res. Not. **2005:23** (2005), 1427–37.

[Sch06] A.H.W. Schmitt, *Geometric Invariant Theory and Decorated Principal Bundles*, Zurich Lectures in Advanced Mathematics, European Mathematical Society (EMS), approx. 396 pp.

[Serre] J.-P. Serre, *The notion of complete reducibility in group theory*, Part II of the *Moursund lectures* 1998, available at http://math.uoregon.edu/resources/serre/, 32 pp.

[Ses01] C.S. Seshadri, *Space of unitary vector bundles on a compact Riemann surface*, Ann. of Math. **85** (1967), 303–36.

[Ses02] C.S. Seshadri, *Quotient spaces modulo reductive algebraic groups*, Ann. of Math. **95** (1972), 511–56.

[S] Ch. Sorger, *Lectures on moduli of principal G-bundles over algebraic curves*, School on Algebraic Geometry (Trieste, 1999), 1–57, ICTP Lect. Notes, 1, Abdus Salam Int. Cent. Theoret. Phys., Trieste, 2000.

[TW] C. Teleman, C. Woodward, *Parabolic bundles, products of conjugacy classes and Gromov-Witten invariants*, Ann. Inst. Fourier **53** (2003), 713–48.

Part III
Research Articles

13

Beilinson type spectral sequences on scrolls

Marian Aprodu

Institute of Mathematics "Simion Stoilow",
P.O.Box 1-764, RO-014700, Bucharest, ROMANIA
and Şcoala Normală Superioară–Bucureşti, Calea Griviţei 21,
RO-010702, Bucharest, ROMANIA
e-mail: Marian.Aprodu@imar.ro

Vasile Brînzănescu

Institute of Mathematics "Simion Stoilow",
P.O.Box 1-764, RO-014700, Bucharest, ROMANIA
and University of Piteşti, Department of Mathematics and Informatics,
RO-110040 Pitesti, ROMANIA
e-mail: Vasile.Brinzanescu@imar.ro

Dedicated to Peter Newstead for his 65th anniversary

Abstract

We construct Beilinson type spectral sequences on scrolls and apply them to vector bundles on rational scrolls. The first application is a cohomological criterion for a vector bundle to be globally generated, Corollary 4.7. The relative canonical bundle can also be described by cohomological conditions, Proposition 4.8.

1 Introduction

In recent years, the theory of derived categories received considerable attention from the mathematical community. Remarkable works have been done in the attempt to understand the way the derived categories reflect the geometry of varieties. In some particular cases (projective bundles, Grassmannianns, quadrics etc), the derived categories have been described explicitly. In other cases, the description can be reduced to some known derived categories. For example, it was shown in [Orl] that if we control the derived category of a projective variety X, then we can control the derived category of any projective bundle on X. Orlov's result relies on a relative Beilinson spectral sequence obtained from a resolution of the diagonal inside the *fibered* product [Orl, p. 855–856].

The aim of this note is to construct slightly different Beilinson type sequences on scrolls, using resolutions of diagonals inside the *usual* product, see Section 4. Working with the usual product instead of the fibered-product has the advantage of giving information on the vanishing of Hochschild cohomology, [Ca]. The precise relationship between resolutions of diagonals and Beilinson type spectral sequences is recalled in Section 3. For the proof of the main result, Theorem 4.1, we use a general Lemma (stated and proved in Section 2) about extending sections in bundles on divisors. In the case of Hirzebruch surfaces, we recover the spectral sequence from [Bch, Section 1]. Section 4 also illustrates potential applications to the description of vector bundles. We obtain sufficient cohomological conditions ensuring that a bundle is globally generated, Corollary 4.7, and we prove that the bundle of relative differentials on a three-dimensional rational scroll is completely determined by the cohomology of suitable twists, Proposition 4.8. On projective spaces, the (classical) Beilinson spectral sequence lead to a number of precise descriptions of moduli spaces of stable bundles. For Hirzebruch surfaces, similar results were obtained in [Bch]. We expect that the same type of techniques could be used for studying moduli of vector bundles on higher-dimensional scrolls, where little is known, cf. [CMR]. Further possible applications are cohomological splitting criteria for rank-2 bundles, as done in [F] for Hirzebruch surfaces.

Acknowledgments. We would like to thank the referee for very useful remarks on the manuscript. MA was partially supported by the ANCS contract CEx05-D11-11/2005 and by a Humboldt Return Fellowship. VB was partially supported by the ANCS contract CEx05-D11-11/2005. Part of this work was done while the second named author was visiting the Mittag-Leffler Institute during the year on Moduli Spaces.

2 Extensions of sections on vector bundles.

In this Section we prove a general fact, which will be of essential use in the sequel. A version of it appeared in an implicit way, in [Bch, Section 1].

Lemma 2.1. *Let Z be a smooth, irreducible variety, Y an effective divisor on Z, F a vector bundle on Z, and $\sigma \in H^0(Y, F_{|Y})$. If G denotes the vector bundle on Z given by the extension:*

$$0 \longrightarrow F \longrightarrow G \longrightarrow \mathcal{O}_Z(Y) \longrightarrow 0,$$

corresponding to the image of σ by the canonical morphism $H^0(Y, F_{|Y}) \xrightarrow{\delta} H^1(Z, F(-Y))$, then there exists $u \in H^0(Z, G)$ such that $u_{|Y} = \sigma$. In *particular, $zero(u) = zero(\sigma) \subset Y$. Moreover, if $H^0(Z, F(-Y)) = 0$, and the extension is not trivial, then u is unique.*

Proof The sheaf G is obviously a vector bundle (it is a local problem, we denote $A = \mathcal{O}_{Z,x}$, and use $\mathrm{Ext}^1(A, A^n) = \bigoplus \mathrm{Ext}^1(A, A) = 0$). By a diagram chase, we remark that the image of σ via the natural map $H^0(Y, F_{|Y}) \to H^0(Y, G_{|Y})$ belongs to

$$\mathrm{Ker}\left(H^0(Y, G_{|Y}) \to H^1(Z, G(-Y))\right) = \mathrm{Im}\left(H^0(Z, G) \to H^0(Y, G_{|Y})\right).$$

In particular, there exists $u \in H^0(Z, G)$ whose restriction to Y equals σ.

For the uniqueness, we remark that, if $H^0(Z, F(-Y)) = 0$ and the map $H^0(Z, \mathcal{O}_Z) \to H^1(Z, F(-Y))$ is non-zero, then $H^0(Z, G(-Y)) = 0$, which shows that the restriction map $H^0(Z, G) \to H^0(Y, G_{|Y})$ is injective. $\square$

3 Resolutions of diagonals and Beilinson spectral sequences

The content of the present Section has the source in the classical construction of Beilinson's spectral sequences on projective spaces, cf. [Be1], and [OSS]. The general construction is well-known, and it was included here for the convenience of the reader, and for the coherence of the exposition.

Let X be a d-dimensional complex manifold, G be a rank-d vector bundle on the product $X \times X$, and $u \in H^0(X \times X, G)$ be a section vanishing exactly along the diagonal Δ_X, i.e. $zero(u) = \Delta_X$ scheme-theoretically. Then there is an exact Koszul complex:

$$0 \to \wedge^d G^\vee \to \cdots \to \wedge^2 G^\vee \to G^\vee \to \mathcal{O}_{X \times X} \to \mathcal{O}_{\Delta_X} \to 0, \qquad (3.1)$$

giving rise by truncation to a complex:

$$0 \to \wedge^d G^\vee \to \cdots \to \wedge^2 G^\vee \to G^\vee \to \mathcal{O}_{X \times X} \to 0.$$

Let $\mathcal{M}$ be a vector bundle on X, and denote $p_1, p_2 : X \times X \to X$ the projections on the first and, respectively, second component, and put

$$C_\mathcal{M}^0 = p_2^* \mathcal{M}, C_\mathcal{M}^{-1} = G^\vee \otimes p_2^* \mathcal{M}, \ldots, C_\mathcal{M}^{-d} = \wedge^d G^\vee \otimes p_2^* \mathcal{M}.$$

We obtain a complex of sheaves on $X \times X$,

$$0 \to C_\mathcal{M}^{-d} \to \cdots \to C_\mathcal{M}^{-1} \to C_\mathcal{M}^0 \to 0,$$

such that $H^i(C^*_\mathcal{M}) = 0$ except for $H^0(C^*_\mathcal{M}) = (p^*_2\mathcal{M})_{|\Delta_X}$, i.e.

$$H^0(C^*_\mathcal{M}) \cong \mathcal{M}$$

via the identification $X \cong \Delta_X$.

There are two spectral sequences abutting to the same limit

$$E^{p,q}_2 = H^p(R^q p_{1*}(C^*_\mathcal{M})) \Rightarrow \mathbf{R}^{p+q} p_{1*}(C^*_\mathcal{M})$$

and

$$'E^{p,q}_2 = R^p p_{1*}(H^q(C^*_\mathcal{M})) \Rightarrow \mathbf{R}^{p+q} p_{1*}(C^*_\mathcal{M}).$$

A word of warning on the notation is necessary. Throughout the above discussion, H^i is the cohomology of the corresponding complexes, and should not be mixed up with sheaf cohomology.

Since $'E^{p,q}_2$ all vanish except for $'E^{0,0}_2 = \mathcal{M}$, we obtain

$$E^{p,q}_1 \Rightarrow \begin{cases} \mathcal{M} & \text{if } p+q = 0 \\ 0 & \text{otherwise.} \end{cases}$$

Moreover, $E^{p,q}_1 = R^q p_{1*}(C^p_\mathcal{M}) = 0$ if $p \geq 1$ or $p \leq -d-1$ or $q \leq -1$ or $q \geq d+1$.

Definition 3.1. The spectral sequence E above is called *the Beilinson spectral sequence of $\mathcal{M}$ associated to the data* (G, u).

Remark 3.2. A distinction between different (G, u) is necessary, as this pair is in general not unique describing the diagonal by equations. The existence of a (G, u) usually fails for an arbitrary manifold X. The whole construction is therefore based on the assumption of the existence of such a pair.

Remark 3.3. Using the base-change formula, we can simplify the situation if the exterior powers of G are all external products, i.e. $\wedge^{-p} G^\vee$ is of type $p^*_1 A_{-p} \otimes p^*_2 B_{-p}$ for all p. Under this assumption, we have $E^{p,q}_1 = H^q(X, B_{-p} \otimes F) \otimes A_{-p}$. This is the case if X is a projective space, [Be1], [OSS], or a rational scroll, Corollary 4.6.

4 Beilinson spectral sequences on scrolls

4.1 General scrolls.

Let C be a smooth, irreducible, projective curve over $\mathbf{C}$, $\mathcal{E}$ be a rank-d vector bundle on C, $X = \mathbf{P}(\mathcal{E}^\vee) = \mathrm{Proj}(\mathrm{Sym}(\mathcal{E}))$ be the projective bundle associated to $\mathcal{E}$, and $\pi : X \to C$, $p : X \times X \to C \times C$ be the natural projections. Put $\mathcal{O}_X(H) = \mathcal{O}_{\mathbf{P}(\mathcal{E}^\vee)}(1)$, and, for any point $y \in C$, put $\mathcal{O}_X(R_y) = \pi^*(\mathcal{O}_C(y))$, where R_y is the fiber of π over y.

We recall that in this situation we have a relative Euler sequence:

$$0 \to \mathcal{O}_X(-H) \to \pi^*(\mathcal{E}^\vee) \to T_{X|C}(-H) \to 0, \tag{4.1}$$

where $T_{X|C}$ denotes the relative tangent bundle. In particular, we can compute the relative canonical bundle

$$\Omega_{X|C}^{d-1} \cong \mathcal{O}_X(-dH) \otimes \pi^*(\det(\mathcal{E})). \tag{4.2}$$

Since $\Delta_C \subset C \times C$ is a divisor, there exist a line bundle L on $C \times C$ and a global section $s_C \in H^0(C \times C, L)$ such that $L = \mathcal{O}_{C \times C}(\Delta_C)$, and Δ_C is the scheme of zeroes of s_C. Denote by

$$Y = X \times_C X \subset X \times X$$

the inverse image of Δ_C by p, which is also the scheme of zeroes of the global section $p^*(s_C)$ of the line bundle $p^*(L)$. For any $x \in X$, denoting $y = \pi(x)$ and identifying $X_x := \{x\} \times X \cong X$, we have

$$p^*(L)|_{X_x} \cong \mathcal{O}_X(R_y).$$

Consider the rank-$(d-1)$ vector bundle on X

$$F = p_1^*(T_{X|C}(-H)) \otimes p_2^*(\mathcal{O}_X(H)). \tag{4.3}$$

Notice that even though the vector bundle F changes while twisting $\mathcal{E}$ by a line bundle, its restriction to Y does not, as it only depends on X.

Orlov observed that, working over Y, similarly to the case of projective spaces, there exists a global section σ of $F_{|Y}$ whose scheme of zeroes is exactly Δ_X, [Orl, p. 855]. From the Koszul complex, he obtains a resolution of $\mathcal{O}_{\Delta_X}$ over $\mathcal{O}_Y$, which eventually yields to the Beilinson sequence [Orl].

We construct a resolution of $\mathcal{O}_{\Delta_X}$ over $\mathcal{O}_{X \times X}$ as follows. Using Lemma 2.1, we obtain a rank-d vector bundle G, given by an extension:

$$0 \longrightarrow F \longrightarrow G \longrightarrow p^*(L) \longrightarrow 0, \tag{4.4}$$

and a global section $u \in H^0(X \times X, G)$ such that $zero(u) = \Delta_X$. Hence we can apply (3.1).

We state now the main result of this note.

Theorem 4.1. *Notation as above. For any vector bundle $\mathcal{M}$ on X, there exists a spectral sequence, depending on $\mathcal{E}$:*

$$E_1^{p,q} = R^q p_{1*} \left(\wedge^{-p} G^\vee \otimes p_2^* \mathcal{M} \right) \Rightarrow \begin{cases} \mathcal{M} & \text{if } p+q = 0 \\ 0 & \text{otherwise} \end{cases},$$

whose terms can be be computed from a long exact sequence:

$$\cdots \to R^q p_{1*} \left(p^*(L^\vee) \otimes p_2^* \mathcal{M}((p+1)H) \right) \otimes \Omega_{X|C}^{-p-1}(-(p+1)H) \to \quad (4.5)$$

$$\to E_1^{p,q} \to H^q(X, \mathcal{M}(pH)) \otimes \Omega_{X|C}^{-p}(-pH) \to$$

$$R^{q+1} p_{1*} \left(p^*(L^\vee) \otimes p_2^* \mathcal{M}((p+1)H) \right) \otimes \Omega_{X|C}^{-p-1}(-(p+1)H) \to E_1^{p,q+1} \to \cdots$$

Moreover,

$$E_1^{0,q} \cong H^q(X, \mathcal{M}) \otimes \mathcal{O}_X, \qquad (4.6)$$

and

$$E_1^{-d,q} \cong R^q p_{1*} \left(p^*(L^\vee) \otimes p_2^* \mathcal{M}((-d+1)H) \right) \otimes \mathcal{O}_X(-H) \otimes \pi^*(\det(\mathcal{E})) \tag{4.7}$$

for all q.

Proof The existence of the spectral sequence $E_1^{p,q}$ follows from Section 3.

To obtain the long exact sequence (4.5), we dualize and take exterior powers in (4.4), and we consider the induced short exact sequences

$$0 \to \wedge^{-p-1} F^\vee \otimes p^*(L^\vee) \to \wedge^{-p} G^\vee \to \wedge^{-p} F^\vee \to 0, \qquad (4.8)$$

for all negative integers p.

The exact sequence (4.5) is then obtained applying p_{1*} to (4.8) tensored with $p_2^* \mathcal{M}$, and using the identifications

$$R^q p_{1*} \left(p^* L^\vee \otimes \wedge^{-p-1} F^\vee \otimes p_2^* \mathcal{M} \right) = \qquad (4.9)$$

$$= R^q p_{1*} \left(p^* L^\vee \otimes p_2^* \mathcal{M}((p+1)H) \right) \otimes \Omega_{X|C}^{-p-1}(-(p+1)H),$$

and, respectively,

$$R^q p_{1*} \left(\wedge^{-p} F^\vee \otimes p_2^* \mathcal{M} \right) = H^q(X, \mathcal{M}(pH)) \otimes \Omega_{X|C}^{-p}(-pH). \qquad (4.10)$$

The last assertion follows directly from (4.5) and (4.2). $\qquad \square$

Remark 4.2. Using the semicontinuity theorem, we have

$$R^q p_{1*} \left(p^*(L^\vee) \otimes p_2^* \mathcal{M}((p+1)H) \right) = 0$$

if

$$H^q(X, \mathcal{M}((p+1)H - R_y)) = 0,$$

for all y.

Remark 4.3. A modified version of Theorem 4.1 can be obtained by interchanging the factors in (4.3), and starting with

$$F = p_1^*(\mathcal{O}_X(H)) \otimes p_2^*(T_{X|C}(-H)).$$

The terms of the new spectral sequence E are computed from a long exact sequence similar to (4.5)

$$R^q p_{1*}\left(p^*(L^\vee) \otimes p_2^*(\mathcal{M} \otimes \Omega_{X|C}^{-p-1}(-(p+1)H))\right) \otimes \mathcal{O}_X((p+1)H) \to$$

$$\tag{4.11}$$

$$\to E_1^{p,q} \to H^q(X, \mathcal{M} \otimes \Omega_{X|C}^{-p}(-pH)) \otimes \mathcal{O}_X(pH) \to \ldots$$

Remark 4.4. The fact that the spectral sequences depend on $\mathcal{E}$ represents an advantage, as we can replace $\mathcal{E}$ by a twist in order to make vanish as many terms as possible, see Section 4.3. For projective spaces, one had sometimes to twist the bundles $\mathcal{M}$ in order to obtain a monad description, [OSS, Ch. II, Section 3.2].

Corollary 4.5. *Notation as in Theorem 4.1. The bundle $\mathcal{M}$ is globally generated if and only if $E_\infty^{-p,p} = 0$ for all $p \geq 1$.*

Proof We remark that the composed map

$$H^0(X, \mathcal{M}) \otimes \mathcal{O}_X = E_1^{0,0} \twoheadrightarrow E_\infty^{0,0} \hookrightarrow \mathcal{M}$$

coincides with the evaluation morphism, hence $\mathcal{M}$ is globally generated if and only if all the other terms of the filtration induced by the spectral sequence vanish.   □

4.2 Rational scrolls.

In the case $C = \mathbf{P}^1$, the Beilinson spectral sequence has a simpler form. Let $H = \mathcal{O}_X(1)$, and $\mathcal{O}_X(R) = \pi^*\mathcal{O}_{\mathbf{P}^1}(1)$ the two generators of the Picard group of X. With the notation above, we have

$$p^*(L) = p_1^*\mathcal{O}_X(R) \otimes p_2^*\mathcal{O}_X(R),$$

hence

$$R^q p_{1*}(p^*(L^\vee) \otimes p_2^*(\mathcal{M}((p+1)H))) =$$
$$H^q(X, \mathcal{M}((p+1)H - R)) \otimes \mathcal{O}_X(-R).$$

Applying Theorem 4.1, we obtain the following (compare to [Bch, Section 1]):

Corollary 4.6. *Let $\mathcal{E} \to \mathbf{P}^1$ be a rank-d vector bundle with $\deg(\mathcal{E}) = n$. Put $X = \mathbf{P}(\mathcal{E}^\vee)$, $H = \mathcal{O}_X(1)$, and let R be the class of a fiber of the ruling. Then, for any vector bundle $\mathcal{M}$ on X, there exists a spectral sequence,*

$$E_1^{p,q} \Rightarrow \begin{cases} \mathcal{M} & \text{if } p + q = 0 \\ 0 & \text{otherwise} \end{cases},$$

whose terms all lie in long exact sequences:

$$\cdots \to H^q(X, \mathcal{M}((p+1)H - R)) \otimes \Omega_{X|\mathbf{P}^1}^{-p-1}(-(p+1)H - R) \to E_1^{p,q} \to$$

$$(4.12)$$

$$\to H^q(X, \mathcal{M}(pH)) \otimes \Omega_{X|\mathbf{P}^1}^{-p}(-pH) \to$$

$$\to H^{q+1}(X, \mathcal{M}((p+1)H - R)) \otimes \Omega_{X|\mathbf{P}^1}^{-p-1}(-(p+1)H - R) \to \cdots$$

Moreover, for all q, we have

$$E_1^{0,q} \cong H^q(X, \mathcal{M}) \otimes \mathcal{O}_X,$$

and

$$E_1^{-d,q} \cong H^q(X, \mathcal{M}((-d+1)H - R)) \otimes \mathcal{O}_X(-H + (n-1)R).$$

An immediate consequence of Corollary 4.5 and Corollary 4.6 is the following cohomological sufficient criterion for globally generatedness.

Corollary 4.7. *Notation as in Corollary 4.6. The bundle $\mathcal{M}$ is globally generated if*

$$H^p(X, \mathcal{M}((p+1)H - R)) = H^p(X, \mathcal{M}(pH)) = 0,$$

for all $p \geq 1$.

4.3 Relative differentials.

We present an application of Corollary 4.6, and prove that the bundle of relative differentials on a three-dimensional rational scroll is completely determined by the cohomology of some of its twists, compare to [OSS, Ch. II, Section 3.2]). The fact of working in dimension three is not relevant, but helps simplifying the computation. We adopt the notation of Section 4, and [Sch, Section 1].

Proposition 4.8. *Let $X \to \mathbf{P}^1$ be a rational scroll defined by a rank-3 vector bundle*

$$\mathcal{E} \cong \mathcal{O}_{\mathbf{P}^1}(e_1) \oplus \mathcal{O}_{\mathbf{P}^1}(e_2) \oplus \mathcal{O}_{\mathbf{P}^1}(e_3)$$

on $\mathbf{P}^1$ with $e_1 \geq e_2 \geq e_3 \geq 0$, and $e_1 + e_2 + e_3 \geq 2$. Then any 2-bundle $\mathcal{V}$ on X with $c_1(\mathcal{V}) = K_X$, verifying the following conditions:

$$h^0(X, \mathcal{V}(H + R)) = 0, \tag{4.13}$$

$$h^1(X, \mathcal{V}(H + R)) = 0, \tag{4.14}$$

$$h^1(X, \mathcal{V}) = 0 \tag{4.15}$$

$$h^1(X, \mathcal{V}(H)) = 0 \tag{4.16}$$

$$h^1(X, \mathcal{V}(-H + R)) = 0, \tag{4.17}$$

is isomorphic to $\Omega^1_{X|\mathbf{P}^1}(-R)$.

Proof The condition $\deg(\mathcal{E}) = e_1 + e_2 + e_3 \geq 2$ ensures that the line bundle H is big; see, for example [Sch, Section 1].

Note that there is a natural isomorphism

$$\mathcal{V} \cong \mathcal{V}^\vee \otimes K_X. \tag{4.18}$$

which will be repeatedly used during the proof.

We apply Corollary 4.6 to the bundle $\mathcal{M} := \mathcal{V}(H + R)$, and prove that

$$\mathcal{M} \cong E_1^{-1,1} \cong \Omega^1_{X|\mathbf{P}^1}(H).$$

By (4.13), and (4.14) it follows that the terms $E_1^{0,q}$ in the Beilinson spectral sequence of $\mathcal{M}$ are zero, for $q = 0, 1$. Since the divisors $aH + bR$ are effective for any positive integers a and b, from (4.13) we obtain also $E_1^{p,0} = 0$ for all p.

From (4.15), and (4.17) it follows that $E_1^{-2,1} = 0$. Applying (4.16), and the Serre duality, we obtain $E_1^{-3,2} = 0$. The spectral sequence is as shown in the figure.

Since $E_\infty^{0,0} = E_1^{0,0} = 0$, it follows that

$$E_\infty^{-1,1} \cong E_1^{-1,1} \cong \Omega^1_{X|\mathbf{P}^1}(H) \subset \mathcal{M}.$$

But $\operatorname{rank}(\mathcal{M}) = \operatorname{rank}(\Omega^1_{X|\mathbf{P}^1}(H))$, $c_1(\mathcal{M}) = c_1(\Omega^1_{X|\mathbf{P}^1}(H))$, and both are locally free. Therefore $\Omega^1_{X|\mathbf{P}^1}(H) \cong \mathcal{M}$. $\qquad\square$

Remark 4.9. Using the Euler sequence (4.1) and the Leray spectral sequence for π, with the assumptions made on e_1, e_2, e_3, one proves that the bundle $\Omega^1_{X|\mathbf{P}^1}(-R)$ satisfies all the conditions of Proposition 4.8.

q

$E_1^{-3,3}$	$E_1^{-2,3}$	$E_1^{-1,3}$	$E_1^{0,3}$
0	$E_1^{-2,2}$	$E_1^{-1,2}$	$E_1^{0,2}$
$E_1^{-3,1}$	0	$E_1^{-1,1}$	0
0	0	0	0

p

Fig. 13.1. The Beilinson spectral sequence in Proposition 4.8.

Remark 4.10. The conditions imposed in the hypothesis might look not so natural at a first glance. However, since they are satisfied by the bundle $\Omega^1_{X|\mathbf{P}^1}(-R)$, they are as natural as they can be. We would like to stress that in this type of problems one cannot hope to obtain a full description of certain bundles by simple conditions on the characteristic classes which are only topological invariants. Supplementary conditions on the cohomology are very often necessary, even for the simpler case of projective spaces [OSS].

Bibliography

[Bel] Beilinson, A. (1978). *Coherent sheaves on* $\mathbf{P}^N$ *and problems of linear algebra,* Funkts. Analysis, **12**, pp. 214–216.

[Bch] Buchdahl, N. P. (1987). *Stable 2-bundles on Hirzebruch surfaces,* Math. Z., **194**, pp. 143–152.

[Ca] Căldăraru, A. and Willerton, S. (2007). *The Mukai pairing, I: a categorical approach,* Preprint arXiv:0707.2052.

[CMR] Costa, L. and Miró-Roig, R. M. (2001). *Moduli spaces of vector bundles on higher-dimensional varieties,* Michigan Math. J., **49**, pp. 605–620.

[Ha] Hartshorne, R. (1977). *Algebraic Geometry,* Grad. Texts in Math. 52, Springer Verlag.

[F] Fulger, M. (2007). *The Beilinson spectral sequence and applications,* Master Thesis, Şcoala Normală Superioară Bucureşti.

[OSS] Okonek, C., Schneider, M. and Spindler, H. (1980). *Vector bundles on complex projective spaces,* Progress in Math. 3, Birkhäuser.

[Orl] Orlov, D. O. (1992). *Projective bundles, monomial transformations, and derived categories of coherent sheaves,* Izv. Ross. Akad. Nauk Ser. Mat., **56**, pp. 852–862.

[Sch] Schreyer, F.-O. (1986). *Syzygies of canonical curves and special linear series,* Math. Ann., **275**, pp. 105–137.

14

Coherent systems on a nodal curve

Usha N. Bhosle

*School of Mathematics, Tata Institute of Fundamental Research,
Homi Bhabha Road, Mumbai 400 005, India.
e-mail:usha@math.tifr.res.in*

Abstract

Let E denote a torsionfree coherent sheaf of rank n, degree d and $V \subset H^0(E)$ be a subspace of dimension k on a nodal curve X. We show that for $k \leq n$ the moduli space of coherent systems (E, V) which are stable for sufficiently large values of a real parameter stabilizes. We study the nonemptiness and properties like irreducibility, smoothness, seminormality for this moduli space G_L.

1 Introduction

Coherent systems on smooth curves have been studied and are being studied extensively ([BG], [BOMN], [KN], [LN1], [LN2], [He], to name a few). A brief survey of coherent systems on smooth curves appears in this volume [Br]. In this paper, we initiate the study of coherent systems on a nodal curve. A coherent system on a nodal curve X of arithmetic genus g is a pair (E, V) where E denotes a torsionfree coherent sheaf of rank n, degree d on X and $V \subset H^0(E)$ is a subspace of dimension k. The (semi)stability condition for coherent systems depends on a real parameter $\alpha > 0$. It is easy to see that if (E, V) is α-semistable then $d \geq 0$. If (E, V) is α-stable, then for $k < n$ one has $d > 0$ and for $k \geq n$ one has $d > 0$ except in case $(E, V) = (\mathcal{O}, H^0(\mathcal{O}))$.

As in the case of smooth curves, for $k \leq n$ the moduli space $G(\alpha; n, d, k)$ of α-stable coherent systems (E, V) stabilises for suffiently large values of the real parameter, we call this moduli space $G_L = G_L(n, d, k)$. Let $U^s(n - k, d)$ be the moduli space of stable torsionfree sheaves of rank $n - k$ and degree d on X. Our main results may be summed up as follows.

Theorem 1.1. *Let $g \geq 2$ or $g = 1, (n - k, d) = 1,\ 0 < k < n,\ d > 0$. Then the moduli space $G_L(n, d, k)$ of coherent systems of type n, d, k, for 'large' α has an open subset G_L^s defined by*

$$G_L^s = \{(E, V) \in G_L \mid E/V \otimes \mathcal{O} \text{ is stable}\}.$$

G_L^s is isomorphic to a fibration over the $U^s(n-k, d)$ with fibres $Gr(k, d+ (n - k)(g - 1))$. It is nonempty if and only if $k \leq d + (n - k)(g - 1)$. If nonempty, it is irreducible of dim $\beta(n, d, k)$ and seminormal. If $(n - k, d) = 1$ then $G_L^s = G_L$. Let

$$G_L' = \{(E, V) \in G_L \mid E \text{ a vector bundle}\}.$$

Then G_L' is nonempty if $k \leq d + (n-k)(g-1)$ and is smooth of dimension $\beta(n, d, k)$ at each point. If it is nonempty and the desingularization of X has genus ≥ 1, then it is irreducible and smooth of dim $\beta(n, d, k)$.

Theorem 1.2. *Let $k = n \geq 2$. Let $\widetilde{G_L}$ be the moduli space of α-semistable coherent systems on the nodal curve X for large α and $\widetilde{G_L'}$ its open subset consisting of (E, V) with E locally free.*

(i) *$\widetilde{G_L}$ is empty if $0 < d < n$, G_L is empty if $0 \leq d \leq n$.*

(ii) *For $d = 0$, $\widetilde{G_L}$ consists of the single point $(\mathcal{O}^{\oplus n}, \mathbf{C}^n)$.*

(iii) *For $d = n$, $\widetilde{G_L}$ is irreducible of dimension n (not of expected dimension). For $g \geq 1$, the closed subset $\widetilde{G_L} - \widetilde{G_L'}$ is nonempty and of dimension $n - 1$.*

(iv) *For $d > n$, $\widetilde{G_L}$ is irreducible and $\widetilde{G_L'}$ is smooth of dimension $dn - n^2 + 1$ (expected dimension).*

2 Coherent systems

2.1 Preliminaries

Let X be a reduced irreducible projective curve of arithmetic genus $g \geq 2$ with at most ordinary nodes as singularities over an algebraically closed field of characteristic 0. For a torsionfree sheaf E on X, let $r(E), d(E), \mu(E)$ denote respectively the rank, degree and slope of E. For torsionfree sheaves E, F, let $Hom(E, F)$ denote the torsionfree sheaf of homomorphisms and let $\mathrm{Hom}\ (E, F) = H^0(X, Hom(E, F))$ denote the (global) homomorphisms from E to F. Similarly, $\mathrm{Ext}^i(E, F)$ and $Ext^i(E, F)$ will denote respectively the Ext groups and the sheaves. Let $E^* = Hom(E, \mathcal{O}_X), E^{**} = (E^*)^*$. Let K denote the dualizing sheaf (a line bundle).

Let A_j and m_j be respectively the local ring and the maximal ideal at y_j. The stalk E_{y_j} of a torsionfree sheaf E at a node y_j is isomorphic to $a_j(E)A_j \oplus b_j(E)m_j$ where $a_j(E), b_j(E)$ are nonnegative integers with $a_j(E) + b_j(E) = r(E)$. We will say that E is of *local type* b, $b = (b_1(E), \cdots, b_m(E))$, m being the number of nodes. Let x_j, z_j be the points of the (partial) desingularization lying over y_j.

Definition 2.1. A coherent system of type (n, d, k) on a nodal curve X is a pair (E, V) where E is a torsionfree sheaf of rank n, degree d and $V \subseteq H^0(E)$ is a subspace of dimension k.

Let α be a real number. The α–slope, α–stability (respectively α–semistability) of a coherent system (E, V) is defined as on the smooth curves ([Br], Definition 2.2). Since $(E, 0) \subset (E, V)$ is a subsystem, it follows that if (E, V) is α–(semi)stable, then $\alpha > (\geq) \; 0$.

A coherent system is α–polystable if it is the direct sum of coherent systems of the same slope. King and Newstead [KN] had constructed moduli spaces of α–semistable coherent systems (Brill-Noether pairs) over any curve.

Theorem 2.2. *Let X be a polarised algebraic curve and α a positive real number. For each (n, d, k), there exists a coarse moduli space $\widetilde{G}(\alpha; n, d, k)$ for families of α–semistable coherent systems of the fixed type (n, d, k).*

The open subscheme $G(\alpha; n, d, k) \subset \widetilde{G}(\alpha; n, d, k)$ consisting of the α–stable coherent systems is the coarse moduli space for α–stable coherent systems. For α–stable (E, V) we have End $(E, V) \cong \mathbb{C}$.

Proposition 2.3. *Let B, C be torsionfree sheaves on the nodal curve X.*
(1) $d(Hom(B, C)) = r(B)d(C) - r(C)d(B) + \sum_j b_j(B)b_j(C)$.
(2) (a) dim $\text{Ext}^1(B, C) = $ dim $H^1(Hom(B, C)) + 2\sum_j b_j(B)b_j(C)$.
(b) If one of B, C is locally free, then $\text{Ext}^1(B, C) = H^1(Hom(B, C))$.
(3) dim $\text{Ext}^i(B, C) = 2\sum_j b_j(B)b_j(C)$ for all $i \geq 2$.

Proof (1) and (2)(a) are proved in [B1], Lemma 2.5.
(2) As seen in [B1], one has $H^1(Hom(B, C)) \subset \text{Ext}^1(B, C)$. By 2(a), both these vector spaces have same dimension and hence must coincide.
(3) If B is locally free, then $Hom(B, C) = B^* \otimes C$ and $\text{Ext}^2(B, C) = H^2(B^* \otimes C) = 0$. If B is not locally free, then there exists an exact

sequence

$$(e) \qquad\qquad 0 \to B \to E \to T \to 0.$$

such that E is locally free, rank $E = r(B)$, $T = \oplus_j T_j \oplus T_0, T_j$ and T_0 are torsion sheaves, T_j is supported at $y_j \forall j, T_0$ is supported outside the singular set of $Y, t_j = dim_{\mathbf{C}} T_j = b_j, t_0 = dim_{\mathbf{C}} T_0$ ([B1], Lemma 2.5). Applying Hom $(-, C)$ to the exact sequence (e), one has the exact sequence

$$\to \operatorname{Ext}^i(E, C) \to \operatorname{Ext}^i(B, C) \to \operatorname{Ext}^{i+1}(T, C) \to \operatorname{Ext}^{i+1}(E, C)$$

for all $i \geq 2$. Since $\operatorname{Ext}^i(E, C) = 0$ for $i \geq 2$, it follows that $Ext^i(B, C) \cong Ext^{i+1}(T, C)$ for $i \geq 2$. For any $\mathcal{O}-$modules N, C, the spectral sequence

$$E_2^{p,q}(N, C) = H^p(Y, Ext_{\mathcal{O}}^q(N, C))$$

converges to the graded functor $\operatorname{Ext}^{p+q}(N, C)$.

The sheaf $Ext^0(T, C) = 0$ as T is a torsion sheaf and C is torsionfree. Since $Ext^q(T, C)$ is a torsion sheaf on the curve, $H^i(Ext^q(T, C)) = 0$ for $i \neq 0$. Hence $E_2^{p,q}(T, C)$ is nonzero only for $p = 0, q \geq 1$. one has

$$Ext^q(T, C) = \oplus_j \operatorname{Ext}_{A_j}^q(T_j, C_{y_j}) \oplus (\oplus_x \operatorname{Ext}_{\mathcal{O}_x}^q(T_x, C_x)),$$

where x varies over (smooth) points in the support of T_0. Since C_x is free over $\mathcal{O}_x$, it follows that $\operatorname{Ext}^q(T_x, \mathcal{O}_x) = 0$ for $q \geq 2$. Since $\operatorname{Ext}^q(k(y_j), A_j) = 0$, $\operatorname{Ext}^q(k(y_j), m_j) = 2k(y_j)$ for $q \geq 2$ and length $(T_j) = b_j(B)$, it follows that $Ext^q(T, C)) = \oplus_j 2b_j(B)b_j(C)k(y_j)$. Thus

$$E_2^{0,q}(T, C) := H^0(Ext^q(T, C)) = \oplus_j 2b_j(B)b_j(C)k(y_j),$$

for all $q \geq 2$. It is easy to see that all the boundary maps $d_2^{p,q}$ are zero. Thus the spectral sequence degenerates and converges to

$$\oplus_j 2b_j(B)b_j(C)k(y_j).$$

Therefore dim $\operatorname{Ext}^q(B, C) = \dim \operatorname{Ext}^{q+1}(T, C) = 2\sum_j b_j(B)b_j(C)$. $\qquad\square$

2.2 *Extensions of Coherent systems*

Let $(E, V), (E', V')$ be two coherent systems. There exists a long exact sequence of global homomorphisms and Ext groups [He]

$$(*) \quad 0 \to \operatorname{Hom}((E', V'), (E, V)) \to \operatorname{Hom}(E', E) \to \operatorname{Hom}(V', H^0(E)/V)$$

$$\to \operatorname{Ext}^1((E', V'), (E, V)) \to \operatorname{Ext}^1(E', E) \xrightarrow{h^*} \operatorname{Hom}(V', H^1(E))$$

$$\to \operatorname{Ext}^2((E', V'), (E, V)) \to \operatorname{Ext}^2(E', E) \to \operatorname{Hom}(V', H^2(E)) = 0$$

Note that Hom $(V', H^2(E)) = 0$ as $H^2(E) = 0$ on any curve. If E' or E is locally free, then one has $\text{Ext}^2(E', E) = 0$ and $\text{Ext}^1(E', E) = H^1(E'^* \otimes E)$. If X is nonsingular, these equalities hold for all torsionfree sheaves E, E'.

By standard results on abelian categories, the space of equivalence classes of extensions

$$0 \to (E_1, V_1) \to (E, V) \to (E_2, V_2) \to 0$$

is isomorphic to $\text{Ext}^1((E_2, V_2), (E_1, V_1))$.

Proposition 2.4. *Let (E_1, V_1) and (E_2, V_2) be coherent systems of type (n_1, d_1, k_1) and (n_2, d_2, k_2). Let*

$$\mathbf{H}_{21}^0 = \text{Hom}((E_2, V_2), (E_1, V_1)), \quad h_{21}^0 = dim \ \mathbf{H}_{21}^0$$

$$\mathbf{H}_{21}^2 = \text{Ext}^2((E_2, V_2), (E_1, V_1)), \quad h_{21}^2 = dim \ \mathbf{H}_{21}^2.$$

Then:

$$(1) \quad dim \ \text{Ext}^1((E_2, V_2), (E_1, V_1)) = h_{21}^0 + h_{21}^2 + C_{21}.$$

Here, $C_{21} := k_2 \chi(E_1) - k_1 k_2 - \chi(Hom(E_2, E_1)).$

$$= n_1 n_2(g-1) - d_1 n_2 + d_2 n_1 + k_2 d_1 - k_2 n_1(g-1) - k_1 k_2 - \sum_j b_j(E_1) b_j(E_2)$$

(2) If at least one of E_1 and E_2 is locally free then

$$\mathbf{H}_{21}^2 \cong [Ker(H^0(E_1^* \otimes K) \otimes V_2 \to H^0(E_1^* \otimes E_2 \otimes K))]^*.$$

(3) Let N_2 be the kernel of the map $V_2 \otimes \mathcal{O} \to E_2$. If E_1 is locally free, then

$$\mathbf{H}_{21}^2 \cong (H^0(E_1^* \otimes N_2 \otimes K))^*.$$

Proof This proposition is a generalization of [BOMN], Propositions 3.1, 3.2 (see also [Br], Proposition 2.7).
(1) The exact sequence in section 2.2 can be written as

$$0 \to \mathbf{H}_{21}^0 \to \text{Hom}(E_2, E_1) \to \text{Hom}(V_2, H^0(E_1)/V_1) \to$$

$$\to \text{Ext}^1((E_2, V_2), (E_1, V_1)) \to \text{Ext}^1(E_2, E_1) \to \text{Hom}(V_2, H^1(E_1)) \to$$

$$\to \mathbf{H}_{21}^2 \xrightarrow{f} \text{Ext}^2(E_2, E_1) \to 0.$$

By Proposition 2.3,

$$\dim \operatorname{Ext}^1(E_2, E_1) = h^1(Hom(E_2, E_1)) + 2\sum_j b_j(E_1)b_j(E_2)$$

and $\dim \operatorname{Ext}^2(E_2, E_1) = 2\sum_j b_j(E_1)b_j(E_2)$. Hence one has

$$h_{21}^0 - h^0(Hom(E_2, E_1)) + (k_2)(h^0(E_1) - k_1) - \dim \operatorname{Ext}^1((E_2, V_2), (E_1, V_1))$$

$$+ h^1(Hom(E_2, E_1)) + 2\sum_j b_j(E_1)b_j(E_2) - k_2(h^1(E_1))$$

$$+ h_{21}^2 - 2\sum_j b_j(E_1)b_j(E_2) = 0.$$

Therefore

$$h_{21}^0 + h_{21}^2 - \chi(Hom(E_2, E_1)) + k_2(\chi(E_1) - k_1)$$

$$- \dim \operatorname{Ext}^1((E_2, V_2), (E_1, V_1)) = 0$$

i.e.

$$\dim \operatorname{Ext}^1((E_2, V_2), (E_1, V_1)) = h_{21}^0 + h_{21}^2 + C_{21},$$

for all torsionfree sheaves E_1, E_2.

(2) If at least one of E_i is locally free, then $Hom(E_2, E_1) = E_2^* \otimes E_1$, $\operatorname{Ext}^2(E_2, E_1) = 0$ as $b_j(E_1)b_j(E_2) = 0$. Then one has (last four terms of the long exact sequence $(*)$)

$$H^1(Hom(E_2, E_1)) \to V_2^* \otimes H^1(E_1) \to \operatorname{Ext}^2((E_2, V_2), (E_1, V_1)) \to 0.$$

Dualizing, one gets

$$\operatorname{Ext}^2((E_2, V_2), (E_1, V_1)))^* \hookrightarrow V_2 \otimes (H^1(E_1))^* \to (H^1(E_2^* \otimes E_1))^*.$$

By Serre duality, this gives

$$0 \to \operatorname{Ext}^2((E_2, V_2), (E_1, V_1))^* \to V_2 \otimes H^0(E_1^* \otimes K) \to H^0(E_1^* \otimes E_2 \otimes K).$$

Therefore $\operatorname{Ext}^2((E_2, V_2), (E_1, V_1)) \cong$ dual of Ker $(V_2 \otimes H^0(E_1^* \otimes K) \to H^0(E_1^* \otimes E_2 \otimes K))$.

(3) Let I be the image of the map $V_2 \otimes \mathcal{O} \to E_2$. If E_1 is locally free, then

$$0 \to N_2 \otimes E_1^* \otimes K \to E_1^* \otimes V_2 \otimes K \to E_1^* \otimes I \otimes K \to 0$$

is exact. Taking global sections, one has the long exact sequence

$$0 \to H^0(N_2 \otimes E_1^* \otimes K) \to H^0(E_1^* \otimes V_2 \otimes K) \to H^0(E_1^* \otimes I \otimes K) \to \cdots$$

Since E_1 is locally free, $H^0(E_1^* \otimes I \otimes K) \subset H^0(E_1^* \otimes E_2 \otimes K)$. The last assertion of the proposition now follows. $\qquad \square$

Corollary 2.5. *Suppose that at least one of E_1, E_2 is locally free and $k_2 > 0, h^0(E_1^* \otimes K) \neq 0$. Then $h_{21}^2 \leq (k_2 - 1)(h^0(E_1^* \otimes K) - 1)$.*

Proof This can be proved as in [BOMN], Lemma 3.3. $\qquad \square$

Brill-Noether number The Brill-Noether number $\beta(n, d, k)$ is defined by

$$\beta(n, d, k) = n^2(g - 1) + 1 - k(k - d + n(g - 1))$$

Lemma 2.6. *For an $\alpha-$stable coherent system (E, V) of type (n, d, k) one has*

$$\dim \mathrm{Ext}^1((E, V), (E, V)) = \beta(n, d, k) - \sum_j b_j(E)^2$$

$$+ \dim \mathrm{Ext}^2((E, V), (E, V)).$$

Proof By Proposition 2.4(1), for $E_1 = E_2 = E$ one has

$$\dim \ \mathrm{Ext}^1((\mathrm{E}, \mathrm{V}), (\mathrm{E}, \mathrm{V})) = \mathrm{h}_{21}^0 + \mathrm{h}_{21}^2 + \mathrm{C}_{21}.$$

Since an $\alpha-$stable coherent system (E, V) is simple, $h_{21}^0 = 1$. Also $C_{21} = k(\chi(E)) - k^2 - \chi(EndE)$. The result follows by substituting these values. $\qquad \square$

Remark 2.7. Note that $\beta(n, d) - \sum_j b_j(E)^2$ is the 'predicted' dimension of the Brill-Noether locus $B(n, d, k) \cap U_b$ in the moduli space U_b of semistable sheaves of local type $b = (b_j(E))_j$ (if nonempty).

3 The moduli space $G(\alpha; n, d, k)$

Let $S := G(\alpha, n, d, k)$ and let $0 := (E, V) \in S$ denote a point defined by an α-stable coherent system. In a neighbourhood of $0 \in S$, the scheme $(S, 0)$ is embedded in a smooth variety $(N, 0)$ of the Zariski tangent space $\mathrm{Ext}^1((E, V), (E, V))$ ([He], Section 3). The ideal J of $(S, 0)$ in $(N, 0)$ satisfies $J \subset m_0^2$, m_0 being the maximum ideal at 0 in N. The subscheme $(S, 0)$ in $(N, 0)$ is defined by a local equation (in a neighbourhood of 0) with values in $\mathrm{Ext}^2(E, V), (E, V))$ ([He, Theorem 3.13]). Thus we have the following.

Proposition 3.1. *(1) The Zariski tangent space to the moduli space of coherent systems at a point defined by (E,V) is isomorphic to*

$$\mathrm{Ext}^1((E,V),(E,V))$$

[He, Theorem 3.12]

(2) If $\mathrm{Ext}^2((E,V),(E,V)) = 0$, then the moduli space of $\alpha-$stable coherent systems is smooth of dimension equal to $\dim \mathrm{Ext}^1((E,V),(E,V))$ in a neighbourhood of (E,V).

Remark 3.2. (1) If E is locally free and the map

$$h^* : \mathrm{Ext}^1(E,E) = H^1(EndE) \to \mathrm{Hom}(V,H^1(E))$$

is surjective, then from the exact sequence $(*)$ of section 2.2, it follows that $\mathrm{Ext}^2((E,V),(E,V)) = 0$. Consequently, the moduli space $G(\alpha; n, d, k)$ is smooth in a neighbourhood of such an (E,V).

(2) For every irreducible component $S' \subset S = G(\alpha; n, d, k)$ passing through $0 := (E,V)$ one has

$$
\begin{aligned}
\dim \mathrm{Ext}^1((\mathrm{E},\mathrm{V}),(\mathrm{E},\mathrm{V})) \;\geq\;& \dim_0 S' \\
\geq\;& \dim \mathrm{Ext}^1((\mathrm{E},\mathrm{V}),(\mathrm{E},\mathrm{V})) \\
& - \dim \mathrm{Ext}^2((\mathrm{E},\mathrm{V}),(\mathrm{E},\mathrm{V})),
\end{aligned}
$$

([He, Corollary 3.14]).

In particular, if

$$\dim_{(E,V)}S = \dim \mathrm{Ext}^1((E,V),(E,V)) - \dim \mathrm{Ext}^2((E,V),(E,V))$$

then S is locally a complete intersection at (E,V) (i.e. in a neighbourhood of (E,V)).

Corollary 3.3. *The dimension of the irreducible component S' of $G(\alpha; n, d, k)$ passing through (E,V) satisfies*

$$\dim \mathrm{Ext}^1((E,V),(E,V)) \geq \dim S' \geq \beta(n,d,k) - \sum_j b_j(E)^2$$

Proof This follows from the inequalities

$$
\begin{aligned}
\dim \mathrm{Ext}^1((E,V),(E,V)) \;\geq\;& \dim_0 S' \\
\geq\;& \dim \mathrm{Ext}^1((E,V),(E,V)) - \\
& - \dim \mathrm{Ext}^2((E,V),(E,V))
\end{aligned}
$$

and Lemma 2.6. $\qquad\square$

Remark 3.4. One has an exact sequence

$$0 \to H^1(Hom(E,E)) \xrightarrow{i} \text{Ext}^1(E,E) \to \sum_j H^0(Ext^1(E_{y_j}, E_{y_j})) \to 0,$$

(Proof of [B1], Lemma 2.5(B)). Dualizing, one gets

$$0 \to \sum_j H^0(Ext^1(E_{y_j}, E_{y_j}))^* \to (\text{Ext}^1(E,E))^* \xrightarrow{i^*} H^1(Hom(E,E))^* \to 0.$$

Let $h^* : \text{Ext}^1(E,E) \to V^* \otimes H^1(E)$ be the map in section 2.2 and let $f^* := h^* \circ i : H^1(Hom(E,E)) \to V^* \otimes H^1(E)$. Let $f : V \otimes H^1(E)^* \to (H^1(Hom(E,E)))^* = H^1(EndE)^*$ be the dual of f^*. By Serre duality, $H^1(E)^* \cong H^0(E^* \otimes K)$ and $H^1(EndE)^* \cong H^0((EndE)^* \otimes K)$. Then the map $f : V \otimes H^0(E^* \otimes K) \to H^0((EndE)^* \otimes K)$ is given as follows. For $v \in H^0(E) = \text{Hom}(\mathcal{O}, E), \ell \in H^0(E^* \otimes K) = \text{Hom}(E, K), g \in H^0(EndE)$, one has

$$(f(v \otimes \ell))(g) = \ell \circ g \circ v.$$

It is easy to check that for E locally free, under the isomorphism $(EndE) \cong (EndE)^*$, the map f is same as the map

$$f' : V \otimes H^0(E^* \otimes K) \to H^0((EndE) \otimes K)$$

given by tensoring of sections.

Remark 3.5. Note that $(EndE)$ is not isomorphic to $(EndE)^*$ if E is not locally free. In fact

$$d(EndE) = \sum_j b_j(E)^2 \neq -\sum_j b_j^2(E) = d((EndE)^*).$$

For example, $E = m_y$, the ideal sheaf of a node y, $End(E) \cong m_y^*$ has degree 1 while $(End(E))^* \cong m_y$ has degree -1.

Definition 3.6. The *Petri map* of a coherent system (E,V) is the map

$$h : V \otimes H^0(E^* \otimes K) \to (\text{Ext}^1(E,E))^*$$

dual to the map $h^* : \text{Ext}^1(E,E) \to V^* \otimes H^1(E)$ of section 2.2.

By Remark 3.4, if E is locally free, the Petri map is the map

$$f' : V \otimes H^0(E^* \otimes K) \to H^0((EndE) \otimes K))$$

given by multiplication of sections.

Proposition 3.7. *If the Petri map of (E, V) is injective and E is locally free then (E, V) is a smooth point of the moduli space $G(\alpha; n, d, k)$ and the moduli space is of dim $\beta(n, d, k)$ at (E, V).*

Proof This follows from Proposition 3.1 and Remark 3.2. Compare with [BOMN], Proposition 3.10; [Br], Proposition 2.8.   □

Remark 3.8. If f is injective, then h is injective. The injectivity of the Petri map h does not imply dim $\text{Ext}^2((E, V), (E, V)) = 0$ if E is not locally free. It only implies that dim $\text{Ext}^2((E, V), (E, V)) = 2 \sum b_j(E)^2$ (from sequence $(*)$ in section 2.2) so that dim $Ext^1((E, V), E(V)) = \beta(n, d, k) + \sum b_j(E)^2$. Thus if h is injective and E is not locally free, then $\dim_{(E,V)} G(\alpha; n, d, k) > \beta(n, d, k)$. Note that $\beta(n, d, k)$ is the dimension of $G(\alpha; n, d, k)$ at a smooth point corresponding to a coherent system with underlying sheaf locally free and Petri map injective. Consequently (E, V) is not a smooth point of $G(\alpha; n, d, k)$ belonging to a component containing such systems.

Corollary 3.9. *If the evaluation map $V \otimes \mathcal{O} \to E$ is injective $(\Rightarrow k \leq n)$ and E is locally free, then the petri map is injective and $G(\alpha; n, d, k)$ is smooth of dim $\beta(n, d, k)$ at (E, V). In particular this happens if $k = 1, E$ locally free.*

Proof There is an exact sequence

$$0 \to V \otimes K \xrightarrow{ev} E \otimes K \to (F \otimes K) \to 0.$$

Tensoring with the locally free sheaf E^* and taking $H^0(\ )$, one gets

$$0 \to V \otimes H^0(E^* \otimes K) \xrightarrow{f'} H^0((EndE) \otimes K) \to \cdots$$

Thus the Petri map f' is injective. By Proposition 3.7, $G(\alpha; n, d, k)$ is smooth of dimension $\beta(n, d, k)$ at (E, V).   □

3.1 Variation of α

Definition 3.10. $\alpha > 0$ is called a *critical value* if there exists a proper subsystem $(E', V') \subset (E, V)$ such that $\frac{k'}{n'} \neq \frac{k}{n}$ but $\mu_\alpha(E', V') = \mu_\alpha(E, V)$.

$\alpha = 0$ is also regarded a critical value.

Lemma 3.11. *Let α_i, α_{i+1} be two consecutive critical values. Then for all $\alpha, \alpha' \in (\alpha_i, \alpha_{i+1})$, a coherent system (E, V) is α−stable if and only if it is α'− stable .*

Proof We may assume that $\alpha_i < \alpha < \alpha' < \alpha_{i+1}$. Suppose that (E, V) is α-stable but not α'-stable. Then there exists a subsystem (E', V') such that $\mu_\alpha(E', V') < \mu_\alpha(E, V)$, but $\mu_{\alpha'}(E', V') > \mu_{\alpha'}(E, V)$. The function $f(x) = \mu_x(E', V') - \mu_x(E, V)$ is continuous in x and $f(\alpha) < 0, f(\alpha') > 0$. Therefore $f(\beta) = 0$ for some β between α, α'. Then β is a critical value lying between α_i, α_{i+1} contradicting the assumption that α_i, α_{i+1} are consecutive. $\qquad\square$

Lemma 3.12. *Assume that α is very small.*
i) If (E, V) is α-stable for small values of α, then E is semistable.
ii) If E is stable, then (E, V) is α-stable for small α.

Proof Easy. $\qquad\square$

Lemma 3.13. *(A) Assume $k < n$, $d = d(E), n = r(E)$.*
(i) If (E, V) is α-semistable, then $\alpha \leq \frac{d}{n-k}$ and hence $d \geq 0$.
(ii) If (E, V) is α-stable, then $\alpha < \frac{d}{n-k}$ and hence $d > 0$.
(B) Let $k \geq n$.
(i) If (E, V) is α-semistable, then $d \geq 0$.
If (E, V) is α-semistable and V generates a subsheaf of E of strictly smaller rank than n, then $\alpha \leq d(n-1)/k$ and hence $d > 0$ for $\alpha > 0$.
(ii) If (E, V) is α-stable, then $d > 0$ except in case

$$(E, V) = (\mathcal{O}, H^0(\mathcal{O})).$$

Proof The proof is similar to that of [BOMN], Lemma 4.2. $\qquad\square$

Remark 3.14. Let $k < n$. By Lemma 3.13, α is bounded. Hence there are finitely many critical values $\alpha_1, \cdots, \alpha_L$, where α_L is the largest critical value (strictly) less than $\frac{d}{n-k}$,

$$0 = \alpha_0 < \alpha_1 < \cdots < \alpha_L < \frac{d}{n-k}.$$

For $\alpha \neq \alpha_i$, the semistable coherent systems are stable. Lemma 3.11 implies that for all $\alpha \in (\alpha_i, \alpha_{i+1})$ the moduli spaces $G(\alpha; n, d, k)$ are the same ([Br], Proposition 2.9).

Corollary 3.15. *Let $k \geq n$. If $\alpha > \frac{d(n-1)}{k}$, then for any α-semistable coherent system (E, V), the torsionfree sheaf E is generically generated by V and the kernel N of the evaluation map $\varphi : V \otimes \mathcal{O} \to E$ has no global sections.*

Proof Recall that E is called generically generated by V if the image of the evaluation map $\varphi : V \otimes \mathcal{O} \to E$ has same rank as $r(E)$. It follows from Lemma 3.13(B)(i) that for $\alpha > \frac{d(n-1)}{k}$, $Im \; \varphi$ has rank n i.e., E is generically generated. The exact sequence

$$0 \to N \to V \otimes \mathcal{O} \overset{\varphi}{\to} E \to T \to 0,$$

where T is a torsion sheaf, gives

$$0 \to H^0(N) \to V \overset{H^0(\varphi)}{\to} H^0(E) \to \cdots.$$

Since $H^0(\varphi)$ is injective, it follows that $H^0(N) = 0$. $\qquad\square$

Proposition 3.16. *Let $k \geq n, \alpha > d(n-1)$. Then (E, V) is $\alpha-$stable if and only if both the following conditions hold.*

(a) E is generically generated by V.

(b) For every proper subsystem $(E', V') \subset (E, V)$ of type (n', d', k'), either $k'/n' < k/n$ or $\{\frac{k'}{n'} = \frac{k}{n}, \frac{d'}{n'} < \frac{d}{n}\}$.

In particular, $\alpha-$stability is independent of α for $\alpha > d(n-1), k \geq n$.

Proof Suppose that (E, V) is $\alpha-$stable. Then (a) holds by Corollary 3.15. Let $(E', V') \subset (E, V)$ be a subsystem. The $\alpha-$stability implies that

$$\frac{d}{n} \geq \frac{d'}{n'} + \alpha(\frac{k'}{n'} - \frac{k}{n}).$$

Replacing E' by a subsheaf generated by V', we may assume that $d' \geq 0$. If (b) does not hold, then either $\frac{k'}{n'} > \frac{k}{n}$ or $\{\frac{k'}{n'} = \frac{k}{n}$ and $\frac{d'}{n'} \geq \frac{d}{n}\}$. In the latter case, (E', V') contradicts α-stability for any α. Since $\alpha > d(n-1)$, in case $\frac{k'}{n'} > \frac{k}{n}$, one has $\alpha(\frac{k'}{n'} - \frac{k}{n}) > d(n-1)\frac{(k'n-n'k)}{nn'} \geq \frac{d}{n}$ (as $n-1 \geq n'$ and $k'n - n'k \geq 1$) so that $d/n < \alpha(\frac{k'}{n'} - \frac{k}{n})$, a contradiction to $\alpha-$stability as $d' > 0$. Thus (b) must hold for (E, V) $\alpha-$stable, $\alpha > d(n-1)$.

Conversely, suppose that (a) and (b) hold. Let $(E', V') \subset (E, V)$ be a subsystem. Since E is generically generated by sections, so is E/E', hence $d(E/E') \geq 0$. Therefore $d' \leq d$. Since $\alpha > d(n-1)$ one has

$$
\begin{aligned}
\frac{d}{n} - \frac{d'}{n'} + \alpha(\frac{k}{n} - \frac{k'}{n'}) \; & > \; \frac{d(n'-n)}{nn'} + \frac{d(n-1)}{nn'}(n'k - nk'). \\
& = \; \frac{d}{nn'}\{(n-1)(n'k - nk') - (n - n')\} \\
& \geq \; 0
\end{aligned}
$$

Thus (E, V) is $\alpha-$stable. $\qquad\square$

Remark 3.17. In view of Proposition 3.16, for $\alpha > d(n-1), k \geq n$, $\alpha-$stability condition is independent of α. Hence there is a critical value α_L

such that $G(\alpha; n, d, k) = G_L$ for $\alpha > \alpha_L$. Thus the α-range is divided into a finite set of intervals bounded by critical values

$$0 = \alpha_0 < \alpha_1 < \cdots < \alpha_L < \infty$$

such that
(i) if α_i, α_{i+1} are two consecutive critical values, the moduli spaces for any $\alpha, \alpha' \in (\alpha_i, \alpha_{i+1})$ coincide.
(ii) For any $\alpha, \alpha' \in (\alpha_L, \infty)$, the moduli spaces coincide.

4 The moduli space G_L for $k \leq n$

In this section, we first study the moduli space $G_L := G_L(n, d, k)$ for $k < n$. The case $k = n$ is dealt with at the end of the section. In view of Remark 3.14, let $\alpha_L < \alpha < \frac{d}{n-k}$. We start with listing a few results from [BG] which hold in the nodal case. Since their proofs are same as in the smooth case, we omit them giving appropriate references. As in [BG], we adapt the more general definition of a coherent system as a triple (E, V, φ) where the pair (E, V) is as before and $\varphi : V \otimes \mathcal{O} \to E$ is a morphism (not necessarily injective).

Proposition 4.1. *(1) There exists $\alpha_M \in [0, \frac{d}{n-k})$ such that for $\alpha > \alpha_M$, an α-semistable coherent system (E, V, φ) of type (n, d, k) determines an extension*

$$0 \to \mathcal{O}^{\oplus k} \xrightarrow{\varphi} E \to F \to 0$$

with F a torsionfree sheaf ([BG], Proposition 3.8). Moreover, F is locally free if and only if E is so. In fact, $\alpha_M = max\{\frac{d(k-1)}{k(n-k+1)}, \frac{d-\frac{n}{k}}{n-k}\}$ suffices.
(2) For any $\alpha \in (0, \frac{d}{n-k})$, if the extension determined by an α-semistable coherent system (E, V, φ) is given by $(e_1, \cdots, e_k) \in H^1(F^* \otimes \mathcal{O}^{\oplus k})$ then the vectors $e_1, \cdots, e_k$ are linearly independent in $H^1(F^*)$ ([BG], Proposition 3.9).
(3) There exists $\alpha'_M \geq \alpha_M$ such that if $\alpha > \alpha'_M$ then an α-semistable coherent system (E, V, φ) of type (n, d, k) determines an extension

$$0 \to \mathcal{O}^{\oplus k} \to E \to F \to 0$$

with F a semistable torsionfree sheaf. One has $h^0(F^*) = 0$ as F^* is semistable and of negative degree ([Br], Propositions 3.3, 3.4).

Corollary 4.2. *Every coherent system* $(E, V, \varphi) \in G_L(n, d, k)$ *determines an extension* $0 \to \mathcal{O}^{\oplus k} \to E \to F \to 0$ *with* F *a semistable torsionfree sheaf with* $h^0(F^*) = 0$.

Proof If $\alpha > \alpha_L > \alpha'_M$, the result holds by Proposition 4.1(3). If $\alpha_L < \alpha'_M$, since α'_M- stability is same as α stability for any $\alpha > \alpha_L$, (E, V, φ) is α-stable for all $\alpha'_M > \alpha > \alpha_L$ too. $\qquad\qquad\square$

Proposition 4.3. *Let* $0 \to \mathcal{O}^{\oplus k} \to E \to F \to 0$ *be an extension with* F *a stable torsionfree sheaf of rank* $n - k$, *degree* $d \geq 0$ *such that the extension is given by linearly independent* $e_1, \cdots, e_k \in H^1(F^*)$. *Then it determines an* α-stable *coherent system of type* (n, d, k) *for* $\alpha > \alpha_L$ *([BG], Proposition 4.2).*

Remark 4.4. (1) Even if (E, V, φ) is α−stable, F may not be α−stable. (2) If F is semistable but not stable, then (E, V, φ) may not be α−semistable.

4.1 The space $BGN^s(n, d, k)$

Denote the family of extensions of type as in the statement of Proposition 4.3 by $BGN^s(n, d, k)$. It is contained in an irreducible bounded family depending on at most $(n-k)^2(g-1) + 1 + k(d + (n-k)g - n) = \beta(n, d, k)$ parameters ([B2], Proof of Theorem 4.9). Corollary 4.2 and Proposition 4.3 imply that there is an inclusion

$$i_\alpha : BGN^s(n, d, k) \overset{i}{\hookrightarrow} G_L(n, d, k).$$

In view of Remark 4.4, this map may not be a surjection. Let

$$G_L^s(n, d, k) := \{(E, V, \varphi) \in G_L \mid E/\varphi(V \otimes \mathcal{O}) \text{ is stable}\}.$$

Corollary 4.2 and Proposition 4.3 say that

$$BGN^s(n, d, k) \cong G_L^s(n, d, k).$$

If $(n - k, d)$ are coprime, then $G_L^s = G_L$ so that $BGN^s(n, d, k) \cong G_L(n, d, k)$.

Proposition 4.5. *The space* $BGN^s(n, d, k)$ *and hence* $G_L^s(n, d, k)$ *is isomorphic to a Grassmannian bundle over the moduli space* $U^s(n-k, d)$ *of stable torsionfree sheaves of rank* $(n - k)$ *and degree* d *with fibre the*

Grassmannian of k-dimensional subspaces in a vector space of dimension $(d + (n - k)(g - 1))$. *In particular,*

$$\dim G_L(n, d, k) = \dim BGN(n, d, k) = fi(n, d, k).$$

Proof This can be proved as in [BG], Proposition 4.4. □

Corollary 4.6. *(1) For* $g \geq 2$, $G_L^s(n, d, k)$ *is nonempty if and only if* $k \leq d + (n - k)(g - 1)$.
(2) If $G_L^s(n, d, k)$ *is nonempty, then it is irreducible.*
If $g \geq 2, (n - k, d) = 1$, *then* $G_L(n, d, k)$ *is nonempty and irreducible if and only if* $k \leq d + (n - k)(g - 1)$.
(3) G_L^s *is seminormal. If* $(n - k, d) = 1$, *then* G_L *seminormal.*

Proof (1) Since $U^s(n - k, d) \neq \varphi$ for $g \geq 2$, ([B2], Theorem 2.5), it follows from Proposition 4.5 that $G_L^s(n, d, k)$ is nonempty if and only if $k \leq d + (n - k)(g - 1)$.
(2) $U^s(n - k, d)$ is irreducible [Re]. It follows from Proposition 4.5 that if $G_L^s(n, d, k)$ is nonempty, then it is irreducible. If $(n - k, d) = 1$, then $U^s(n - k, d) = U(n - k, d)$ and so $G_L(n, d, k) = G_L^s(n, d, k)$.
(3) The moduli space $U^s(n - k, d)$ is seminormal [Su]. Since seminormality is a local property and a product of a seminormal variety with a normal variety is seminormal [GT], it follows from Proposition 4.5 that $G_L^s(n, d, k)$ is a seminormal variety. □

Our next result needs GPBs (generalized parabolic bundles) on the desingularization Z of X. Let x_j, z_j be the points of Z lying over the node $y_j, j = 1, \cdots m$ of X. Recall that a generalized parabolic bundle (GPB) of rank r, degree d on Z is a tuple (E, q_j) where E is a vector bundle of rank r, degree d on Z and $q_j : E_{x_j} \oplus E_{z_j} \rightarrow Q_j(E)$ is a quotient, $Q_j(E)$ a k-vector space of dimension $r, j = 1, \cdots, m$. There is a (semi)stability notion for GPBs and there exist coarse moduli spaces for semistable GPBs which are normal projective varieties. There is a surjective morphism from the moduli space of GPBs on Z to the moduli space of torsionfree sheaves on X which is an isomorphism over the open subset corresponding to stable vector bundles on X.

Proposition 4.7. *Assume that* $g \geq 2$ *or* $g = 1, (n - k, d) = 1$. *Let*

$$G_L' = \{(E, V) \in G_L \mid E \text{ is locally free}\}.$$

(1) If $k \leq d + (n - k)(g - 1)$, *then* G_L' *is nonempty and is smooth of dimension* $\beta(n, d, k)$ *at each point.*

(2) Assume that the genus of the desingularization of X is at least 1. Then $G'_L \neq \varphi$ if and only if $k \leq d + (n-k)(g-1)$. If is nonempty, G'_L is irreducible and smooth of dimension $\beta(n, d, k)$ (compare [Br], Theorem 3.7).

Proof By Corollary 4.2, the evaluation map $V \otimes \mathcal{O} \to E$ is injective for $(E, V) \in G_L$. Then by Corollary 3.9, G_L is smooth of dimension $\beta(n, d, k)$ at $(E, V) \in G'_L$. It follows that every component of G'_L is smooth of dimension $\beta(n, d, k)$ (G'_L could possibly be a disjoint union of smooth open sets).

Let $G'^s_L = G'_L \cap G^s_L$. It is isomorphic to a Grassmannian bundle over the moduli space $U'^s(n-k, d)$ of stable vector bundles of rank $(n-k)$ and degree d with fibre the Grassmannian of k-dimensional subspaces in a vector space of dimension $d + (n-k)(g-1)$. Hence it is irreducible, smooth of dimension $\beta(n, d, k)$ and is contained in one component of G'_L. Since $U'^s(n-k, d)$ is nonempty for $g \geq 2$ or $g = 1, (n-k, d) = 1$ [B2], it follows that G'^s_L is nonempty if and only if $k \leq d + (n-k)(g-1)$. In view of these considerations, it suffices to prove that all irreducible components of $G'_L - G'^s_L$ have dimension strictly less than $\beta(n, d, k)$, we shall check this in the following.

We first construct a bounded family containing all the *BGN* extensions with F a vector bundle of rank $(n-k)$, of positive degree and $h^0(F^*) = 0$. Since we do not know how to do this directly on X, we use GPBs (generalized parabolic bundles) on the desingularization Z of X. Let $F_1 = F \otimes K_X$. Note that if F is semistable, then $F_1 = F \otimes K_X$ is semistable and $d(F) > 0$ implies that $h^0(F^*) = 0$. If $h^0(F^*) = 0$, then by Serre duality, $h^1(F_1) = 0$. The vector bundle F_1 corresponds to a unique GPB $(E_1, q_j(E_1))$ on Z. One has

$$0 \to F_1 \to \pi_*(E_1) \to \oplus Q \to 0,$$

where $Q = \oplus_j Q_j(F_1)$ is a torsion sheaf supported on nodes. The associated cohomology exact sequence shows that $h^1(F_1) = 0$ implies that $h^1(E_1) = 0$. It is known that all vector bundles E_1 with $h^1(E_1) = 0$ occur in a bounded family $\mathcal{E}$ parametrized by a scheme T whose irreducible components have dim $\leq (n-k)^2(g_Z - 1) + 1, g_Z \geq 1$. Moreover, if $T_s \subset T$ is the open subset corresponding to stable bundles, then all components of $T - T_s$ have dimension strictly less than $(n-k)^2(g_Z - 1) + 1$. Let $Gr_j = Gr(\mathcal{E}|_{x_j \times T} \oplus \mathcal{E}|_{z_j \times T}, n-k) \to T$ be the Grassmannian bundle of $n-k$ dimensional quotients of the vector bundle $\mathcal{E}|_{x_j \times T} \oplus \mathcal{E}|_{z_j \times T}$, on T and Gr the fibre product of Gr_j over T. Then the GPBs $(E_1, q_j(E_1))$

and therefore vector bundles F_1 (and hence F) occur in a bounded family parametrized by Gr. The family parametrized by Gr contains non locally free torsionfree sheaves too. The vector bundles in the family parametrized by Gr are parametrized by a fibre bundle $Gr' \subset Gr$ such that the fibres of $Gr' \to T$ are open subsets of fibres of Gr. Let S' (respectively S'') be the open subsets of Gr' corresponding to stable (respectively semistable) vector bundles F, then $S' \subset S''$. The extensions $0 \to \mathcal{O}^k \to E \to F \to 0$ with F semistable are then parametrized by a projective bundle P over S''. Then $P \mid_{S''-S'}$ parametrizes the extensions with F strictly semistable. There is an open subset P' of $P \mid_{S''-S'}$ which parametrizes all BGN extensions with F strictly semistable.

Recall that $T_s \subset T$ is the subset corresponding to stable E_1. If $(E_1, q_j(E_1))$ is a GPB which gives a vector bundle F_1 on Y and if E_1 is stable, then F_1 is stable. Therefore $Gr' \mid_{T_s}$ parametrizes a family of stable vector bundles F_1 on X. Thus $Gr' \mid_{T_s} \subset S'$. Since $(T - T_s)$ has all irreducible components of dimension $< (n-k)^2(g_z - 1) + 1$, the dimension of all components of $Gr' - Gr' \mid_{T_s} < (n-k)^2 + (g-1) + 1$. Since $S'' - S' \subset Gr' - S' \subset Gr' - Gr' \mid_{T_s}$; dim $S'' - S' < (n-k)^2 + (g-1) + 1$ (by dim we mean dimension of each component). Hence dim $P' = $ dim $P \mid_{S''-S'} < \beta(n,d,k)$. It follows that all components in $G'_L - G'^s_L$ have dimension strictly less than $\beta(n,d,k)$. This completes the proof of the proposition. $\qquad\Box$

Thus we have proved the following theorem.

Theorem 4.8. *Let $g \geq 2$ or $g = 1, (n-k,d) = 1, 0 < k < n, d > 0$. Then the moduli space $G_L(n,d,k)$ of coherent systems of type n, d, k, for 'large' α has an open subset G^s_L defined by*

$$G^s_L = \{(E,V) \in G_L \mid E/V \otimes \mathcal{O} \text{ is stable}\}.$$

G^s_L is isomorphic to a fibration over the $U^s(n-k,d)$ with fibres $Gr(k, d + (n-k)(g-1))$. It is nonempty if and only if $k \leq d + (n-k)(g-1)$. If nonempty, it is irreducible of dim $\beta(n,d,k)$ and seminormal. If $(n-k,d) = 1$ then $G^s_L = G_L$. Let

$$G'_L = \{(E,V) \in G_L \mid E \text{ a vector bundle}\}.$$

Then G'_L is nonempty if $k \leq d+(n-k)(g-1)$ and is smooth of dimension $\beta(n,d,k)$ at each point. If it is nonempty and the desingularization of X has genus ≥ 1, then it is irreducible and smooth of dim $\beta(n,d,k)$.

Proposition 4.9. *There is a* $1 - 1$ *correspondence between*

(a) isomorphism classes of BGN extensions of type (n, d, k) *with* F *stable but* E *not semistable and*

(b) the points in $G_L(n, d, k)$ *represented by triples* (E, V, φ) *such that* $F = E/\varphi(V \otimes \mathcal{O}_X)$ *is stable and the coherent system* (E, V) *is* α-*unstable for* $\alpha < \alpha_1$ *(or equivalently not* α-*semistable for some* $\alpha \geq 0$).

Thus, if

$$G_{L,0}^s(n, d, k) = \{(E, V) \in G_L \mid E/V \otimes \mathcal{O} \text{ stable}, (E, V) \ \alpha - \text{stable},$$

$$\alpha \in (0, \frac{d}{n - k})\}.$$

then

$$\{BGN^s(n, d, k)_{E \text{ unstable}}\} \leftrightarrow \{G_L(n, d, k) - G_{(L,0)}^s(n, d, k)\}.$$

Proof Same as [BG], Theorem 4.6. □

Remark 4.10. The stability of F is not used in proof of the implication $(b) \Rightarrow (a)$. Thus $(E, V) \in G_L$ but unstable for $\alpha < \alpha_1$ implies that E is not semistable (however it implies that F is semistable).

Remark 4.11. For $g = 0$ or $g = 1, (n - k, d) \neq 1$, $U^s(n - k, d)$ is empty, hence Theorem 1.1 may fail. The cases $g = 0, 1$ for nonsingular curve X were studied in detail by Lange and Newstead ([LN1], [LN2]). For $g = 0, X = \mathbb{P}^1$, a nonsingular curve. When X is a nodal curve of $g = 1$, the case $(n - k, d) \neq 1$ needs detailed investigation. Atiyah's classification of vector bundles on smooth elliptic curves fails in the nodal case.

4.2 Moduli space G_L for $k = n$

Theorem 4.12. *Let* $k = n \geq 2$. *Let* $\widetilde{G_L}$ *be the moduli space of* α-*semistable coherent systems on a nodal curve* X *for sufficiently large* α *and* $\widetilde{G'_L}$ *its open subset consisting of* (E, V) *with* E *locally free.*

(i) $\widetilde{G_L}$ *is empty if* $0 < d < n$, G_L *is empty if* $0 \leq d \leq n$.

(ii) *For* $d = 0$, $\widetilde{G_L}$ *consists of the single point* $(\mathcal{O}^{\oplus n}, \mathbf{C}^n)$.

(iii) *For* $d = n$, $\widetilde{G_L}$ *is irreducible of dimension* n *(not of expected dimension). For* $g \geq 1$, *the closed subset* $\widetilde{G_L} - \widetilde{G'_L}$ *is nonempty and of dimension* $n - 1$.

(iv) *For* $d > n$, $\widetilde{G_L}$ *is irreducible and* $\widetilde{G'_L}$ *is smooth of dimension* $dn - n^2 + 1$ *(expected dimension).*

Proof The theorem follows exactly as [BOMN], Theorem 5.6, we only indicate the modifications needed. For $d = n$, the irreducibility of $\widetilde{G_L}$ can be shown and the dimension of $\widetilde{G_L}$ can be computed as in [BOMN]. For $g = 0$, $\widetilde{G_L} = \widetilde{G'_L}$. For $d = n, g \geq 1$ and for each node y_j, , the coherent systems $(\mathcal{O}(y_j), H^0(\mathcal{O}(y_j)) \oplus (\mathcal{O}(P_1), H^0(\mathcal{O}(P_1)) \oplus \cdots \oplus (\mathcal{O}(P_{n-1}), H^0(\mathcal{O}(P_{n-1})))$, where $P_i, i = 1, \cdots, n-1$ are points of X, give distinct elements of $\widetilde{G_L} - \widetilde{G'_L}$. Hence $\dim \widetilde{G_L} - \widetilde{G'_L} \geq n - 1$. Since $\widetilde{G_L} - \widetilde{G'_L}$ is a proper subset of $\widetilde{G_L}$, it has dimension $n - 1$. For $d > n$, the smoothness of G'_L follows from Corollary 3.9. $\qquad\square$

Bibliography

[B1] Bhosle U. N. Maximal subsheaves of torsionfree sheaves on nodal curves. J. London Math. Soc. (2) 74 (2006) 59–74.

[B2] Bhosle U. N. Brill-Noether theory on nodal curves. Internat. J. Math., Vol. 18, No. 10 (2007) 1133–1150.

[BG] Bradlow S.B. and García-Prada O. An application of coherent systems to a Brill - Noether problem. J. Reine Angew. Math. 551 (2002) 123–143.

[BOMN] Bradlow S.B., García-Prada O. Muncoz V. and Newstead P.E. Coherent systems and Brill-Noether theory. Internat. J. Math. Vol. 14, No.7 (2003) 683–783.

[Br] Bradlow S.B. Coherent systems: A brief survey. Moduli spaces and vector bundles, ed. L. Brambila-Paz, S.Bradlow, O. Garcia-Prada, and S. Ramanan, London Math. Soc. Lecture Notes Series 359, Cambridge University Press.

[Gi] Giesekar, D. Stable curves and special divisors: Petri's conjecture. Invent math. 66 (1982) 251–275.

[GT] Greco S., Traverso, C. On seminormal schemes. Compositio Math., Vol. 40, No. 3 (1980) 325–365.

[He] He M. Espaces de modules de systémes cohérents. Internat. J. Math. 9 (1998) 545–598.

[KN] King A., Newstead P.E. Moduli of Brill-Noether pairs on algebraic curves. Internat. J. Math. 6 (1995) 733–748.

[LN1] Lange H., Newstead, P. E. Coherent systems of genus 0. Internat. J. Math. Vol. 15, No. 4 (2004) 409–424.

[LN2] Lange H., Newstead P.E. Coherent systems on elliptic curves. Internat. J. Math. Vol. 16, No. 7 (2005) 787–805.

[Re] Rego C.J. Compactification of the space of vector bundles on a singular curve. Comment Math. Helvetici 57 (1982) 226–236.

[Su] Sun X.: Degeneration of moduli spaces and generalized theta functions. J. Algebraic Geom. 9 , No. 3, (2000) 459–527.

15

Brill-Noether bundles and coherent systems on special curves

L. Brambila-Paz

CIMAT

Apdo. Postal 402, C.P. 36240. Guanajuato, Gto, México
e-mail: lebp@cimat.mx

Angela Ortega

Instituto de Fisico-Matematicas
Universidad Michoacana, Morelia, Mich., México
e-mail: ortega@ifm.umich.mx

Dedicated to Peter Newstead on the occasion of his 65th birthday.

Abstract

Let X be a smooth algebraic curve of genus $g \geq 2$ and Clifford index γ. For any $n \geq 2$ we define positive numbers $d_u(n, g, \gamma)$ and $d_\ell(n, g, \gamma)$, depending on n, g and γ and prove that if $d < d_\ell(n, g, \gamma)$ and $k > n$ the Brill-Noether locus $B(n, d, k)$ and the moduli spaces $G(\alpha : n, d, k)$ of α-stable coherent systems are empty for all $\alpha > 0$. Moreover, for $d_\ell(n, g, \gamma) \leq d < d_u(n, g, \gamma)$, $B(n, d, k)$ is non-empty if and only if $G(\alpha : n, d, k)$ is non-empty for any $\alpha > 0$. Furthermore, if $G(\alpha : n, d, k) \neq \emptyset$ for some $\alpha > 0$ then the maximal critical value α_L of the parameter α is 0. We prove that the lower bound $d_\ell(n, g, \gamma)$ is sharp. We give examples where the Brill-Noether number is negative and Brill-Noether bundles over special curves that define α-stable coherent systems for all $\alpha > 0$ exist.

1 Introduction

Let X be a smooth complex algebraic curve of genus $g \geq 2$ over $\mathbb{C}$ and $\mathcal{M}(n, d)$ the moduli space of stable bundles of rank n and degree d over X.

Brill Noether theory is concerned with the study of the subvarieties of the moduli space $\mathcal{M}(n, d)$ consisting of stable bundles having at least a specified number of independent sections. For line bundles the basic questions, concerning non-emptiness, connectedness, irreducibility, dimension, singularities, cohomology classes, etc., have been completely answered when the underlying curve is general (see [ACGH]).

For higher rank these questions are far from being answered even for a generic curve. The most complete results about non-emptiness, connectedness, irreducibility, dimension, singularities are those in [BGN], for the case $d \leq n$. In [M1] Mercat solved the problems of non-emptiness, dimension and singularities when $n < d \leq 2n$ and Teixidor i Bigas has answered these questions when the curve is generic and n, d and the number of sections satisfy some relations ([T]).

That the Brill-Noether bundles are related to coherent systems follows directly from the definitions. Recall that a coherent system of type (n, d, k) is a pair (E, V) where E is a vector bundle of rank n, degree d and V is a linear subspace of $H^0(X, E)$ of dimension k. The definition of stability of coherent systems depends on a real parameter α, which corresponds to the choice of linearization of a group action. This notion of stability leads to a family of moduli spaces. As α varies, the moduli spaces change only when α passes through one of a discrete set of points, called critical values. The relation between two consecutive moduli spaces is given by the so called "flips" (see [BGMN]). If $k < n$, the range of the parameter α is a finite interval and the family of moduli spaces has only a finite number of distinct members. If $k \geq n$ the range of the parameter is infinite; however, there is only a finite number of distinct moduli spaces. Denote by $G_0(n, d, k)$ the first member of the moduli spaces family and by $G_L(n, d, k)$ the last one.

It is well known that there is a map from the moduli space $G_0(n, d, k)$ to the semistable Brill-Noether locus and its image includes the entire stable Brill-Noether locus. Our purpose here is to study, on special curves, the relationship between bundle stability and the coherent systems stability for any $\alpha > 0$. By *special curve* we mean a curve with Clifford index $\gamma < [\frac{g-1}{2}]$.

In order to state our results we recall the following definitions.

Denote by $B(n, d, k)$ the Brill-Noether locus defined by vector bundles $E \in \mathcal{M}(n, d)$ such that $\dim H^0(X, E) \geq k$. A vector bundle E in $B(n, d, k)$ is called a Brill-Noether bundle of type (n, d, k). The expected dimension of $B(n, d, k)$ is the Brill-Noether number $\beta(n, d, k) := n^2(g - 1) + 1 - k(k - d + n(g - 1))$.

For any real number $\alpha > 0$ denote by $G(\alpha : n, d, k)$ the moduli space of α-stable coherent systems of type (n, d, k). Every irreducible component of $G(\alpha : n, d, k)$ has dimension of at least the Brill-Noether number $\beta(n, d, k)$ (see [BGMN, Corollary 3.14]).

For coherent systems (E, V) of type (n, d, k) with $k > n$ define $U(n, d, k)$, $G_{gg}(n, d, k)$ and $G_g(n, d, k)$ as:

$$U(n, d, k) := \{(E, V) : (E, V) \text{ is } \alpha\text{ - stable for all } \alpha > 0 \text{ and } E \text{ is stable}\},$$

$$G_{gg}(n, d, k) := \{(E, V) : (E, V) \text{ is generically generated with } H^0(I_E^*) = 0\},$$

and

$$G_g(n, d, k) := \{(E, V) : (E, V) \text{ is generated with } H^0(E^*) = 0\},$$

where I_E is the image of the evaluation map $V \otimes \mathcal{O} \to E$. By a *generated triple* we mean a triple (n, d, k) where $G_g(n, d, k) \neq \emptyset$. Note that $G(\alpha : n, d, k)$ can be non-empty for all $\alpha > 0$, but $U(n, d, k) = \emptyset$. However, if $U(n, d, k) \neq \emptyset$ then $G(\alpha : n, d, k) \neq \emptyset$ for all $\alpha > 0$ and $B(n, d, k) \neq \emptyset$. From the openness of α-stability, $U(n, d, k)$ is an open subset of $G_L(n, d, k)$. Moreover, if $G_L(n, d, k)$ is irreducible and $U(n, d, k) \neq \emptyset$, $U(n, d, k)$ is irreducible and hence $B(n, d, k)$ is irreducible. Our aim in this paper is to study $U(n, d, k)$, in particular non-emptiness, when d is given in terms of the Clifford index γ of the curve X and $k > n$.

Let X be a curve of genus g and Clifford index γ. For any $n \geq 2$ define the positive number $d_u(n, g, \gamma)$ by

$$d_u(n, g, \gamma) := \begin{cases} n + g - 1 + \frac{g-1}{n-1} & \text{if } \gamma \geq g - n \\ 2n + \gamma + \frac{\gamma}{n-1} & \text{if } \gamma < g - n. \end{cases} \tag{1.1}$$

Note that if $g \leq n$ then $d_u(n, g, \gamma) \leq 2n$. Brill-Noether bundles and coherent systems of slope ≤ 2 have been studied in [BGN] and [M1]; and [BGMN] and [BGMMN2] respectively. So we will assume that $g > n$.

Recall from [BGMN] that if $k > n$, there exists a value $\alpha_L \geq 0$ of the parameter α such that for all $\alpha, \alpha' > \alpha_L$, $G(\alpha : n, d, k) = G(\alpha' : n, d, k)$. In section 3 we prove:

Theorem 1.1. *Let X be a curve of genus g with Clifford index γ. Assume $d < d_u(n, g, \gamma)$. Then $B(n, d, k) \neq \emptyset$ if and only if $G(\alpha : n, d, k) \neq \emptyset$ for $\alpha > 0$. Furthermore, if $G(\alpha : n, d, k) \neq \emptyset$ for some $\alpha > 0$, then*

   (i) $G(\alpha : n, d, k) \neq \emptyset$ *for all $\alpha > 0$;*
   (ii) $\alpha_L = 0$;
   (iii) $G(n, d, k) := G_L(n, d, k) = U(n, d, k) = G_{gg}(n, d, k) \neq \emptyset$.

On general curves it was proved in [B-P, Theorem 3.9] that $B(n, d, k)$ and $G(\alpha : n, d, k)$ are empty for all $\alpha > 0$ if $\beta(1, d, n+1) < 0$. For curves of genus g, Clifford index γ and $n \geq 2$ we define the positive number $d_\ell(n, g, \gamma)$ as

$$d_\ell(n, g, \gamma) := \begin{cases} n + g - 1 & \text{if } \gamma \geq g - n \\ 2n + \gamma & \text{if } \gamma < g - n \end{cases} \tag{1.2}$$

and prove in section 3 the following theorem:

Theorem 1.2. *Let X be a curve of genus g and Clifford index γ. If $d < d_\ell(n, g, \gamma)$ and $k > n$, $B(n, d, k) = \emptyset$. Moreover, $G(\alpha : n, d, k)$ is empty for all $\alpha > 0$.*

In section 4 we use the dual span correspondence to prove that the lower bound $d_\ell(n, g, \gamma)$ is sharp.

Let $(G, W) \in G_g(n_G, d, n_G + n)$ be a generated coherent system of type $(n_G, d, n_G + n)$. Denote by $M_{W,G}$ the dual of the kernel of the evaluation map $W \otimes \mathcal{O} \to G$. The coherent system $(M_{W,G}, W^*)$ is called the *dual span* of (G, W). For general curves Butler, in [Bu], conjectures that if (G, W) is α-stable for $\frac{1}{\alpha} \gg 0$ and is general in $G_g(n_G, d, n_G + n)$ then $M_{W,G}$ is stable. From Theorem 1.2 we have that

- if $d < d_\ell(n, g, \gamma)$, then $M_{W,G}$ is not stable. Therefore $(M_{W,G}, W^*) \notin U(n, d, n_G + n)$.

It was proved in [Bu] and [M1] (see also [EL]) that if $d \geq 2gn_G$ and $W = H^0(G)$, then $M_{W,G}$ is stable. As a corollary of Theorem 1.1 we have the following theorem (see Theorem 4.1):

Theorem 1.3. *Let X be a curve of genus g and Clifford index γ and $(G, W) \in G_g(n_G, d, n_G + n)$. If $d_\ell(n, g, \gamma) \leq d < d_u(n, g, \gamma)$ then $(M_{W,G}, W^*) \in U(n, d, n_G + n)$.*

Let $(G, W) \in G_g(n_G, d, n_G + n)$ and $Grass(n_G, W^*)$ be the Grassmannian of n_G-dimensional linear subspaces of W^*. Denote by $TGrass$ the tangent bundle of $Grass(n_G, W^*)$. Any $(G, W) \in G_g(n_G, d, n_G + n)$ induces a morphism $\varphi_{W,G} : X \to Grass(n_G, W^*)$. Such morphisms have been an intensive object of study, in particular when $n_G = 1$, since they reflect the geometry of X. From our results on coherent systems we have (see Theorem 4.3) that if $d_\ell(n, g, \gamma) \leq d < d_u(n, g, \gamma)$ and G is stable then

- $\varphi_{W,G}^*(TGrass)$ is semistable. Moreover, if $n_G = 1$, $\varphi_{W,G}^*(TGrass)$ is stable.

From [RV, Lemma 1.5] we have (see Corollary 4.4):

Corollary 1.4. *Let X be a curve of genus g with Clifford index γ and $G_g(s, d, n + s) \neq \emptyset$. If $d_\ell(n, g, \gamma) \leq d < d_u(n, g, \gamma)$ and $d' \geq 0$ then $U(n, d + nd', n + s) \neq \emptyset$.*

In section 5 we consider particular curves and prove (see Corollary 5.1):

Corollary 1.5. *Let X be a smooth plane curve of degree $d \geq 5$. Then $U(2, d, 3) \neq \emptyset$. Moreover, the Brill-Noether number $\beta(2, d, 3) < 0$ for all $d \geq 5$.*

The *Clifford dimension* of X is the positive number

$$n := \min\{h^0(L) - 1 | L \text{ computes the Clifford index of } X\}.$$

If $n \geq 2$, a line bundle L which achieves the minimum is very ample ([ELMS, Lemma 1.1]) and defines an embedding $\varphi_L : X \to \mathbf{P}(H^0(L)^*)$ of X in $\mathbf{P}(H^0(L)^*)$. We call such an embedding a *Clifford embedding*.

For any $n \geq 3$ curves of Clifford dimension n with

- genus $g = 4n - 2$ and
- Clifford index $\gamma = 2n - 3$

exist (see [ELMS]). For such curves the Brill-Noether number $\beta(n, g - 1, n + 1)$ is negative and we have (see Corollary 5.2),

Corollary 1.6.   (i) $B(n, g - 1, n + 1) \neq \emptyset$;
   (ii) $G(\alpha : n, g - 1, n + 1) \neq \emptyset$ for all $\alpha > 0$;
   (iii) $\alpha_L = 0$;
   (iv) $U(n, g - 1, n + 1) = G_g(n, g - 1, n + 1) = G_L(n, g - 1, n + 1) \neq \emptyset$.
Moreover, if φ_L is a Clifford embedding, $\varphi_L^*(T\mathbf{P})$ is stable.

We give further examples where $U(n, d, n + 1) \neq \emptyset$ but the Brill-Noether number is negative. The examples are based on the existence of curves lying on K3 surfaces with linear systems g_d^n computing the Clifford index.

Notation

We will denote by K the canonical bundle over X, by I_E the image of the evaluation map $V \otimes \mathcal{O} \to E$, $H^i(X, E)$ by $H^i(E)$, $\dim H^i(X, E)$ by $h^i(E)$, the rank of E by n_E and the degree of E by d_E.

2 General results

The reader is invited to go to the survey and lecture notes on coherent systems and Brill-Noether theory in this volume. We also refer the reader to [M2], [BDGW], [Br] and [BGMN] for basic properties and results of Brill-Noether bundles of rank ≥ 2 and coherent systems on algebraic curves.

In this section we recall the main definitions and results that we will use.

For any real number $\alpha > 0$, define the α-slope of the coherent system (E, V) of type (n, d, k) by

$$\mu_\alpha(E, V) := \mu(E) + \alpha \frac{k}{n},$$

where $\mu(E) := d/n$ is the slope of the vector bundle E. A coherent subsystem $(F, W) \subset (E, V)$ is a coherent system such that F is a subsheaf of E and $W \subseteq V \cap H^0(F)$. A coherent system (E, V) is α-stable if for all proper coherent subsystems (F, W) we have $\mu_\alpha(F, W) < \mu_\alpha(E, V)$.

Denote the moduli space of α-stable coherent systems of type (n, d, k) by $G(\alpha : n, d, k)$ and by $\beta(n, d, k)$ the Brill-Noether number $\beta(n, d, k) := n^2(g - 1) + 1 - k(k - d + n(g - 1))$.

Recall that from the definition of α-stability and stability of a vector bundle we have that if (E, V) is a coherent system of type (n, d, k) then

(i) if $(E, V) \in G(\alpha : n, d, k)$ and E is stable then (E, V) is α'-stable for all $0 < \alpha' < \alpha$;

(ii) if E is stable and for all coherent subsystems $(F, W) \subset (E, V)$ one has that $\frac{\dim W}{n_F} \leq \frac{k}{n}$, then (E, V) is α-stable for all $\alpha > 0$.

Coherent systems with $k \leq n$ have already been studied in [BG] [BGMN] and [BGMMN1]. Precise conditions for the moduli space $G(\alpha : n, d, k)$ to be non-empty were given in [BGMN, Theorem 3.3]. It was also shown there that each non-empty moduli space has an irreducible component of the expected dimension.

Let (E, V) be a coherent system of type (n, d, k) with $k > n$. We shall say that (E, V) is generically generated if the image I_E of the evaluation map $V \otimes \mathcal{O} \to E$ has rank n. That is, we have the exact sequence

$$0 \longrightarrow I_E \longrightarrow E \longrightarrow \tau \longrightarrow 0, \tag{2.1}$$

where τ is a torsion sheaf. We shall say that (E, V) is generated if $\tau = 0$ and globally generated if in addition $V = H^0(E)$.

Let (E, V) be a generically generated coherent system. For any quotient coherent system (Q, W) we have the following commutative diagram

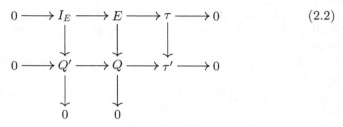

$$\hspace{11cm}(2.2)$$

where Q' is a vector bundle and τ' is a torsion sheaf.

Proposition 2.1. *Let (E, V) be a coherent system of type (n, d, k) with $k > n$.*

  (i) *If (E, V) is generically generated and $H^0(I_E^*) = 0$, then any quotient coherent system (Q, W) is generically generated and $H^0(I_Q^*) = 0$.*

 (ii) *If (E, V) is generated, then (Q, W) is generated.*

(iii) *If (E, V) is generated and $H^0(E^*) \neq 0$, then $(E, V) \cong (\mathcal{O}^s, H^0(\mathcal{O}^s)) \oplus (G, W)$ where $H^0(G^*) = 0$, $W = H^0(G) \cap V$ and (G, W) is generated with $1 \leq n_G < n$.*

(iv) *There exists a generated coherent subsystem (G, W) of (E, V) such that $\dim W > n_G$ and $H^0(G^*) = 0$.*

Proof (i) The image of I_E in Q is precisely I_Q, hence I_Q is a quotient of I_E and the result follows from the cohomology of the dual of diagram (2.2), since $Q' = I_Q$. Part (ii) is trivial. Part (iii) follows from the fact that if E is generated with $H^0(E) \neq 0$ and $H^0(E^*) \neq 0$ then $E \cong \mathcal{O}^s \oplus G$. Part (iv) follows by applying part (iii) to the coherent system (I_E, V). $\qquad\square$

We finish this section by recalling that the Clifford index is a numerical measure for line bundles to have the greatest number of independent sections for its degree. That is, the *Clifford index* for a line bundle L over X is given by $\mathrm{Cliff}(L) := d_L - 2(h^0(L) - 1)$ and the *Clifford index of X* is defined by

$$\gamma := \gamma(X) = \min\{\mathrm{Cliff}(L) \mid h^0(L) \geq 2, \; h^1(L) \geq 2\}.$$

We say that L *computes the Clifford index* if $\mathrm{Cliff}(L) = \gamma$ with both $h^0(L) \geq 2$ and $h^1(L) \geq 2$.

It is known that the Clifford index γ is bounded by $0 \leq \gamma \leq [\frac{g-1}{2}]$. For general curves $\gamma = [\frac{g-1}{2}]$ and the curve X is hyperelliptic if and only if $\gamma(X) = 0$. Brill-Noether bundles and coherent systems over hyperelliptic curves have been considered in [BMNO] and [BGMMN2] respectively (see also Remark 4.2).

3   Proof of Theorems 1.1 and 1.2

Let X be a smooth algebraic curve of genus g with Clifford index γ and $d_u(n, g, \gamma)$ and $d_\ell(n, g, \gamma)$ as in (1.1) and (1.2) respectively. We assume that $g > n$ and $k > n$. We shall prove Theorems 1.1 and 1.2 by means of a sequence of propositions. Most of the propositions are generalizations of the corresponding propositions for a general curve given in [B-P] and the proofs are similar.

Remark 3.1. Paranjape and Ramanan in [PR, Proposition 3.3] proved that the degree of any generated vector bundle E with $H^0(E^*) = 0$ has a lower bound. The lower bound for generically generated bundles E with $H^0(I_E^*) = 0$, will be given by the bound for I_E. That is, if E is generically generated and $H^0(I_E^*) = 0$ then

$$d_E \geq d_{I_E} \geq \begin{cases} n_E + g & \text{if} \quad h^1(\det I_E) = 0 \\ n_E + g - 1 & \text{if} \quad h^1(\det I_E) = 1 \\ 2n_E + \gamma & \text{if} \quad h^1(\det I_E) \geq 2 \end{cases}$$

since $n_E = n_{I_E}$.

First we prove a lemma that we will need.

Lemma 3.1. *Let (E, V) be a coherent system of type (n, d, k) with $k > n$ over X. Let (F, W) be a generically generated coherent system with $H^0(I_F^*) = 0$. If $n_F < n$ and $d \leq d_u(n, g, \gamma)$ then $\mu(F) \geq \mu(E)$; and if $d < d_u(n, g, \gamma)$, then $\mu(F) > \mu(E)$.*

Proof From Remark 3.1

$$\mu(F) \geq \mu(I_F) \geq \begin{cases} 1 + \frac{g}{n_F} & \text{if} \quad h^1(\det I_F) = 0 \\ 1 + \frac{g-1}{n_F} & \text{if} \quad h^1(\det I_F) = 1 \\ 2 + \frac{\gamma}{n_F} & \text{if} \quad h^1(\det I_F) \geq 2. \end{cases}$$

If $g - n \leq \gamma$, then

$$\mu(E) = \frac{d}{n} \leq \frac{d_u(n, g, \gamma)}{n} = 1 + \frac{g-1}{n-1} \leq \mu(F)$$

for $h^1(\det I_F) \leq 1$. If $h^1(\det I_F) \geq 2$ we have

$$\mu(E) \leq 1 + \frac{g-1}{n-1} \leq 2 + \frac{\gamma}{n-1} \leq 2 + \frac{\gamma}{n_F} \leq \mu(F),$$

since $g - n \leq \gamma$.

In the case $g - n > \gamma$,

$$\mu(E) \leq \frac{d_u(n,g,\gamma)}{n} = 2 + \frac{\gamma}{n-1} \leq \mu(F)$$

for $h^1(\det I_F) \geq 2$. If $h^1(\det I_F) \leq 1$, we have that

$$\mu(E) \leq 2 + \frac{\gamma}{n-1} \leq 1 + \frac{g-1}{n-1} \leq 1 + \frac{g-1}{n_F} \leq \mu(F),$$

since $g - n > \gamma$. Therefore, $\mu(F) \geq \mu(E)$; and if $d < d_u(n,g,\gamma)$, then $\mu(F) > \mu(E)$. □

Proposition 3.2. *Let (E,V) be a coherent system of type (n,d,k) with $k > n$ and $d < d_u(n,g,\gamma)$. Then $(E,V) \in G_{gg}(n,d,k)$ if and only if E is stable.*

Proof Suppose that (E,V) is generically generated with $H^0(I_E^*) = 0$ and Q a quotient bundle. By Proposition 2.1 one has that Q is generically generated and $H^0(I_Q^*) = 0$. By Lemma 3.1 $\mu(E) < \mu(Q)$, since $n_Q < n$. Therefore E is stable.

Conversely, assume E is stable and suppose $H^0(I_E^*) \neq 0$. From Proposition 2.1, we have that $I_E \cong \mathcal{O}^s \oplus G$ with $s \geq 1$ and G satisfying $H^0(G^*) = 0$. We have then that the coherent subsystem $(G, H^0(G) \cap V)$ is generated with $n_G < n$. From Lemma 3.1 it follows that $\mu(G) > \mu(E)$, which contradicts stability of E. Therefore, $s = 0$ and hence $H^0(I_E^*) = 0$. Similarly if $n_{I_E} < n$ we get a contradiction. Hence E is generically generated and $H^0(I_E^*) = 0$. □

Remark 3.2. Note that if $(E,V) \in G_{gg}(n,d,k)$ and $d = d_u(n,g,\gamma)$, then E is semistable. Indeed, if Q is a quotient bundle, as in the proof of Proposition 3.2, $\mu(Q) \geq \mu(E)$.

Proposition 3.3. *Let (E,V) be a coherent system of type (n,d,k) with $k > n$ and $d < d_u(n,g,\gamma)$. The following statements are equivalent*

   (i) *(E,V) is generically generated and $H^0(I_E^*) = 0$;*
  (ii) *E is stable;*
 (iii) *(E,V) is α-stable for $\alpha > 0$.*

Proof $(i) \Leftrightarrow (ii)$ is Proposition 3.2.

$(ii) \Rightarrow (iii)$. Let $(F, W) \subset (E, V)$ be a coherent subsystem of (E, V) with $n_F < n_E$. If $\dim(W) \geq n_F + 1$ then by Proposition 2.1(iv) there exists a generated coherent subsystem (F', W') of (F, W) with $H^0(F'^*) = 0$. From Lemma 3.1 we obtain $\mu(F') > \mu(E)$ which contradicts stability of E. Hence, $\dim W \leq n_F$ and with the stability of E, (E, V) is α-stable for any $\alpha > 0$.

$(iii) \Rightarrow (i)$. Suppose that $I_E \cong \mathcal{O}^s \oplus G$ with $s \geq 1$ and $H^0(G^*) = 0$. From Proposition 2.1 and Lemma 3.1 it follows that $\mu(G) > \mu(E)$. Note that the subsystem $(G, H^0(G) \cap V) \subset (E, V)$ is of type $(n_G, d_G, k - s)$. From the α-stability of (E, V) for $\alpha > 0$ we get

$$\alpha \left(\frac{k - s}{n_G} - \frac{k}{n} \right) < \mu(E) - \mu(G) < 0.$$

Hence, $n(k - s) < n_G k$, and $k(n - n_G) < ns$ which is a contradiction since $s \leq n - n_G$ and $k > n$. Therefore, $s = 0$ and then, $H^0(I_E^*) = 0$.

Suppose now $n_{I_E} < n$ and let $(I_E, V) \subset (E, V)$ be the coherent subsystem defined by I_E. From the α-stability of (E, V) we obtain the following inequality

$$\mu(I_E) - \mu(E) < \alpha \left(\frac{k}{n} - \frac{k}{n_{I_E}} \right).$$

If $n_{I_E} < n$, then $\mu(I_E) - \mu(E) < 0$ which is a contradiction since from Lemma 3.1 we have that $\mu(I_E) \geq \mu(E)$. Therefore $n_{I_E} = n$ and hence (E, V) is generically generated. $\square$

We are now able to prove Theorems 1.1 and 1.2.

Proof of Theorem 1.1. From Proposition 3.3 one has that $B(n, d, k) \neq \emptyset$ if and only if $G(\alpha : n, d, k) \neq \emptyset$ for $\alpha > 0$. Parts $(i), (ii)$ and (iii) follow also from Proposition 3.3 once we notice that 3.3 (iii) is for any $\alpha > 0$. $\square$

Proof of Theorem 1.2. Suppose that $(E, V) \in G(\alpha : n, d, k)$ with $d < d_\ell(n, g, \gamma)$. From Proposition 3.3, (E, V) is generically generated and $H^0(I_E^*) = 0$. If $\gamma \geq g - n$ then $d < n + g - 1 = d_\ell(n, g, \gamma)$. But by Remark 3.1 $d \geq n + g - 1$ if $h^1(\det I_E) \leq 1$ and $d \geq g + n$ if $h^1(\det I_E) \geq 2$, obtaining a contradiction. Similarly if $\gamma < g - n$ we get a contradiction. Therefore, $G(\alpha : n, d, k) = \emptyset$ for all $\alpha > 0$. $\square$

4 Dual span

Let (G, W) be a generated coherent system of type $(n_G, d, n_G + n)$ with $n \geq 1$ and $H^0(G^*) = 0$. If we denote by $M_{W,G}$ the dual of the kernel of the evaluation map $W \otimes \mathcal{O} \to G$, we have the exact sequence

$$0 \to M_{W,G}^* \to W \otimes \mathcal{O} \to G \to 0. \qquad (4.1)$$

If $W = H^0(G)$ we will denote $M_{W,G}$ by M_G. The coherent system $(M_{W,G}, W^*)$ is called the *dual span* of (G, W). For such coherent systems we have:

Theorem 4.1. *If $d_\ell(n, g, \gamma) \leq d < d_u(n, g, \gamma)$, then $M_{W,G}$ is stable and the coherent system $(M_{W,G}, W^*)$ is α-stable for all $\alpha > 0$. Moreover, if $d < d_\ell(n, g, \gamma)$ then $M_{W,G}$ is not stable.*

Proof The theorem follows directly from Propositions 3.2 and 3.3 since $M_{W,G}$ is generated with $H^0(M_{W,G}^*) = 0$ and $d < d_u(n, g, \gamma)$. The second part follows from Theorem 1.2. $\qquad \square$

Remark 4.1. Note that if

$$\max\{d_\ell(n_G, g, \gamma), d_\ell(n, g, \gamma)\} \leq d < \min\{d_u(n_G, g, \gamma), d_u(n, g, \gamma)\}$$

then G and $M_{W,G}$ are stable (see [Bu, Conjecture 2]).

From Theorems 4.1 and 1.1 we have:

Theorem 4.2. *Let X be a curve of genus g and Clifford index γ and $G_g(n_s, d, n_s + n) = \emptyset$. If $d_\ell(n, g, \gamma) \leq d < d_u(n, g, \gamma)$ then*

(i) $B(n, d, n_s + n) \neq \emptyset$;
(ii) $G(\alpha : n, d, n_s + n) \neq \emptyset$ for all $\alpha > 0$;
(iii) $\alpha_L = 0$;
(iv) $G(n, d, n_s + n) := G_L(n, d, n_s + n) = U(n, d, n_s + n) = G_{gg}(n, d, n_s + n) \neq \emptyset$.

Remark 4.2.   (i) Note that if $(E, V) \in G_{gg}(n, d, k)$ and $d = d_u$ (n, g, γ), then E is semistable (see Remark 3.2). Moreover, if $k = n + 1$ and (E, V) is generated, from [B-P, Proposition 2.5], (E, V) is α-stable for all $\alpha > 0$.

(ii) If $\gamma = 0$, then $d_\ell(n, g, \gamma) = d_u(n, g, \gamma) = 2n$. The line bundles that compute the Clifford index are precisely the powers of the hyperelliptic line bundle $\widetilde{L}$. Note that $h^0(\widetilde{L}^{\otimes s}) = s + 1$ for $1 \leq s \leq g$. Furthermore, $M_{\widetilde{L}}$ is semistable (see [PR]) and $(M_{\widetilde{L}^{\otimes s}}, H^0(\widetilde{L}^{\otimes s}))$ is α-stable for all $\alpha > 0$ (see [BGMMN2]).

A generated coherent system $(G, W) \in G_g(n_G, d, n_G + n)$ defines a map $\varphi_{W,G} : X \to Grass(n_G, W^*)$ where $Grass(n_G, W^*)$ is the Grassmannian of n_G-dimensional linear subspaces of W^*. Note that $Grass(n_G, W^*) = Grass(n, W)$. Denote by $TGrass$ the tangent bundle of $Grass(n_G, W^*)$. From the above results on coherent systems we have the following theorem.

Theorem 4.3. *Let X be a curve of genus $g \geq 2$ with Clifford index γ and $(G, W) \in G_g(n_G, d, n_G + n)$. If G is stable and $d_\ell(n, g, \gamma) \leq d < d_u(n, g, \gamma)$ then $\varphi_{W,G}^*(TGrass)$ is semistable. Moreover, if $n_G = 1$, $\varphi_{W,G}^*(TGrass)$ is stable.*

Proof Let S and Q be the universal subbundle and quotient bundle, respectively, over $Grass(n_G, W^*)$. Such bundles fit in the following exact sequence

$$0 \to S \to W^* \otimes \mathcal{O}_{Grass} \to Q \to 0 \qquad (4.2)$$

over $Grass(n_G, W^*)$. It is well known that the tangent bundle $TGrass$ can be described as $TGrass \cong S^* \otimes Q$. Since $\varphi_{W,G}^*(S) = G^*$ and $\varphi_{W,G}^*(Q) = M_{W,G}$,

$$\varphi_{W,G}^*(TGrass) = M_{W,G} \otimes G.$$

By the hypothesis G is stable and from Theorem 4.1 $M_{W,G}$ is stable, therefore $\varphi_{W,G}^*(TGrass)$ is semistable.

In particular if G is a line bundle, (G, W) defines the map $\varphi_{W,G} : X \to \mathbf{P}(W^*) = \mathbf{P}^n$. Hence $\varphi_{W,G}^*(T\mathbf{P}^n)$ is stable since

$$\varphi_{W,G}^*(T_\mathbf{P}) \cong M_{W,G} \otimes G.$$

$\square$

For generated triples $(s, d, n + s)$ with $d_\ell(n, g, \gamma) \leq d < d_u(n, g, \gamma)$, we have from Theorem 4.2 and [RV, Lemma 1.5] the following corollary.

Corollary 4.4. *Let X be a curve of genus $g \geq 2$ with Clifford index γ and $G_g(s, d, n + s) \neq \emptyset$. If $d_\ell(n, g, \gamma) \leq d < d_u(n, g, \gamma)$ and $d' \geq 0$ then $U(n, d + nd', n + s) \neq \emptyset$.*

Let X be a curve of genus g and Clifford index γ; and L_0 a line bundle computing the Clifford index. If $d_0 = d_{L_0}$ and $n_0 = h^0(L_0) - 1$ then $\gamma = d_0 - 2n_0$. Assume $\gamma < g - n_0$. Then, d_0 is precisely the lower bound $d_\ell(n_0, g, \gamma) = d_0$. For such values d_0 and n_0 we have the following corollary.

Corollary 4.5. $B(n_0, d_0, n_0 + 1) \neq \emptyset$. *Moreover $U(n_0, d, n_0 + 1) \neq \emptyset$ for all the values of $d = \gamma + rn_0$ with $r \geq 2$.*

Proof As a straightforward consequence of the definition of Clifford index we have that a line bundle computing the Clifford index is globally generated. Hence the triple $(1, d_0, n_0 + 1)$ is generated. Since $d_\ell(n_0, g, \gamma) = d_0$ and $d_0 < d_u(n_0, g, \gamma)$ the corollary follows from Corollary 4.4.   □

Assume that $\gamma \geq g - n$ with $n \geq 2$. In this case the lower bound is $d_\ell := d_\ell(n, g, \gamma) = n + g - 1$ and we have the following corollary.

Corollary 4.6. *Let X be a curve with Clifford index γ. If $\gamma \geq g - n > 0$ then $B(n, d_\ell, n + 1) \neq \emptyset$. Moreover, for $d' \geq 0$, $U(n, d_\ell + nd', n + 1) \neq \emptyset$.*

Proof The Brill-Noether number $\beta(1, g - n - 1, 1) = g - (n + 1) \geq 0$, hence $B(1, g - n - 1, 1) \neq \emptyset$ and $\dim B(1, g - n - 1, 1) \geq g - n - 1$. Denote by A the subset

$$A := \{L \in B(1, g - n - 1, 1) : h^0(L(x)) \geq 2 \text{ for some } x \in X\}.$$

The bundles $L(x)$ are in the locus $B(1, g - n, 2)$. By Clifford's Theorem $2 \leq g - n$ and since $\gamma > 0$, X is non-hyperelliptic. Hence, the upper bound for the dimension of $B(1, g - n, 2)$ given by Martens (see e.g. [ACGH, page 191]) is $g - n - 3$, therefore the complement $B(1, g - n - 1, 1) - A \neq \emptyset$.

If $L \in B(1, g - n - 1, 1) - A$,

$$h^0(L) = 1 \quad \text{and} \quad h^0(L) = h^0(L(x)) \tag{4.3}$$

for all $x \in X$. Let $M = K \otimes L^*$ where K is the canonical bundle. By Serre duality $M \in B(1, d_\ell, n + 1)$. Moreover, M is generated since from (4.3), $h^1(M) = h^1(M(-x))$ for all $x \in X$.

The result now follows from Corollary 4.4, since

$$(M, H^0(M)) \in G_g(1, d_\ell, n + 1).$$

□

For triples $(1, d, n + 1)$ with $d_\ell(n, g, \gamma) < d$ we have the following corollary.

Corollary 4.7. *Let X be a curve of genus $g \geq 3$ and Clifford index γ. If $n \geq 2$ and either $0 < g - n \leq \gamma$ or $0 < \gamma < g - n$ and $g - n < \gamma(\frac{n}{n-1})$ then $U(n, g + n, n + 1) \neq \emptyset$.*

Proof We use the same technique as in the proof of Corollary 4.6. Let C be the subset

$$C := \{L \in Pic^{g-n-2}(X) : h^0(L(x)) \geq 1 \text{ for some } x \in X\}.$$

In this case $L(x) \in B(1, g-n-1, 1)$ and since dim $B(1, g-n-1, 1) \leq g-2$, the complement $Pic^{g-n-2}(X) - C \neq \emptyset$ for $n \geq 2$.

For $L \in Pic^{g-n-2}(X) - C$ we have that

$$h^0(L) = h^0(L(x)) = 0 \tag{4.4}$$

for all $x \in X$.

If $M = K \otimes L^*$ with $L \in Pic^{g-n-2}(X) - C$ then $M \in B(1, g+n, n+1)$ and $(M, H^0(M)) \in G_g(1, g+n, n+1)$ since from (4.4), $h^1(M) = h^1(M(-x)) = 0$ for all $x \in X$.

If $g - n \leq \gamma$ then $d_u(n, g, \gamma) = n + g - 1 + \frac{g-1}{n-1}$ and $d_\ell(n, g, \gamma) < g + n < d_u(n, g, \gamma)$ since $n < g$. If $0 < \gamma < g - n$, then $d_u(n, g, \gamma) = 2n + \gamma(\frac{n}{n-1})$ and $g + n < d_u(n, g, \gamma)$ if and only if $g - n < \gamma(\frac{n}{n-1})$. Therefore $d_\ell(n, g, \gamma) < g + n < d_u(n, g, \gamma)$ and the corollary follows from Theorem 4.2. $\square$

5 Coherent systems over special curves

In this section we give examples of coherent systems on special curves for which Theorem 4.1 applies and for which the Brill-Noether number is negative.

Define the Clifford dimension of X as

$$n = \min\{h^0(L) - 1 | L \text{ computes the Clifford index of } X\}.$$

A line bundle L which achieves the minimum is said to compute the Clifford dimension. If $n \geq 2$, any line bundle computing the Clifford dimension is very ample ([ELMS, Lemma 1.1]) and defines an embedding $\varphi_L : X \to \mathbf{P}(H^0(L)^*)$ of X in $\mathbf{P}(H^0(L)^*)$. We call such an embedding a *Clifford embedding*.

The curves of Clifford dimension 2 are exactly the smooth plane curves of degree $d \geq 5$.

Let X be a smooth plane curve of degree $d \geq 5$ and Clifford index γ. Hence $d = \gamma + 4$. Note that the line bundle L with $h^0(L) = 3$ corresponding to the embedding of X in $\mathbf{P}^2$ computes the Clifford index and it is generated by its sections. Thus the triple $(2, d, 3)$ is generated. Since $g = \frac{(d-2)(d-1)}{2}$, we have that $d < g + 2$ and $d_u(2, g, \gamma) = 4 + 2\gamma = 2d - 4$. Hence, $d < d_u(2, g, \gamma)$ if and only if $4 < d$, therefore from

Theorem 4.1 the dual span M_L is always stable if $d \geq 5$. From Theorem 4.2 we have the following result.

Corollary 5.1. *Let X be a smooth plane curve of degree $d \geq 5$. Then $U(2, d, 3) \neq \emptyset$. Moreover, the Brill-Noether number $\beta(2, d, 3) < 0$ for $d \geq 5$.*

The curves of Clifford dimension 3 are complete intersections of pairs of cubics in $\mathbf{P}^3$. The square of the line bundle that embeds such a curve in $\mathbf{P}^3$ is the canonical bundle. Furthermore, the genus of X is 10, the Clifford index is $\gamma = 3$ and if L computes the Clifford dimension, then L has degree 9 and $d_\ell(3, 10, 3) = 9 < d_u(3, 10, 3)$. Applying Theorem 4.2 we see that for curves of Clifford dimension 3

- $B(3, 9, 4) \neq \emptyset$;
- $G(\alpha : 3, 9, 4) \neq \emptyset$ for all $\alpha > 0$

and the Brill-Noether number $\beta(3, 9, 4) = -6$.

Actually, from [ELMS, Theorem 4.3] there exist curves of Clifford dimension $n \geq 3$ with:

- genus $g = 4n - 2$;
- Clifford index $\gamma = 2n - 3$.

Moreover, the degree of a line bundle L computing the Clifford dimension is $d_L = g - 1$. For such curves $\gamma < g - n$, hence $d_u(n, d, \gamma) = 2n + \gamma + \frac{\gamma}{n-1}$ and $d_\ell(n, g, \gamma) = d_L < d_u(n, g, \gamma)$. From Theorem 4.2 we have:

Corollary 5.2. *If X is a curve of Clifford dimension $n \geq 3$, genus $g = 4n - 2$ and Clifford index $\gamma = 2n - 3$ then*

(i) $B(n, g - 1, n + 1) \neq \emptyset$;
(ii) $G(\alpha : n, g - 1, n + 1) \neq \emptyset$ for all $\alpha > 0$;
(iii) $\alpha_L = 0$;
(iv) $U(n, g - 1, n + 1) = G_g(n, g - 1, n + 1) = G_L(n, g - 1, n + 1) \neq \emptyset$;
(v) if φ_L is a Clifford embedding, $\varphi_L^*(T\mathbf{P})$ is stable.

Moreover, $\beta(n, g - 1, n + 1) < 0$.

In addition to the curves of Clifford dimension $n \geq 3$, existence of curves lying on K3 surfaces with linear systems g_d^n computing the Clifford index was proved in [M].

From [M] we know that curves of genus $g \geq 2$ with a simple g_{g-1}^n computing the Clifford index $\gamma = g - 1 - 2n$ exist if and only if

$$4(n - 1) \leq g - 1 \leq 4n.$$

For such curves generated triples $(1, g-1, n+1)$ with $d_\ell(n, g, \gamma) = g-1 < d_u(n, g, \gamma)$ exist. Therefore, Theorem 4.2 holds. Furthermore, the Brill Noether number satisfies

$$\beta = \beta(n, g - 1, n + 1) = -n^2 - 2n + g - 1$$

and, since $4n - 4 \le g - 1 \le 4n$, we have that $\beta \le -n^2 + 2n < 0$ for $n > 2$.

Let $n \ge 3$ and $g \ge 2$. Assume $4(n - 1) \le d_1 \le g - 1$. From [M] we have that if

$$2(d_1 - 2n) \le g - 1 < \frac{3(d_1 - 1)(d_1 + 3)}{16(n - 1)}$$

then there is a curve X of genus g with a very ample $g_{d_1}^n$ computing the Clifford index γ, that is, $\gamma = d_1 - 2n$. For such a curve the triples $(1, d_1, n + 1)$ are generated and $d_\ell(n, g, \gamma) = d_1 < d_u(n, g, \gamma)$. Therefore, Theorem 4.1 holds.

Acknowledgements. Both authors acknowledge Peter Newstead for his continuous teachings; the first author, additionally, thanks him for his help and support over these years. We also thank the referee for her/his comments and suggestions. The first author was supported by CONACYT grant 48263-F and the second author by CONACYT grant 61011-J1

Bibliography

[ACGH] E. Arbarello, M. Cornalba, P.A. Griffiths and J. Harris, *Geometry of Algebraic curves* Vol **1** Springer-Verlag, New York 1985.

[Br] Bradlow S.B. Coherent systems: A brief survey. Moduli spaces and vector bundles, ed. L. Brambila-Paz, S.Bradlow, O. Garcia-Prada, and S. Ramanan, London Math. Soc. Lecture Notes Series 359, Cambridge University Press.

[BDGW] S.B. Bradlow, G. Daskalopoulos, O. Garcia-Prada and R. Wentworth, Stable augmented bundles over Riemann surfaces. *Vector bundles in Algebraic Geometry*, Durham 1993, ed N.J. Hitchin, P.E. Newstead and W.M. Oxbury, LMS Lecture Notes Series **208**, 15–67.

[BG] S.B. Bradlow and O. Garcia-Prada, An application of coherent systems to a Brill-Noether problem, J. Reine Angew Math. **551** (2002), 123–143.

[BGMN] S.B. Bradlow, O. Garcia-Prada, V. Muñoz and P.E. Newstead, Coherent systems and Brill-Noether theory. *Internat. J. Math.* **14** (2003), 683–733.

[BGMMN1] S.B. Bradlow, O. Garcia-Prada, V. Mercat, V. Muñoz and P.E. Newstead, On the geometry of moduli spaces of coherent systems on algebraic curves, *Internat. J. Math.* **18** (2007), 411–453.

[BGMMN2] S.B. Bradlow, O. Garcia-Prada, V. Mercat, V. Muñoz and P.E. Newstead, Moduli spaces of coherent systems of small slope on algebraic curves, Comm. in Algebra (To appear)

[B-P] L. Brambila-Paz, Non-emptiness of moduli spaces of coherent systems, *Intenat. J. Math.* **18**, (2008), 777–799.

[BGN] L. Brambila-Paz, I. Grzegorczyk and P.E. Newstead, Geography of Brill-Noether loci for small slopes, *Jour. Alg. Geom.* **6** (1997), 645–669.

[BMNO] L. Brambila-Paz, V. Mercat, P.E. Newstead, and F. Ongay, Nonemptiness of Brill-Noether loci, *Internat. J. Math.* **11** (2000), 737–760.

[Bu] D.C. Butler, Birational maps of moduli of Brill-Noether pairs, preprint, arXiv:math.AG/ 9705009.

[EL] L. Ein and R. K. Lazarsfeld, *Stability and restrictions of Picard bundles with an application to the normal bundles of elliptic curves*, in Complex Projective Geometry (ed. G. Ellingsrud, C. Peskine, G. Sacchiero and S. A. Strømme), London Math. Soc. Lecture Notes Series **179**, Cambridge Univ. Press, 1992, 149–156.

[ELMS] D. Eisenbud, H. Lange, G. Martens, F.-O. Schreyer, The Clifford dimension of a projective curve, *Compositio Math* **72**, No. 2 (1989), 173–204.

[H] M. He, Espaces de modules de systèmes cohérents, *Internat. J. Math.* **9** (1998) 545–598.

[KN] A. King and P.E. Newstead, Moduli of Brill-Noether pairs on algebraic curves *Internat. J. Math.* **6** (1995) 733–748.

[LeP] J. Le Potier, Faisceaux semistables et systèmes cohérents, *Vector bundles in Algebraic Geometry*, Durham 1993, ed N.J. Hitchin, P.E. Newstead and W.M. Oxbury, LMS Lecture Notes Series **208**, 179–239.

[M] G. Martens, Linear series computing the Clifford index of a projective curve, *Abh. Math. Sem. Uni. Hamburg* **76** (2006), 115–130.

[M1] V. Mercat, Le problème de Brill-Noether pour des fibrés stables de petite pente, *J. Reine Angew. Math.* **506** (1999), 1–41.

[M2] V. Mercat, Le problème de Brill-Noether: présentation. http://www.liv.ac.uk/ newstead/bnt.html.

[PR] K. Paranjape and S. Ramanan, On the canonical ring of a curve, *Algebraic geometry and Commutative Algebra in Honor of Masayoshi Nagata* (1987) 503–516.

[RV] N. Raghavendra and P.A. Vishwanath, Moduli of pairs and generalised theta divisors, *Tôhoku Math. J.* **46** (1994) 321–340.

[T] M. Teixidor i Bigas, Brill-Noether Theory for stable vector bundles, Duke Math. J. **62**, (1991), 385–400.

16

Higgs bundles in the vector representation

Nigel Hitchin

Mathematical Institute,
24-29 St Giles, Oxford OX1 3LB, UK
e-mail: hitchin@maths.ox.ac.uk

Dedicated to Peter Newstead on the occasion of his 65th birthday.

1 Introduction

A *Higgs bundle*, as introduced by the author 20 years ago [H1], consists of a holomorphic bundle E over a compact Riemann surface Σ together with a section $\Phi \in H^0(\Sigma, \operatorname{End} E \otimes K)$ (called the Higgs field) satisfying a certain stability condition. This can be generalized to a holomorphic principal G-bundle P over Σ together with a section Φ of the bundle $\operatorname{Ad}(P) \otimes K$. The moduli space of such pairs has a very rich geometry and featured recently in the physical derivation by A.Kapustin and E.Witten of the central ideas of the geometric Langlands programme [KW]. A further paper by Kapustin [K] initiates a parallel discussion where the Higgs field Φ takes values in $R(P) \otimes K$ for a representation more general than the adjoint. In particular, he considers the case where $G = SU(n)$ and R consists of k copies of the n-dimensional vector representation. This means that we have a rank n vector bundle E with trivial determinant and $\Phi = (\varphi_1, \ldots, \varphi_k) \in H^0(\Sigma, E \otimes K) \otimes \mathbf{C}^k$.

The study of the algebraic geometry of these moduli spaces already exists in the literature under the name of *stable k-pairs* [BDGW], [BDW]. These papers, however, generally differ from the current problem in two ways. One is the type of moduli space considered. For historical reasons, perhaps based on the original problem of the abelian vortex equations [JT], the moduli spaces considered in the literature are generally compact. For example, the noncompact hyperkähler Higgs bundle moduli space for the adjoint representation would be replaced by the one-parameter family of compact Kähler quotients by the natural circle action. The second issue is that the gauge-theoretic equations which arise when Φ takes values in $E \otimes K \otimes \mathbf{C}^k$ depend only on the conformal structure of Σ and not on a choice of metric, as is the usually-treated case. An exception is [BGM] which contains in a very general context the results necessary to prove existence for the equations we need.

However, to get a feel for the issues involved and the differences encountered in straying from the adjoint representation, we study here the simplest rank 2 situation in a style which closely follows the original paper [H1]. We prove a basic existence theorem and then study features of the moduli space. The most notable one is the analogue of the integrable system: for a Higgs bundle in the adjoint representation, the invariant polynomials on $\mathfrak{g}$ define a proper map from the moduli space to a vector space, whose generic fibre is an abelian variety. The two-dimensional representation of $SU(2)$ has no symmetric invariant polynomials but instead it has a skew form. Using this we define a proper map to $H^0(\Sigma, K^2) \otimes \Lambda^2 \mathbf{C}^k$. Here the generic fibre for $k > 2$ is finite. On the other hand, for $k > 3$ the image satisfies constraints. These Plücker relations link the moduli space to the space of maps from Σ to a Grassmannian, which was the original point of view of [BDW].

The author wishes to thank A.Kapustin for introducing him to the subject and S.Bradlow and O.García Prada for valuable conversations.

2 The equations

Let E be a rank 2 C^∞ hermitian vector bundle on a compact oriented surface Σ of genus g, with a unitary trivialization of $\Lambda^2 E$. The space $\mathcal{A}$ of $SU(2)$-connections on E is an affine space modelled on $\Omega^1(\Sigma, \mathfrak{g})$ where $\mathfrak{g}$ denotes the adjoint bundle of skew-adjoint trace zero endomorphisms of E. As is well-known, this is formally a symplectic manifold and the moment map for the gauge group $\mathcal{G}$ is the curvature F_A. If Σ is now given a complex structure then labelling a connection by its $(0, 1)$ part d_A'' (the holomorphic structure on E defined by the connection) we can regard $\mathcal{A}$ as an infinite dimensional Kähler manifold. In this case $\Omega^{1,0}(\Sigma, E)$ also has by integration a well-defined hermitian form, so that the pairs $(A, \varphi) \in \mathcal{A} \times \Omega^{1,0}(\Sigma, E)$ such that $d_A'' \varphi = 0$ formally constitute a complex submanifold of a Kähler manifold and which is therefore Kähler.

The moment map for the action of the group of $SU(2)$ gauge transformations $\mathcal{G}$ on this manifold is

$$F_A + \varphi \otimes \varphi^* - \frac{1}{2}\langle \varphi, \varphi \rangle 1 \in \Omega^2(\Sigma, \mathfrak{g}).$$

Here $\varphi \otimes \varphi^* \in \operatorname{End} E$ is defined by $\varphi \otimes \varphi^*(e) = \langle e, \varphi \rangle \varphi$. If we replace E by $E \otimes \mathbf{C}^k$ and use a hermitian form on $\mathbf{C}^k$, then the zeros of the moment map are given by

$$F_A + \sum_1^k \left(\varphi_i \otimes \varphi_i^* - \frac{1}{2} \langle \varphi_i, \varphi_i \rangle 1 \right) = 0 \qquad (2.1)$$

This, together with $d''_A \varphi_i = 0$, provides the Higgs bundle equations we shall study. For brevity we shall write $R(\Phi)$ for the term involving the Higgs field.

Definition 2.1. *(cf [BDGW]) The k-pair (E, Φ) with*

$$\Phi = (\varphi_1, \ldots, \varphi_k) \in H^0(\Sigma, E \otimes K) \otimes \mathbf{C}^k$$

is stable if, whenever the φ_i all take values in a line bundle $L \otimes K \subset E \otimes K$, then $\deg L < 0$.

Note that if $g \leq 1$, then $L \otimes K$ has negative degree and hence no sections. On the other hand, the classification of rank 2 bundles for $g \leq 1$ tells us that all sections of $E \otimes K$ lie in a subbundle. It follows that our stable k-pairs only exist for $g \geq 2$.

Proposition 2.2. *If (E, Φ) is the k-pair defined by a solution to the equations (2.1) then either the pair is stable or else $\Phi = 0$ and A reduces to a flat $U(1)$-connection.*

Proof Suppose the φ_i all lie in $L \otimes K$. Put on L a hermitian connection whose curvature is dv where $d = \deg L$ and v is a Kähler form. Let $s \in H^0(\Sigma, L^* \otimes E)$ be the inclusion, then $d''_B s = 0$ where B is the induced connection on $L^* \otimes E$. Furthermore each $\varphi_i = \lambda_i s$. The Weitzenböck formula gives

$$\int_\Sigma \langle d'_B s, d'_B s \rangle = \int_\Sigma \langle F_B s, s \rangle. \qquad (2.2)$$

From equation (2.1) $F_B = -R(\Phi) - dv$ so

$$\langle F_B s, s \rangle = -\sum_1^k |\langle s, \varphi_i \rangle|^2 + \frac{1}{2} \sum_i^k \langle \varphi_i, \varphi_i \rangle \langle s, s \rangle - d\langle s, s \rangle v$$

$$= -\frac{1}{2} \sum_1^k |\lambda_i|^2 \|s\|^4 - d\|s\|^2 v.$$

Since the left hand side of (2.2) is non-negative we must have $d < 0$, unless $\lambda_i = 0$ and $d = 0$ in which case $d'_B s = 0$ and s is covariant constant. This reduces the holonomy of the connection to $U(1)$ and from (2.1) and $\varphi_i = 0$ it is flat. $\qquad \square$

3 Existence

We now prove the converse – if (E, Φ) is stable then it admits a solution of the equations. We follow [H1], which itself is based on Donaldson's original argument. See [BG] for a similar argument in the present context.

Theorem 3.1. *Let (A_0, Φ_0) define a stable k-pair, then there exists a C^∞ $SL(2, \mathbf{C})$ automorphism of E, unique up to an $SU(2)$ gauge transformation, which takes (A_0, Φ_0) to a solution of equation (2.1).*

Proof We start with the generic case:

Lemma 3.2. *If (E, Φ_0) is a stable k-pair then for $k \geq 2$ there is an open dense set of $\Phi \in H^0(\Sigma, E \otimes K) \otimes \mathbf{C}^k$ for which the φ_i do not all lie in any subbundle.*

Note that any of the φ_i can be identically zero, so that a stable ℓ-pair is a stable k-pair for $\ell \leq k$. So a stable pair for $k < 2$ is also the limit of one of these generic pairs.

Proof: We need to find a pair φ_1, φ_2 such that $\varphi_1 \wedge \varphi_2$ is not identically zero. If all $\varphi \in H^0(\Sigma, E \otimes K)$ lie in a line bundle $L \otimes K$, then, by the stability of Φ_0, $d = \deg L$ is negative. But then $\dim H^0(\Sigma, L \otimes K) \leq g$. By Riemann-Roch $\dim H^0(\Sigma, E \otimes K) \geq 2g - 2$ and so for $g > 2$ we have a contradiction. Since $\dim H^0(\Sigma, E \otimes K) \geq 2$ we have two sections φ_1, φ_2, not everywhere linearly dependent and so $\varphi_1 \wedge \varphi_2 \neq 0$. When $g = 2$, a line bundle of degree 0 or 1 has at most one section, so the theorem holds in this case too.

To start the proof we also need a crucial pointwise estimate. If $\Phi \in \mathbf{C}^2 \otimes \mathbf{C}^k$, ω is a skew form on $\mathbf{C}^2$ and $e_1, \ldots, e_k$ a basis for $\mathbf{C}^k$, define $\Lambda\Phi = \sum_{i,j} \omega(\varphi_i, \varphi_j)e_i \wedge e_j \in \Lambda^2\mathbf{C}^k$. If $\Lambda\Phi = 0$ then $\varphi_i = \lambda_i e$ and

$$R(\Phi) = \sum_1^k |\lambda_i|^2 \left(e \otimes e^* - \frac{1}{2}\|e\|^2 \right).$$

This is non-vanishing unless all φ_i vanish. Thus, restricted to the unit sphere $\|\Phi\| = 1$, the function $\|R(\Phi)\|^2 + \|\Lambda\Phi\|^2 \geq m > 0$ and, since $\Lambda(\Phi)$ and $R(\Phi)$ are quadratic in Φ we get an inequality

$$\|R(\Phi)\|^2 + \|\Lambda\Phi\|^2 \geq m\|\Phi\|^4 \tag{3.1}$$

Now let $\mathcal{G}^c$ be the group of L_2^2 $SL(2, \mathbf{C})$ gauge transformations and consider a $\mathcal{G}^c$-orbit of (A_0, Φ_0). Let (A_n, Φ_n) denote a minimizing sequence for $\|F_A + R(\Phi)\|_{L^2}^2$. Since $d_A'' \Phi = 0$, the Weitzenböck formula gives

$$0 \leq \int_\Sigma \langle d_B' \Phi, d_B' \Phi \rangle = \int_\Sigma \langle F_B(\Phi), \Phi \rangle = \int_\Sigma \langle F_A(\Phi), \Phi \rangle + (2g-2) \int_\Sigma \|\Phi\|^2 v.$$

Since $\|F_A + R(\Phi)\|_{L^2}^2$ is bounded for the sequence (A_n, Φ_n), this gives

$$\|R(\Phi_n)\|_{L^2}^2 \leq c_1 \int_\Sigma \langle R(\Phi_n) \Phi_n, \Phi_n \rangle \leq c_2 \|\Phi_n\|_{L^2}^2. \tag{3.2}$$

Now since $\mathcal{G}^c$ preserves the skew form on E, $\Lambda\Phi \in \Lambda^2 \mathbf{C}^k \otimes H^0(\Sigma, K^2)$ is constant on the orbit. Thus from (3.1)

$$\|R(\Phi_n)\|_{L^2}^2 + k_1 \geq m \|\Phi_n\|_{L^4}^4 \geq k_2 \|\Phi_n\|_{L^2}^4$$

where the last inequality comes from the Schwarz inequality. Together with (3.2) this gives a uniform L^2 bound on Φ_n and again from (3.2) on $R(\Phi_n)$. Since $\|F_{A_n} + R(\Phi_n)\|_{L^2}^2$ is also bounded this provides an L^2 bound on F_{A_n}. By Uhlenbeck's weak compactness theorem this means that, after applying $SU(2)$ gauge transformations, we may assume A_n converges weakly in L_1^2 to a connection A.

As in [H1], elliptic regularity and the L_1^2 bounds on A_n applied to $d_{A_n}'' \Phi_n = 0$ yields L_1^2 bounds on Φ_n and hence weak convergence. Furthermore, the element $\psi_n \in \mathcal{G}^c$ which takes (A_n, Φ_n) to (A_0, Φ_0) has a non-zero weak limit ψ. As a section of $E_n^* \otimes E_0$ it is holomorphic, and since both E_n and E_0 are of the same degree, it is either an isomorphism or maps to a line bundle in E_0. But since it takes Φ_n to Φ_0 and Φ_0 was chosen to be generic, so that its components do *not* all lie in a line sub-bundle, ψ must be an isomorphism and the limit (A, Φ) lies on the same orbit as (A_0, Φ_0). Thus (A, Φ) realizes the minimum of $\|F_A + R(\Phi)\|_{L^2}^2$ on this orbit and from the moment map interpretation, its value at this critical point must be zero.

To complete the proof we must pass from the generic case to any stable k-pair, again following [H1]. Our sequence (A_n, Φ_n) is now a sequence of solutions in the generic case in the same equivalence class of holomorphic structure on E. Since $F_{A_n} + R(\Phi_n) = 0$ we can use the equality (3.2) and from it and (3.1) the important relation

$$\|\Phi\|_{L^2}^4 \leq c_1 + c_2 \|\Lambda\Phi\|^2 \tag{3.3}$$

If (E, Φ_0) is the initial stable k-pair then $\Phi_n \to \Phi_0$ in $H^0(\Sigma, E \otimes K) \otimes \mathbf{C}^k$ and so $\Lambda\Phi_n \to \Lambda\Phi_0$ in the finite-dimensional space $\Lambda^2 \mathbf{C}^k \otimes H^0(\Sigma, K^2)$

and hence in any norm. From (3.3) and (3.2) we have an L^2 bound on $R(\Phi_n)$ and hence on $F_{A_n} = -R(\Phi_n)$. The resulting weak convergence preserves the equations (2.1) so the limit is also a solution. All that remains is to show that the limit of the isomorphisms ψ_n from E_n to E is also an isomorphism. But (E, Φ_0) is stable by assumption and the limit is stable by Proposition 2.2, and a map between stable k-pairs is an isomorphism. □

4 The moduli space

Theorem 3.1 enables us to describe the moduli space of stable k-pairs as a gauge-theoretic moduli space – the space of solutions of equation (2.1) modulo $SU(2)$ gauge transformations. The linearization of this equation is

$$(\dot{A}^{01}, \dot{\Phi}) \mapsto d'_A \dot{A}^{01} + r(\Phi)\dot{\Phi}^* + r^*(\Phi)\dot{\Phi} \in \Omega^2(\Sigma, \mathfrak{g})$$

where $\dot{\Phi}^* \in \Omega^{0,1}(\Sigma, E^*) \otimes \mathbf{C}^k$ is defined by $\dot{\Phi}$ using the hermitian metric, and

$$r(\Phi)\dot{\Phi}^* = \sum_1^k \left(\varphi_i \otimes \dot{\varphi}_i^* - \frac{1}{2}\langle \varphi_i, \dot{\varphi}_i \rangle 1 \right),$$

$$r^*(\Phi)\dot{\Phi} = \sum_1^k \left(\dot{\varphi}_i \otimes \varphi_i^* - \frac{1}{2}\langle \dot{\varphi}_i, \varphi_i \rangle 1 \right).$$

The linearization of the holomorphicity condition $d''_A \Phi = 0$ is $d''_A \dot{\Phi} + \dot{A}^{01}\Phi = 0$.

An infinitesimal gauge transformation $\psi \in \Omega^0(\Sigma, \mathfrak{g})$ defines the infinitesimal variation $\dot{A}^{01} = d''_A \psi, \dot{\Phi} = \psi\Phi$. Define

$$d_1\psi = (d''_A \psi, \psi\Phi) \in \Omega^{0,1}(\Sigma, \mathfrak{g}^c) \oplus \Omega^{1,0}(\Sigma, E \otimes \mathbf{C}^k)$$

and

$$d_2(\dot{A}^{01}, \dot{\Phi}) = (d'_A \dot{A}^{01} + r(\Phi)\dot{\Phi}^* + r^*(\Phi)\dot{\Phi}, d''_A \dot{\Phi} + \dot{A}^{01}\Phi) \in \Omega^2(\Sigma, \mathfrak{g} \oplus E \otimes \mathbf{C}^k)$$

then $d_2 d_1 = 0$ and we obtain a three-step elliptic complex. The group H^1 of the complex consists of infinitesimal variations of solutions to the equations modulo infinitesimal gauge transformations and, if H^0 and H^2 vanish, standard methods give a smooth finite-dimensional moduli space if $\mathcal{G}$ acts freely.

Introducing the adjoint of d_2 and putting the two real spaces involving $\mathfrak{g}$ into a single complex one gives the complex-linear elliptic operator:

$$d_2^* + d_1 : \Omega^0(\Sigma, \mathfrak{g}^c) \oplus \Omega^0(\Sigma, E \otimes \mathbf{C}^k) \to \Omega^{0,1}(\Sigma, \mathfrak{g}^c) \oplus \Omega^{1,0}(\Sigma, E \otimes \mathbf{C}^k).$$

This has index $-(2k+3)(g-1)$.

Suppose $(d_2^* + d_1)(\psi_1, \psi_2) = 0$, then

$$d_A'' \psi_1 + r^*(\Phi)\psi_2 = 0, \qquad d_A' \psi_2 + \text{tr}(\psi_1 r(\Phi)) = 0$$

where in the second expression we regard $r(\Phi)$ as a linear map from $E \otimes K \otimes \mathbf{C}^k$ to $\mathfrak{g}^c$. Integrating over Σ the identity $d\langle d_A'' \psi_1, \psi_1 \rangle = \langle d_A' d_A'' \psi_1, \psi_1 \rangle - \langle d_A'' \psi_1, d_A'' \psi_1 \rangle$ and using Stokes' theorem gives

$$0 = \| \text{tr}(\psi_1 r(\Phi)) \|_{L^2}^2 + \| d_A'' \psi_1 \|_{L^2}^2$$

and so $d_A'' \psi_1 = 0, \text{tr}(\psi_1 r(\Phi)) = 0$. The second condition implies that $\psi_1 \varphi_i = 0$ for all i. Now use the Weitzenböck formula for the holomorphic section ψ_1 of $\mathfrak{g}^c$ and we get

$$0 \le \int_\Sigma \langle d_A' \psi_1, d_A' \psi_1 \rangle = - \int_\Sigma \langle [R(\Phi), \psi_1], \psi_1 \rangle.$$

But if $\psi_1 \varphi_i = 0$, the right hand side is $-\|\psi_1 \Phi^*\|_{L^2}^2$ and so vanishes. This means that ψ_1 is covariant constant. Considering $\psi_1 - \psi_1^*$ or $i(\psi_1 + \psi_1^*)$ we get a trace zero covariant constant skew-hermitian transformation ψ and since $\psi_1 \varphi_i = 0 = \psi_1 \varphi_i^*$, $\psi \varphi_i = 0$ which implies $\psi = 0$ or the $\varphi_i = 0$. In the latter case E is flat and the argument shows that if ψ is non-zero it reduces the connection to a flat $U(1)$ connection.

Using Proposition 2.2 we see that H^0 and H^2 of the complex are zero if (A, Φ) defines a stable k-pair and non-zero only if $\Phi = 0$ and the connection is reducible to a flat $U(1)$ connection.

The moduli space $\mathcal{M}_k$ of solutions to the equations can be constructed by the normal analytical methods. It is smooth for the stable pairs so long as the group of gauge transformations $\mathcal{G}$ acts freely. Now only the identity in $SU(2)$ fixes a vector in $\mathbf{C}^2$, so if Φ is non-zero the action of $\mathcal{G}$ is free. When $\Phi = 0$, which is the moduli space of flat $SU(2)$-connections, the element -1 acts trivially. Thus this $3g-3$-dimensional subvariety forms the singular locus of $\mathcal{M}_k$.

Remark: We know from [R] that in rank two, degree zero, there is no universal bundle and no Picard bundle to realize $H^0(\Sigma, E \otimes K)$. Thus the stable bundle moduli space has no normal bundle in $\mathcal{M}_k$, and must for that reason alone be singular.

5 Invariants of the Higgs field

When the Higgs field is in the adjoint representation, an important role is played by evaluating the invariant polynomials on the Lie algebra on Φ. For $SU(2)$ this is the quadratic differential $\operatorname{tr}\Phi^2$. In the vector representation there are no such invariant polynomials but the invariant skew form ω led us to define

$$\Lambda(\Phi) \in \Lambda^2 \mathbf{C}^k \otimes H^0(\Sigma, K^2).$$

Since this is invariant under the action of $\mathcal{G}$, we get a map

$$p : \mathcal{M}_k \to \Lambda^2 \mathbf{C}^k \otimes H^0(\Sigma, K^2).$$

The inequality (3.3), together with the Uhlenbeck compactness theorem, shows that p is *proper*. In contrast to the adjoint case, the map p is, for $k > 3$, not surjective. This is because when we evaluate at $x \in \Sigma$ the $\varphi_i(x)$ lie in the two-dimensional space E_x and so $\Lambda(\Phi) = \sum_{i,j} \omega(\varphi_i, \varphi_j) e_i \wedge e_j$ is decomposable. Thus

$$\Lambda(\Phi) \wedge \Lambda(\Phi) = 0 \in \Lambda^4 \mathbf{C}^k \otimes H^0(\Sigma, K^4).$$

We consider the various non-zero values of k next.

5.1 $k = 1$

When $k = 1$, p is zero and this means in particular that $\mathcal{M}_1$ is compact. More generally, $p^{-1}(0)$ is compact and contains $\mathcal{M}_1$. This is the analogue of the nilpotent cone in the adjoint case: it consists of the k-pairs where all the φ_i lie in a line bundle.

5.2 $k = 2$

In this case p maps to the $3g - 3$-dimensional vector space $H^0(\Sigma, K^2)$. Given $q \in H^0(\Sigma, K^2)$ with $4g - 4$ simple zeros, the pairs (E, Φ) which map to q consist of maps

$$\mathbf{C}^2 \to E \otimes K$$

whose image is one-dimensional at the zeros of q. Dualizing, this means that $E \otimes K^{-1}$ is the subsheaf of $\mathcal{O} \oplus \mathcal{O}$ which is the kernel of a map to a skyscraper sheaf supported on the zeros of q – this is a *Hecke modification* of the trivial rank 2 bundle. A more geometrical description is to say that the ruled surface, the projective bundle $\mathrm{P}(E) \to \Sigma$, is obtained from $\mathbf{P}^1 \times \Sigma$ by choosing over each zero of q a point in $\mathbf{P}^1$, blowing up that point and blowing down the fibre which now has self-intersection -1. It follows that the fibre of $p : \mathcal{M}_2 \to H^0(\Sigma, K^2)$ over such a q is $(\mathbf{P}^1)^{4g-4}$. Note that

$$\dim \mathcal{M}_2 = 7g - 7 = \dim H^0(\Sigma, K^2) \times (\mathbf{P}^1)^{4g-4}.$$

Since p is proper and its image contains a dense open set of $H^0(\Sigma, K^2)$, it is surjective.

Remark: If $g = 2$ and E is stable then $\dim H^0(\Sigma, E \otimes K) = 2g - 2 = 2$, and the quadratic differential q is determined, up to a scale, by E itself. In fact in this case the moduli space of (semi)-stable bundles is $\mathbf{P}^3$, [NR] the linear system $|2\Theta|$ on the Jacobian $J^1(\Sigma)$. The theta-divisor is Σ, with normal bundle K, so that a divisor of 2Θ meets it in a divisor of K^2.

5.3 $k = 3$

Here p maps to the vector space $\Lambda^2 \mathbf{C}^3 \otimes H^0(\Sigma, K^2) \cong \mathbf{C}^3 \otimes H^0(\Sigma, K^2)$ and the relations are trivial. Given $(q_1, q_2, q_3) \in \mathbf{C}^3 \otimes H^0(\Sigma, K^2)$, suppose that the q_i have no common zeros, which is possible for any genus $g > 1$. Then they define a map

$$b : \Sigma \to \mathbf{P}^2$$

such that $b^* \mathcal{O}(1) \cong K^2$. Then if T is the tangent bundle of $\mathbf{P}^2$, $b^* T$ is a rank 2 vector bundle and $\Lambda^2 b^* T \cong b^* \mathcal{O}(3) \cong K^6$. Hence the bundle

$$E = b^* T \otimes K^{-3}$$

has trivial determinant. Moreover the Euler sequence on $\mathbf{P}^2$

$$0 \to \mathcal{O}(-1) \to \mathcal{O} \otimes \mathbf{C}^3 \to T(-1) \to 0$$

pulls back to a map $\mathbf{C}^3 \to E \otimes K$. It is straightforward to check that this is the stable 3-pair which maps to (q_1, q_2, q_3).

For $k = 3$ we see again the p has an open set in its image and hence is surjective. In this case the fibre is a point, and we also observe that

$$\dim \mathcal{M}_3 = 9g - 9 = \dim(\mathbf{C}^3 \otimes H^0(\Sigma, K^2)).$$

5.4 $k = 4$

For $k = 4$ we have for $\Lambda \Phi \in \Lambda^2 \mathbf{C}^4 \otimes H^0(\Sigma, K^2)$ the single relation

$$\Lambda \Phi \wedge \Lambda \Phi = 0 \in \Lambda^4 \mathbf{C}^4 \otimes H^0(\Sigma, K^4).$$

This means that the map $\Lambda^2 \mathbf{C}^4 \to H^0(\Sigma, K^2)$, if there is no common zero, defines a map $c : \Sigma \to \mathbf{P}^5$ whose image lies in the 4-dimensional Klein quadric $Q \subset \mathbf{P}^5$. The Klein quadric parametrizes the lines in a projective space $\mathbf{P}^3$ and therefore also in its dual – the choice is one of identifying the two families of planes in Q. Put another way, a surjective map

$$\mathbf{C}^4 \to E \otimes K$$

has kernel $F \otimes K^{-1}$ with $\Lambda^2 F$ trivial. There is a dual map $\mathbf{C}^4 \to F \otimes K$. It follows that the generic fibre of p here consists of two points.

Note that

$$\begin{aligned}
\dim \mathcal{M}_4 &= 11g - 11 = (18g - 18) - (7g - 7) \\
&= \dim(\Lambda^2 \mathbf{C}^4 \otimes H^0(\Sigma, K^2)) - \dim H^0(\Sigma, K^4).
\end{aligned}$$

5.5 $k > 4$

The two case indicates what the general situation is: the quadratic relations $\Lambda(\Phi) \wedge \Lambda(\Phi) = 0 \in \Lambda^4 \mathbf{C}^k \otimes H^0(\Sigma, K^4)$ are the relations for the Plücker embedding of the Grassmannian $G(2, k)$ of rank 2 quotients of $\mathbf{C}^k$ in $\mathbf{P}^{k(k-1)/2}$. A generic point in the image of p defines a map

$$f : \Sigma \to G(2, k)$$

such that if U is the universal rank two bundle over $G(2, k)$ then $\Lambda^2 f^* U \cong K^2$. Without the latter constraint this was the context for the first discussion of stable k-pairs in [BDW].

Bibliography

[BDGW] S.Bradlow, G.D.Daskalopoulos, O.García-Prada, & R.Wentworth, *Stable augmented bundles over Riemann surfaces*, in "Vector bundles in algebraic geometry (Durham, 1993)", 15–67, London Math. Soc. Lecture Note Ser. **208**, Cambridge Univ. Press, Cambridge (1995).

[BDW] A.Bertram, G.Daskalopoulos & R.Wentworth, *Gromov invariants for holomorphic maps from Riemann surfaces to Grassmannians*. J. Amer. Math. Soc. **9** (1996) 529–571.

[BG] S.Bradlow & O.García-Prada, *A Hitchin-Kobayashi correspondence for coherent systems on Riemann surfaces*, J. London Math. Soc. **60** (1999) 155–170.

[BGM] S.Bradlow, O.García-Prada & I.Mundet i Riera, *Relative Hitchin-Kobayashi correspondences for principal pairs*, Q. J. Math. **54** (2003) 171–208.

[H1] N.J.Hitchin, *The self-duality equations on a Riemann surface*. Proc. London Math. Soc. **55** (1987), 59–126.

[H2] N.J.Hitchin, *Stable bundles and integrable systems,* Duke Math. J. **54** (1987), 91–114.

[JT] A.Jaffe & C.Taubes *Vortices and monopoles,* Progress in Physics **2,** Birkhäuser, Boston, Mass. (1980).

[KW] A. Kapustin & E. Witten, *Electric-magnetic duality and the geometric Langlands program,* Commun. Number Theory Phys. **1** (2007), 1–236.

[K] A. Kapustin, *Holomorphic reduction of $N = 2$ gauge theories, Wilson -'t Hooft operators, and S-duality,* **hep-th/0612119** (2006).

[NR] M.S.Narasimhan & S.Ramanan, *Moduli of vector bundles on a compact Riemann surface,* Ann. of Math. **89** (1969) 14-51.

[R] S.Ramanan, *The moduli spaces of vector bundles over an algebraic curve,* Math. Ann. **200** (1973), 69–84.

17

Moduli spaces of torsion free sheaves on nodal curves and generalisations - I

C.S. Seshadri

Chennai Mathematical Institute
H1, SIPCOT Information Technology Park
Padur Post, Siruseri-603 103
Tamil Nadu, India.
e-mail: css@@cmi.ac.in

The contents of this paper are essentially the same as in the unpublished note [13], but for some more comments on *ramified H-bundles*, H a semi-simple algebraic group. The work of Ivan Kausz [4] is closely related to [13].

§1. Introduction

Let X be an irreducible projective curve whose only singularities are nodes, say only one at $x_0 \in X$. We take the base field as the field $\mathbb{C}$ of complex numbers. The moduli spaces of (semi-stable) torsion free sheaves on X provide a good generalisation of the moduli spaces of vector bundles on smooth projective curves. They provide compactifications of the moduli spaces of (semi-stable) vector bundles on X and have good specialisation properties i.e. when a smooth projective curve specialises to X, these objects specialise well. An interesting question is to generalise the moduli spaces of torsion free sheaves in the context of reductive or semi-simple algebraic groups H i.e. to have moduli spaces on X which are compactifications of the moduli spaces of semi-stable principal H-bundles (in the sense of A. Ramanathan [8]). There is some progress now due to the works of A. Schmitt and Usha Bhosle ([2],[11]).

Let F be a torsion free sheaf of rank n on X. The starting point of this paper is the observation (consequence of some considerations in [7], [12]) that locally at x_0, F is the *invariant direct image* of a vector bundle on a cyclic ramified covering; or equivalently F can be represented by a *ramified G-bundle* $P \longrightarrow X$ i.e. $G = GL(n)$ operates on P with finite isotropies, $X = P$ mod G and the action of G on P is free at points which lie over $x \in X$, $x \neq x_0$. Intuitively speaking, even though F may not be locally free at x_0, we can work with locally free sheaves by going to a finite ramified cyclic covering locally at x_0. This representation of F endows it with additional structures. Let $j : Y \longrightarrow X$ be the normalisation of X, x_1 and x_2 the points of Y which lie over x_0 and E the vector bundle $j^*(F)$ mod torsion on Y. These additional structures on F can be expressed by means of parabolic structures on E at x_1, x_2. This is rather natural, as in the smooth case the representation of a vector bundle E as the *invariant direct image* of a vector bundle on a ramified covering leads to parabolic structures on E. However, what is rather remarkable is that one finds (Remarks 6 and 7 below) that these additional structures can be identified with the additional structures provided by the (generalised) *Gieseker moduli space* over F i.e. the fibre over F of the canonical morphism of the Gieseker moduli space to the moduli space of torsion free sheaves ([7], [12]). This leads to the conjecture that the moduli space of torsion free sheaves with these additional structures is isomorphic to the Gieseker moduli space (Remark 7 below).

We can now define *ramified H-bundles* $P \longrightarrow X$, say when H is a semi-simple algebraic group, in the same way as we did above for the case of the linear group G (P is a principal H-bundle over $X \smallsetminus x_0$ and is represented, locally at x_0, by a principal H-bundle on a ramified cyclic covering). The ramified H bundles can also be expressed by means of *parabolic structures* on the normalisation Y of X. In this case the parabolic structures turn out to be subtler and they involve degenerations of H to groups which need not be semi-simple (Bruhat-Tits theory). This will be carried out in a future publication.

A. Schmitt and Usha Bhosle have shown that there is a moduli space of *singular H-bundles* on X with good specialisation properties ([2],[11]). From their results it follows that if a smooth curve $\underline{X}$ specialises to X and $P \longrightarrow \underline{X}$ is a semi-stable principal H-bundle (in the sense of A. Ramanathan), then it specialises to a ramified H-bundle on X (Remark 9, below).

Thus it is natural to conjecture that a compactification of the moduli space of semi-stable principal H-bundle on X can be expressed by means of ramified H-bundles on X.

In the sequel we have to refine the concept of ramified G or H bundles to what are called *distinguished G or H bundles* and we work with them.

§1. Distinguished G-spaces

Let X be as above. We fix a 1-parameter family of smooth curves specialising to X i.e. we take $\underline{X} \longrightarrow S$ to be a flat projective morphism such that $S = \operatorname{Spec} A$ (A a discrete valuation ring (d.v.r.) with residue field $\mathbb{C}$. Depending on the context S may be taken as a smooth affine curve or A to be complete or 1-dimensional open disc) such that for s generic in S, the fibre $\underline{X}_s$ is smooth and for the closed point $s_0 \in S$, $\underline{X}_{s_0} \simeq X$.

Recall that if F_s ($s \neq s_0$) is a torsion free $\mathcal{O}_{X_s}$-module (of finite rank), then it can be extended to a flat family $F = \{F_s\}$, $s \in S$, of torsion free sheaves over S. To see this, one first extends F_s to a flat family F' over S and then the double dual $(F')^{**}$ does the job. On the other hand given a torsion free $\mathcal{O}_X$-module F_0 on X, it is not difficult to see (possibly by going to a finite covering of S) that there is a flat family $\{F_s\}$, $s \in S$, as above extending F_0 (easy consequence of flatness of Quot schemes over S etc.).

Let us then represent a torsion free $\mathcal{O}_X$-module F_0 on X as a specialisation of a flat family $F = \{F_s\}$, $s \in S$, such that F_s is locally free for $s \neq s_0$. Consider now x_0 as a point of $\underline{X}$. Locally at x_0 (as a complete local ring or complex analytically), $\underline{X}$ has the following representation (consequence of the deformation properties of a double point singularity) in the 3-dimensional affine space (x, y, t).

$$xy = t^r \qquad (1)$$

the morphism $\underline{X} \longrightarrow S$ (locally at x_0) being given by $(x, y, t) \longmapsto t$, t being the uniformising parameter of A. Hence (as remarked in [7], [12]) if D is the 2-dimensional ball with coordinates (u, v) and Γ is the cyclic group of order r, represented by the r-th roots of unity operating on D by

$$\zeta(u, v) = (\zeta u, \overline{\zeta} v), \quad \zeta \in \Gamma, \overline{\zeta} \text{ complex conjugate of } \zeta \qquad (2)$$

then a neighbourhood C of x_0 is represented by the quotient $D \mod \Gamma$. The canonical morphism $f : D \longrightarrow C$ is given by

$$x = u^r, \quad y = v^r. \qquad (3)$$

Recall that ([7],[12]) we have a Γ-vector bundle V on D (i.e. the action of Γ on D lifts to V on D) such that

$$F|_C = f_*^\Gamma(V) \quad (\Gamma\text{-invariant direct image}). \qquad (4)$$

In fact, if f' is the restriction of f to $D \smallsetminus \{0\}$ (we see that Γ acts freely on $D \smallsetminus (0)$ and the quotient $(D \smallsetminus (0)) \mod \Gamma$ is $C \smallsetminus x_0$), we see that $(f')^*(F)$ extends to D as a Γ-vector bundle (the extension is locally free since we are in the smooth 2-dimensional case) which is the required V. One knows that V is defined by a representation i.e. if V_0 is the fibre of V at (0), then V_0 is a Γ-module, the diagonal action of Γ on $D \times V_0$ endows it with the structure of a vector bundle over D and V is Γ-isomorphic to $D \times V_0$.

Let $n = \operatorname{rk} V$ and $G = GL(n) \simeq \operatorname{Aut} V_0$. We call V_0 the *standard G-module*. Let $\varrho : \Gamma \longrightarrow G$ be the representation which defines the Γ-vector bundle V. Then $D \times G$ has a canonical *structure of a Γ-G bundle* (i.e. the action of Γ on D lifts to the (trivial) principal G-bundle $D \times G \longrightarrow D$ over D) and the associated vector bundle for the G-module V_0 is the vector bundle V on D. We denote this $(\Gamma - G)$ bundle over D by P_D. We have commuting actions by Γ and G on P_D (we take the action of G to be on the right and the action of Γ on the left) and if we set $P_C = P_D \mod \Gamma$ (geometric quotient) we have a canonical action of G on P_C (on the right). We denote by $p_C : P_C \longrightarrow C$ the canonical quotient morphism. We see that p_C induces a principal G-fibration over $C \smallsetminus x_0$ and at a point over x_0 the isotropy subgroup is a conjugate of $\varrho(\Gamma)$ in G. Hence p_C is what we call a *ramified G-bundle*. A morphism $P \longrightarrow Z$ is called a *ramified G-bundle* if we have an action of G (say on the right) such that Z is the geometric quotient $P \mod G$ and all the isotropy groups for the G-action are finite. Isomorphisms of ramified bundles over Z are G-isomorphisms which induce the identity map on Z. Consider the diagonal action of G on $P_C \times V_0$. Then this is a *G-vector bundle* over P_C (i.e. the action of G on P_C lifts to this vector bundle). We denote this G-vector bundle on P_C by $\underline{\underline{V}}_0$. We see that

$$(p_C)^G_*(\underline{\underline{V}}_0) = F|_C. \tag{5}$$

In fact, more generally if W is a G-module, then W is also a Γ-module through $\varrho : \Gamma \longrightarrow G$. Then $D \times W$ is canonically a Γ-vector bundle on D, which we denote by W'. Let $\underline{W}$ denote the G-vector bundle $P_C \times W$ over P_C. We see that

$$f^\Gamma_*(W') = (p_C)^G_*(\underline{W}). \tag{6}$$

We could call the objects on the RHS above as sheaves on C *associated to the G-modules* W for the ramified G-bundle $P_C \longrightarrow C$ (similar to the notion for principal G-bundles).

Let us now note a property of the Γ-covering $f : D \longrightarrow C$. Consider the diagram

$$
\begin{array}{ccccc}
(u,v) \in D & \xrightarrow{\ f\ } & C & (x,y,t) & x = u^r \\
& \searrow & \downarrow & \downarrow & y = v^r \\
& & S & T & t = uv
\end{array} \tag{7}
$$

The morphism $D \longrightarrow S$ is the composite of f and the morphism $C \longrightarrow S$ so that it is given by $(u,v) \longrightarrow t = uv$. Then we see that

$$
\Gamma \text{ leaves } D_s \text{ (fibre over `}s\text{') stable and } D_s \bmod \Gamma = C_s. \tag{8}
$$

Let $F_1 = F|_{\underline{X} \smallsetminus x_0}$. Then F_1 is a vector bundle on $\underline{X} \smallsetminus x_0$ and is defined by a principal G-bundle $p_1 : P_1 \longrightarrow \underline{X} \smallsetminus x_0$. We have an analytic G-morphism

$$
P_C|_{C \smallsetminus x_0} \simeq P_1|_{C \smallsetminus x_0}. \tag{9}
$$

By patching through this isomorphism, we obtain a ramified G-bundle (á priori analytic)

$$
p : P \longrightarrow \underline{X} \tag{10}
$$

such that $p|_C = p_C$ and $p|_{\underline{X} \smallsetminus x_0} = p_1$. Further we note that

$$
p_*^G(\underline{V}_0) = F \tag{11}
$$

where $\underline{V}_0$ is the G-vector bundle $P \times V_0$ on P. By comparison theorems of analytic and algebraic geometry (Serre type of theorems) $p : P \longrightarrow \underline{X}$ acquires a canonical algebraic structure. Consider

$$
\begin{array}{ccc}
P & \longrightarrow & \underline{X} \\
& \searrow & \downarrow \\
& & S
\end{array} \tag{12}
$$

where $P \longrightarrow S$ is the composite of p and $\underline{X} \longrightarrow S$. Then by (8) the morphism $p_s : P_s \longrightarrow \underline{X}_s$ (induced by p on the fibres over $s \in S$) gives a flat family of ramified G-bundles and for $s \neq s_0$, $p_s : P_s \longrightarrow \underline{X}_s$ is a principal G-fibration which specialises to the ramified G-bundle $p_0 = p_{s_0} : P_0 (= P_{s_0}) \longrightarrow \underline{X}_{s_0} = X$. Note that since geometric quotients by G commute with base change in characteristic zero, we have

$$
(p_s^G)_*(\underline{W}_s) = (p_*^G(\underline{W}))_s. \tag{13}
$$

By the preceding discussions (especially (8)), we have the following concrete description for the ramified G-bundle $p_0 : P_0 \longrightarrow X$ over a neighbourhood of x_0. We *drop the suffix zero* below i.e. we call P_0, p_0, D_0 etc. simply as P, p, D etc.

$$
\left\{
\begin{array}{l}
D = \{(u,v)|uv = 0\}, \quad C = \{(x,y)|xy = 0\} \\
\text{The cyclic group } \Gamma \text{ of order } r \text{ acts on } D \text{ as follows.} \\
\text{Fix a generator } \vartheta \text{ of } \Gamma. \text{ Then } \vartheta(u,v) = (\zeta^{-1}u, \zeta v) \text{ where } \zeta \text{ is a} \\
\text{primitive } r\text{th root of unity.} \\
f : D \longrightarrow C, u^r = x, \ y^r = v \ D \mod \Gamma = C, C \text{ a neighbourhood} \\
\text{of } x_0 \ (u,v,x,y \text{ sufficiently small}). \\
\text{If } P_C = p^{-1}(C), \ P_C = (D \times G) \mod \Gamma \ (\Gamma\text{-action through} \\
\varrho : \Gamma \longrightarrow G).
\end{array}
\right.
\tag{14}
$$

Definition 1:

(i) A *distinguished G-bundle* over X is a ramified G-bundle $p : P \longrightarrow X$ having the description (14) above locally at x_0 and (14) is referred to as the *local data* for the distinguished G-bundle.

(ii) In a similar manner we define a *distinguished H-bundle* $p : P \longrightarrow X$ for any semi-simple group (in fact any group). The above considerations give

Proposition 1:

(i) Given a torsion free sheaf F on X, there is a distinguished G-bundle $p : P \longrightarrow X$ such that $(p_*)^G(\underline{V}) = F$ i.e. F is the sheaf on X associated to the standard G-module V for $P \longrightarrow X$.

(ii) Given a principal G-bundle $P_s \longrightarrow \underline{X}_s$ $(s \neq s_0)$, it specialises to a distinguished G-bundle $p : P \longrightarrow X$.

Remark 1: The assertion (2) of Proposition 1 above extends to the case of any semi-simple group H (or any group) by arguments, as above, provided the following holds (the point to observe is that, as in the case of G, if D is a 2-dimensional ball with a Γ-action as above and we have a Γ-H principal bundle in $D \smallsetminus (0)$, then it extends to a Γ-H principal bundle in D given by a representation $\varrho : \Gamma \longrightarrow H$):

$$
\left\{
\begin{array}{l}
\text{Given a principal } H\text{-bundle } P_s \longrightarrow \underline{X}_s \ (s \neq s_0) \text{ it can be extended} \\
\text{to a principal } H\text{-bundle. } p' : P' \longrightarrow U, \text{ where } U \text{ is open in } \underline{X} \\
\text{and the complement of } U \text{ in } \underline{X} \text{ consists only of a finite number of} \\
\text{points of } \underline{X}_{s_0} \simeq X.
\end{array}
\right.
\tag{15}
$$

Remark 2: The assertion (1) of Proposition 1 can be seen directly and quite easily without using the fact that F is the specialisation of a vector bundle on a smooth projective curve specialising to X. We have only to show that (see (6) above) we have a ramified covering $f : D \longrightarrow C$ with notations as in (14) above such that

$$F|_C = (f_*^{\Gamma})(V'),$$

$$V' = \Gamma- \text{ vector bundle or sheaf } D \times V, \ V \ a \ \Gamma- \text{ module.} \qquad (16)$$

Let us recall that locally at x_0, F is the direct sum of a finite number of copies of the structure sheaf $\mathcal{O}_{x_0}$ and a finite number of copies of the maximal ideal m of $\mathcal{O}_{x_0}$ and the number of copies of m is called the *type* of F (at x_0). Thus to prove (16) we can assume that F is either $\mathcal{O}_{x_0}$ or m (locally at x_0). In the first case we take for $\varrho : \Gamma \longrightarrow \text{Aut } V$ (here V is of rank one) the trivial character. In the second case, any non-trivial character would do; in fact, if e is a basis element of V, we have (with the notations as in (14) above):

$$\vartheta.e \text{ (i.e. } \varrho(\vartheta)e) = \zeta^s e, \ \ 1 \leq s < r.$$

Then we check easily that the RHS of (16) above is generated by (the Γ-invariant elements) $u^s e$ and $v^{r-s}e$ as a $\mathcal{O}_{x_0}$ module. From this (16) above follows easily.

Remark 3: One notes that for the argument in Remark 2 above, we could have taken $\Gamma = \mathbb{Z}$ mod 2 and ϱ as the trivial or non-trivial character. Thus for the assertion of Remark 2, ramified double coverings would do. However, to have the specialisation property i.e. the assertion (2) of Proposition 1, we may require a general Γ.

§2 Connection with parabolic structures

We have seen in Proposition 1 above that given a torsion free sheaf F of rank n on X, it is of the form $p_*^G(\underline{V})$ where $p : P \longrightarrow X$ is a distinguished G-bundle $(G = GL(n))$ and $\underline{V}$ the G-vector bundle $P \times V$ on P, V being the standard G-module. Thus we get a functor:

$$\left\{ \begin{array}{ccc} \text{(Distinguished } G\text{-bundles over } X) & \longrightarrow & \text{(Torsion free sheaves} \\ & & \text{of rk } n \text{ on } X) \\ P & \longmapsto & (p_*^G)(\underline{V}) \end{array} \right.$$

$$\qquad (17)$$

where the LHS denotes the category of distinguished G-bundles over X and similarly for the RHS. This is not an equivalence of categories and the distinguished G-bundle P can be viewed as providing additional structures on F. We shall see now that these additional structures can be expressed by means of parabolic structures on the normalisation of X.

Let $j : Y \longrightarrow X$ be the normalisation of X and x_1, x_2 the points of Y lying over x_0. Let $Q' \longrightarrow Y$ be the scheme theoretic pull-back of $P \longrightarrow X$ by j and $Q = Q'_{red}$ (we see that Q' and Q coincide set theoretically and $Q = Q'_{red}$ over the open subset $Y \smallsetminus \{x_1, x_2\}$). In fact Q' is *not* reduced when ϱ is not trivial (see the comments following (20) below). Let $E = j^*(F)$ mod torsion. We see easily (again see the comments below) that the vector bundle E on Y is the bundle associated to the standard G-module V for the ramified G-bundle $Q \longrightarrow Y$. One knows that the ramified G-bundle $Q \longrightarrow Y$ can be equivalently described by parabolic structures for E at the points x_1, x_2. We see also that P can be obtained from Q by an identification $Q_{x_1} \longrightarrow Q_{x_2}$ (Q_{x_i} fibre of Q at x_i, $i = 1, 2$) which is a G-morphism. Note that $Q_{x_i} \simeq \varrho(\Gamma) \smallsetminus G$ ($i = 1, 2$). Hence this identification gives an element of the centraliser of $\varrho(T)$ in G. We shall now describe all these via parabolic structures of E at x_1, x_2.

We shall now use the local data (14) above. Let D_1, D_2 be the discs (with centre (0)) which are the components of D. The coordinates of D_1, D_2 are given by u, v respectively and the action of Γ on D_i is given by

$$\vartheta u = \zeta^{-1} u, \ \vartheta v = \zeta v,$$

ϑ a generator of Γ and ζ a primitive rth root of unity. (18)

Then $C_i = D_i$ mod Γ and can be identified with a neighbourhood of $x_i \in Y$, $i = 1, 2$. Let $f_i : D_i \longrightarrow C_i$ be the canonical map $i = 1, 2$. The coordinates of C_1, C_2 are x, y respectively. Now V is a G-module ($G = \text{Aut } V$). It is also a Γ-module through $\varrho : \Gamma \longrightarrow G$. Now $D_i \times V$ is a Γ-vector bundle over D_i ($i = 1, 2$) and the Γ-action is given by:

$$\vartheta \cdot (u, z) = (\zeta^{-1} u, \varrho(\vartheta) z), \vartheta \cdot (v, z) = (\zeta u, \varrho(\vartheta) z), z \in V. \quad (19)$$

From the previous considerations ((5), (6) above and definition of E), we have ($i = 1, 2$):

$$\begin{cases} (f_i)_*^\Gamma (D_i \times V) = E|_{C_i} = (q_{C_i})_*^G (\underline{V}), \\ q_{C_i} = \text{restriction of } q : Q \longrightarrow Y \text{ over the open set over } C_i \end{cases} \quad (20)$$

Let us understand (20) above a little better. Recall that $P_D = D \times G$ and $P_C = (D \times G) \mod \Gamma$. The base change of $P_C \longrightarrow C$ by the closed immersion $C_1 \longrightarrow C$ identifies with $Q'_{C_1} \longrightarrow C_1$ (where Q', as defined above, is the base change of $P \longrightarrow X$ by $j : Y \longrightarrow X$ and $Q'_{C_1} = Q'|_{C_1}$, C_1 identified with a neighbourhood of x_1 in Y). Then $Q'_{C_1} \longrightarrow C_1$ is a "ramified" G-bundle and if F' denotes the coherent sheaf on C_1 associated to the standard G-module, F' coincides with the restriction of F to C_1 (by the commutativity properties of taking G-invariants with respect to base change). Hence $E = F' \mod$ torsion on C_1. We claim that the coherent sheaf F' is *not* locally free if the type of $F \geq 1$ (i.e. not locally free). To prove the claim it suffices to consider the case when locally at x_0, F is defined by the maximal ideal. Then it is easily checked $F' = $ direct sum of the maximal ideal at x_1 and a torsion sheaf of rk 1 supported at x_1. This in particular implies that Q_{C_1} is *not* reduced for otherwise $Q'_{C_1} = Q_{C_1}$ and the coherent sheaf on C_1 associated to the standard G-module would be locally free, which proves the assertion made above that Q' is *not* reduced if ϱ is non-trivial.

Let us suppose, for simplicity, that on the whole of X, $F = $ maximal ideal at x_0. Then we see that E on Y is the ideal sheaf defined by the two points at x_1, x_2. We have $\deg E = -2$ and $\deg F = -1$ i.e. $\deg E = \deg F - \text{type } F$ and from this argument it follows easily that in the general case we have the same relation $\deg E = \deg F - \text{type } F$.

Let D'_1 be the closed subscheme of D obtained as base change of $D \longrightarrow C$ by $C_1 \longrightarrow C$ i.e. we have the commutative diagram:

$$
\begin{array}{ccc}
D'_1 & \longrightarrow & D \\
\downarrow & & \downarrow \\
C_1 & \longrightarrow & C
\end{array}
\quad (D'_1)_{red} = D_1.
$$

Then $P_{D'_1} = D'_1 \times G$ is the base change of $P_D \longrightarrow D$ by $D'_1 \longrightarrow D$ and by the commutativity properties of taking Γ-invariants with respect to base change, we see that $P_{D'_1} \mod \Gamma = Q'_{C_1}$, $D'_1 \mod \Gamma = C_1$. We see that D'_1 is *not* reduced. In fact the complete local ring of D'_1 at (0) is given by $\mathbb{C}[[u, v]]/(uv, y)$ and if $\overline{v}$ is the image of v, $\overline{v} \neq 0$ and $\overline{v}^r = 0$.

We can find a basis $\{e_i\}$, $1 \leq i \leq n$, of V such that the action of Γ on E is given as follows (with the notations as in (14) and (19)):

$$\vartheta \cdot e_i = \zeta^{a_i} e_i; \quad 0 \leq a_1 \leq a_2 \leq \cdots \leq a_r < r. \tag{21}$$

Let $\mathcal{O}_{D_1}$ represent the local ring of D_1 at (0). We see that $\mathcal{O}_{C_1} = \Gamma$-invariants of $\mathcal{O}_{D_1}$, can be identified with the local ring of C_1 (or Y) at x_1. The Γ-vector bundle $D_1 \times V$ is represented locally at (0) by the free $\mathcal{O}_{D_1}$-module $\mathcal{O}_{D_1} \otimes V$ with basis e_i (i.e. $1 \otimes e_i$), $1 \leq i \leq n$. We see easily that its Γ-invariant elements form a free submodule over $\mathcal{O}_{C_1}$ with basis $\{g_i\}$, $1 \leq i \leq n$, given by

$$\begin{cases} g_i = u^{a_i} e_i = x^{\alpha_i'} e_i, \ \alpha_i' = a_i/r, \\ 0 \leq \alpha_1' \leq \alpha_2' \leq \cdots \leq \alpha_n' < 1 \end{cases} \tag{22}$$

or equivalently we can say that locally at x_1, E is a free $\mathcal{O}_{Y,x_1}$-module with basis $\{g_i\}$, $1 \leq i \leq n$. Let A be a $\Gamma - \mathcal{O}_{D_1}$ automorphism of $\mathcal{O}_{D_1} \otimes V_0$. We have

$$Ae_i = \sum_{j=1}^n a_{ij}(u)e_j; \ \ a_{ij} \in \mathcal{O}_{D_1}, \ A = (a_{ij}). \tag{23}$$

Then we get

$$\begin{cases} Ag_i = \sum_{j=1}^n b_{ij} g_j; \ \text{with} \\ b_{ij} = a_{ij}(u)u^{a_i - a_j}, \ \ B = (b_{ij}). \end{cases} \tag{24}$$

Since A is a Γ-$\mathcal{O}_{D_1}$ automorphism (or equivalently, $(f_1)_*^\Gamma$ is a functor), we see that it induces an $\mathcal{O}_{C_1}$ automorphism B of the $\mathcal{O}_{C_1}$-module of Γ-invariant elements of $\mathcal{O}_{D_1} \otimes V$, so that $b_{ij} \in \mathcal{O}_{C_1}$. Thus we get

$$\begin{cases} b_{ij} = b_{ij}(u^r) = b_{ij}(x) \in \mathcal{O}_{C_1} \\ b_{ij}(0)(i.e.b_{ij}(x_1)) = 0 \ \text{if} \ a_i > a_j \end{cases} \tag{25}$$

We see that A is a $(\Gamma - \mathcal{O}_{D_1})$ automorphism if and only if the automorphism B induced on the basis of Γ-invariant elements $\{g_j\}$, $1 \leq j \leq n$, satisfy the conditions (25) above. Observe that $\{g_j\}$ can be identified with a basis of sections of E locally at x_1. Let $\{\overline{g}_i\}$ be the basis elements of the fibre E_{x_1} of E at x_1 induced by $\{g_i\}$.

We set the following notations:

$$\begin{cases} k_1, k_2, \cdots, k_s \text{ integers such that} \\ a_1 = a_2 = \cdots = a_{k_1} < a_{k_1+1} = \cdots = a_{k_1+k_2} < a_{k_1+k_2+1} < \cdots \\ \alpha_i \text{ the distinct elements among } \alpha_i' = a_i/r. \\ \text{We have } 0 \leq \alpha_1 < \alpha_2 < \cdots < \alpha_s < 1 \\ n = k_1 + \cdots + k_s. \end{cases} \tag{26}$$

Let us consider the *strictly* decreasing flag $\mathcal{F}_{x_1} = \{\mathcal{F}_{x_1}^i\}$ on E_{x_1} defined as follows:

$$\begin{cases} E_{x_1} = \mathcal{F}_{x_1}^1, \ \ \mathcal{F}_{x_1}^2 = \text{subspace spanned by } \overline{g}_j, \ j \geq k_1 + 1 \\ \mathcal{F}_{x_1}^3 = \text{subspace spanned by } \overline{g}_j, \ j \geq k_1 + k_2 + 1, \\ \mathcal{F}_{x_1}^s = \text{subspace spanned by } \overline{g}_j, \ j \geq k_1 + \cdots + k_{s-1} + 1, \\ \mathcal{F}_{x_1}^{s+1} = (0). \end{cases} \tag{27}$$

The important point is that A above is a Γ-automorphism if and only if the automorphism induced by B on E_{x_1} leaves the flag stable. Recall the definition of parabolic structures ([5]):

Definition 2:

(a) The flag $\mathcal{F}_{x_1} = \{\mathcal{F}_{x_1}^i\}$, $1 \leq i \leq s+1$, on E_{x_1} together with the weights $\alpha_1, \cdots, \alpha_s$ in (26) and (27) above, is called a *parabolic structure* of E at x_1. The weight of $\mathcal{F}_{x_1}^i$ is α_i. We set

$$\operatorname{Gr} E_{x_1} = \operatorname{Gr} \mathcal{F}_{x_1} = \bigoplus \mathcal{F}_{x_1}^i / \mathcal{F}_{x_1}^{i+1}, \operatorname{Gr}^i \mathcal{F}_{x_1} = \mathcal{F}_{x_1}^i / \mathcal{F}_{x_1}^{i+1},$$

we call $\{k_i\}$, $1 \leq i \leq s$, the multiplicities of the parabolic structure. Note that $k_i = \operatorname{rk} \operatorname{Gr}^i \mathcal{F}_{x_1}$. We see that $s =$ length of the flag.

(b) Recall that a *quasi-parabolic structure* of E at x_1 is the underlying structure when we ignore the weights.

Remark 4: Recall that we have defined $g_i = u^{a_i} e_i$ (see (2.2) above). We can change the coordinate u to $u\delta$, where δ is a unit in $\mathcal{O}_{D_1}$. If we set $g_i' = (u\delta)^{a_i} e_i$, we see that $\overline{g}_i' = \lambda^{a_i} \overline{g}_i$, where $\lambda \in \mathbb{C}$, $\lambda \neq 0$. Thus we see that once a basis $\{e_i\}$ of V is fixed, the basis element $\overline{g}_i$ is determined upto a scalar multiple λ^{a_i}, $\lambda \in \mathbb{C}$, $\lambda \neq 0$.

Remark 5: The group Γ acts on the tangent space T_1 (one dimensional) of D_1 at (0). This action is canonically given by a character $\chi : \Gamma \longrightarrow \mathbb{C}^*$ (since $\operatorname{Aut} T \simeq \mathbb{C}^*$ canonically) and hence if we fix the generator $\vartheta \in \Gamma$ (as we have) by its value at ϑ. In our case we see that $\chi(\vartheta) = \zeta$ (as in (14) and (21) above) since the action of Γ on the dual space T_1^* is given by χ^{-1} and we have $\vartheta \cdot u = \zeta^{-1} u$. The action of Γ on V with respect to the basis $\{e_i\}$ (see (21) above) is to be interpreted as

$$\vartheta \cdot e_i = \chi^{a_i}(\vartheta) e_i$$

i.e. we write V as a direct sum of one dimensional Γ-modules defined by the characters χ^{a_i} $0 \leq a_1 \leq a_2 \leq \cdots \leq a_n < r$. This gives a canonical interpretation of the integers $\{a_i\}$. We set $\operatorname{Gr}_\chi V = \oplus \operatorname{Gr}^i V$, where $\operatorname{Gr}^i V =$ isotypical component defined by χ^{a_i}.

We shall now see that we have a canonical identification of $\operatorname{Gr}_\chi V$ with $\operatorname{Gr} \mathcal{F}_{x_1}$, modulo a well-defined equivalence relation. The assignment $e_i \longmapsto \overline{g}_i$ defines a graded (linear) isomorphism.

$$\psi_1 : \operatorname{Gr}_\chi V \longrightarrow \operatorname{Gr} \mathcal{F}_{x_1} \tag{28}$$

which is canonically defined modulo the following equivalence relation i.e. if ψ_1 and ψ' are two such isomorphisms (see Remark 4), we have

$$\psi_1(y) = \lambda^{a_i} \psi_1'(y), \quad y \in \mathrm{Gr}^i V, \lambda \in \mathbb{C}, \ \lambda \neq 0. \tag{29}$$

In a similar manner we can define a parabolic structure on the fibre E_{x_2}. However, we note the following difference. The action of Γ on the tangent space T_2 of D_2 at (0) is given by the character χ^{-1} (χ as in Remark 5) i.e. T_2 is dual to T_1 (as Γ-modules). If we take $\chi' = \chi^{-1}$, then we see that $\chi^{a_i} = (\chi')^{b_i}$, $\chi^{-a_i} = \chi^{b_i}$, $0 \leq b_i < r$, with $b_i = r - a_i$ $(b_i/r = 1 - \alpha_i)$ if $a_i \neq 0$ and $b_i = a_i$ if $a_i = 0$. We rearrange b_i in increasing order $0 \leq b_1 \leq \cdots \leq b_n < r$. Let V^* be the Γ-dual of V. We now see easily that $\mathrm{Gr}_{\chi'} V \simeq \mathrm{Gr}_\chi V^*$. Thus when we carry out a parabolic structure on E_{x_2} similar to the case of E_{x_1}, we get an isomorphism

$$\psi_2 : \mathrm{Gr}_\chi V^* \simeq \mathrm{Gr}\, \mathcal{F}_{x_2} \tag{30}$$

which is canonically determined modulo the equivalence relation defined as in the case of ψ_1 above. We see that the weights of $\mathcal{F}_{x_2}$ on E_{x_2} are *dual to* $\alpha_1, \cdots, \alpha_s$ i.e. the dual weights are $1 - \alpha_s, 1 - \alpha_{s-1}, \cdots, 1 - \alpha_1$ if $\alpha_1 \neq 0$ and $0, 1 - \alpha_s, \cdots, 1 - \alpha_2$ if $\alpha_1 = 0$. Now the flag $\mathcal{F}_{x_2}$ and E_{x_2} induces a canonical flag $\mathcal{F}_{x_2}^*$ (called the *dual flag*) on the dual $E_{x_2}^*$ of E_{x_2}. The weights of $\mathcal{F}_{x_2}^*$ are dual to those of $\mathcal{F}_{x_2}$ and hence coincide with $\alpha_1, \cdots, \alpha_s$. We see easily that we can identify $\mathrm{Gr}\, \mathcal{F}_{x_2}^*$ with $\mathrm{Gr}\, \mathcal{F}_{x_2}$ by a shifting of degrees as follows:

$$\begin{cases} \mathrm{Gr}^i \mathcal{F}_{x_2}^* = \mathrm{Gr}^{s+1-i} \mathcal{F}_{x_2}; \ 1 \leq i \leq s, \text{ if } \alpha_1 \neq 0 \\ \mathrm{Gr}^1 \mathcal{F}_{x_2}^* = \mathrm{Gr}^1 \mathcal{F}_{x_2}; \mathrm{Gr}^i \mathcal{F}_{x_2}^* = \mathrm{Gr}^{s+2-i} \mathcal{F}_{x_2} \text{ if } \alpha_1 = 0. \end{cases} \tag{31}$$

Thus we have a graded linear isomorphism $\psi_2^* : \mathrm{Gr}\, V \longrightarrow \mathrm{Gr}\, \mathcal{F}_{x_2}^*$ determined upto an equivalence relation, as mentioned above. Thus we get a graded linear isomorphism:

$$\varphi : \mathrm{Gr}\, \mathcal{F}_{x_1} \longrightarrow \mathrm{Gr}\, \mathcal{F}_{x_2}^* \tag{32}$$

determined upto an equivalence relation as in (29) above.

Definition 3: Let E be a vector bundle of $\mathrm{rk}\, n$ of Y. Then a *distinguished parabolic (DP) structure* on E at x_1, x_2 consists of the following data:

(a) $\mathcal{F} = (\mathcal{F}_{x_2}, . \mathcal{F}_{x_2})$, where $\mathcal{F}_{x_1}$ and $\mathcal{F}_{x_2}$ define parabolic structures on E_{x_1} and E_{x_2} respectively such that the weights of $\mathcal{F}_{x_2}$ are dual to the weights $\alpha_1, \cdots, \alpha_s$ of $\mathcal{F}_{x_1}$. Let k_i = multiplicity of α_i so that s = length of the flag $\mathcal{F}_{x_1}$.

(b) a graded linear isomorphism:

$$\varphi : \operatorname{Gr} \mathcal{F}_{x_1} \longrightarrow \operatorname{Gr} \mathcal{F}_{x_2}^*$$

where $\mathcal{F}_{x_2}^*$ is the flag dual to $\mathcal{F}_{x_2}$ (see (31) and (32) above) and the weights of $\mathcal{F}_{x_2}^*$ are dual to those of $\mathcal{F}_{x_2}$ and hence coincide with $\alpha_1, \cdots, \alpha_s$. Further, φ is determined upto an equivalence relation defined as follows:

$\varphi \sim \varphi'$ if there exist $\lambda_i \in \mathbb{C}$, $\lambda_i \neq 0$, $1 \leq i \leq s$, such that $\varphi(y) = \lambda_i \varphi'(y)$, $y \in \operatorname{Gr}^i \mathcal{F}_{x_1}$ (note the difference in the equivalence relation from that in (29) above).

Definition 4: A *distinguished quasi-parabolic* (DQP) *structure* on a vector bundle E on Y at x_1, x_2 consists of the following data (essentially we forget the non-zero weights of a (DP) structure on E):

(a) $\mathcal{F} = (\mathcal{F}_{x_1}, \mathcal{F}_{x_2})$, where $\mathcal{F}_{x_1}$ and $\mathcal{F}_{x_2}$ denote (strictly decreasing) flags on E_{x_1} and E_{x_2} respectively (as in Definition 3 above) with the same length s and multiplicity k_i, $1 \leq i \leq s$, i.e.

$$k_i = \operatorname{rk} \operatorname{Gr}^i \mathcal{F}_{x_1} = \operatorname{rk} \operatorname{Gr}^i \mathcal{F}_{x_2}^*, \quad 1 \leq i \leq s$$

(b) we attach a "weight" to $\mathcal{F}_{x_1}$ (resp. $\mathcal{F}_{x_2}$) and call it zero or non-zero. We suppose that

$$\operatorname{wt} \mathcal{F}_{x_1} (= E_{x_1}) = 0 \Longleftrightarrow \operatorname{wt} \mathcal{F}_{x_2} = 0.$$

We call k_1 the *multiplicity* of the zero weight and denote it by k. Let $l = n - k$ ($\mathcal{F}_{x_1}^i$ (resp. $\mathcal{F}_{x_2}^i$) is supposed to be of non-zero weight for $i \geq 2$)

(c) a graded linear isomorphism

$$\varphi : \operatorname{Gr} \mathcal{F}_{x_1} \longrightarrow \operatorname{Gr} \mathcal{F}_{x_2}^* \quad (\mathcal{F}_{x_2}^* \text{ flag dual to } \mathcal{F}_{x_2})$$

determined upto an equivalence relation as in Definition 3 above. Let E be a (DP) vector bundle on Y. Then we see that

$$\text{par.deg.(parabolic degree of) } E =$$

$$= \deg E + \sum_{i=1}^{s} k_i \alpha_i + \sum (1 - k_i)\alpha_i \text{ if } \alpha_1 \neq 0$$

$$= \deg E + \sum_{i=2}^{s} k_i \alpha_i + \sum_{i=2}^{s} k_i (1 - \alpha_i) \text{ if } \alpha_1 = 0.$$

Hence

$$\text{par.deg. } E \;=\; \deg E + n \text{ if } \alpha_1 \neq 0$$
$$=\; \deg E + (n - k_1) \text{ if } \alpha_1 \neq 0.$$

Thus for a (DQP) vector bundle E on Y, we define

$$\text{Quasi-par.deg. } E = \deg E + (n - k) = \deg E + l. \qquad (33)$$

Remark 5: We have thus established a functor

(Distinguished $G-$bundles on X) $\longrightarrow$ (DP vector bundles E on Y of rk n)

which induces a map from the set of isomorphism classes of distinguished G-bundles to the set of isomorphism classes of distinguished parabolic vector bundles on Y of rk n. This map is surjective but not bijective, essentially because of the equivalence relation in (b) of Definition 3 which is finer than that in (29) above.

Let us summarize our discussions above. We defined a functor (see (17) above)

(Distinguished G-bundles on X) $\longrightarrow$ (Torsion free sheaves of rk n on X)

$$P \longmapsto (p_*^G)(\underline{V}) = F, \ V \text{ standard } G \text{ module.}$$

This representation of F can be viewed as giving additional structures on F, which have been interpreted as (DP) and (DQP) structures on the vector bundle $E = j^*(F)$ mod torsion on the normalisation Y of X. In particular this gives a functor

(Distinguished G-bundles on X) $\longrightarrow$ (DQP vector bundles E of rk n on Y).
$$\qquad (34)$$

We would expect to recover the torsion free sheaf F from E. Let us first observe that if $E = j^*(F)$ mod torsion and F comes from a distinguished G-bundle, as above, we have

$$\deg F = \text{par.deg. (or quasi} - \text{par.deg.)} E. \qquad (35)$$

To see this, observe that par.deg. $E = \deg E + l$, where l is $(n - k_1)$ (see (33) above), where k_1 is the multiplicity of the zero weight of the parabolic or quasi-parabolic structure which is also the multiplicity of the trivial representation of Γ in Γ-module V. One knows from the local representation of F via ramified coverings (see Remark 2 and (16) above) that $(n - k_1)$ is the type of F.

We shall now associate canonically a torsion free sheaf F on X to a vector bundle E on Y (or rk n) with a (DQP) structure $\mathcal{F}$ such that

$\deg F = \text{quasi} - \text{par.deg.} E$. Let us recall ([6]) that F can be represented by a vector bundle E' on Y and a *linear map* $\delta : E'_{x_1} \longrightarrow E'_{x_2}$ (called a *triple* in [6]) such that

$$\deg F = \deg E', \quad \text{type of } F = \text{rk}(\text{Coker } \delta). \tag{36}$$

In fact the association $F \longmapsto (E', \delta)$ is an equivalence of categories. We note that we have fixed an order on the points of Y lying over x_0. The vector bundle E' is defined as a Hecke modification of $E = j^*(F)$ mod torsion (recall that F can also be defined by a GPB structure on E' in the sense of [1], the GPB structure associated to the triple is just the graph of δ). More precisely, this is done as follows in our case:

Case 1: $s = 1$ ($s =$ length of the flags defined by $\mathcal{F}$) and the weight of $\mathcal{F}_{x_1} = E_{x_1}$ is zero.

In this case the map φ defined in (c) of Definition 4 reduces to an isomorphism $\varphi : E_{x_1} \longrightarrow E_{x_2}$. One knows that this defines canonically a *vector bundle* F on X. We see that $\deg F = \deg E = \text{quasi-par.deg.} E$.

Case 2: Weight of $\mathcal{F}_{x_1}$ is non-zero.

Let $E \longrightarrow E'$ be the Hecke modification such that $E_{x_2} \longrightarrow E'_{x_2}$ is the zero map and $E_\alpha \longrightarrow E'_x$ is an isomorphism for $x \neq x_2$. Then $\deg E' = \deg E + n$. Let $\delta : E'_{x_1} \longrightarrow E'_{x_2}$ be the zero map. Then δ defines a torsion free sheaf F on X such that $\deg F = \deg E' = \deg E + n = par.deg.E$, and the the type of F is n.

Case 3: $s \geq 2$ and the weight of $\mathcal{F}_{x_1}$ is zero

We take the Hecke modification $\lambda : E \longrightarrow E'$ such that the kernal of $\lambda_{x_2} : E_{x_2} \longrightarrow E'_{x_2}$ is $\mathcal{F}^2_{x_2}$ and $\lambda_x : E_x \longrightarrow E'_x$ is an isomorphism for $x \neq x_2$. The (DQP) structure on E defines an isomorphism $\varphi_1 : E_{x_1}/\mathcal{F}^2_{x_1} \longrightarrow E_{x_2}/\mathcal{F}^2_{x_2}$. This induces a map $\varphi'_1 : E_{x_1} \longrightarrow E_{x_2}/\mathcal{F}^2_{x_2}$. We see that the "composite" $\lambda_{x_2} \cdot \varphi'_1$ defines a well-defined map $\delta : E'_{x_1} \longrightarrow E'_{x_2}$ ($E'_{x_1} = E_{x_1}$) whose image identifies with $E_{x_2}|_{\mathcal{F}^2_{x_2}}$. This defines a torsion free sheaf F on X. We see rk Coker $\delta = \text{rk}\,\mathcal{F}^2_{x_2} = l$. We see that $\deg F = \text{quasi-par.deg. } E$ and type of $F = l$.

Thus we have defined a functor

$$\text{(DQP vector bundles on } Y) \longrightarrow \text{(Torsion free sheaves on } X). \tag{37}$$

Taking the composite of the functors (34) and (37), we get the functor (17).

Remark 6: Let us now determine the set S of all isomorphism classes of (DQP) vector bundles $(E, \mathcal{F})$ on Y such that the underlying torsion free sheaf defined as above in (36) is fixed. More precisely, we fix E and then S is the set of all (DQP) structures on E for which the Hecke modification $E \longrightarrow E'$ and the linear map $\delta : E'_{x_1} \longrightarrow E'_{x_2}$ as defined above, are fixed.

Case 1: $\underline{\delta = 0}$. Let us fix flags $\tau_1 = \mathcal{F}_{x_1}$, $\tau_2 = \mathcal{F}^*_{x_2}$ on E_{x_1} and $E^*_{x_2}$ respectively and take an isomorphism $\varphi : \operatorname{Gr} \mathcal{F}_{x_1} \longrightarrow \operatorname{Gr} \mathcal{F}^*_{x_2}$ modulo the equivalence relation as in Definition 4 above. Let S_{τ_1, τ_2} be the set of all isomorphisms φ when we fix τ_1, τ_2. We see that $S_{\tau_1, \tau_2} \simeq PGL(k_1) \times \cdots \times PGL(k_s)$, $k_i =$ multiplicity i.e. rk $\operatorname{Gr}^i \mathcal{F}_{x_1}$. Fix a parabolic subgroup Q in G corresponding to the flag type τ_1 (or τ_2) (fixing a maximal torus in G, we take Q containing this torus). We see that $(\tau_1, \tau_2) \in G/Q \times G/Q$. Let us denote by S_Q the set of all $(\tau_1, \tau_2, \varphi)$ when we fix the flag type. We fix the parabolic subgroup. Then we see that we have a map

$$S_Q \longrightarrow G/Q \times G/Q \tag{38}$$

and the fibre over (τ_1, τ_2) identifies with S_{τ_1, τ_2}. We see that S_{τ_1, τ_2} identifies essentially with the Levi subgroup of Q i.e. we substitute the projective groups for the linear groups in the decomposition of the Levi group of Q as a product of linear groups. We see that S is the disjoint union of S_Q when Q varies over all parabolic subgroups containing T (we have also to include the case $Q = GL(n)$, which corresponds to the case when the length of the flag is 1. Then (τ_1, τ_2) reduces to one point and $S_{\tau_1, \tau_2} \simeq PGL(n)$). Recall that the S_Q as above provide a stratification of the wonderful compactification of $PGL(n)$ ([3]). Thus S can be identified set theoretically with the *wonderful compactification of $PGL(n)$*. Note that $n =$ type of F (torsion free sheaf defined by δ).

Case 2: $\underline{\delta \neq 0}$.

Suppose δ is an isomorphism. This means that $E' = E$, the torsion free sheaf F on X is a vector bundle and S reduces to one point.

Suppose then that δ is *not* an isomorphism. Then for the DQP structure on E, there is also a zero weight and the isomorphism

$$\operatorname{Gr}^1(\varphi) : \operatorname{Gr}^1 \mathcal{F}_{x_1} \longrightarrow \operatorname{Gr}^1 \mathcal{F}^*_{x_2}$$

is fixed. Then we have a similar consideration as in case 1 by taking flags on $\mathcal{F}^2_{x_1}$ and $(\mathcal{F}^*_{x_2})^2$. Then we see easily that S identifies with the *wonderful compactification of $PGL(l)$* ($l =$ type of F-torsion free sheaf defined by δ).

Definition 5:

(a) A distinguished quasi-parabolic vector bundle E on Y is said to be *stable* if the torsion free sheaf F on X associated (by the functor (36) above) is stable.

(b) Let $(DQP)^s(n,d)$ denote the set of isomorphism classes of stable (DP) vector bundles on Y of rk n and quasi-parabolic deg d.

Remark 7: We have a canonical map (defined by (36) above)

$$g : (DQP)^s(n,d) \longrightarrow U^s(n,d)$$

where $U^s(n,d)$ denotes the moduli space of torsion free sheaves on X of rk n and deg d. By Remark 6, if $F \in U^s(n,d)$ and its type is l, the fibre over F for the map identifies (set theoretically) with the wonderful compactification of $PGL(l)$. On the other hand, if $G^s(n,d)$ denotes the moduli space of stable Gieseker vector bundles of rk n and deg d, there is a canonical morphism $G^s(n,d) \longrightarrow U^s(n,d)$ such that the fibres of this morphism have the same description as for g above ([7],[12]). Thus it is natural to conjecture:

$$\begin{cases} (DQP)^s(n,d) \text{ has a natural structure of a quasi-projective variety} \\ \text{and canonically identifies with } G^s(n,d). \end{cases}$$
(39)

There is some progress due to Ivan Kausz in the proof of this conjecture ([4]). It seems likely that one could construct a moduli space of semi-stable Gieseker bundles $G^{ss}(n,d)$ from this point of view, probably using the (DP) structures.

Remark 8: Let H be a semi-simple algebraic group (or more generally a reductive algebraic group). As we have seen before ((2) of Definition 1), we have the notion of a *distinguished H-bundle $p : P \longrightarrow X$*. The pull-back of P to Y defines a *ramified H-bundle $q : Q \longrightarrow Y$* and P can be equivalently described by Γ-structures of Q locally at x_1, x_2 and an identification $Q_{x_1} \longrightarrow Q_{x_2}$. By these Γ-structures we mean the local descriptions of Q over the neighbourhoods C_1, C_2 of x_1, x_2 respectively i.e. we take the diagonal action of Γ on $D_i \times H$ (the action on H through a representation $\varrho : \Gamma \longrightarrow H$) and we have $Q|_{C_i} \simeq (D_i \times H)$ mod Γ, $i = 1, 2$. In the case of the linear group G, the Γ-structures are equivalently described by *parabolic structures* on Y as we saw in Definition 2. In the case of a general semi-simple group, the interpretation of Γ-structures as parabolic structures is a little more subtle. The set of $(\Gamma - H)$ automorphisms of $D_1 \times H$ identifies with the set $H(D_1)^\Gamma$ of Γ-invariant regular maps $D_1 \longrightarrow H$. The computations (24) and (25) above show that

$$\begin{cases} G(D_1)^\Gamma \simeq \begin{cases} \text{the subset of } G(C_1) \text{ consisting of elements } B \text{ such that} \\ B((0)) \text{ belongs to a parabolic subgroup of } G. \end{cases} \end{cases}$$
(40)

This is the reason for the interpretation of Γ-structures (or equivalently ramified G-bundles over Y) as parabolic structures again for principal G-bundles over Y (or equivalently parabolic structures for vector bundles on Y). Let Q_{x_1} denote the local ring of y at x_1 and k_{x_1} its quotient field. Then for a semi-simple H, one checks that $H(D_1)^\Gamma$ (D_1 sufficiently small) determines canonically a conjugacy class of "lattices" or bounded subgroups of $H(k_{x_1})$ (see [14]) and then by Bruhat-tits theory we have the following: For a semi-simple (or reductive) group H, we have

$$
\left\{
\begin{aligned}
&H(D_1)^\Gamma \simeq \text{Subgroup of } \mathcal{H}(C_1), \text{ where } \mathcal{H} \longrightarrow C_1 \text{ is a group scheme} \\
&\qquad \text{such that the generic fibre is } H \text{ but the fibre over } (0) \\
&\qquad \text{could be non-reductive.}
\end{aligned}
\right.
$$

$$(41)$$

For example when H is the orthogonal group, $H(D_1)^\Gamma$ leaves invariant a quadratic form B defined over C_1 which over the generic point over C_1 is non-degenerate (with the associated orthogonal group $\simeq H$) but at (0) (i.e. $B(0)$ could be degenerate). Thus the interpretation of ramified H-bundles $Q \longrightarrow Y$ (or Γ-structures) as structures over Y would involve principal bundles or *torsors* associated to group schemes $\mathcal{H} \longrightarrow Y$ such that the generic fibre is isomorphic to H but the fibres over some closed points (in our case x_1, x_2) could be non-reductive.

As for the interpretation for the identification $Q_{x_1} \longrightarrow Q_{x_2}$, let us note the following for the case of linear group G. Let $L = $ Centraliser of $\varrho(\Gamma)$ in G. Let K be the Levi subgroup of the parabolic subgroup of G (i.e. the parabolic subgroup of G figuring on the RHS of (39) above) which defines the parabolic structure, say at $x_1 \in Y$ or equivalently at (0) of C_1, coming from the Γ-vector bundles $D_1 \times V$. Then we have an isomorphism.

$$\psi_1 : L \longrightarrow K(\text{canonically defined upto an equivalence relation}).$$

$$(42)$$

This is equivalent to the map (denoted by the same letter) ψ_1 defined in (28) above. In the case of a semi-simple group H, we can again take $L = $ Centraliser of $\varrho(\Gamma)$ in H. Possibly the isomorphism ψ_1 generalises as well and the identification $Q_{x_1} \longrightarrow Q_{x_2}$ via a map which generalises the map φ in Definition 3 and Definition 4.

Remark 9: Let $\underline{X} \longrightarrow S$ be a flat projective morphism as in the beginning of §1 and $P_s \longrightarrow \underline{X}_s$ be a principal H-bundle, s being the generic point of S. Suppose that P_s is stable (or semi-stable) in the sense of A.

Ramanathan [8]. Then A. Schmitt and Usha Bhosle have shown ([2], [11]) that P_s can be extended to a principal H-bundle over an open subset U of $\underline{X}$ such that $\underline{X} \smallsetminus U$ consists of only a finite number of points on the closed fibre X_0 of $\underline{X} \longrightarrow S$. This is a consequence of their result that P_s can be extended as a family of semi-stable *singular H-bundles* over the whole of $\underline{X}$ such that on the closed fibre it is an *honest singular H-bundle*. Then by Remark 1, we see that P_s specialises to a distinguished H-bundle on X_0. This gives a "properness" criterion for distinguished H-bundles.

Thus it seems natural to conjecture that one can construct a proper moduli space via distinguished H-bundles on a nodal curve, having good specialisation properties.

Remark 10: One would like to formulate the definition of stability of torsion free sheaves F in the language of principal G-bundles and their reductions, as is done by A. Ramanathan ([8]) so that it could lead to a suitable generalisation in the semi-simple case. Let us even suppose that F is a vector bundle V and $P \longrightarrow X$ the associated principal G-bundle. A reduction of P to a parabolic subgroup of G is equivalent to giving an exact sequence of vector bundles

$$0 \longrightarrow L \longrightarrow V \longrightarrow M \longrightarrow 0 \qquad (43)$$

and Ramanathan's definition of semi-stability is equivalent to saying that for the sub-bundle $L \otimes M^*$ (resp. quotient bundle $M \otimes L^*$) of $V \otimes V^*$, we have

$$\deg(L \otimes M^*) \leq 0 \Longleftrightarrow \deg(M \otimes L^*) \geq 0$$
$$\Longleftrightarrow \mu(L) \leq \mu(V). \qquad (44)$$

However, to say that V is semi-stable, we should have $\mu(L) \leq \mu(V)$ for exact sequence as in (43) such that L and M are *torsion free*, not merely locally free. Then (43) does *not* define a reduction of the structure group of the principal bundle on the whole of X but only on $X \smallsetminus x_0$. It may happen that if W is the maximal torsion subsheaf of $V \otimes V^*$, which extends the image of $L \otimes M^*$ in $V \otimes V^*$ on $X \smallsetminus x_0$ to the whole of X, then $\deg W$ is strictly positive, in particular $V \otimes V^*$ may not be semi-stable. In fact, let $\operatorname{rk} V = 2$ and V be semi-stable of degree zero. We can certainly construct examples such that L and M are respectively of rank one, degree zero and of type one. Let L' (resp. M') denote $j^*(L)$ (resp. $j^*(M)$ mod . torsion (j is the map $Y \longrightarrow X$, Y being the normalisation of X). Then one sees that the line sub-bundle of V (resp. V^*) generated by L' (resp. M') is of the form $L'(x_1 + x_2)$ (resp. $M'(x_1 + x_2)$). We have $\deg L' = \deg M' = -1$, so that $\deg L'(x_1 + x_2) = \deg M'(x_1 + x_2) = 1$. Then we see that

$$N = (L'(x_1 + x_2) \otimes M'(x_1 + x_2))(-x_1 - x_2)$$

goes down to a torsion free subsheaf of $V \otimes V^*$, which is indeed W. Since $\deg N = 0$, we see that $\deg W = 1$. This proves the assertion and shows that the definition of semi-stability has to be carefully formulated to generalise the definition of Ramanathan. In the case when $\operatorname{rk} V$ is arbitrary and $\deg L = \deg M = 0$, the degree of the torsion subsheaf W of $V \otimes V^*$, defined as above, seems to be equal to (type L) $\cdot$ (type M) (in fact, in our case type L = type M, since V is locally free). Thus it seems that we should put in a correction factor depending on the types for defining semi-stability.

Recall (Remark 9) that one has the notion of singular H-bundles and concepts of stability for these bundles on X (due to A. Schmitt [9], [10], [11]) and moduli spaces with good specialisation properties have been constructed. Probably using these definitions one could arrive at notions of stability for distinguished H-bundles in the spirit of A. Ramanathan.

Remark 11: Let us fix a covering $f : D \longrightarrow C$, $C = D$ mod Γ of a neighbourhood C of x_0, $x_0 \in X$ (as in (14) above) or to be more precise we fix only the homomorphism f^* induced by f from the local ring of C at (0) to that of D at (0). One can ask the question whether there is a global covering $g : Z \longrightarrow X$ with an action of a finite group π such that $X = Z$ mod π and g extends the local covering f (or rather f^*). This means that given $z \in Z$ such that $g(z) = x_0$, the isotropy subgroup at z is isomorphic to Γ and we have a neighbourhood D' of z which is a Γ-covering over a neighbourhood C' of x_0 such that the map $D' \longrightarrow C'$ is of the same type as $D \longrightarrow C$. We can also ask whether $g : Z \longrightarrow X$ could be unramified over points outside x_0 but this is not necessary for our purpose.

Suppose then that we have a covering $g : Z \longrightarrow X$ as desired above. Let $p : P \longrightarrow X$ be a distinguished H-bundle with local data as in (14) (i.e. $P|_C = (D \times H)$ mod Γ, the homomorphism $\varrho : \Gamma \longrightarrow H$ need not be fixed). Then we see that there is a $(\pi - H)$ principal bundle P_Z over Z (i.e. the action of π on Z lifts to an action on P_Z and commutes with the (right) action of the H-principal bundle P_Z over Z) such that $P_Y = P_Z$ mod π (extending the local construction of $P \longrightarrow X$). This would be very useful in the global construction of moduli spaces associated to distinguished H-bundles over X.

Suppose now that $W \longrightarrow Y$ is a double cover (i.e. $Y = W \mod \Gamma$, $\Gamma = \mathbb{Z}/(2)$) and that w_1, w_2 are two points of W which are fixed points under the action of Γ and map onto x_1, x_2 respectively. Let Z be the non-normal curve obtained by identifying w_1, w_2. Then we see we have a canonical map $Z \longrightarrow X$ and a commutative diagram

$$
\begin{array}{ccc}
W & \longrightarrow & Z \\
\downarrow & & \downarrow \\
Y & \longrightarrow & X
\end{array}
\tag{45}
$$

We see that Z is a required choice for the case $\Gamma = \mathbb{Z}/(2)$.

More generally, suppose that $Z \longrightarrow X$ exists with the required properties. Let W be the normalisation of Z so that we have a commutative diagram as above. Let z_0 be a point of Z lying over x_0. We can find w_1, w_2 in W lying over z_0 and project to x_1, x_2 respectively. We see easily that the group π acts on W, $Y = W \mod \pi$ and we have a subgroup (isomorphic to Γ) which fixes w_1, w_2 and in fact coincides with the isotropic groups at these points. Hence this gives a necessary condition for the existence of Z via Galois coverings over Y.

Bibliography

[1] Usha N. Bhosle - Generalized parabolic bundles and applications to torsion free sheaves on nodal curves, *Arkiv för mathematik*, 30 (1992).

[2] Usha N. Bhosle - Tensor fields and singular principal bundles, *International Mathematics Research Notices*, (2004), 3057–3077.

[3] De Concini and C. Procesi - Complete symmetric varieties I, Invariant theory, *Proceedings, Montecatini F. Gheradelli Ed.*, Lect. Notes Math. 996, Springer-Verlag, Heidelberg-Berlin- New York (1983) 1–44.

[4] Ivan Kausz - Twisted vector bundles on pointed nodal curves, *Proc. Indian Acad. Sci.*, 115 (2005), no.2, 147–165.

[5] V.B. Mehta and C.S. Seshadri - Moduli of vector bundles on curves with parabolic structures, *Math. Ann.*, 248 (1980), 205–239.

[6] D.S. Nagaraj and C.S. Seshadri - Degenerations of moduli spaces of vector bundles on curves I, *Proc. Indian Acad. Sci., Math. Sci.*, 107 (1997) 101–137.

[7] D.S. Nagaraj and C.S. Seshadri - Degenerations of the moduli spaces of vector bundles on curves II (Generalized Gieseker moduli spaces), *Proc. Indian Acad. Sci., (Math. Sci.)*, 109 (1999) 165–201.

[8] A. Ramanathan - Moduli of principal bundles over algebraic curves I, *Proc. Indian Acad. Sci., Math. Sci.*, 106 (1996), no.3, 301–328.

[9] Alexander H.W. Schmitt - Singular principal bundles over higher dimensional manifolds and their moduli spaces, *International Mathematics Research Notices*, (2002), no.23, 1183–1209.

[10] Alexander H.W. Schmitt - A closer look at semistability for singular principal bundles, (2004), 62, 3327–3366.

[11] Alexander H.W. Schmitt - Moduli spaces for semi-stable honest singular bundles on a nodal curve which are compatible with degeneration - A remark on Bhosle's paper "Tensor fields and singular principal bundles", *International Mathematics Research Notices*, (2005), no.23, 1427–1437.

[12] C.S. Seshadri - Degenerations of the moduli spaces of vector bundles on curves, *ICTP Lecture Notes, School on algebraic geometry*, 26 July – 13 August 1999, 209–265.

[13] C.S. Seshadri - Moduli of torsion free sheaves and *G*-spaces on nodal curves (*written in 2002, unpublished*).

[14] C.S. Seshadri – Remarks on parabolic structures, *talk at Madrid conference in honour of S. Romanan* (to appear).

Printed in the United States
by Baker & Taylor Publisher Services